London Mathematical Society Lecture Note Series: 404

Operator Methods for Boundary Value Problems

SEPPO HASSI
University of Vaasa

HENDRIK S. V. DE SNOO
University of Groningen

FRANCISZEK HUGON SZAFRANIEC
Jagiellonian University, Kraków

CAMBRIDGE UNIVERSITY PRESS
Cambridge, New York, Melbourne, Madrid, Cape Town,
Singapore, São Paulo, Delhi, Mexico City

Cambridge University Press
The Edinburgh Building, Cambridge CB2 8RU, UK

Published in the United States of America by Cambridge University Press, New York

www.cambridge.org
Information on this title: www.cambridge.org/9781107606111

First published 2012

Printed and bound in the United Kingdom by the MPG Books Group

A catalogue record for this publication is available from the British Library

Library of Congress Cataloguing in Publication data

ISBN 978-1-107-60611-1 Paperback

Contents

Contributors

Yury Arlinskiĭ
Department of Mathematical Analysis
East Ukrainian National University
Kvartal Molodyozhny 20-A, Lugansk 91034, Ukraine
yury_arlinskii@yahoo.com

Damir Z. Arov
Division of Applied Mathematics and Informatics
Institute of Physics and Mathematics
South-Ukrainian National Pedagogical University
65020 Odessa, Ukraine
arov_damir@mail.ru

Jussi Behrndt
Technische Universität Graz
Institut für Numerische Mathematik
Steyrergasse 30, 8010 Graz, Austria
behrndt@tugraz.at

Vladimir Derkach
Department of Mathematics
Donetsk National University
Universitetskaya 24, 83055 Donetsk, Ukraine
derkach.v@gmail.com

Gerd Grubb
Department of Mathematical Sciences, University of Copenhagen
Universitetsparken 5, DK-2100 Copenhagen, Denmark
grubb@math.ku.dk

Seppo Hassi
Department of Mathematics and Statistics
University of Vaasa
P.O. Box 700, FI-65101 Vaasa, Finland
sha@uwasa.fi

Mikael Kurula
University of Twente, Department of Applied Mathematics
P.O. Box 217, 7500 AE Enschede, The Netherlands
mikael.kurula@abo.fi
Matthias Langer
Department of Mathematics and Statistics, University of Strathclyde
26 Richmond Street, Glasgow G1 1XH, United Kingdom
m.langer@strath.ac.uk
Mark M. Malamud
Department of Mathematics
Donetsk National University
Universitetskaya 24, 83055 Donetsk, Ukraine
mmm@telenet.dn.ua
Arjan van der Schaft
Johann Bernoulli Institute for Mathematics and Computer Science
University of Groningen
P.O. Box 407, 9700 AK Groningen, The Netherlands
a.j.van.der.schaft@rug.nl
Hendrik S.V. de Snoo
Johann Bernoulli Institute for Mathematics and Computer Science
University of Groningen
P.O. Box 407, 9700 AK Groningen, The Netherlands
desnoo@math.rug.nl
Olof J. Staffans
Åbo Akademi University, Department of Mathematics
FIN-20500 Åbo, Finland
olof.staffans@abo.fi
Franciszek Hugon Szafraniec
Instytut Matematyki, Wydział Matematyki i Informatyki
Uniwersytet Jagielloński
ul. Łojasiewicza 6, 30 348 Kraków, Poland
umszafra@cyf-kr.edu.pl
Rudi Wietsma
Department of Mathematics and Statistics
University of Vaasa
P.O. Box 700, FI-65101 Vaasa, Finland
rwietsma@uwasa.fi
Hans Zwart
University of Twente, Department of Applied Mathematics
P.O. Box 217, 7500 AE Enschede, The Netherlands
h.j.zwart@math.utwente.nl

Preface

The theory of unbounded operators, which dates back to the early 1930s, was developed by J. von Neumann and M.H. Stone. Of course the earlier work of H. Weyl on boundary eigenvalue problems and of T. Carleman on singular integral operators should be mentioned as stepping stones for the general abstract treatment. One of the underlying ideas was to put quantum mechanics on a rigorous mathematical foundation. J. von Neumann used a distinction between symmetric (Hermitian) and selfadjoint (hypermaximal) operators and referred to E. Schmidt for this. Once this distinction was made, it was natural to determine all selfadjoint extensions of a symmetric operator (necessarily with equal deficiency indices). J. von Neumann gave such a description for densely defined symmetric operators by means of his well known formulas. This description requires the knowledge of the deficiency spaces of the symmetric operator. Another approach involving abstract boundary conditions was developed by J.W. Calkin in his 1937 Harvard doctoral dissertation, which was written under the direction of Stone, who suggested the topic. Unfortunately Calkin's work on boundary value problems did not receive the attention it deserved; probably because he never returned to it after his mathematical work related to World War II.

A revival of interest in applications of this approach to boundary value problems is due to M.G. Kreĭn, M.I. Vishik, M.S. Birman, and R. S. Phillips in the 1950s and, later, to G. Grubb, F.S. Rofe-Beketov, and M.L. Gorbachuk. The notion of boundary triplet was proposed in the 1970s by A.N. Kochubei and V.M. Bruk. The boundary triplet idea was used to introduce the associated Weyl function: an abstract analog of the classical Titchmarsh-Weyl m-function. This function formed a connection with the classical work of M.A. Naimark and M.G. Kreĭn from the early 1940s about exit space extensions. A densely defined symmetric operator with possibly unequal deficiency indices always has selfadjoint extensions once exit space extensions are allowed. In fact, Kreĭn had discovered the formula bearing his name describing the resolvents of the selfadjoint extensions, possibly in exit spaces; this procedure involves an operator-valued analytic function, the so-called Q function, which equals the above Weyl function up to an additive constant. In the same period, during the 1970s, the theory was facilitated by the systematic introduction of the notion of linear relation (multivalued linear operator), so that it was not necessary to restrict attention to densely defined operators any longer. The background of this extension employed an idea of von Neumann, which was to use the graph to analyze an unbounded operator.

In Kreĭn's formula the Weyl function and a possibly multivalued parameter family play similar roles. A geometric interpretation of the Kreĭn formula gave rise to the notion of boundary relation by treating both the Weyl function and the parameter family on an even basis. The notion of boundary relation is based on an abstract form of Green's or Lagrange's identity, and this is where isometric and unitary relations in Kreĭn spaces come into play. Simultaneously, Kreĭn space methods received increasing attention in system theory. Recently, system theoretic equivalents for boundary triplets and boundary relations have been developed. The theory of boundary relations and triplets offers another approach to extension theory and operator theoretic methods to concrete boundary value problems. Various forms of boundary relations have appeared recently in conjunction with boundary value problems for elliptic operators (Dirichlet-to-Neumann mappings) and with local point interactions.

The idea of the present volume arose during a workshop at the Lorentz Center in Leiden, the Netherlands (December 14–December 18, 2009), entitled *Boundary relations, extension theory, and applications* and organized by the present editors. Several of the talks were devoted to recent developments in boundary triplet methods, system theory, and partial differential equations. This volume presents more or less a coherent presentation of these new developments with a historical review, beginning with Calkin's work from the late 1930s. Different aspects of boundary value problems are treated in contributions written by specialists in the field from their own point of view. We thank the Lorentz Center for their support in organizing the workshop and their hospitality during the conference. Moreover, we thank Sam Harrison of Cambridge University Press for his pleasant and constructive cooperation.

Seppo Hassi, Henk de Snoo and Franek Szafraniec

1

John Williams Calkin: a short biography

Seppo Hassi, Henk de Snoo and Franciszek Hugon Szafraniec

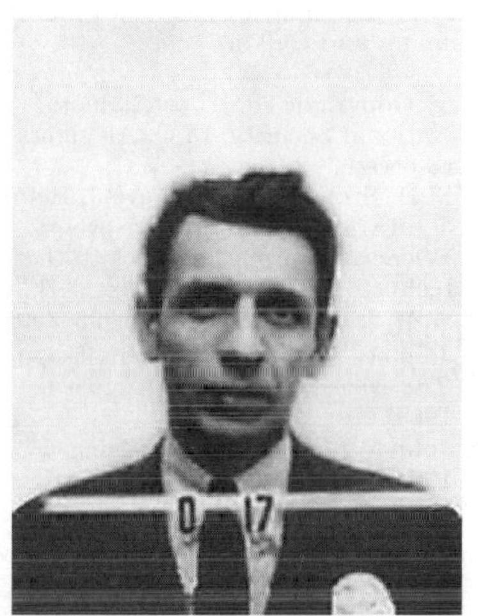

All employees at Los Alamos had Project Y security badges. Calkin's photograph[1] is one of these badge photos.

John Williams Calkin (October 11, 1909–August 5, 1964). A native of New Rochelle, N.Y., he graduated with honors in mathematics from Columbia University in 1933. He was awarded his MA in 1934 and his PhD in June 1937 by Harvard University. His PhD dissertation, *Applications of the Theory of Hilbert Space to Partial Differential Equations; the Self-Adjoint Transformations in Hilbert Space Associated with a Formal Partial Differential Operator of the Second Order and Elliptic Type*, was written under the direction and at the suggestion of M.H. Stone; fruitful conversations with J. von Neumann were acknowledged. In the fall of 1937 Calkin went to the Institute for Advanced Study in Princeton on a year's fellowship to work with O. Veblen and J. von Neumann. He later served as an assistant professor at the University of New Hampshire and the Illinois Institute of Technology in Chicago. During World War II Calkin was part of a mine warfare operations analysis group with J.L. Doob, J. von Neumann, and M.H. Stone[2]. Von Neumann and Calkin worked on shock waves and damage by explosives; they were sent to England to learn of the progress under way there. When it appeared that their special knowledge would be useful for the Manhattan Project, they moved to Los Alamos. The Los Alamos Laboratory, or Project Y, had come into existence in early 1943 for a single purpose: to design and build an atomic bomb. Calkin remained there until 1946 when he accepted a Guggenheim fellowship at the California Institute of Technology. He later taught at the Rice Institute in Houston before returning to Los Alamos in 1949 as a member of the theoretical division. In 1958, he accepted a consulting appointment at New York University and at Brookhaven National Laboratory and, in 1961, was named head, and then chairman of the Applied Mathematics Department. He died[3] [4] in Westhampton, N.Y., at the age of 54.

Calkin's publications date back to the time before he joined the war effort. We have not been able to find other published mathematical work. The list consists of the following papers:

- Abstract self-adjoint boundary conditions, *Proc. Nat. Acad. Sci.*, **24** (1938), 38–42.
- General self-adjoint boundary conditions for certain partial differential

operators, *Proc. Nat. Acad. Sci.*, **25** (1939), 201–206.

- Abstract symmetric boundary conditions, *Trans. Amer. Math. Soc.*, **45** (1939), 369–442.
- Abstract definite boundary value problems, *Proc. Nat. Acad. Sci.*, **26** (1940), 708–712.
- Functions of several variables and absolute continuity I, *Duke Math. J.*, **6** (1940), 170–186.
- Symmetric transformations in Hilbert space, *Duke Math. J.*, **7** (1940), 504–508.
- Two-sided ideals and congruences in the ring of bounded operators in Hilbert space. *Ann. Math.*, **42** (1941), 839–873.

Today, Calkin is mostly remembered for the algebras bearing his name; the relevant work dates back to 1941. However, it is clear that this work was directly influenced by his earlier work on boundary value problems. After World War II Calkin did not return to his earlier work, but remained involved in applied mathematics and physics both at Los Alamos and at Brookhaven National Laboratory.

The environmentalist Brant Calkin[5] (1934) writes us about his father: "My most vivid memories of him were while we were living at Los Alamos, the atomic bomb laboratory which we first went to in 1943. There was of course no opportunity to visit him at work, since all the offices were behind security fences patrolled by the army. It was there that I can remember the long lasting and raucous poker games which we often hosted. What games those must have been. Players included Stan Ulam, Nick Metropolis, Carson Mark, and occasionally, I think, Enrico Fermi and John von Neumann. The latter, someone my dad had worked with at Princeton. Much earlier, we lived in Chicago, where my father haunted obscure and mostly black night clubs where he loved the jazz, and where he met singers who later became famous. His love of jazz resulted in the music which we mostly heard in our house at that time. His contacts were esoteric, and in the late 50s he invited me to join him at a New York City jazz club. There, he greeted a famous drummer by first name and with whom he had played chess. He was not a car fan, but he had fondness for them. In Chicago, he once walked out to buy some groceries and came back with a used car."

Another glimpse of the man Calkin is provided by S.M. Ulam[6]. When Ulam and his wife traveled to New Mexico by train he was infinitely surprised to see Calkin, whom he had known from Chicago, waiting at the whistle-stop to drive them to Los Alamos: "He was a tall, pleasant-looking, man with more savoir-faire than most mathematicians", and Ulam continues describing life in Los Alamos, including discussions with von Neumann and Calkin.

[2] M.H. Stone, Review: Steve J. Heims, John von Neumann and Norbert Wiener, from mathematics to the technologies of life and death, *Bull. A.M.S. (N.S.)*, **8**, (1983), 395–399.

[3] John W. Calkin, a mathematician, *The New York Times*, August 6, 1964.

[4] John W. Calkin, *Phys. Today*, **17**, 1964, 102–105.

[5] Brant Calkin, personal communication.

[6] S.M. Ulam, *Adventures of a mathematician*, University of California Press, 1991.

2

On Calkin's abstract symmetric boundary conditions

Seppo Hassi and Rudi Wietsma

Abstract J.W. Calkin introduced in the late 1930s the concept of a reduction operator in order to investigate (maximal) symmetric extensions of symmetric operators. A precise interpretation of reduction operators can be given using Kreĭn space terminology; this simultaneously connects his work with recent investigations on boundary triplets. Here an overview of his main results on reduction operators is given, with emphasis on the more involved unbounded case. In particular, by extending a domain decomposition of reduction operators from [Calkin, 1939a] into a graph decomposition and using Kreĭn space methods, simplified proofs for most of his main results are given.

2.1 Introduction

In order to investigate boundary value problems appearing in connection with ordinary and partial differential equations, see e.g. [Calkin, 1939b], Calkin introduced in the end of the 1930s the concept of a reduction operator in [Calkin, 1939a]; Definition 2.1 below contains a slightly reformulated form of his notion.

Definition 2.1 Let S be a closed linear operator in a Hilbert space $(\mathfrak{H}, (\cdot,-)_{\mathfrak{H}})$ and assume that S is densely defined. Then a closed linear operator U with domain in the graph of S^* and with range in a Hilbert space $(\mathfrak{K}, (\cdot,-)_{\mathfrak{K}})$ is said to be a *reduction operator* for S^* if:

(i) $\overline{\operatorname{dom} U} = \operatorname{gr} S^*$;
(ii) there exists a unitary operator W in $(\mathfrak{K}, (\cdot,-)_{\mathfrak{K}})$ such that

$$(\operatorname{gr} U)^{\perp} = \{\{S^*f, -f, WU\{f, S^*f\}\} : \{f, S^*f\} \in \operatorname{dom} U\}.$$

Here $\mathfrak{K}$ is called the *range space* of U and W is called the *rotation* associated with U.

The assumption that S is densely defined in Definition 2.1 means that S^* exists as an operator; this prevents the use of linear relations. Without this condition all the theory that follows holds without essential changes. However, as in Calkin's original paper, here only the densely defined case is considered.

Next observe that condition (ii) in Definition 2.1 implies the following (abstract) Green's (or Lagrange's) identity:

$$(f, S^*g)_{\mathfrak{H}} - (S^*f, g)_{\mathfrak{H}} = (WU\{f, S^*f\}, U\{g, S^*g\})_{\mathfrak{K}}, \qquad (2.1)$$

where $\{f, S^*f\}, \{g, S^*g\} \in \operatorname{dom} U$. This equality shows in particular that $\ker U = \operatorname{gr} S$ and, hence, S is a closed symmetric operator.

The main problems investigated in [Calkin, 1939a] can be roughly formulated as follows:

(i) investigate the cardinality and type of maximal symmetric extensions of S whose graph is contained in the domain of a reduction operator for S^*;
(ii) determine necessary and sufficient conditions on a subspace belonging to the range of the reduction operator for its pre-image under the reduction operator to be maximal symmetric;
(iii) describe the pathology of unbounded reduction operators.

It is the purpose of this chapter to give an overview of Calkin's answers to the above questions. It should be noted that concepts which are essentially special cases of Calkin's reduction operator have been introduced in later literature. For instance, the widely known concept of a boundary triplet (or boundary value space), see [Gorbachuk and Gorbachuk, 1991] and the references therein, can be considered as a bounded reduction operator with a suitable choice of the rotation W. In particular, this means that Calkin's results for bounded reduction operators, i.e. for reduction operators with $\operatorname{dom} U = S^*$, have in special cases been rediscovered in later works; for further details see Chapter 7 and the references therein. In this overview the main focus will be on the more complicated unbounded case. In order to make Calkin's results more accessible, they are here reformulated using Kreĭn space terminology, which is a natural framework for his investigation. Furthermore, simplified proofs are included for the main statements by extending the central result from [Calkin, 1939a]. In addition, it is shown how the concept of a reduction operator is connected to modern day equivalent objects used in the extension theory of symmetric operators.

The contents of the chapter are now outlined. Section 2 contains some preliminary results on operators (relations) in Kreĭn spaces. In Section 3 reduction operators are investigated: their connection to (unitary) boundary triplets is made explicit, they are characterized and a classification of them is introduced. In Section 4, Calkin's answers to the above mentioned questions are presented.

The idea for writing this overview of Calkin's 1939 paper entitled "Abstract symmetric boundary conditions" came from Vladimir Derkach's lecture "Survey of boundary triplets and boundary relations" given during a workshop at the Lorentz Center in Leiden, the Netherlands (December 14–18, 2009). That lecture included a short survey of Calkin's work as can be found in Chapter 7.

2.2 Preliminaries

The notion of a Kreĭn space is recalled including a short introduction to relations in such spaces, see [Azizov and Iokhvidov, 1989] for details. In particular, definitions of (nonstandard) unitary operators in Kreĭn spaces and unitary boundary triplets are given; these objects, which are used in the extension theory of symmetric operators, are later shown to be closely related to reduction operators.

Kreĭn spaces Let $\mathfrak{j}$ be a *fundamental symmetry*, also called a *signature operator*, in the Hilbert space $(\mathfrak{H}, (\cdot, -))$, i.e., $\mathfrak{j}$ is an everywhere defined operator satisfying $\mathfrak{j}^* = \mathfrak{j} = \mathfrak{j}^{-1}$. Then the space $\mathfrak{H}$ equipped with the indefinite inner product $(\mathfrak{j}\,\cdot, -)$, i.e. $(\mathfrak{H}, (\mathfrak{j}\,\cdot, -))$, is called a *Kreĭn space.* Note that the fundamental symmetry $\mathfrak{j}$ induces an orthogonal decomposition $\mathfrak{H}^+ + \mathfrak{H}^-$ of $\mathfrak{H}$ by $\mathfrak{H}^\pm = \ker(I \mp \mathfrak{j})$ and that $(\mathfrak{H}^\pm, \pm(\mathfrak{j}\cdot, -))$ are Hilbert spaces; this orthogonal decomposition of $\mathfrak{H}$ is called the *canonical decomposition of* $(\mathfrak{H}, (\cdot, -))$ *induced by* $\mathfrak{j}$.

Example 2.2 For a Hilbert space $(\mathfrak{H}, (\cdot, -))$ define $\mathfrak{j}_\mathfrak{H}$ on $\mathfrak{H}^2$ as

$$\mathfrak{j}_\mathfrak{H}\{f, f'\} = \{-\mathrm{i}f', \mathrm{i}f\}, \quad \{f, f'\} \in \mathfrak{H}^2. \tag{2.2}$$

Then $\mathfrak{j}_\mathfrak{H} = (\mathfrak{j}_\mathfrak{H})^{-1} = (\mathfrak{j}_\mathfrak{H})^*$, i.e., $\mathfrak{j}_\mathfrak{H}$ is a fundamental symmetric in $(\mathfrak{H}^2, (\cdot, -))$. Consequently, $(\mathfrak{H}^2, (\mathfrak{j}_\mathfrak{H}\,\cdot, -))$ is a Kreĭn space. Note that if $\mathfrak{H}^+ + \mathfrak{H}^-$ is the canonical decomposition of $\mathfrak{H}^2$ induced by $\mathfrak{j}_\mathfrak{H}$, then

$$\mathfrak{H}^+ = \ker(I - \mathfrak{j}_\mathfrak{H}) = \{\{f, \mathrm{i}f\} : f \in \mathfrak{H}\},$$
$$\mathfrak{H}^- = \ker(I + \mathfrak{j}_\mathfrak{H}) = \{\{f, -\mathrm{i}f\} : f \in \mathfrak{H}\}.$$

Let $\mathfrak{L}$ be a subspace of the Hilbert space $(\mathfrak{H}, (\cdot, -))$ with fundamental symmetry j and let $[\cdot, -] := (\mathrm{j}\cdot, -)$. Then $\mathfrak{L}$ is called a *positive*, *negative*, *nonnegative*, *nonpositive* or *neutral* subspace of $(\mathfrak{H}, (\mathrm{j}\cdot, -))$ if $[f, f] > 0$, $[f, f] < 0$, $[f, f] \geq 0$, $[f, f] \leq 0$ or $[f, f] = 0$ for every $f \in \mathfrak{L} \setminus \{0\}$, resp. (Note that Calkin uses the notion of W-symmetry for neutrality, see [Calkin, 1939a, Definition 1.3].) A neutral, nonnegative or nonpositive subspace is called *maximal neutral*, *maximal nonnegative* or *maximal nonpositive* if it has no neutral, nonnegative or nonpositive extension, resp. In particular, maximal semi-definite (i.e., neutral, nonnegative or nonpositive) subspaces are closed. If $P^{\pm}$ denotes the orthogonal projection onto $\mathfrak{H}^{\pm}$ in $\mathfrak{H}$ with respect to $(\cdot, -)$ for a canonical decomposition $\mathfrak{H}^{+} + \mathfrak{H}^{-}$ of $(\mathfrak{H}, (\mathrm{j}\cdot, -))$, then a nonnegative or nonpositive subspace $\mathfrak{L}$ is maximal if and only if $P^{+}\mathfrak{L} = \mathfrak{H}^{+}$ or $P^{-}\mathfrak{L} = \mathfrak{H}^{-}$, resp.

The *orthogonal complements* of $\mathfrak{L}$ with respect to $(\cdot, -)$ and $[\cdot, -] := (\mathrm{j}\,\cdot, -)$ are denoted by $\mathfrak{L}^{\perp}$ and $\mathfrak{L}^{[\perp]}$: they are the closed subspaces defined as

$$\mathfrak{L}^{\perp} = \{f \in \mathfrak{H} : (f, g) = 0 \text{ for all } g \in \mathfrak{L}\},$$
$$\mathfrak{L}^{[\perp]} = \{f \in \mathfrak{H} : [f, g] = 0 \text{ for all } g \in \mathfrak{L}\}.$$

Clearly, $\mathfrak{L}^{[\perp]} = \mathrm{j}\mathfrak{L}^{\perp} = (\mathrm{j}\mathfrak{L})^{\perp}$. With the above notation, a subspace $\mathfrak{L}$ is neutral if and only if $\mathfrak{L} \subseteq \mathfrak{L}^{[\perp]}$. For a neutral subspace $\mathfrak{L}$ of $(\mathfrak{H}, (\mathrm{j}\,\cdot, -))$ the *abstract first von Neumann formula* holds:

$$\mathfrak{L}^{[\perp]} = \operatorname{clos}(\mathfrak{L}) \dotplus (\mathfrak{L}^{[\perp]} \cap \mathfrak{H}^{+}) \dotplus (\mathfrak{L}^{[\perp]} \cap \mathfrak{H}^{-}), \tag{2.3}$$

see [Azizov and Iokhvidov, 1989, Chapter 1 : 4.20]. Note that the first von Neumann formula is nothing else than the canonical decomposition for the Kreĭn space $(\mathfrak{L}^{[\perp]} \ominus \operatorname{clos}\mathfrak{L}, (\mathrm{j}\,\cdot, -))$ induced by the canonical decomposition $\mathfrak{H}^{+} + \mathfrak{H}^{-}$ of $(\mathfrak{H}, (\mathrm{j}\cdot, -))$. For a neutral subspace $\mathfrak{L}$ its defect numbers $n_{+}(\mathfrak{L})$ and $n_{-}(\mathfrak{L})$ are defined as

$$\begin{aligned} n_{+}(\mathfrak{L}) &= \dim(\mathfrak{L}^{[\perp]} \cap \mathfrak{H}^{-}) = \dim(\mathfrak{H}^{-} \ominus P^{-}\mathfrak{L}); \\ n_{-}(\mathfrak{L}) &= \dim(\mathfrak{L}^{[\perp]} \cap \mathfrak{H}^{+}) = \dim(\mathfrak{H}^{+} \ominus P^{+}\mathfrak{L}). \end{aligned} \tag{2.4}$$

A neutral subspace is called *hyper-maximal neutral* if it is maximal nonnegative and maximal nonpositive, see [Azizov and Iokhvidov, 1989, Chapter 1: Definition 4.15]. Equivalently, a neutral subspace $\mathfrak{L}$ of the Kreĭn space $(\mathfrak{H}, (\mathrm{j}\,\cdot, -))$ is hyper-maximal neutral if and only if $\mathfrak{L} = \mathfrak{L}^{[\perp]}$. In other words, $\mathfrak{L}$ is hyper-maximal neutral if and only if it induces the following orthogonal decomposition of the Hilbert space $(\mathfrak{H}, (\cdot, -))$:

$$\mathfrak{H} = \mathfrak{L} \oplus \mathrm{j}\,\mathfrak{L}. \tag{2.5}$$

Operators in Hilbert spaces Let $(\mathfrak{H}_1, (\cdot,-)_1)$ and $(\mathfrak{H}_2, (\cdot,-)_2)$ be Hilbert spaces. Then H is called a *(linear) relation* from $\mathfrak{H}_1$ to $\mathfrak{H}_2$ if it is a subspace of $\mathfrak{H}_1 \times \mathfrak{H}_2$; here and in what follows, linear operators are usually identified with their graphs. In particular, H is *closed* if and only if its graph, $\operatorname{gr} H$, is closed (as a subspace of $\mathfrak{H}_1 \times \mathfrak{H}_2$). The symbols $\operatorname{dom} H$, $\operatorname{ran} H$, $\ker H$, and $\operatorname{mul} H$ stand for the domain, range, kernel, and the multi-valued part of H, respectively. In particular, $\operatorname{mul} H = \{0\}$ if and only if H is an operator. For a subspace $\mathfrak{L}$ of $\operatorname{dom} H$, $H(\mathfrak{L})$ denotes the subspace

$$H(\mathfrak{L}) = \{f' \in \mathfrak{H}_2 : \ \{f, f'\} \in H \ \text{ for some } f \in \mathfrak{L}\}.$$

The inverse H^{-1} of a relation H is defined as

$$H^{-1} = \{\{f, f'\} \in \mathfrak{H}_2 \times \mathfrak{H}_1 : \{f', f\} \in H\}$$

and its adjoint H^* is defined via

$$H^* = \{\{f, f'\} \in \mathfrak{H}_2 \times \mathfrak{H}_1 : (f', g)_1 = (f, g')_2, \quad \{g, g'\} \in H\}. \tag{2.6}$$

From this definition it follows that $\ker H^* = (\operatorname{ran} H)^{\perp_2}$ and $\operatorname{mul} H^* = (\operatorname{dom} H)^{\perp_1}$. In particular, this shows that H^* is an operator if and only if $\overline{\operatorname{dom} H} = \mathfrak{H}_1$.

An operator U from the Hilbert space $(\mathfrak{H}_1, (\cdot,-)_1)$ to the Hilbert space $(\mathfrak{H}_2, (\cdot,-)_2)$ is called *isometric* or *unitary* if $U \subseteq U^{-*}$ or $U = U^{-*}$, respectively [1]. An isometric operator is said to be *maximal isometric* if it has no isometric extension. Similarly, an operator (or relation) S in the Hilbert space $(\mathfrak{H}, (\cdot,-))$ is called *symmetric* or *selfadjoint* if

$$S \subseteq S^* \quad \text{or} \quad S = S^*,$$

respectively, and a symmetric operator (or relation) is said to be *maximal symmetric* if it has no symmetric extension. For a symmetric operator (or relation) S in the Hilbert space $(\mathfrak{H}, (\cdot,-))$ the notation $\widehat{\mathfrak{N}}_\lambda(S^*)$ is used to denote its defect spaces:

$$\widehat{\mathfrak{N}}_\lambda(S^*) = \{\{f_\lambda, \lambda f_\lambda\} : f_\lambda \in \ker(S^* - \lambda)\}, \quad \lambda \in \mathbb{C}.$$

With this notation the *first von Neumann formula* holds:

$$S^* = S \dotplus \widehat{\mathfrak{N}}_\lambda(S^*) \dotplus \widehat{\mathfrak{N}}_{\bar\lambda}(S^*), \quad \lambda \in \mathbb{C} \setminus \mathbb{R}, \tag{2.7}$$

cf. (2.3). In fact, the direct sums in (2.7) are orthogonal for $\lambda = \pm\,\mathrm{i}$. The defect numbers $n_+(S)$ and $n_-(S)$ for S are defined as

$$n_+(S) = \dim \widehat{\mathfrak{N}}_{\bar\lambda}(S^*) \quad \text{and} \quad n_-(S) = \dim \widehat{\mathfrak{N}}_\lambda(S^*), \quad \lambda \in \mathbb{C}_+.$$

[1] For a relation U, the symbol U^{-*} is a shorthand notation for $(U^*)^{-1} = (U^{-1})^*$.

There exist connections between (maximal) symmetric or selfadjoint relations, (maximal) isometric or unitary relations and (maximal) neutral or hyper-maximal neutral subspaces. In Calkin's paper the connection between neutral subspaces of a Kreĭn space and isometric operators, by means of what are now called angular operators (see e.g. [Azizov and Iokhvidov, 1989, Chapter 1: §8]), is used very frequently. Also the connection between symmetric operators and neutral subspaces of Kreĭn spaces was used, see in particular [Calkin, 1939a, Theorem 2.7]. Since this connection is also frequently used here, it is formulated below.

Proposition 2.3 *Let $(\mathfrak{H}, (\cdot, -))$ be a Hilbert space and let $\mathfrak{j}_{\mathfrak{H}}$ be the fundamental symmetry in $\mathfrak{H}^2$ as in Example 2.2. Then S is a (closed, maximal) symmetric or selfadjoint operator (or relation) in $(\mathfrak{H}, (\cdot, -))$ if and only if $\operatorname{gr} S$ is a (closed, maximal) neutral or hyper-maximal neutral subspace of the Kreĭn space $(\mathfrak{H}^2, (\mathfrak{j}_{\mathfrak{H}} \cdot, -))$, respectively. Moreover,*

$$S^* = (\operatorname{gr} S)^{[\perp]},$$

where $[\perp]$ stands for the orthogonal complement with respect to $(\mathfrak{j}_{\mathfrak{H}} \cdot, -)$.

Unitary operators and boundary triplets Let H be an operator (or relation) from the Hilbert space $(\mathfrak{H}_1, (\cdot, -)_1)$ to the Hilbert space $(\mathfrak{H}_2, (\cdot, -)_2)$ and let $\mathfrak{j}_i$ be a fundamental symmetry in $(\mathfrak{H}_i, (\cdot, -)_i)$, $i = 1, 2$. Then the adjoint $H^{[*]}$ of H as a mapping from the Kreĭn space $(\mathfrak{H}_1, (\mathfrak{j}_1 \cdot, -)_1)$ to the Kreĭn space $(\mathfrak{H}_2, (\mathfrak{j}_2 \cdot, -)_2)$ is defined via

$$H^{[*]} = \{\{f, f'\} \in \mathfrak{H}_2 \times \mathfrak{H}_1 : (\mathfrak{j}_1 f', g)_1 = (\mathfrak{j}_2 f, g')_2, \quad \{g, g'\} \in H\}.$$

This Kreĭn space adjoint $H^{[*]}$ of H and the Hilbert space adjoint H^* of H in (2.6) are connected by

$$H^{[*]} = \mathfrak{j}_1 H^* \mathfrak{j}_2. \tag{2.8}$$

By means of the Kreĭn space adjoint an operator (or relation) U from the Kreĭn space $(\mathfrak{H}_1, (\mathfrak{j}_1 \cdot, -)_1)$ to the Kreĭn space $(\mathfrak{H}_2, (\mathfrak{j}_2 \cdot, -)_2)$ is called an *isometric* or a *unitary* operator (or relation) if

$$U^{-1} \subseteq U^{[*]} \quad \text{or} \quad U^{-1} = U^{[*]}, \tag{2.9}$$

respectively. Different from the Hilbert space situation, there exist isometric and unitary relations (with non-trivial kernels). Here, however, only unitary operators are encountered. In this connection note that combining the equalities $\ker H^* = (\operatorname{ran} H)^{\perp_2}$ and $\operatorname{mul} H^* = (\operatorname{dom} H)^{\perp_1}$

with (2.8) and (2.9) yields that for a unitary relation U between Kreĭn spaces the following equalities hold:

$$\ker U = (\operatorname{dom} U)^{[\perp]_1} \quad \text{and} \quad \operatorname{mul} U = (\operatorname{ran} U)^{[\perp]_2}, \tag{2.10}$$

where $[\perp]_i$ is the orthogonal complement with respect to $(\mathfrak{j}_i\,\cdot,-)_i$, for $i = 1, 2$. In particular, a unitary operator has a dense range and a unitary operator has a trivial kernel if and only if it is densely defined.

By the definitions in (2.9) a unitary operator is always closed, and an operator and its inverse are simultaneously isometric or unitary. Clearly, an operator U is isometric in the sense of (2.9) if and only if

$$(\mathfrak{j}_1 f, g)_1 = (\mathfrak{j}_2 U f, U g)_2, \quad f, g \in \operatorname{dom} U. \tag{2.11}$$

Furthermore, an isometric operator U is unitary if and only if for $g \in \mathfrak{K}_1$ and $g' \in \mathfrak{K}_2$, $(\mathfrak{j}_1 f, g)_1 = (\mathfrak{j}_2 U f, g')_2$ for all $f \in \operatorname{dom} U$, implies that $g \in \operatorname{dom} U$ and $g' = Ug$. In other words, unitary operators are a special type of maximal isometric operators.

Equation (2.11) shows that isometric operators map neutral, nonnegative and nonpositive subspaces of $(\mathfrak{H}_1, (\mathfrak{j}_1\,\cdot,-)_1)$ onto neutral, nonnegative and nonpositive subspaces of $(\mathfrak{H}_2, (\mathfrak{j}_2\,\cdot,-)_2)$, respectively. In general, isometric or unitary operators do not map closed subspaces onto closed subspaces or maximal semi-definite subspaces onto maximal semi-definite subspaces. However, if a unitary operator has a closed domain, or, equivalently, a closed range, then it preserves all such properties. In fact, unitary operators with closed domain, i.e. bounded unitary operators, also preserve the defect numbers of neutral subspaces, cf. Theorem 2.20 below.

Next the general definition of a unitary boundary triplet is recalled, cf. [Derkach et al., 2006, Definition 3.1], and its connection to unitary operators between Kreĭn spaces is explained.

Definition 2.4 Let S be a closed symmetric relation in a Hilbert space $(\mathfrak{H}, (\cdot,-)_{\mathfrak{H}})$ and let $(\mathcal{H}, (\cdot,-)_{\mathcal{H}})$ be an auxiliary Hilbert space. Then the triplet $\{\mathcal{H}, \Gamma_0, \Gamma_1\}$, where $\Gamma_i : S^* \subseteq \mathfrak{H}^2 \to \mathcal{H}$ is a linear operator for $i = 0, 1$, is called a *unitary boundary triplet* for S^*, if

(i) $\overline{\operatorname{dom}\Gamma} = S^*$ and $\overline{\operatorname{ran}}\,\Gamma = \mathcal{H}^2$, where $\Gamma = \Gamma_0 \times \Gamma_1$;
(ii) Green's identity holds: for every $\{f, f'\}, \{g, g'\} \in \operatorname{dom}\Gamma$,

$$(f', g)_{\mathfrak{H}} - (f, g')_{\mathfrak{H}} = (\Gamma_1\{f, f'\}, \Gamma_0\{g, g'\})_{\mathcal{H}} - (\Gamma_0\{f, f'\}, \Gamma_1\{g, g'\})_{\mathcal{H}};$$

(iii) if $g, g' \in \mathfrak{H}$ and $k, k' \in \mathcal{H}$ are such that for all $\{f, f'\} \in \operatorname{dom} \Gamma$

$$(f', g)_{\mathfrak{H}} - (f, g')_{\mathfrak{H}} = (\Gamma_1\{f, f'\}, k)_{\mathcal{H}} - (\Gamma_0\{f, f'\}, k')_{\mathcal{H}},$$

then $\{g, g'\} \in \operatorname{dom} \Gamma$ and $\{k, k'\} = \Gamma\{g, g'\}$.

If S is a densely defined operator, so that its adjoint S^* is also an operator, then $\{f, f'\}$ and $\{g, g'\}$ appearing in Definition 2.4 can be replaced by f and g, respectively. Green's identity in that case becomes:

$$(S^* f, g)_{\mathfrak{H}} - (f, S^* g)_{\mathfrak{H}} = (\Gamma_1 f, \Gamma_0 g)_{\mathcal{H}} - (\Gamma_0 f, \Gamma_1 g)_{\mathcal{H}},$$

cf. (2.1). Note that, with the notation as in (2.2), Definition 2.4 means that Γ is a unitary operator from $(\mathfrak{H}^2, (\mathfrak{j}_{\mathfrak{H}} \cdot, -))$ to $(\mathcal{H}^2, (\mathfrak{j}_{\mathcal{H}} \cdot, -))$. This, in particular, implies that $\ker \Gamma = S$, cf. (2.10) and Proposition 2.3.

Remark 2.5 If S is a symmetric operator and there exists a unitary boundary triplet for S^*, then $n_{\pm}(S) = \dim \mathcal{H}^{\mp}$, where $\mathcal{H}^+ + \mathcal{H}^-$ is the canonical decomposition of $\mathcal{H}^2$ induced by $\mathfrak{j}_{\mathcal{H}}$. Since it is easy to see that $\dim \mathcal{H}^+ = \dim \mathcal{H}^-$, unitary boundary triplets only exist for the adjoints of symmetric relations with equal defect numbers. For symmetric relations with unequal defect numbers either D-boundary triplets or boundary relations have to be used, see [Mogilevskiĭ, 2006] or [Derkach et al., 2006, Proposition 3.7], respectively.

2.3 Reduction operators

In the first subsection the basic properties of reduction operators are given and it is shown how reduction operators can be interpreted as unitary operators and/or unitary boundary triplets. In the second subsection a graph decomposition of reduction operators is presented which will be the central tool in Section 2.4. In the third subsection a classification of reduction operators is introduced and characterized. In the fourth and final subsection bounded reduction operators are shortly considered.

Reduction operators, unitary operators and boundary triplets
Let U be a reduction operator for S^* as in Definition 2.1, then as a direct consequence of its definition U has the following three basic properties:

(i) $\ker U = S$ and, hence, S is a symmetric operator in $(\mathfrak{H}, (\cdot, -))$;
(ii) $\overline{\operatorname{ran}} U = \mathfrak{K}$;
(iii) $I + W^2 = 0$,

see [Calkin, 1939a, Theorems 1.1, 1.2]. The third of the above properties allows one to connect reduction operators to unitary operators acting between Kreĭn spaces and, hence, to unitary boundary triplets. Namely, define the operator $\mathfrak{j}_W$ in $\mathfrak{K}$ as

$$\mathfrak{j}_W g = \mathrm{i}Wg, \quad g \in \mathfrak{K}.$$

Then $WW^* = I$ and $WW = -I$, see Definition 2.1 and (iii) above, yield that $W^{-1} = -W = W^*$ and, hence, $\mathfrak{j}_W = (\mathfrak{j}_W)^{-1} = (\mathfrak{j}_W)^*$. I.e., $\mathfrak{j}_W$ is a fundamental symmetry in the Hilbert space $(\mathfrak{K}, (\cdot, -))$ and, hence, $(\mathfrak{K}, (\mathfrak{j}_W \cdot, -))$ is a Kreĭn space.

Remark 2.6 Let U be a reduction operator for S^* as in Definition 2.1. In what follows $\mathfrak{j}_W$ always denotes the fundamental symmetry in the range space $(\mathfrak{K}, (\cdot, -))$ associated with the rotation W: $\mathfrak{j}_W = \mathrm{i}W$. The canonical decomposition of $\mathfrak{K}$ induced by $\mathfrak{j}_W$ is denoted by $\mathfrak{K}_W^+ + \mathfrak{K}_W^-$ and for the orthogonal projections onto $\mathfrak{K}_W^+$ and $\mathfrak{K}_W^-$ in $(\mathfrak{K}, (\cdot, -))$ the notation P_W^+ and P_W^- is reserved, respectively.

With the above notation (see also Example 2.2), condition (ii) in Definition 2.1 can be rewritten as

$$(\mathrm{gr}\, U)^\perp = (\mathfrak{j}_\mathfrak{H} \times -\mathfrak{j}_W)(\mathrm{gr}\, U). \tag{2.12}$$

Thus, $\mathrm{gr}\, U$ is a hyper-maximal neutral subspace of the Kreĭn space $(\mathfrak{H}^2 \times \mathfrak{K}, ((\mathfrak{j}_\mathfrak{H} \times -\mathfrak{j}_W) \cdot, -))$, see [Calkin, 1939a, Theorem 2.11]. This observation yields that reduction operators can be interpreted as unitary operators between Kreĭn spaces, and vice versa.

Proposition 2.7 *Let U be a reduction operator for S^* as in Definition 2.1. Then U is a unitary operator with $\ker U = S$ from the Kreĭn space $(\mathfrak{H}^2, (\mathfrak{j}_\mathfrak{H} \cdot, -))$ to the Kreĭn space $(\mathfrak{K}, (\mathfrak{j}_W \cdot, -))$. Conversely, if $\mathfrak{H}$ is a Hilbert space, U is a unitary operator from the Kreĭn space $(\mathfrak{H}^2, (\mathfrak{j}_\mathfrak{H} \cdot, -))$ to the Kreĭn space $(\mathfrak{K}, [\cdot, -])$, $\mathfrak{j}$ is a fundamental symmetry for $(\mathfrak{K}, [\cdot, -])$ and $S = \ker U$, then U is a reduction operator for S^* with range space $\mathfrak{K}$ and associated rotation $W := -\mathrm{i}\mathfrak{j}$.*

Proof The assertions follow from the discussion preceding the statement and the fact that U is a unitary relation between the Kreĭn spaces $(\mathfrak{K}_1, [\cdot, -]_1)$ and $(\mathfrak{K}_2, [\cdot, -]_2)$ if and only if $\mathrm{gr}\, U$ is a hyper-maximal neutral subspace of the Kreĭn space $(\mathfrak{K}_1 \times \mathfrak{K}_2, [\cdot, -]_{1,-2})$, where

$$[\{f, f'\}, \{g, g'\}]_{1,-2} = [f, g]_1 - [f', g']_2, \quad f, g \in \mathfrak{K}_1, f', g' \in \mathfrak{K}_2,$$

see e.g. [Wietsma, 2012, Proposition 3.3]. □

Since reduction operators are unitary they satisfy the equality

$$U^{-1} = \mathrm{j}_{\mathfrak{H}} U^* \mathrm{j}_W, \tag{2.13}$$

see Proposition 2.7 and (2.8), (2.9). Note that U^{-1} has a non-trivial multi-valued part if and only if U is a reduction operator for the adjoint of a non-trivial symmetry. Since unitary operators between Kreĭn spaces map neutral subspaces onto neutral subspace, see (2.11), Proposition 2.7 together with Proposition 2.3 shows that a reduction operator for S^* maps the graphs of symmetric extensions of S onto neutral subspaces of $(\mathfrak{K}, (\mathrm{j}_W \cdot, -))$ and, conversely; see [Calkin, 1939a, Section 1.3].

If U is a reduction operator for the adjoint of the symmetric operator S with range space $\mathfrak{K}$ and rotation W, then the defect numbers of S are directly connected with the dimensions of the Kreĭn space $(\mathfrak{K}, (\mathrm{j}_W \cdot, -))$:

$$n_+(S) = \dim \mathfrak{K}_W^- \quad \text{and} \quad n_-(S) = \dim \mathfrak{K}_W^+, \tag{2.14}$$

see [Calkin, 1939a, Theorem 3.7]. In view of Remark 2.5, this implies that reduction operators can only in certain cases be interpreted as unitary boundary triplets.

Corollary 2.8 *Let U be a reduction operator for S^* and assume that $n_+(S) = n_-(S)$. Then there exists a Hilbert space $(\mathcal{H}, (\cdot, -))$ and an everywhere defined unitary operator U_t from the Kreĭn space $(\mathfrak{K}, (\mathrm{j}_W \cdot, -))$ onto the Kreĭn space $(\mathcal{H}^2, (\mathrm{j}_{\mathcal{H}} \cdot, -))$ such that $\Gamma := U_t U$ induces a unitary boundary triplet for S^*. Conversely, if $\{\mathcal{H}, \Gamma_0, \Gamma_1\}$ is a unitary boundary triplet for S^*, then $\Gamma = \Gamma_0 \times \Gamma_1$ is a reduction operator for S^* with range space $\mathcal{H}^2$ and associated rotation $W := -\,\mathrm{i}\mathrm{j}_{\mathcal{H}}$.*

Proof By the assumption $n_+(S) = n_-(S)$ one has $\dim \mathfrak{K}_W^+ = \dim \mathfrak{K}_W^-$, see (2.14). Let $(\mathcal{H}, (\cdot, -))$ be any Hilbert space of dimension $\dim \mathfrak{K}_W^+$ and let $\mathcal{H}^+ + \mathcal{H}^-$ be the canonical decomposition of $\mathcal{H}^2$ induced by $\mathrm{j}_{\mathcal{H}}$, see Example 2.2. Then $\dim \mathcal{H}^+ = \dim \mathcal{H}^- = \dim \mathfrak{K}_W^+$ and, hence, there exist Hilbert space unitary operators U_+ and U_- from $\mathfrak{K}_W^+$ to $\mathcal{H}^+$ and $\mathfrak{K}_W^-$ to $\mathcal{H}^-$ and, consequently, $U_t = U_+ \times U_-$ is an everywhere defined Kreĭn space unitary operator as in the statement. Finally, general composition results imply that $U_t U$ is a unitary operator from the Kreĭn space $(\mathfrak{H}^2, (\mathrm{j}_{\mathfrak{H}} \cdot, -))$ to the Kreĭn space $(\mathcal{H}^2, (\mathrm{j}_{\mathcal{H}} \cdot, -))$. In other words, $\Gamma := U_t U$ induces a unitary boundary triplet for S^*, see Definition 2.4 and the discussion following it.

Recall that if $\{\mathcal{H}, \Gamma_0, \Gamma_1\}$ is a unitary boundary triplet, then $\Gamma = \Gamma_0 \times \Gamma_1$ is a unitary operator from $(\mathfrak{H}^2, (\mathrm{j}_{\mathfrak{H}} \cdot, -))$ to $(\mathcal{H}^2, (\mathrm{j}_{\mathcal{H}} \cdot, -))$, see the discussion following Definition 2.4. Hence, the converse assertion follows directly from Proposition 2.7. □

A graph decomposition of reduction operators Here a graph decomposition of reduction operators is presented; this result, which in [Calkin, 1939a, Theorem 3.5] is formulated only for the domain of a reduction operator, plays a central role in the rest of this article.

As a first preparation for proving the above mentioned result Calkin showed that it suffices to consider only reduction operators for the trivial symmetry, see [Calkin, 1939a, Theorem 3.1]. Namely, if U is a reduction operator for S^* as in Definition 2.1, then the operator U_o defined by

$$U_o\{f,f'\} = U\{f,f'\}, \quad \{f,f'\} \in \operatorname{dom} U_o := (\operatorname{gr} S)^{\perp} \tag{2.15}$$

is unitary from the Kreĭn space $(\mathfrak{H}^2 \ominus (\ker U + \mathrm{j}_{\mathfrak{H}} \ker U), (\mathrm{j}_{\mathfrak{H}}\,\cdot, -))$ to the Kreĭn space $(\mathfrak{K}, (\mathrm{j}_W\,\cdot, -))$ with $\ker U_o = \{0\}$. Conversely, if U_o is a unitary operator from $(\mathfrak{H}^2 \ominus (\ker U + \mathrm{j}_{\mathfrak{H}} \ker U), (\mathrm{j}_{\mathfrak{H}}\,\cdot, -))$ to $(\mathfrak{K}, (\mathrm{j}\,\cdot, -))$, then U defined via

$$\operatorname{gr} U = \operatorname{gr} S \times \{0\} + \operatorname{gr} U_o \tag{2.16}$$

is a reduction operator for S^* with range space $\mathfrak{K}$ and rotation $W := -\mathrm{i}\mathrm{j}$.

As the second step Calkin introduced equivalence classes of reduction operators and showed that each equivalence class possesses a nonnegative selfadjoint representative, see [Calkin, 1939a, Theorem 3.6]; this representative was in fact obtained from the polar decomposition of U_o as in (2.15). As a consequence of the connection between reduction operators and nonnegative selfadjoint operators (which are also unitary), Calkin showed that the domain of a reduction operator can be decomposed in a specific manner, see [Calkin, 1939a, Theorem 3.5]. Here the interesting part of that statement is reproduced in an expanded form, cf. [Wietsma, 2012, Theorem 5.1]. Using this more general form, the proofs for several of Calkin's statements can be essentially simplified.

Theorem 2.9 *Let U be a reduction operator for S^*. Then the domain and range of U have the following decompositions:*

$$\begin{aligned} \operatorname{dom} U &= \operatorname{gr} S \oplus \mathfrak{K}_d \oplus \mathfrak{N}_d \oplus (\mathrm{j}_{\mathfrak{H}} \mathfrak{N}_d \cap \operatorname{dom} U); \\ \operatorname{ran} U &= \mathfrak{K}_r \oplus \mathfrak{N}_r \oplus (\mathrm{j}_W \mathfrak{N}_r \cap \operatorname{ran} U), \end{aligned} \tag{2.17}$$

where

(i) $(\mathfrak{K}_d, (\mathrm{j}_{\mathfrak{H}}\,\cdot, -))$ *and* $(\mathfrak{K}_r, (\mathrm{j}_W\,\cdot, -))$ *are Kreĭn spaces such that* $U(\mathfrak{K}_d) = \mathfrak{K}_r$*, and, moreover,* $\mathrm{j}_{\mathfrak{H}} \mathfrak{K}_d = \mathfrak{K}_d$ *and* $\mathrm{j}_W \mathfrak{K}_r = \mathfrak{K}_r$*;*

(ii) $\operatorname{gr} S + \mathfrak{N}_d \subseteq \operatorname{dom} U$ *and* $\mathfrak{N}_r \subseteq \operatorname{ran} U$ *are hyper-maximal neutral subspaces of* $(\mathfrak{H}^2 \ominus \mathfrak{K}_d, (\mathrm{j}_{\mathfrak{H}} \cdot, -))$ *and* $(\mathfrak{K} \ominus \mathfrak{K}_r, (\mathrm{j}_W \cdot, -))$, *respectively, such that*

$$U(\mathfrak{N}_d) = \mathrm{j}_W \mathfrak{N}_r \cap \operatorname{ran} U \quad \textit{and} \quad U(\mathrm{j}_{\mathfrak{H}} \mathfrak{N}_d \cap \operatorname{dom} U) = \mathfrak{N}_r \tag{2.18}$$

and, moreover,

$$\operatorname{clos}(\mathrm{j}_{\mathfrak{H}} \mathfrak{N}_d \cap \operatorname{dom} U) = \mathrm{j}_{\mathfrak{H}} \mathfrak{N}_d \quad \textit{and} \quad \operatorname{clos}(\mathrm{j}_W \mathfrak{N}_r \cap \operatorname{ran} U) = \mathrm{j}_W \mathfrak{N}_r.$$

Proof To prove the statement assume without loss of generality that S is the trivial symmetry, see (2.15), and let $\mathfrak{K}$ be the range space of U and W its associated rotation. Recall that $\mathrm{j}_{\mathfrak{H}}$ as in (2.2) is a fundamental symmetry in $(\mathfrak{H}^2, (\cdot, -))$ and that $\mathrm{j}_W = \mathrm{i}W$ is a fundamental symmetry in $(\mathfrak{K}, (\cdot, -))$.

Step 1: Let $U_t|U|$ be the polar decomposition of U as a closed operator from the Hilbert space $(\mathfrak{H}^2, (\cdot, -))$ to the Hilbert space $(\mathfrak{K}, (\cdot, -))$. Then $|U|$ is a nonnegative selfadjoint operator in $(\mathfrak{H}^2, (\cdot, -))$ and U_t is a closed isometric operator from $(\mathfrak{H}^2, (\cdot, -))$ to $(\mathfrak{K}, (\cdot, -))$ with $\overline{\operatorname{ran}}\, U_t = \overline{\operatorname{ran}}\, U = \mathfrak{K}$ and $\operatorname{dom} U_t = \overline{\operatorname{ran}}\, U^* = (\ker \operatorname{clos}(U))^{\perp} = \mathfrak{H}^2$, i.e. U_t is a unitary operator from $(\mathfrak{H}^2, (\cdot, -))$ to $(\mathfrak{K}, (\cdot, -))$.

Step 2: Next it is shown that $|U|$ is a unitary operator in the Kreĭn space $(\mathfrak{H}^2, (\mathrm{j}_{\mathfrak{H}} \cdot, -))$; then U_t is also a unitary operator from the Kreĭn space $(\mathfrak{H}^2, (\mathrm{j}_{\mathfrak{H}} \cdot, -))$ to the Kreĭn space $(\mathfrak{K}, (\mathrm{j}_W \cdot, -))$, i.e.

$$|U|^{-1} = \mathrm{j}_{\mathfrak{H}} |U|^* \mathrm{j}_{\mathfrak{H}} = \mathrm{j}_{\mathfrak{H}} |U| \mathrm{j}_{\mathfrak{H}} \quad \text{and} \quad U_t \mathrm{j}_{\mathfrak{H}} = \mathrm{j}_W U_t. \tag{2.19}$$

To see that (2.19) holds, recall that since U is unitary it satisfies $U^{-1} = \mathrm{j}_{\mathfrak{H}} U^* \mathrm{j}_W = \mathrm{j}_{\mathfrak{H}} |U| U_t^{-1} \mathrm{j}_W$, see (2.13). Applying U_t on both sides of the preceding equality yields

$$|U|^{-1} = (U_t^{-1} U)^{-1} = U^{-1} U_t = \mathrm{j}_{\mathfrak{H}} |U| U_t^{-1} \mathrm{j}_W U_t. \tag{2.20}$$

Taking adjoints on both sides and using the selfadjointness of $|U|$, $\mathrm{j}_{\mathfrak{H}}$ and j_W yields $|U|^{-1} = U_t^{-1} \mathrm{j}_W U_t |U| \mathrm{j}_{\mathfrak{H}}$. Combining this with (2.20) gives $|U|^{-2} = \mathrm{j}_{\mathfrak{H}} |U|^2 \mathrm{j}_{\mathfrak{H}} = (\mathrm{j}_{\mathfrak{H}} |U| \mathrm{j}_{\mathfrak{H}})^2$. Since the nonnegative square root of a nonnegative operator is unique, this yields $|U|^{-1} = \mathrm{j}_{\mathfrak{H}} |U| \mathrm{j}_{\mathfrak{H}}$ and (2.19).

Step 3: Let $\{E_t\}_{t \in \mathbb{R}}$ and $\{F_t\}_{t \in \mathbb{R}}$ be the spectral families of the nonnegative selfadjoint operators $|U|$ and $|U|^{-1}$ in $(\mathfrak{H}^2, (\cdot, -))$, respectively. Then $F_t = I - E_{(1/t)-}$ for $t > 0$. Moreover, $\mathfrak{N}_d$, $\mathfrak{K}_d$, and $\mathfrak{M}_r$ defined as

$$\mathfrak{N}_d = \operatorname{ran} E_{1-}, \quad \mathfrak{K}_d = \ker(|U| - I), \quad \mathfrak{M}_r = \operatorname{ran} F_{1-} \tag{2.21}$$

are closed subspaces of $(\mathfrak{H}^2, (\cdot, -))$ such that

$$\operatorname{dom} |U| = \mathfrak{N}_d \oplus \mathfrak{K}_d \oplus |U|^{-1}(\mathfrak{M}_r), \quad \operatorname{ran} |U| = \mathfrak{M}_r \oplus \mathfrak{K}_d \oplus |U|(\mathfrak{N}_d). \tag{2.22}$$

The equality $|U|^{-1} = \mathfrak{j}_{\mathfrak{H}}|U|\mathfrak{j}_{\mathfrak{H}}$ proven in Step 2 combined with the stated connection between the spectral measures of $|U|$ and $|U|^{-1}$ implies that

$$I - E_{(1/t)-} = \mathfrak{j}_{\mathfrak{H}} E_t \mathfrak{j}_{\mathfrak{H}}, \quad t > 0. \tag{2.23}$$

This yields $E_1 - E_{1-} = \mathfrak{j}_{\mathfrak{H}}(E_1 - E_{1-})\mathfrak{j}_{\mathfrak{H}}$ and, hence, $\ker(|U| - I) = \operatorname{ran}(E_1 - E_{1-})$ is $\mathfrak{j}_{\mathfrak{H}}$-invariant. Consequently, $(\mathfrak{K}_d, (\mathfrak{j}_{\mathfrak{H}}\cdot, -))$ is a Kreĭn space, since $\mathfrak{K}_d$ is closed by definition, see (2.21). From (2.23) it also follows that $\mathfrak{j}_{\mathfrak{H}}\operatorname{ran}(I - E_1) = \operatorname{ran}(E_{1-}\mathfrak{j}_{\mathfrak{H}}) = \mathfrak{N}_d$. Since $\operatorname{ran}(I - E_1) \cap \operatorname{dom}|U| = |U|^{-1}(\mathfrak{M}_r)$, this implies that

$$|U|^{-1}(\mathfrak{M}_r) = \mathfrak{j}_{\mathfrak{H}}\mathfrak{N}_d \cap \operatorname{dom}|U|.$$

Combining this with the observation that

$$\mathfrak{H}^2 = \overline{\operatorname{dom}}\,|U| = \mathfrak{N}_d \oplus \mathfrak{K}_d \oplus \operatorname{clos}(|U|^{-1}(\mathfrak{M}_r))$$

and the proven fact that $\mathfrak{K}_d$ is $\mathfrak{j}_{\mathfrak{H}}$-invariant shows that $\mathfrak{N}_d$ is a hyper-maximal neutral subspace of the Kreĭn space $(\mathfrak{H}^2 \ominus_1 \mathfrak{K}_d, (\mathfrak{j}_{\mathfrak{H}}\cdot, -))$, see (2.5), and that $\operatorname{clos}(|U|^{-1}(\mathfrak{M}_r)) = \mathfrak{j}_{\mathfrak{H}}\mathfrak{N}_d$.

Step 4: A similar argument as above shows that $\mathfrak{M}_r$ in (2.22) is a hyper-maximal neutral subspace of the Kreĭn space $(\mathfrak{H}^2 \ominus \mathfrak{K}_d, (\mathfrak{j}_{\mathfrak{H}}\cdot, -))$, that $|U|(\mathfrak{N}_d) = \mathfrak{j}_{\mathfrak{H}}\mathfrak{M}_r \cap \operatorname{ran}|U|$ and that $\operatorname{clos}(|U|(\mathfrak{N}_d)) = \mathfrak{j}_{\mathfrak{H}}\mathfrak{M}_r$. Consequently, by (2.19) and (2.22) the statement holds with $\mathfrak{N}_d$ and $\mathfrak{K}_d$ as (2.21) and with $\mathfrak{N}_r := U_t(\mathfrak{M}_r)$ and $\mathfrak{K}_r := U_t(\mathfrak{K}_d)$. □

Remark 2.10 From the proof of Theorem 2.9 one can read off the central mapping properties of a reduction operator (i.e. of a nondensely defined Kreĭn space unitary operator) U:

(i) U has dense domain in S^* with $\ker U = S$ and in its polar decomposition $U = U_t|U|$, U_t is unitary both in the Hilbert and the Kreĭn space sense, and $|U|$ is a nonnegative selfadjoint operator which is also unitary in the Kreĭn space sense.

(ii) $U : \mathfrak{K}_d \to \mathfrak{K}_r$ is unitary in the Kreĭn space sense and, moreover $U^0 := U \restriction \mathfrak{K}_d = U_t \restriction \mathfrak{K}_d$, i.e. U^0 is also unitary in the Hilbert space sense.

(iii) The (possible) unbounded behavior of U is only on the subspace $\mathfrak{N}_d \oplus (\mathfrak{j}_{\mathfrak{H}}\mathfrak{N}_d \cap \operatorname{dom} U)$. To describe this consider the restriction $U' := U \restriction (\mathfrak{N}_d \oplus \mathfrak{j}_{\mathfrak{H}}\mathfrak{N}_d)$, which maps into the Kreĭn space $(\mathfrak{N}_r \oplus \mathfrak{j}_W\mathfrak{N}_r, (\mathfrak{j}_W\cdot, -))$. Since $\mathfrak{N}_d \subseteq \operatorname{dom} U'$ is a closed subspace, (2.18) implies that U' admits an off-diagonal block representation with respect to the decompositions

$\mathfrak{N}_d \oplus \mathfrak{j}_{\mathfrak{H}}\mathfrak{N}_d$ and $\mathfrak{N}_r \oplus \mathfrak{j}_W\mathfrak{N}_r$ which (in view of (2.13)) is given by

$$U' = \begin{pmatrix} 0 & C^{-*}\mathfrak{j}_{\mathfrak{H}} \\ \mathfrak{j}_W C & 0 \end{pmatrix}, \quad C = \mathfrak{j}_W U \upharpoonright \mathfrak{N}_d, \tag{2.24}$$

where C is a closed contractive operator from $(\mathfrak{N}_d, (\cdot, -))$ to $(\mathfrak{N}_r, (\cdot, -))$ with $\operatorname{dom} C = \mathfrak{N}_d$, $\overline{\operatorname{ran}}\, C = \mathfrak{N}_r$ and $\ker C = \{0\}$. Note also that the (expansive) operator $C^{-*} : \mathfrak{N}_d \to \mathfrak{N}_r$ is unbounded if and only if U' or, equivalently, U is unbounded. Consequently, all the major statements concerning unbounded reduction in this paper can be most easily understood having the block representation (2.24) in mind.

(iv) Next observe that $U_t^{-1}U' = |U| \upharpoonright (\mathfrak{N}_d \oplus \mathfrak{j}_{\mathfrak{H}}\mathfrak{N}_d)$ is a unitary operator in the Kreĭn space $(\mathfrak{H}^2, (\mathfrak{j}_{\mathfrak{H}} \cdot, -))$, because U_t is a bounded unitary operator from $(\mathfrak{H}^2, (\mathfrak{j}_{\mathfrak{H}} \cdot, -))$ to $(\mathfrak{K}, (\mathfrak{j}_W \cdot, -))$. In fact, since U_t maps $\mathfrak{N}_d$ onto $\mathfrak{N}_r$ and $\mathfrak{j}_{\mathfrak{H}}\mathfrak{N}_d$ onto $\mathfrak{j}_W\mathfrak{N}_r$, cf. (2.19), item (iii) implies that there exists a closed (expansive) operator B in the Hilbert space $(\mathfrak{N}_d, (\cdot, -))$ with $\overline{\operatorname{dom}}\, B = \mathfrak{N}_d = \operatorname{ran} B$ and $\ker B = \{0\}$, such that

$$U_t^{-1}U' = \begin{pmatrix} 0 & B\mathfrak{j}_{\mathfrak{H}} \\ \mathfrak{j}_{\mathfrak{H}} B^{-*} & 0 \end{pmatrix}, \tag{2.25}$$

where the adjoint of B is taken in $(\mathfrak{N}_d, (\cdot, -))$ and the block representation is with respect to the decomposition $\mathfrak{N}_d \oplus \mathfrak{j}_{\mathfrak{H}}\mathfrak{N}_d$ of $\mathfrak{H}^2$, see (2.5).

(v) Using Proposition 2.7 a converse to Theorem 2.9 can be established: For some decomposition of $\mathfrak{H}^2$ of the form (2.17) define Kreĭn space unitary operators $U^0 : \mathfrak{K}_d \to \mathfrak{K}_r$ (bounded) and $U' : \mathfrak{N}_d \oplus \mathfrak{j}_{\mathfrak{H}}\mathfrak{N}_d \to \mathfrak{N}_d \oplus \mathfrak{j}_{\mathfrak{H}}\mathfrak{N}_d$ (unbounded) via (2.25), where B is a closed densely defined operator with dense range and $\ker B = \{0\}$, then U defined by (2.16) with $U_o = U^0 \oplus U'$ is a unitary operator in $(\mathfrak{H}^2, (\mathfrak{j}_{\mathfrak{H}} \cdot, -))$ and, hence, can be interpreted as a reduction operator for S^* by Proposition 2.7. In fact, in what follows the selection $\mathfrak{K}_d = \{0\}$ and $U^0 = 0$ is mainly used.

One consequence of Theorem 2.9, which is not explicitly stated in Calkin's paper, is the following result.

Corollary 2.11 *Let U be a reduction operator for S^*. Then there exists a maximal symmetric extension S_m of S such that $S_m^* \subseteq \operatorname{dom} U$.*

Proof Let $\mathfrak{M}_m$ be any maximal neutral subspace of $(\mathfrak{K}_d, (\mathfrak{j}_{\mathfrak{H}} \cdot, -))$; cf. Theorem 2.9 (i). Then, clearly,

$$\operatorname{gr} S_m := \ker U + \mathfrak{N}_d + \mathfrak{M}_m$$

is a maximal neutral subspace of $(\mathfrak{H}^2, (\mathfrak{j}_{\mathfrak{H}} \cdot, -))$ and, consequently, S_m is a maximal symmetric extension of S, see Proposition 2.3. Moreover,

$$S_m^* = (\operatorname{gr} S_m)^{[\perp]} = \ker U + \mathfrak{N}_d + \mathfrak{M}_m^{[\perp]} \cap \mathfrak{K}_d$$

shows that $S_m^* \subseteq \operatorname{dom} U$. □

A classification of reduction operators Here Calkin's classification of reduction operators into two disjoint classes is presented and some characterizations for those classes are given. The reason for introducing this classification is connected with the type of maximal symmetric operators that are contained in the domain of a reduction operator; see Theorems 2.21, 2.22 below.

Definition 2.12 [Calkin, 1939a, Definition 3.5] A reduction operator U for S^* is said to be of *type I* if at least one of the subspaces

$$\widehat{\mathfrak{N}}_{\mathrm{i}}(S^*) \cap \operatorname{dom} U \quad \text{and} \quad \widehat{\mathfrak{N}}_{-\mathrm{i}}(S^*) \cap \operatorname{dom} U$$

contains no infinite-dimensional closed subspace. Otherwise U is said to be of *type II*.

Using the following characterization of compact operators, Calkin connected the above definition to compact operators.

Lemma 2.13 [Calkin, 1939a, Lemma 3.1] *Let T be a closed densely defined operator from the Hilbert space $(\mathfrak{H}_1, (\cdot, -)_1)$ to the Hilbert space $(\mathfrak{H}_2, (\cdot, -)_2)$. Then T is compact if and only if each closed subspace contained in* $\operatorname{ran} T$ *is finite-dimensional.*

Proof First notice that T is compact or, equivalently, T^* is compact if and only if $|T^*| = (TT^*)^{\frac{1}{2}}$ is compact by the polar decomposition $T^* = V|T^*|$. Since $\operatorname{ran}|T^*| = \operatorname{ran} T$, it suffices to prove the statement for nonnegative selfadjoint operators T. In addition, one can assume that $\ker T = \{0\}$ and that $\dim \mathfrak{H}_1 = \infty$.

Now to prove the necessity, let T be compact and assume that $\mathfrak{M} \subseteq \operatorname{ran} T$ is a closed subspace. Since T is bounded, $\mathfrak{N} := T^{-1}(\mathfrak{M})$ is also closed and $T_0 := T \upharpoonright \mathfrak{N}$ is a closed bounded operator from $\mathfrak{N}$ onto $\mathfrak{M}$. By the closed graph theorem T_0^{-1} is bounded. If $\mathfrak{U} \subseteq \mathfrak{M}$ is a closed bounded set, $T_0^{-1}(\mathfrak{U})$ is also a closed bounded set in $\mathfrak{N}$ and, hence the set $\mathfrak{U}$ must be compact by compactness of T. Therefore, $\dim \mathfrak{M} < \infty$.

Conversely, assume that $\operatorname{ran} T$ contains only finite-dimensional closed subspaces. Let $E(\cdot)$ be the spectral measure of the nonnegative selfadjoint operator T. Then $\mathfrak{M}_n := \operatorname{ran} E([a_n, \infty))$, where $a_n > 0$, is a closed subspace which reduces T. The operator $T_n = T \upharpoonright \mathfrak{M}_n$ is selfadjoint in $\mathfrak{M}_n$ and boundedly invertible, since $a_n > 0$ is a lower bound for T_n. Thus, $\operatorname{ran} T_n = \mathfrak{M}_n \subseteq \operatorname{ran} T$ and hence $\dim \mathfrak{M}_n < \infty$. Therefore, T_n being a selfadjoint operator in a finite-dimensional space has only finitely many eigenvalues with finite multiplicity. By letting $a_n \to 0$ this clearly implies that T is a compact operator. □

Remark Lemma 2.13 was used by Calkin to analyze the type of reduction operators and to obtain answers to the problems stated in the introduction. In his famous paper [Calkin, 1941] used Lemma 2.13 as a tool to characterize all two-sided ideals contained in the algebra of bounded everywhere defined operators on a separable Hilbert space; see Theorem 2.32. The property that all closed subspaces contained in the range are finite-dimensional has been studied subsequently also in the Banach space setting; see e.g. [Cross, 1995] and the references therein.

The type of a reduction operator was characterized in [Calkin, 1939a, Theorem 3.13] using the domain decomposition in Theorem 2.9.

Theorem 2.14 *Let U be a reduction operator for S^* and let $\mathfrak{K}_d$ and $\mathfrak{N}_d$ be as in Theorem 2.9. Then U is of type I if and only if*

(i) *$|U| = (U^*U)^{\frac{1}{2}}$ induces a compact operator in $\mathfrak{N}_d \Leftrightarrow C$ in (2.24), or equivalently B^{-1} in (2.25), is a compact operator;*
(ii) *at least one of the subspaces $\widehat{\mathfrak{N}}_{\mathrm{i}}(S^*) \cap \mathfrak{K}_d$ and $\widehat{\mathfrak{N}}_{-\mathrm{i}}(S^*) \cap \mathfrak{K}_d$ is finite-dimensional, i.e, $\mathfrak{K}_d$ (and therefore also $\mathfrak{K}_r$) in Theorem 2.9 is a Pontryagin space.*

Proof By Theorem 2.9 and items (ii) and (iii) in Remark 2.10 the defect subspaces of S contained in the domain of U can be described as follows:

$$\begin{aligned}\widehat{\mathfrak{N}}_{\mathrm{i}}(S^*) \cap \operatorname{dom} U &= (\widehat{\mathfrak{N}}_{\mathrm{i}}(S^*) \cap \mathfrak{K}_d) \oplus \{f + \mathfrak{j}_{\mathfrak{H}} f : f \in \operatorname{ran} C^*\};\\ \widehat{\mathfrak{N}}_{-\mathrm{i}}(S^*) \cap \operatorname{dom} U &= (\widehat{\mathfrak{N}}_{-\mathrm{i}}(S^*) \cap \mathfrak{K}_d) \oplus \{f - \mathfrak{j}_{\mathfrak{H}} f : f \in \operatorname{ran} C^*\},\end{aligned} \tag{2.26}$$

see (2.24). Note that in order to obtain (2.26) it was used that $\mathfrak{K}_d = \mathfrak{K}_d \cap \widehat{\mathfrak{N}}_{\mathrm{i}}(S^*) + \mathfrak{K}_d \cap \widehat{\mathfrak{N}}_{-\mathrm{i}}(S^*)$, which holds because $\mathfrak{K}_d$ is $\mathfrak{j}_{\mathfrak{H}}$-invariant. The equations in (2.26) show that U is of type I if and only if each closed subspace contained in $\operatorname{ran} C^*$ is finite-dimensional and in addition (ii) holds. By Lemma 2.13 the first condition is satisfied if and only if C^* or, equivalently, $C = (\mathfrak{j}_W U_t |U|) \upharpoonright \mathfrak{N}_d$ is a compact operator. □

The formulas (2.26) in the proof of Theorem 2.14 show that

$$n_{\pm}(S) = \dim (\widehat{\mathfrak{N}}_{\mp\mathrm{i}}(S^*) \cap \mathfrak{K}_d) + \dim \mathfrak{N}_d. \tag{2.27}$$

If U is unbounded, then $\dim \mathfrak{N}_d = \infty$ and, thus, $n_+(S) = n_-(S) = \infty$ and, moreover, $\widehat{\mathfrak{N}}_{\mathrm{i}}(S^*) \cap \operatorname{dom} U$ and $\widehat{\mathfrak{N}}_{-\mathrm{i}}(S^*) \cap \operatorname{dom} U$ contain both at least finite-dimensional closed subspaces, since in this case $\operatorname{ran} C^*$ is a non-closed infinite-dimensional subspace. Observe, also that

$$\operatorname{dom} U \cap \mathfrak{H}^{\pm} = \operatorname{dom} U \cap \widehat{\mathfrak{N}}_{\pm\mathrm{i}}(S^*), \tag{2.28}$$

see Example 2.2. Denote

$$k_{\pm}(\mathfrak{K}_d) = \dim (\mathfrak{H}^{\pm} \cap \mathfrak{K}_d) = \dim (\widehat{\mathfrak{N}}_{\pm\mathrm{i}}(S^*) \cap \mathfrak{K}_d).$$

If U is of type I, then by Theorem 2.14 at least one of the numbers $k_+(\mathfrak{K}_d)$ and $k_-(\mathfrak{K}_d)$ is finite, and the pair of numbers

$$\begin{cases} (0, k_+(\mathfrak{K}_d) - k_-(\mathfrak{K}_d)), & \text{if } k_+(\mathfrak{K}_d) > k_-(\mathfrak{K}_d); \\ (k_-(\mathfrak{K}_d) - k_+(\mathfrak{K}_d), 0), & \text{otherwise}, \end{cases} \tag{2.29}$$

is called the *characteristic index* of U, see [Calkin, 1939a, Definition 3.6]; the small difference in the present definition is explained by its later use (cf. e.g. [Calkin, 1939a, Theorem 3.15]). This notion is used when describing maximal symmetric extensions of S contained in the domain of a reduction operator of type I.

Next proceed to characterizations of reduction operators of type II. It is useful to start with a characterization which is not explicitly stated in Calkin's paper, cf. Corollary 2.11.

Proposition 2.15 *Let U be a reduction operator for S^*. Then U is of type II if and only if there exists a closed symmetric extension T of S with $n_+(T) = \infty = n_-(T)$ such that*

$$S \subset T \subset T^* \subseteq \operatorname{dom} U.$$

Proof To prove the sufficiency, let T be a closed symmetric relation as in the statement. The assumption $n_+(T) = \infty = n_-(T)$ implies that

$$T^* \cap \widehat{\mathfrak{N}}_{\mathrm{i}}(S^*) \quad \text{and} \quad T^* \cap \widehat{\mathfrak{N}}_{-\mathrm{i}}(S^*)$$

are infinite-dimensional closed subspaces. Since $T^* \subseteq \operatorname{dom} U$, it follows that U is of type II.

To prove the converse let S_m be a maximal symmetric extension of S such that $S_m^* \subseteq \operatorname{dom} U$, see Corollary 2.11, and assume without loss of generality that $n_-(S_m) = 0$. Moreover, let K be the angular operator of $\operatorname{gr} S_m$ with respect to $\mathfrak{H}^+ \supseteq \widehat{\mathfrak{N}}_{\mathrm{i}}(S^*)$:

$$\operatorname{gr} S_m = \{ f + Kf : f \in \mathfrak{H}^+ \},$$

see Example 2.2. Next let $\mathfrak{D}^+$ be an infinite-dimensional closed subspace of $\widehat{\mathfrak{N}}_{\mathrm{i}}(S^*) \cap \operatorname{dom} U \subseteq \mathfrak{H}^+$, which exists because U is by assumption a reduction operator of type II. Then, since the angular operator is a closed isometric operator from the Hilbert space $(\mathfrak{H}^+, (\mathrm{j}_{\mathfrak{H}} \cdot, -))$ to the Hilbert space $(\mathfrak{H}^-, -(\mathrm{j}_{\mathfrak{H}} \cdot, -))$, $K(\mathfrak{D}^+)$ is an infinite-dimensional closed subspace of $\mathfrak{H}^-$, which is by construction contained in the domain of U, since $S_m^* \subseteq \operatorname{dom} U$. I.e., $K(\mathfrak{D}^+) \subseteq \widehat{\mathfrak{N}}_{-\mathrm{i}}(S^*) \cap \operatorname{dom} U$, see (2.28). Consequently, the restriction T of S_m defined via

$$\operatorname{gr} T = \{\, f + Kf : f \in \mathfrak{H}^+ \ominus \mathfrak{D}^+\} \tag{2.30}$$

satisfies $T^* = S_m^* + \mathfrak{D}^+ = S_m^* + K(\mathfrak{D}^+) \subseteq S^* \cap \operatorname{dom} U$. Moreover, $n_+(T) \geq \dim K(\mathfrak{D}^+) = \infty$ and $n_-(T) \geq \dim \mathfrak{D}^+ = \infty$. □

Next the type of a reduction operator is characterized by means of maximal symmetric relations which are contained in its domain. The argument for the sufficiency of the condition in the next theorem combines [Calkin, 1939a, Lemma 4.1] and part of [Calkin, 1939a, Theorem 4.4].

Theorem 2.16 *Let U be a reduction operator for S^* and assume that $n_+(S)$, $n_-(S) \leq \aleph_0$. Then U is of type II if and only if there exist maximal symmetric extensions $S_1 \subseteq \operatorname{dom} U$ and $S_2 \subseteq \operatorname{dom} U$ of S such that $n_+(S_1) \neq n_+(S_2)$ or $n_-(S_1) \neq n_-(S_2)$.*

Proof The necessity is obtained directly from Proposition 2.15, since $T \subseteq T^* \subseteq \operatorname{dom} U$ has defect numbers $n_+(T) = n_-(T) = \infty$ and all maximal symmetric extensions of T are automatically contained in $\operatorname{dom} U$.

To prove the sufficiency assume without loss of generality that S is the trivial symmetry. Then $\mathfrak{H}^+ + \mathfrak{H}^-$, where $\mathfrak{H}^+ := \widehat{\mathfrak{N}}_{\mathrm{i}}(S^*)$ and $\mathfrak{H}^- := \widehat{\mathfrak{N}}_{-\mathrm{i}}(S^*)$, is a canonical decomposition of $(\mathfrak{H}^2, (\mathrm{j}_{\mathfrak{H}} \cdot, -))$, see Example 2.2. Since $n_+(S), n_-(S) \leq \aleph_0$, there are three distinct cases to consider:

(a) $n_-(S_1) = 0 = n_-(S_2)$, $n_+(S_1) < n_+(S_2)$ and $n_+(S_1) < \infty$;
(b) $n_+(S_1) = 0 = n_+(S_2)$, $n_-(S_1) < n_-(S2)$ and $n_-(S_1) < \infty$;
(c) $n_-(S_1) = 0 \neq n_-(S_2)$ and $n_+(S_1) \neq 0 = n_+(S_2)$.

In the proof of the above cases the (closed) angular operators of $\operatorname{gr} S_1$ and $\operatorname{gr} S_2$ with respect to $\mathfrak{H}^+$ are denoted by K_1 and K_2:

$$\operatorname{gr} S_i = \{f^+ + K_i f^+ : f^+ \in P^+(\operatorname{gr} S_i)\}, \quad i = 1, 2. \tag{2.31}$$

Then $n_-(S_i) = \dim(\operatorname{dom} K_i)^\perp$ and $n_+(S_i) = \dim(\operatorname{ran} K_i)^\perp$, $i = 1, 2$.

(a) & (b) By symmetry it suffices to consider only the case (a).

Step 1: The mapping $X := K_2 K_1^{-1}$ is a closed isometric operator in the Hilbert space $(\mathfrak{H}^-, -(\mathrm{j}_{\mathfrak{H}}\,\cdot, -))$, which, because of the assumptions in (a), satisfies

$$\dim(\operatorname{dom} X)^{\perp_1} < \dim(\operatorname{ran} X)^{\perp_1} \quad \text{and} \quad \dim(\operatorname{dom} X)^{\perp_1} < \infty.$$

Consequently, there exists a finite-dimensional isometric extension Y of X such that $\operatorname{dom} Y = \mathfrak{H}^-$ and $\operatorname{ran} Y \neq \mathfrak{H}^-$. If $\operatorname{ran}(I - Y)$ does not contain an infinite-dimensional closed subspace, then $I - Y$ is a compact operator by Lemma 2.13. Hence, either $0 \in \rho(Y)$ or $0 \in \sigma_p(Y)$ and, since $\operatorname{ran} Y \neq \mathfrak{H}^-$, this yields $\ker Y \neq \{0\}$ (cf. the Fredholm alternative); this is impossible since Y is an everywhere defined isometric operator in a Hilbert space. Hence, $\operatorname{ran}(I - Y)$ and, thus, also $\operatorname{ran}(I - X)$ contain an infinite-dimensional closed subspace. Next observe that the assumption $S_1, S_2 \subseteq \operatorname{dom} U$ and the formulas (2.31) imply that $\operatorname{ran}(K_1 - K_2) \subseteq \operatorname{dom} U \cap \mathfrak{H}^-$ and therefore

$$\operatorname{ran}(I - X) = \operatorname{ran}((K_1 - K_2)K_1^{-1}) \subseteq \operatorname{ran}(K_1 - K_2) \subseteq \operatorname{dom} U \cap \mathfrak{H}^-.$$

Hence, $\operatorname{dom} U \cap \mathfrak{H}^-$ contains an infinite-dimensional closed subspace.

Step 2: Let $\mathfrak{N}$ be an infinite-dimensional closed subspace of $\operatorname{dom} U \cap \mathfrak{H}^-$. Since $n_+(S_1) = \dim(\mathfrak{H}^- \ominus P^-(\operatorname{gr} S_1)) < \infty$, $\mathfrak{N} \cap P^-(\operatorname{gr} S_1)$ is also an infinite-dimensional closed subspace of $\operatorname{dom} U \cap \mathfrak{H}^-$. Since $\mathfrak{N} \subseteq \operatorname{dom} U$, this implies that $K_1^{-1}(\mathfrak{N} \cap P^-(\operatorname{gr} S_1))$ is an infinite-dimensional closed subspace of $\operatorname{dom} U \cap \mathfrak{H}^+$. Due to (2.28) this proves that U is of type II.

(c) If the assumptions in (c) hold, then the fact that $\operatorname{dom} U \cap \mathfrak{H}^+$ and $\operatorname{dom} U \cap \mathfrak{H}^-$ contain an infinite-dimensional closed subspace follows by arguments as in Step 1, applied to $X := K_2^{-1}K_1$ and $X := K_1K_2^{-1}$. □

Corollary 2.17 *Let U be a reduction operator for S^* and assume that $n_+(S), n_-(S) \leq \aleph_0$. Then U is of type I if and only if for any two maximal symmetric extensions $S_1 \subseteq \operatorname{dom} U$ and $S_2 \subseteq \operatorname{dom} U$ of S one has $n_+(S_1) = n_+(S_2)$ and $n_-(S_1) = n_-(S_2)$.*

This section is concluded with an example of a unitary boundary triplet which is also a reduction operator of type I.

Example 2.18 Let $\{\mathcal{H}, \Gamma_0, \Gamma_1\}$ be a unitary boundary triplet for S^* whose Weyl function $M(\lambda)$ is an everywhere defined compact operator in $(\mathcal{H}, (\cdot, -))$ for $\lambda \in \mathbb{C} \setminus \mathbb{R}$. Recall, that the corresponding γ-field γ_λ satisfies

$$(\lambda - \overline{\lambda})\gamma_{\lambda}^*\gamma_\lambda = M(\lambda) - M(\lambda)^*,$$

cf. [Derkach et al., 2006, (5.3) & Section 5.3]. This shows that $\gamma_\lambda^*\gamma_\lambda$, and hence also, γ_λ is a compact operator (cf. proof of Lemma 2.13). Then $\mathfrak{N}_\lambda(T)$, $\operatorname{gr} T = \operatorname{dom}\Gamma$, as the range of γ_λ contains only finite-dimensional closed subspaces, $\lambda \in \mathbb{C} \setminus \mathbb{R}$; i.e. in this case $\Gamma := \Gamma_0 \times \Gamma_1$ is a reduction operator of type I. Note that this situation occurs, for instance, in the case of partial differential equations, see Chapters 6 and 7 as well as the references therein.

Bounded reduction operators Recall that if U is a reduction operator for the adjoint of a symmetric operator, then

$$U^{-1} = \mathrm{j}_{\mathfrak{H}} U^* \mathrm{j}_W,$$

see (2.13). The above equality implies that $\operatorname{dom} U$ is closed if and only if $\operatorname{ran} U$ is closed, see [Calkin, 1939a, Theorem 3.2]. In particular, since the range of a reduction operator is always dense in its range space $\mathfrak{K}$, this implies that a reduction operator U is bounded if and only if $\operatorname{ran} U = \mathfrak{K}$.

Different from the case of unbounded reduction operators, the type of bounded reduction operators for the adjoint of a symmetric operator is completely determined by the defect numbers of the symmetry.

Proposition 2.19 [Calkin, 1939a, Theorem 3.12] *Let U be a bounded reduction operator for S^*. Then U is of type I if and only if $n_+(S) < \infty$ or $n_-(S) < \infty$.*

Proof Since $\operatorname{dom} U$ is closed, the closed subspaces $\widehat{\mathfrak{N}}_{\mathrm{i}}(S^*)$ and $\widehat{\mathfrak{N}}_{-\mathrm{i}}(S^*)$ are contained in $\operatorname{dom} U$. Therefore, U is of type I if and only if at least one of these two subspaces is finite-dimensional; in other words, if and only if at least one of the defect numbers $n_+(S)$ or $n_-(S)$ is finite. □

Bounded reduction operators for the adjoint of a symmetric operator S preserve (in a certain sense) all the essential properties of symmetric extensions of S, see [Calkin, 1939a, Theorem 4.1].

Theorem 2.20 *Let U be a bounded reduction operator for S^*. If T is a symmetric extension of S such that $T \subseteq \operatorname{dom} U$, then*

$$U(T^*) = (U(T))^{[\perp]_W} \text{ and } U(\operatorname{clos}(T)) = \operatorname{clos}(U(T)),$$

where $[\perp]_W$ is the orthogonal complement with respect to $(\mathrm{j}_W \cdot, -)$ with $\mathrm{j}_W := \mathrm{i}W$. Moreover,

$$n_+(T) = n_+(U(T)) \text{ and } n_-(T) = n_-(U(T)).$$

In particular, T is a (closed, maximal) symmetric or selfadjoint operator *in $(\mathfrak{H}, (\cdot, -))$ if and only if $U(T)$ is a (closed, maximal) neutral or hypermaximal neutral subspace of $(\mathfrak{K}, (\jmath_W \cdot, -))$.*

The proof of this statement is fairly simple; it is based on the following equivalence, which holds due to the closedness of $\operatorname{dom} U = S^*$:

$$S \subseteq \operatorname{clos}(T) \subseteq \operatorname{dom} U \quad \Longleftrightarrow \quad S \subseteq T^* \subseteq \operatorname{dom} U.$$

The next theorem contains information about the maximal symmetric extensions of S contained in the domain of a bounded reduction operator for S^*, see [Calkin, 1939a, Theorem 4.1 & p. 409].

Theorem 2.21 *Let U be a bounded reduction operator for S^* and assume that $n_+(S), n_-(S) \leq \aleph_0$. Then the following statements hold:*

(i) *if S_m is a maximal symmetric extension of S, then $S_m \subseteq \operatorname{dom} U$;*

(ii) *if S is not a maximal symmetric operator, then the class of all maximal symmetric extensions S_m of S such that $S_m \subseteq \operatorname{dom} U$ has the cardinal number of the continuum.*

Moreover, if U is a bounded reduction operator of type I, then

(I) *if $S_m \subseteq \operatorname{dom} U$ is a maximal symmetric extension of S, then the pair $(n_+(S_m), n_-(S_m))$ coincides with the characteristic index of U,*

and if U is a bounded reduction operator of type II, then

(II) *for every $0 \leq p \leq \aleph_0$ there exists a maximal symmetric extension S_m of S with $n_+(S_m) = 0$ and $n_-(S_m) = p$ or $n_+(S_m) = p$ and $n_-(S_m) = 0$ such that $S_m \subseteq \operatorname{dom} U$.*

Proof (i) If T is an extension of S, then $T^* \subseteq S^* = \operatorname{dom} U$. Hence, if T is a symmetric extension of S, then $T \subseteq T^* \subseteq \operatorname{dom} U$, which shows that (i) holds.

(ii) If S is not a maximal symmetric operator, then $n_+(S) \geq 1$ and $n_-(S) \geq 1$. Since $S^* \subseteq \operatorname{dom} U$, see (i), the fact that (ii) holds is obvious.

(I) The invariance property of the defect numbers of S_m holds by Corollary 2.17. The fact that $(n_+(S_m), n_-(S_m))$ is the characteristic index of U is obtained from (2.27) and (2.29), since now $\dim \mathfrak{N}_d < \infty$.

(II) This follows from the fact that $n_+(S) = \infty = n_-(S)$, see Proposition 2.19, and the fact that $S^* \subseteq \operatorname{dom} U$, see (i). □

2.4 Maximal symmetric extensions and unbounded reduction operators

Here Calkin's answers to the three questions posed in the introduction are given. More precisely, in the first subsection some of the properties of maximal symmetric extensions of S contained in the domain of an unbounded reduction operator for S^* are described; note that the action of the reduction operator is not used at all there. In the second subsection the maximal symmetric extensions of S contained in the domain of an unbounded reduction operator for S^* are characterized by means of neutral subspaces in the range of the reduction operator. Therein an essential role is played by a graph decomposition of reduction operators together with angular operators. In the third subsection the pathology of unbounded reduction operators is indicated; therein the graph decomposition from Theorem 2.9 plays again a central role.

Maximal symmetric extensions The properties of maximal symmetric extensions of S contained in the domain of an unbounded reduction operator for S^* are described, cf. [Calkin, 1939a, Section 4.3]. In particular, it is shown that for unbounded reduction operators (of type I and II) the properties listed in Theorem 2.21 still hold except for the fact that not every maximal symmetric extension is contained in the domain. In fact, there exist maximal symmetric extensions whose graph intersected with the domain of a unbounded reduction operator is the graph of the original symmetry, see Corollary 2.25 below. Recall, that if U is an unbounded reduction operator then $n_{\pm}(S) = \infty$, see (2.27).

Theorem 2.22 *Let U be an unbounded reduction operator for S^* and assume that $n_+(S), n_-(S) \leq \aleph_0$. Then the following statements hold:*

(i) *the class of all maximal symmetric extensions S_m of S such that $S_m \subseteq \operatorname{dom} U$ has the cardinal number of the continuum;*
(ii) *the class of all maximal symmetric extensions S_m of S such that $S_m \not\subseteq \operatorname{dom} U$ and $\operatorname{clos}(S_m \cap \operatorname{dom} U) = S_m$ has the cardinal number of the continuum.*

Moreover, if U is an unbounded reduction operator of type I, then

(I) *if $S_m \subseteq \operatorname{dom} U$ is a maximal symmetric extension of S, then the pair $(n_+(S_m), n_-(S_m))$ coincides with the characteristic index of U,*

and if U is an unbounded reduction operator of type II, then

(II) *for every $0 \leq p \leq \aleph_0$ there exists a maximal symmetric extension S_m of S with $n_+(S_m) = 0$ and $n_-(S_m) = p$ or $n_+(S_m) = p$ and $n_-(S_m) = 0$ such that $S_m \subseteq \operatorname{dom} U$.*

Proof (i) Since U is an unbounded reduction operator, (2.27) shows that $\widehat{\mathfrak{N}}_{\mathrm{i}}(S^*) \cap \operatorname{dom} U$ and $\widehat{\mathfrak{N}}_{-\mathrm{i}}(S^*) \cap \operatorname{dom} U$ contain both at least one element. Therefore, as in the proof of Proposition 2.15, cf. (2.30), it is seen that there exists a closed symmetric relation T in $(\mathfrak{H}, (\cdot, -))$ with $n_\pm(T) \geq 1$ such that

$$S \subseteq T \subseteq T^* \subseteq \operatorname{dom} U. \tag{2.32}$$

From this observation (i) follows.

(ii) If S_m is a maximal symmetric operator as in statement (i) and in addition $S_m^* \subseteq \operatorname{dom} U$ (take $S_m \supseteq T$ in (2.32)), then T_m defined via

$$\operatorname{gr} T_m = \operatorname{gr} S + \mathrm{j}_{\mathfrak{H}}(\operatorname{gr} S_m \ominus \operatorname{gr} S)$$

satisfies the conditions in (ii). This one can easily see if S is assumed to be trivial and S_m to be selfadjoint, because in that case

$$\operatorname{dom} U = \operatorname{gr} S_m \oplus (\mathrm{j}_{\mathfrak{H}}(\operatorname{gr} S_m) \cap \operatorname{dom} U),$$
$$\mathfrak{H}^2 = \overline{\operatorname{dom} U} = \operatorname{gr} S_m \oplus \mathrm{j}_{\mathfrak{H}}(\operatorname{gr} S_m).$$

The general case follows by similar arguments.

(I) The invariance property of the defect numbers of S_m holds again by Corollary 2.17. Hence, it suffices to construct one maximal symmetric extension of S with defect numbers given by the characteristic index of U. Let $\mathfrak{M}_m$ be some maximal neutral subspace of the Kreĭn space $(\mathfrak{K}_d, (\mathrm{j}_{\mathfrak{H}}\cdot, -))$, see (2.17). Since U is of type I, $\dim \mathfrak{M}_m = \min\{k_+(\mathfrak{K}_d), k_-(\mathfrak{K}_d)\} < \infty$ and the pair of defect numbers of $\mathfrak{M}_m$ in $\mathfrak{K}_d$, see (2.4), coincides with the characteristic index of U. Now clearly the subspace $\operatorname{gr} S \oplus \mathfrak{M}_m \oplus \mathfrak{N}_d$ is a maximal neutral subspace contained in $\operatorname{dom} U$, see Theorem 2.9, and by construction it determines the graph of a maximal symmetric extension of S with the defect numbers as claimed.

(II) This follows immediately from Proposition 2.15. □

In fact, the proof shows that in (i) & (II) of Theorem 2.22 the condition $S_m \subseteq \operatorname{dom} U$ can be replaced by the stronger condition $S_m^* \subseteq \operatorname{dom} U$.

Theorem 2.22 implies that if $\mathfrak{K}_d = \{0\}$ and C (or B^{-1}) in Remark 2.10 is a compact operator, then every maximal symmetric extension of S contained in the domain of U is selfadjoint, because by Theorem 2.9 (ii) $\operatorname{gr} S + \mathfrak{N}_d \subseteq \operatorname{dom} U$ is the graph of a selfadjoint extension of S and by Theorem 2.14 U is a reduction operator of type I. If C is not compact, then by Theorem 2.22 (II) there are also non-selfadjoint maximal symmetric extensions of S contained in the domain of U with arbitrary defect.

The next result is not contained in Calkin's paper; it complements Theorem 2.21 by showing the existence of different types of reduction operators for any symmetric operator having infinite defect numbers.

Proposition 2.23 *Let S be a symmetric operator in the Hilbert space $(\mathfrak{H},(\cdot,-))$ with defect numbers $n_+(S) = n_-(S) = \aleph_0$. Then for every $0 \le p \le \aleph_0$ there exists reduction operators U for S^* of type I with the characteristic index $(p,0)$ or with the characteristic index $(0,p)$, and there exist reduction operators U for S^* of type II.*

Proof Without loss of generality one can assume that $S = \ker U$ is the trivial symmetry, see (2.15). By Remark 2.10 (v) it suffices to construct a reduction operator U for $\{0\}^*$ in the separable (infinite-dimensional) Hilbert space $(\mathfrak{H},(\cdot,-))$ with the fundamental symmetry $\mathfrak{j}_{\mathfrak{H}}$ in $\mathfrak{H}^2$ as in (2.2), such that U has the stated properties. For this purpose let S_m be a maximal symmetric operator in $(\mathfrak{H},(\cdot,-))$ with defect numbers $(p,0)$ or $(0,p)$ and let $\mathfrak{N}_d = \operatorname{gr} S_m$. Then it is easy to see that $\mathfrak{H}^2$ admits the following decomposition (cf. (2.5)):

$$\mathfrak{H}^2 = \mathfrak{N}_d \oplus (\mathfrak{N}_d^{[\perp]} \cap \mathfrak{j}_{\mathfrak{H}}\mathfrak{N}_d^{[\perp]}) \oplus \mathfrak{j}_{\mathfrak{H}}\mathfrak{N}_d.$$

Here the subspace $\mathfrak{K}_d := (\mathfrak{N}_d^{[\perp]} \cap \mathfrak{j}_{\mathfrak{H}}\mathfrak{N}_d^{[\perp]})$ is uniformly definite (when $\mathfrak{K}_d \neq \{0\}$) of dimension p.

Let B be a closed unbounded operator in the Hilbert space $(\mathfrak{N}_d,(\cdot,-))$ with $\overline{\operatorname{dom} B} = \mathfrak{N}_d = \operatorname{ran} B$ and $\ker B = \{0\}$. Then U defined on $\mathfrak{H}^2 = \mathfrak{K}_d \oplus \mathfrak{N}_d \oplus \mathfrak{j}_{\mathfrak{H}}\mathfrak{N}_d$ via the block formula

$$U = I_{\mathfrak{K}_d} \oplus \begin{pmatrix} 0 & B\mathfrak{j}_{\mathfrak{H}} \\ \mathfrak{j}_{\mathfrak{H}}B^{-*} & 0 \end{pmatrix}$$

is an unbounded reduction operator for $\{0\}^*$ by Remark 2.10. Since $\min\{k_+(\mathfrak{K}_d), k_-(\mathfrak{K}_d)\} = 0$, Theorem 2.14 now implies that U is of type I if B is a compact operator and of type II otherwise. Moreover, if U is of type I, then by construction S_m is a maximal symmetric extension of $\{0\}$ such that its graph, $\mathfrak{N}_d$, is contained in $\operatorname{dom} U$. Hence the characteristic index of U is in that case equal to $(p,0)$ or $(0,p)$, respectively, see Theorem 2.22. □

Theorem 2.22 shows that not all the graphs of maximal symmetric extensions of S are contained in the domain of an unbounded reduction operator for S^*. In fact, using a result going actually back to J. von Neumann, it can be shown that there exists a maximal symmetric extension of S such that the intersection of its graph with the domain of the reduction operator coincides with the graph of S itself. The following more general result was communicated to Calkin by J. von Neumann and it is mentioned in [Calkin, 1939a, p. 416] without providing a proof.

Theorem 2.24 *Let U be an unbounded reduction operator for S^* and assume that $n_+(S) = n_-(S) = \aleph_0$. Then there exists an unbounded reduction operator U_1 for S^* such that*

$$\operatorname{dom} U \cap \operatorname{dom} U_1 = S.$$

Sketch of the proof Without loss of generality assume that $S = \{0\}$. Moreover, for simplicity assume that $\mathfrak{K}_d = \{0\}$ in (2.17), so that with respect to the decomposition $\mathfrak{N}_d \oplus \mathfrak{j}_{\mathfrak{H}} \mathfrak{N}_d$ of $(\mathfrak{H}^2, (\cdot, -))$ the reduction operator U for $\{0\}^*$ is given by

$$U = \begin{pmatrix} 0 & B\mathfrak{j}_{\mathfrak{H}} \\ \mathfrak{j}_{\mathfrak{H}} B^{-*} & 0 \end{pmatrix},$$

where B is some closed unbounded operator in the (separable) Hilbert space $(\mathfrak{N}_d, (\cdot, -))$ with $\overline{\operatorname{dom} B} = \mathfrak{N}_d = \operatorname{ran} B$ and $\ker B = \{0\}$; see Remark 2.10.

By a well-known result of von Neumann, cf. [Fillmore and Williams, 1971, Theorem 3.6], one concludes that there exists a bounded selfadjoint operator K in $(\mathfrak{N}_d, (\cdot, -))$ with $\overline{\operatorname{ran} K} = \mathfrak{N}_d$ and $\operatorname{dom} B \cap \operatorname{ran} K = \{0\}$. Then $\mathfrak{L} := \{f + \mathfrak{j}_{\mathfrak{H}}\, iKf : f \in \mathfrak{N}_d\}$ is a hyper-maximal neutral subspace in $(\mathfrak{H}^2, (\mathfrak{j}_{\mathfrak{H}} \cdot, -))$ with $\mathfrak{L} \cap \operatorname{dom} U = \{0\}$.

Next apply [Fillmore and Williams, 1971, Theorem 2.2] to conclude that $\operatorname{ran} K + \operatorname{dom} B$ is an operator range. Therefore, again there exists a closed compact (selfadjoint) operator D in $(\mathfrak{N}_d, (\cdot, -))$ with $\overline{\operatorname{ran} D} = \mathfrak{N}_d$ and $\ker D = \{0\}$, such that $\operatorname{ran} D \cap (\operatorname{ran} K + \operatorname{dom} B) = \{0\}$. Then the subspace $\mathfrak{D}^+ := \{f + \mathfrak{j}_{\mathfrak{H}} f : f \in \operatorname{ran} D\}$ of $(\mathfrak{H}^2, (\mathfrak{j}_{\mathfrak{H}} \cdot, -))$ is uniformly positive with $\operatorname{clos}(\mathfrak{D}^+) = \mathfrak{H}^+$ and, moreover, $(\mathfrak{L} + \mathfrak{D}^+) \cap \operatorname{dom} U = \{0\}$. To complete the proof (with the above choice of U), a reduction operator U_1 with $\operatorname{dom} U_1 = \mathfrak{L} + \mathfrak{D}^+$ is now constructed.

Since $\mathfrak{D}^+ \subseteq \mathfrak{H}^+$, $\mathfrak{D}^+ = \{f + \mathfrak{j}_{\mathfrak{H}} f : f \in \mathfrak{D}\}$ for a subspace $\mathfrak{D}$ of $\mathfrak{L}$, see (2.5). In fact, since $\mathfrak{D}^+$ is by definition an operator range, also $\mathfrak{D}$ is an operator range. Therefore, there exists a closed operator D_t in the Hilbert space $(\mathfrak{L}, (\cdot, -))$ such that $\overline{\operatorname{dom} D_t} = \mathfrak{L} = \operatorname{ran} D_t$, $\ker D_t = \{0\}$, and $\operatorname{dom} D_t = \mathfrak{D}$. Using D_t define U_1 with respect to the orthogonal decomposition $\mathfrak{L} \oplus \mathrm{j}_{\mathfrak{H}}\mathfrak{L}$ of $(\mathfrak{H}^2, (\cdot, -))$, see (2.5), as:

$$U_1(f + \mathrm{j}_{\mathfrak{H}} f') = D_t f' + \mathrm{j}_{\mathfrak{H}} D_t^{-*} f, \quad f \in \mathfrak{L},\ f' \in \operatorname{dom} D_t.$$

By Remark 2.10 U_1 is a reduction (or unitary) operator in $(\mathfrak{H}^2, (\mathrm{j}_{\mathfrak{H}} \cdot, -))$ and, in addition, one has $\operatorname{dom} U_1 = \mathfrak{L} + \mathrm{j}_{\mathfrak{H}}\mathfrak{D} = \mathfrak{L} + \mathfrak{D}^+$. □

Corollary 2.25 [Calkin, 1939a, Theorem 4.6] *Let U be an unbounded reduction operator for S^* and assume that $n_+(S), n_-(S) \leq \aleph_0$. Then there exists a maximal symmetric extension S_m of S such that*

$$S_m \cap \operatorname{dom} U = S.$$

Proof This is a consequence of Theorem 2.24 and Theorem 2.22 (i). □

Maximal symmetric extensions and angular operators Here the neutral subspaces in the range of a reduction operator are described such that their pre-image is a maximal symmetric relation, cf. [Calkin, 1939a, Section 4.4]. Therefore, with the notation as in Remark 2.6, the following graph decompositions of a reduction operator U are used:

$$\begin{aligned} U &= \{\{f, f'\} \in U : f \in \widehat{\mathfrak{N}}_{\mathrm{i}}(S^*)\} + \{\{f, f'\} \in U : f' \in \mathfrak{K}_W^-\} \\ &= \{\{f, f'\} \in U : f \in \widehat{\mathfrak{N}}_{-\mathrm{i}}(S^*)\} + \{\{f, f'\} \in U : f' \in \mathfrak{K}_W^+\}, \end{aligned} \tag{2.33}$$

see [Calkin, 1939a, Theorem 3.9]. These graph decompositions can be deduced from the fact that the graph of a reduction operator can be interpreted as a hyper-maximal neutral subspace, see (2.12). Combining these decompositions with the fact that the domain and range of a reduction operator contain the orthogonal complement of a maximal neutral subspace, see Theorem 2.9 and Corollary 2.11, yields:

$$\begin{aligned} P^+U^{-1}(\operatorname{ran} U \cap \mathfrak{K}_W^+) &= \mathfrak{H}^+, \quad P_W^+ U(\widehat{\mathfrak{N}}_{\mathrm{i}}(S^*)) = \mathfrak{K}_W^+; \\ P^-U^{-1}(\operatorname{ran} U \cap \mathfrak{K}_W^-) &= \mathfrak{H}^-, \quad P_W^- U(\widehat{\mathfrak{N}}_{-\mathrm{i}}(S^*)) = \mathfrak{K}_W^-. \end{aligned} \tag{2.34}$$

Let K_+ and K_- be the angular operators of the subspaces $U(\widehat{\mathfrak{N}}_{\mathrm{i}}(S^*))$ and $U(\widehat{\mathfrak{N}}_{-\mathrm{i}}(S^*))$ (occurring in (2.33)) with respect to $\mathfrak{K}_W^+$ and $\mathfrak{K}_W^-$:

$$\begin{aligned} U(\widehat{\mathfrak{N}}_{\mathrm{i}}(S^*)) &= \{f^+ + K_+ f^+ : f^+ \in \mathfrak{K}_W^+\}; \\ U(\widehat{\mathfrak{N}}_{-\mathrm{i}}(S^*)) &= \{f^- + K_- f^- : f^- \in \mathfrak{K}_W^-\}. \end{aligned}$$

Then the following result gives an answer to the second problem stated in the introduction, cf. [Calkin, 1939a, Theorems 4.8, 4.11, 4.12].

Theorem 2.26 *Let U be a reduction operator for S^*, let $\mathfrak{N}$ be a neutral subspace of the Kreĭn space $(\mathfrak{K}, (\mathfrak{j}_W \cdot, -))$ and let K be the angular operator of $\mathfrak{N}$ with respect to $\mathfrak{K}_W^+$:*

$$\mathfrak{N} = \{f^+ + Kf^+ : f^+ \in P_W^+ \mathfrak{N}\}.$$

Then the symmetric extension $S_m := U^{-1}(\mathfrak{N} \cap \operatorname{ran} U)$ of S satisfies

$$n_+(S_m) = 0 \quad \Leftrightarrow \quad \operatorname{ran} U \cap \mathfrak{K}_W^- \subseteq \operatorname{ran}(K_+ - K);$$
$$n_-(S_m) = 0 \quad \Leftrightarrow \quad \operatorname{ran} U \cap \mathfrak{K}_W^+ \subseteq \operatorname{ran}(K_- - K^{-1}).$$

Proof It is shown that $n_+(S_m) = 0$ if and only if $\operatorname{ran} U \cap \mathfrak{K}_W^- \subseteq \operatorname{ran}(K_+ - K)$; the other case is analogous. Hence, assume that $\operatorname{ran} U \cap \mathfrak{K}_W^- \subseteq \operatorname{ran}(K_+ - K)$. Then for every $g^- \in \operatorname{ran} U \cap \mathfrak{K}_W^-$, there exists a $g_i \in U(\widehat{\mathfrak{N}}_{\mathrm{i}}(S^*))$ such that $g^- + g_i \in \mathfrak{N} \cap \operatorname{ran} U$. Indeed, by assumption

$$g^- = (K_+ - K)g^+ = (I + K_+)g^+ - (I + K)g^+ \tag{2.35}$$

for some $g^+ \in \mathfrak{K}_W^+$, which implies the claim with $g_i = -(I + K_+)g^+$. Now, equivalently, for every element $f^- \in U^{-1}(\operatorname{ran} U \cap \mathfrak{K}_W^-)$ there exists an element $f_i \in \widehat{\mathfrak{N}}_{\mathrm{i}}(S^*)$ such that $f^- + f_i \in U^{-1}(\mathfrak{N} \cap \operatorname{ran} U) = S_m$. Since $P^- U^{-1}(\operatorname{ran} U \cap \mathfrak{K}_W^-) = \mathfrak{H}^-$, see (2.34), the above arguments shows that $P^- \operatorname{gr} S_m = \mathfrak{H}^-$, i.e., $n_+(S_m) = 0$.

The converse assertion follows by reversing the arguments. □

Observe, that $\operatorname{ran}(K_+ - K) \subseteq \operatorname{ran} U \cap \mathfrak{K}_W^-$ if and only if $\mathfrak{N} \subseteq \operatorname{ran} U$, and similarly $\operatorname{ran}(K_- - K^{-1}) \subseteq \operatorname{ran} U \cap \mathfrak{K}_W^+$ if and only if $\mathfrak{N} \subseteq \operatorname{ran} U$; see (2.35). It should also be noted that using (the proof of) Theorem 2.26 one can also describe maximal symmetry of the closure $\operatorname{clos}(U^{-1}(\mathfrak{N} \cap \operatorname{ran} U))$ for an arbitrary neutral subspace $\mathfrak{N}$ of the Kreĭn space $(\mathfrak{K}, (\mathfrak{j}_W \cdot, -))$, see [Calkin, 1939a, Theorem 4.9].

The next result, which cannot be found in [Calkin, 1939a], is included here to give an explicit connection between Theorem 2.26 and [Derkach et al., 2006, Theorem 4.13]. Therefore observe that for a unitary boundary triplet $\{\mathcal{H}, \Gamma_0, \Gamma_1\}$ with associated Weyl function $M(\lambda)$, the subspaces $U(\widehat{\mathfrak{N}}_{\mathrm{i}}(S^*))$ and $U(\widehat{\mathfrak{N}}_{-\mathrm{i}}(S^*))$ occurring in (2.33) coincide with $\operatorname{gr} M(\mathrm{i})$ and $\operatorname{gr} M(-\mathrm{i})$.

Theorem 2.27 *Let $\{\mathcal{H}, \Gamma_0, \Gamma_1\}$ be a unitary boundary triplet for S^* with the Weyl function $M(\cdot)$, and let $A_0 = \ker \Gamma_0$. Then*

(i) *the range of Γ can be decomposed by means of $M(\pm\mathrm{i})$ as follows*

$$\operatorname{gr} M(\mathrm{i}) + \operatorname{ran} \Gamma \cap \mathfrak{K}_W^- = \operatorname{ran} \Gamma = \operatorname{gr} M(-\mathrm{i}) + \operatorname{ran} \Gamma \cap \mathfrak{K}_W^+; \tag{2.36}$$

(ii) *if* K_0 *is the angular operator for the neutral subspace* $\mathcal{H}\times\{0\}$ *then*

$$P_{\mathcal{H}\times\{0\}}\mathrm{ran}\,(K_\pm - K_0^{\pm 1}) = \mathrm{dom}\,M(\pm\mathrm{i});$$

(iii) *maximal symmetry of* A_0 *can be characterized as follows:*

$$n_+(A_0)=0 \quad\Leftrightarrow\quad \mathrm{ran}\,\Gamma_0 = \mathrm{dom}\,M(\mathrm{i});$$
$$n_-(A_0)=0 \quad\Leftrightarrow\quad \mathrm{ran}\,\Gamma_0 = \mathrm{dom}\,M(-\mathrm{i}).$$

Proof (i) Consider $\Gamma=\Gamma_0\times\Gamma_1$ as a reduction operator for S^*, see Corollary 2.8. Then (i) is obtained directly from (2.33).

(ii) Clearly, $\Theta_0=\{0\}\times\mathcal{H}$ is a (hyper-maximal) neutral subspace of $(\mathcal{H}^2,(\mathfrak{j}_\mathcal{H}\cdot,-))$ and $A_0=\Gamma^{-1}(\Theta_0\cap\mathrm{ran}\,\Gamma)$. The angular operator K_+ of the subspace $\Gamma(\widehat{\mathfrak{N}}_\mathrm{i}(S^*))=M(\mathrm{i})$ with respect to $\mathfrak{K}_W^+$ is given by

$$K_+\begin{pmatrix}h\\ \mathrm{i}h\end{pmatrix}=\begin{pmatrix}\mathrm{i}(I+\mathrm{i}\,M(\mathrm{i}))(\mathrm{i}\,I+M(\mathrm{i}))^{-1}h\\ (I+\mathrm{i}\,M(\mathrm{i}))(\mathrm{i}\,I+M(\mathrm{i}))^{-1}h\end{pmatrix}.$$

The angular operator K_0 of Θ_0 with respect to $\mathfrak{K}_W^+$ is clearly $K_0=(-I_\mathcal{H})\oplus I_\mathcal{H}$. Therefore,

$$(K_+-K_0)\begin{pmatrix}h\\ \mathrm{i}h\end{pmatrix}=\begin{pmatrix}2\,\mathrm{i}(\mathrm{i}\,I+M(\mathrm{i}))^{-1}h\\ 2(\mathrm{i}\,I+M(\mathrm{i}))^{-1}h\end{pmatrix},$$

which shows that $\mathrm{ran}\,(K_+-K_0)=\{\{f,-\mathrm{i}f\}: f\in\mathrm{dom}\,M(\mathrm{i})\}$. Similarly it is seen that $\mathrm{ran}\,(K_- - K_0^{-1})=\{\{f,\mathrm{i}f\}: f\in\mathrm{dom}\,M(-\mathrm{i})\}$.

(iii) Now Theorem 2.26 is applied. By (ii) $\mathrm{ran}\,\Gamma\cap\mathfrak{K}_W^-\subseteq\mathrm{ran}\,(K_+-K_0)$ if and only if $P_{\mathcal{H}\times\{0\}}\mathrm{ran}\,\Gamma\cap\mathfrak{K}_W^-\subseteq\mathrm{dom}\,M(\mathrm{i})$. In view of (2.36), this happens if and only if $\mathrm{ran}\,\Gamma_0=P_{\mathcal{H}\times\{0\}}\mathrm{ran}\,\Gamma=\mathrm{dom}\,M(\mathrm{i})$.

Similar arguments show that $\mathrm{ran}\,U\cap\mathfrak{K}_W^+\subseteq\mathrm{ran}\,(K_- - K^{-1})$ if and only if $\mathrm{ran}\,\Gamma_0=\mathrm{dom}\,M(-\mathrm{i})$, which completes the proof. □

Remark 2.28 Similarly, by applying Theorem 2.26 to the subspace $\Theta_1=\mathcal{H}\times\{0\}$ one obtains the following equivalences: $n_\pm(A_1)=0 \Leftrightarrow \mathrm{ran}\,\Gamma_1=\mathrm{ran}\,M(\pm\mathrm{i})$, where $A_1=\ker\Gamma_1$. If, in particular, $\{\mathcal{H},\Gamma_0,\Gamma_1\}$ is a generalized boundary triplet for S^* with the Weyl function $M(\lambda)$, and $\Theta=\Theta^*$ is an arbitrary selfadjoint operator in $\mathcal{H}$, then $\{\mathcal{H},\Gamma_0,\Gamma_1-\Theta\Gamma_0\}$ is a unitary boundary triplet for S^* with Weyl function $M(\lambda)-\Theta$, see Chapter 7. Therefore, the selfadjoint boundary condition

$$\Gamma_1\widehat{f}-\Theta\Gamma_0\widehat{f}=0$$

determines a symmetric extension $A_\Theta=\Gamma^{-1}(\Theta\cap\mathrm{ran}\,\Gamma)$ of S for which

$$n_\pm(A_\Theta)=0 \quad\Leftrightarrow\quad \mathrm{ran}\,(\Gamma_1-\Theta\Gamma_0)=\mathrm{ran}\,(M(\pm\mathrm{i})-\Theta).$$

A similar characterization for $n_\pm(A_\Theta) = 0$ in a more general setting will be discussed elsewhere.

Pathology of unbounded reduction operators In this subsection two results are presented which show how unbounded reduction operators change the defect numbers of (maximal) symmetric operators contained in the domain. For this purpose the graph decomposition of reduction operators in Theorem 2.9 is used together with the following general property of unbounded operators in Hilbert spaces.

Lemma 2.29 [Calkin, 1939a, Lemma 4.2] *Let B be a closed unbounded operator between the Hilbert spaces $(\mathfrak{H}_1, (\cdot,-)_1)$ and $(\mathfrak{H}_2, (\cdot,-)_2)$ such that $\overline{\operatorname{dom} B} = \mathfrak{H}_1$, $\operatorname{ran} B = \mathfrak{H}_2$ and $\ker B = \{0\}$. Then there exists an infinite-dimensional closed subspace $\mathfrak{L}$ of $\mathfrak{H}_1$ such that*

$$\mathfrak{L} \cap \operatorname{dom} B = \{0\} \quad \textit{and} \quad \mathfrak{H}_2 = \operatorname{clos} B^{-*}(\mathfrak{L}^{\perp_1}).$$

Proof The existence of a closed subspace $\mathfrak{L}$ with $\mathfrak{L} \cap \operatorname{dom} B = \{0\}$ in the case of a separable Hilbert space is obtained from the already mentioned result of J. von Neumann, cf. [Fillmore and Williams, 1971, Theorem 3.6]. For non-separable Hilbert spaces, it can be deduced for instance from a description of operator ranges in [Fillmore and Williams, 1971, Theorem 1.1].

As to the second assertion, note that the assumptions imply that $\operatorname{ran} B^* = \mathfrak{H}_1$ and $\ker B^* = \{0\}$. Hence, if $g \in \mathfrak{H}_2$ is orthogonal to $B^{-*}(\mathfrak{L}^{\perp_1})$, then

$$0 = (g, B^{-*}f)_2 = (B^{-1}g, f)_1, \quad f \in \mathfrak{L}^{\perp_1}.$$

This implies that $B^{-1}g \in \operatorname{dom} B \cap \mathfrak{L} = \{0\}$. Consequently, $g = 0$ and $\mathfrak{H}_2 = \operatorname{clos} B^{-*}(\mathfrak{L}^{\perp_1})$. □

Combining Lemma 2.29 with Theorem 2.9 yields the following statement, see [Calkin, 1939a, Lemma 4.3 & Theorems 4.13, 4.14].

Theorem 2.30 *Let U be an unbounded reduction operator for S^*. Then there exists a maximal symmetric extension S_m of S such that*

(i) $S_m \subseteq \operatorname{dom} U$;
(ii) $\operatorname{clos}(U(S_m))$ *is a maximal neutral subspace of* $(\mathfrak{K}, (\mathrm{j}_W \cdot, -))$;
(iii) *for every* $0 \le p \le \aleph_0$ *there exists a closed symmetric relation* S_p *such that* $S \subseteq S_p \subseteq S_m$, $\operatorname{clos}(U(S_p)) = \operatorname{clos}(U(S_m))$, *and*

$$n_+(S_p) = n_+(S_m) + p \quad \textit{and} \quad n_-(S_p) = n_-(S_m) + p.$$

Proof To prove the statement assume without loss of generality that $S = \{0\}$. Then, with the notation as in Theorem 2.9 and Remark 2.10, recall that $B := C^{-*}$, where C is as defined in (2.24), is a closed Hilbert space operator from $(\mathfrak{N}_d, (\cdot, -))$ to $(\mathfrak{N}_r, (\cdot, -))$ which satisfies the conditions of Lemma 2.29. Hence for every $0 \leq p \leq \aleph_0$ there exists a p-dimensional closed subspace $\mathfrak{L}_p$ of $\mathfrak{N}_d$ such that $\mathfrak{L}_p \cap \operatorname{dom} B = \{0\}$ and

$$\mathfrak{N}_r = \operatorname{clos} B^{-*}(\mathfrak{L}_p^{\perp_1} \cap \mathfrak{N}_d) = \operatorname{clos} C(\mathfrak{N}_d \ominus \mathfrak{L}_p).$$

Now let $\mathfrak{M}_d$ be any maximal neutral subspace of $(\mathfrak{K}_d, (\mathfrak{j}_{\mathfrak{H}} \cdot, -))$, see Theorem 2.9, and define S_p via

$$\operatorname{gr} S_p = \mathfrak{N}_d \ominus \mathfrak{L}_p + \mathfrak{M}_d.$$

Then $\operatorname{clos} U(S_p) = \operatorname{clos}(\mathfrak{j}_W C(\mathfrak{N}_d \ominus \mathfrak{L}_p) + U(\mathfrak{M}_d)) = \mathfrak{j}_W \mathfrak{N}_r + U(\mathfrak{M}_d)$ is a maximal neutral subspace of $(\mathfrak{K}, (\mathfrak{j}_W \cdot, -))$, because the subspace $U(\mathfrak{M}_d)$ is a maximal neutral subspace of $(\mathfrak{K}_r, (\mathfrak{j}_W \cdot, -))$, see Theorem 2.9 and Remark 2.10. Consequently, the statement holds with $S_m := S_0$. □

In particular, since the reduction operator and its inverse have the same behavior, one can show analogous to Theorem 2.30 that there exists a maximal neutral subspace $\mathfrak{N}_m$ of $(\mathfrak{K}, (\mathfrak{j}_W \cdot, -))$ such that

(i) $\mathfrak{N}_m \subseteq \operatorname{ran} U$;
(ii) $\operatorname{clos} U^{-1}(\mathfrak{N}_m)$ is the graph of a maximal symmetric operator;
(iii) for every $0 \leq p \leq \aleph_0$ there exists a closed neutral subspace $\mathfrak{N}_p$ of $\mathfrak{N}_m$ such that $\operatorname{clos} U^{-1}(\mathfrak{N}_p) = \operatorname{clos} U^{-1}(\mathfrak{N}_m)$ and

$$n_+(\mathfrak{N}_p) = n_+(\mathfrak{N}_m) + p \quad \text{and} \quad n_-(\mathfrak{N}_p) = n_-(\mathfrak{N}_m) + p.$$

Calkin also showed that for unbounded reduction operators there exist maximal symmetric operators whose graphs are contained in the domain of the reduction operator such that their image is a closed neutral subspace whose defect numbers are different from the original maximal symmetric operator, cf. [Calkin, 1939a, Lemma 4.4 & Theorem 4.15].

Theorem 2.31 *Let U be an unbounded reduction operator for S^*. Then for every $0 \leq p \leq \aleph_0$ there exists a maximal symmetric extension S_m of S with $S_m \subseteq \operatorname{dom} U$ such that $\mathfrak{N} := U(S_m)$ is a closed neutral subspace with*

$$n_+(\mathfrak{N}) = n_+(S_m) + p \quad \textit{and} \quad n_-(\mathfrak{N}) = n_-(S_m) + p.$$

Proof Again for simplicity it is assumed that S is the trivial symmetry, that $\mathfrak{K}_d = \{0\}$, and that $U_t = I$ in Theorem 2.9, so that U is a reduction operator for $\{0\}^*$ with range space $\mathfrak{H}^2$ and associated rotation $-\mathrm{i}\mathrm{j}_{\mathfrak{H}}$ of the form as given in (2.25):

$$U = \begin{pmatrix} 0 & B\mathrm{j}_{\mathfrak{H}} \\ \mathrm{j}_{\mathfrak{H}}B^{-*} & 0 \end{pmatrix},$$

where B is a closed unbounded operator in the Hilbert space $(\mathfrak{N}_d, (\cdot, -))$ satisfying $\overline{\operatorname{dom} B} = \mathfrak{N}_d = \operatorname{ran} B$ and $\ker B = \{0\}$. Since B^* is unbounded, there exists a p-dimensional closed subspace $\mathfrak{L}_p$ of $\mathfrak{N}_d$ such that $\operatorname{dom} B^* \cap \mathfrak{L}_p = \{0\}$ and

$$\mathfrak{N}_d = \operatorname{clos}\{B^{-1}f : f \subset \mathfrak{N}_d \ominus \mathfrak{L}_p\},$$

see Lemma 2.29. Hence,

$$Cf = Bf, \quad f \in \operatorname{dom} C = \{g \in \operatorname{dom} B : \; Bg \in \mathfrak{N}_d \ominus \mathfrak{L}_p\},$$

considered as an operator from $\mathfrak{N}_d$ to $\mathfrak{H} \ominus \mathfrak{L}_p$ is a closed operator which satisfies $\overline{\operatorname{dom} C} = \mathfrak{N}_d$, $\operatorname{ran} C = \mathfrak{N}_d \ominus \mathfrak{L}_p$ and $\ker C = \{0\}$. Now define the operator V from $\mathfrak{H}^2$ to $(\mathfrak{N}_d \ominus \mathfrak{L}_p)^2$ by

$$V(f + \mathrm{j}_{\mathfrak{H}} f') = Cf' + \mathrm{j}_{\mathfrak{H}} C^{-*} f, \quad f \in \mathfrak{N}_d, \; f' \in \operatorname{dom} C.$$

Then by definition $\operatorname{dom} V \subseteq \operatorname{dom} U$ and from Remark 2.10 one concludes that V is a reduction operator for $\{0\}^*$ with range space $(\mathfrak{N}_d \ominus \mathfrak{L}_p)^2$ and associated rotation $-\mathrm{i}\mathrm{j}_{\mathfrak{N}_d \ominus \mathfrak{L}_p}$. Let WK be the polar decomposition of C. Then K is a (nonnegative) selfadjoint operator in $(\mathfrak{N}_d, (\cdot, -))$ with $\operatorname{dom} K = \operatorname{dom} C$ and, hence, the subspace $\mathfrak{L}$ of $(\mathfrak{H}^2, (\mathrm{j}_{\mathfrak{H}} \cdot, -))$ defined by

$$\mathfrak{L} := \{Kf + \mathrm{j}_{\mathfrak{H}} if : f \in \operatorname{dom} K\}$$

is a hyper-maximal neutral and it is contained in $\operatorname{dom} V = \mathfrak{N}_d \times \operatorname{dom} C$.

By definition of K, KC^{-1} is a closed operator from $(\mathfrak{N}_d \ominus \mathfrak{L}_p, (\cdot, -))$ to $(\mathfrak{N}_d, (\cdot, -))$ with domain $\mathfrak{N}_d \ominus \mathfrak{L}_p$. Moreover, KB^{-1} coincides with KC^{-1} when the latter is considered as a mapping in $(\mathfrak{N}_d, (\cdot, -))$, because $\operatorname{dom} K = \operatorname{dom} C$ and $C \subseteq B$. Therefore $S := B^{-*}KB^{-1}$ is a closed symmetric operator in $(\mathfrak{N}_d, (\cdot, -))$ with domain $\mathfrak{N}_d \ominus \mathfrak{L}_p$, i.e. S is a bounded symmetric operator with $n_\pm(S) = p$. Now the proof is completed by observing that $\mathfrak{L} \subseteq \operatorname{dom} U$ and that $\mathfrak{N} := U(\mathfrak{L}) = \{f - \mathrm{j}_{\mathfrak{H}} iSf : f \in \operatorname{dom} S\}$ is a neutral subspace with $n_+(\mathfrak{N}) = n_+(S)$ and $n_-(\mathfrak{N}) = n_-(S)$. □

Observe, that all the results in this section apply to unitary boundary triplets (and, hence, in particular to generalized boundary triplets), which are not ordinary boundary triplets, see Chapter 7.

Two-sided ideals of bounded operators Here Calkin's characterization of all two-sided ideals contained in the algebra $\mathfrak{B} := \mathfrak{B}(\mathfrak{H})$ of bounded operators in a separable Hilbert space $(\mathfrak{H}, (\cdot, -))$ is presented. Basic examples of proper two-sided ideals in $\mathfrak{B}$ are the ideal of finite-rank operators and the ideal $\mathfrak{B}_\infty$ of compact operators. Calkin derived his characterization of all two-sided ideals directly from Lemma 2.13, which makes a connection between his main two papers [Calkin, 1939a] and [Calkin, 1941]. This result from [Calkin, 1941] is repeated here only to make the connection between these two papers explicit.

Theorem 2.32 [Calkin, 1941, Theorem 1.4] *Let $\mathfrak{I}$ be an arbitrary two-sided ideal in $\mathfrak{B}$. Then either $\mathfrak{I} = \mathfrak{B}$ or $\mathfrak{I} \subseteq \mathfrak{B}_\infty$.*

Proof Let $\mathfrak{I}$ be an ideal which is not contained in $\mathfrak{B}_\infty$. Then there is a bounded operator $T \in \mathfrak{I}$ whose range contains an infinite-dimensional closed subspace $\mathfrak{M}$, see Lemma 2.13. Since T is bounded, $T^{-1}(\mathfrak{M})$ and, hence also $\mathfrak{U} := T^{-1}(\mathfrak{M}) \ominus \ker T$, is a closed subspace. Hence, $T : \mathfrak{U} \to \mathfrak{M}$ is bounded with bounded inverse, in particular, $\dim \mathfrak{U} = \dim \mathfrak{M} = \dim \mathfrak{H} = \aleph_0$ by separability of $\mathfrak{H}$. Hence, there are partial isometries $X : \mathfrak{H} \to \mathfrak{U}$ and $Y : \mathfrak{H} \to \mathfrak{M}$ with $\operatorname{dom} X = \operatorname{dom} Y = \mathfrak{H}$ and $\operatorname{ran} X = \mathfrak{U}$, $\operatorname{ran} Y = \mathfrak{M}$. Then the operator $B := Y^*AX$ belongs to $\mathfrak{I}$ and, by construction, $\operatorname{dom} B = \mathfrak{H} = \operatorname{ran} B$ and $\ker B = \{0\}$. Thus B^{-1} is also bounded, which implies that $I = B^{-1}B \in \mathfrak{I}$. Therefore, $\mathfrak{I} = \mathfrak{B}$. □

References

Azizov, T. Ya., and Iokhvidov, I. S. 1989. *Linear operators in spaces with an indefinite metric*, John Wiley and Sons, New York.

Calkin, J. W. 1939a. Abstract symmetric boundary conditions, *Trans. Amer. Math. Soc.*, **45**, 369–442.

Calkin, J. W. 1939b. General self-adjoint boundary conditions for certain partial differential operators, *Proc. N.A.S.*, **25**, 201–206.

Calkin, J. W. 1941. Two-Sided ideals and congruences in the ring of bounded operators in Hilbert space, *Annals of Mathematics*, **42**, 839–873.

Cross, R. W. 1995. Closed operators and the Calkin property, *Proceedings of the Royal Irish Academy*, **95A**, 109–111.

Derkach, V. A., Hassi, S., Malamud, M. M., and de Snoo, H. S. V. 2006. Boundary relations and their Weyl families, *Trans. Amer. Math. Soc.*, **358**, 5351–5400.

Fillmore, P. A., and Williams, J. P. 1971. On operator ranges, *Advances in Math.*, **7**, 254–281.

Gorbachuk, V.I., and Gorbachuk, M.L. 1991. *Boundary value problems for operator differential equations*, Kluwer Academic Publishers Group, Dordrecht.

Mogilevskii, V. 2006. Boundary triplets and Krein type resolvent formula for symmetric operators with unequal defect numbers, *Methods of Functional Analysis and Topology*, **12**, 258–280.

Wietsma, H. L., 2012. Representations of unitary relations between Kreĭn spaces, *Integral Equations Operator Theory*, **72**, 309–344.

3

Boundary triplets and maximal accretive extensions of sectorial operators

Yury Arlinskiĭ

Abstract This chapter is a survey of results related to the problem of a description of all maximal accretive extensions for a densely defined sectorial operator; the problem in more generality was originally posed by R. Phillips. We also treat maximal sectorial extensions. Our approach uses the concepts of boundary pairs and boundary triplets.

3.1 Introduction

A linear operator S in a complex Hilbert space $\mathfrak{H}$ is called *accretive* [Kato, 1995] if $\operatorname{Re}(Su, u) \geq 0$ for all $u \in \operatorname{dom} S$. An accretive operator S is called *maximal accretive* (m-accretive) if one of the following equivalent conditions is satisfied [Kato, 1995; Lyantse, 1954; Phillips, 1959a,b, 1969]:

- the operator S is closed and has no accretive extensions in $\mathfrak{H}$;
- $\rho(S) \cap \Pi_- \neq \emptyset$, where Π_- denotes the open left half-plane;
- the operator S is densely defined and closed, and S^* is accretive;
- the operator $-S$ generates a one-parameter contractive semigroup $T(t) = \exp(-tS)$, $t \geq 0$.

The resolvent set $\rho(S)$ of an m-accretive operator contains the open left half-plane and

$$\|(S - zI)^{-1}\| \leq \frac{1}{|\operatorname{Re} z|}, \quad \operatorname{Re} z < 0.$$

The class of m-accretive operators plays an essential role in differential equations, scattering theory, stochastic processes, passive linear systems and, for instance, hydrodynamics. It should be noted that Phillips calls an operator A *dissipative* if $-A$ is accretive. Nowadays the term dissipative is used for operators A with $\mathrm{Im}\,(Af,f) \geq 0$.

The Phillips problem is to find all m-accretive extensions of a densely defined accretive operator; cf. [Phillips, 1959a,b, 1969]. Phillips proved that any closed densely defined accretive operator has an m-accretive extension. Phillips proposed to use the geometry of spaces with indefinite inner product and this method has been further developed in [Evans and Knowles, 1985, 1986] for positive definite ordinary differential operators and in [Mil'yo and Storozh, 1991, 1993] in the abstract setting for positive definite symmetric operators with finite defect numbers. Another approach of Phillips is connected with fractional-linear transformations, which reduces the problem to a parametrization of all contractive extensions of a not everywhere defined contraction. The solution of the last problem has been obtained in [Crandall, 1969].

Two operators S and T are said to form a *dual pair* $\{S,T\}$ if

$$(Sf,g) = (f,Tg), \quad f \in \mathrm{dom}\, S, \quad g \in \mathrm{dom}\, T.$$

A linear operator $\widetilde{S}$ satisfying the conditions

$$\widetilde{S} \supset S \quad \text{and} \quad \widetilde{S}^* \supset T,$$

is called *an extension of a dual pair* $\{S,T\}$. This property is equivalent to $S \subset \widetilde{S} \subset T^*$ for densely defined closed operators S and T. An example of a dual pair is provided by the minimal operators associated with a differential expression and its formal adjoint. The existence of m-accretive extensions of a dual pair of densely defined accretive operators was proved in [Phillips, 1969]. Fractional-linear transformations of a dual pair of accretive operators lead to aparametrization of all contractive extensions of a dual pair of non-densely defined contractions. This problem was solved in [Shmul'yan and Yanovskaya, 1981; Arsene and Geondea, 1982; Davis et al., 1982].

A linear operator S in a Hilbert space $\mathfrak{H}$ is called *α-sectorial* (or sectorial for short) with vertex at $z = 0$ and semi-angle $\alpha \in [0, \pi/2)$ [Kato, 1995] if the numerical range of S

$$W(S) := \{(Su,u) \in \mathbb{C} : \; u \in \mathrm{dom}\, S, \; \|u\| = 1\}$$

is contained in the sector $\mathcal{S}(\alpha) := \{z \in \mathbb{C} : |\arg z| \leq \alpha\}$; equivalently

$$|\mathrm{Im}\,(Sf,f)| \leq (\tan\alpha)\mathrm{Re}\,(Sf,f), \quad f \in \mathrm{dom}\, S.$$

The case of $\alpha = 0$ corresponds to nonnegative symmetric or self-adjoint operators.

If S is sectorial (α-sectorial) and S is m-accretive then it is called m-sectorial (m-α-sectorial). The resolvent set of an m-sectorial operator S contains the set $\mathbb{C} \setminus \mathcal{S}(\alpha)$,

$$\|(S - zI)^{-1}\| \le \frac{1}{\operatorname{dist}(z, \mathcal{S}(\alpha))}, \quad z \in \mathbb{C} \setminus \mathcal{S}(\alpha),$$

and the one-parameter semigroup $T(t) = \exp(-tS)$, $t \ge 0$, admits a holomorphic contractive continuation into the interior of the sector $\mathcal{S}(\pi/2 - \alpha)$.

We are interested in a description of all m-accretive and m-sectorial extensions of a densely defined sectorial operator S. An important special case of this problem is a parametrization of all nonnegative self-adjoint extensions for a nonnegative symmetric operator. This problem was studied by J. von Neumann, K. Friedrichs and later by M.G. Kreĭn in his famous papers [Kreĭn, 1947a,b]. Kreĭn's approach is based on a fractional-linear transformation and a parametrization of all self-adjoint contractive extensions for a non-densely defined Hermitian contraction. Kreĭn's results were essentially completed in [Birman, 1956] for a symmetric operator with positive lower bound. Birman used the approach proposed in [Vishik, 1952]. It should be noted that for a second order partial elliptic differential operator in a bounded domain Vishik used Green's identity and constructed special boundary spaces and corresponding boundary operators. The methods of Vishik and Birman for dual pairs of boundedly invertible operators and for positive definite operators with subsequent applications to elliptic partial differential operators were developed in [Grubb, 1968, 1970, 1971, 1973, 1974, 1983, 2006]. The boundary triplets approach, which goes back to [Calkin, 1939], has been applied to the description of all nonnegative selfadjoint and proper (quasi-selfadjoint) m-accretive extensions of a nonnegative operator ($S \subset \widetilde{S} \subset S^*$) in [Kochubei, 1979; Gorbachuk and Gorbachuk, 1984; Mikhailets, 1980] (see also [Gorbachuk et al., 1989]), and later in [Derkach et al., 1988, 1989; Arlinskiĭ, 1988; Storozh, 1990] with essential additions connected to a Weyl function of the boundary triplet [Derkach and Malamud, 1991, 1995] (see the recent survey [Arlinskiĭ and Tsekanovskiĭ, 2009] and references therein). In [Mil'yo and Storozh, 1999a,b, 2002; Pipa and Storozh, 2006] criteria for maximal accretiveness have been established for some classes of perturbations of a positive definite operator with applications to differential-boundary operators.

The Kreĭn and Phillips approach via fractional-linear transformations and a description of all quasi-selfadjoint contractive extensions $\widetilde{A}$ of a non-densely defined Hermitian contraction A ($\widetilde{A} \supset A$, $\widetilde{A}^* \supset A$) have been used in [Arlinskiĭ and Tsekanovskiĭ, 1984a,b, 1988; Kolmanovich and Malamud, 1985; Hassi, Malamud and de Snoo, 2004]. In the case of proper extensions the actions of the extensions are known and the problem is reduced to the determination of domains. An intrinsic description and parametrization (in analogy to von Neumann's formulas for quasi-self-adjoint extensions) of the domains of all m-accretive and m-sectorial quasi-self-adjoint extensions of nonnegative S are given in [Arlinskiĭ, 1995; Arlinskiĭ, Kovalev and Tsekanovskiĭ, 2012].

In the case of a non-symmetric operator neither actions nor domains of extensions are known. The results of [Arlinskiĭ, 1996]–[Arlinskiĭ, 2006] are based on the existence of at least one m-sectorial extension, the Friedrichs extension, cf. [Kato, 1995]. It may occur that this m-sectorial extension is unique. Otherwise there are infinitely many extensions and among them there exists a special m-sectorial extension (namely the Kreĭn-von Neumann extension), possessing properties similar to the one in the nonnegative situation [Kreĭn, 1947a]. Under some additional condition, which is satisfied for coercive sectorial operators, it is possible to describe all m-sectorial and m-accretive extensions by means of a boundary pair and two parameters. The use of special boundary triplets allows us determine the actions and domains.

The aim of this paper is to demonstrate the efficiency of the approach via boundary pairs and boundary triplets in the problems outlined above.

3.2 Preliminaries

Notations The domain, the range, and the null-space of a linear operator T are denoted by $\operatorname{dom} T$, $\operatorname{ran} T$, and $\ker T$ respectively; $\overline{\operatorname{dom}}\, T$ and $\overline{\operatorname{ran}}\, T$ stands for the closures. The Banach space of bounded linear operators acting between Hilbert spaces $\mathfrak{H}$ and $\mathfrak{K}$ is denoted by $\boldsymbol{B}(\mathfrak{H}, \mathfrak{K})$ and the Banach algebra $\boldsymbol{B}(\mathfrak{H}, \mathfrak{H})$ by $\boldsymbol{B}(\mathfrak{H})$. For a contraction $T \in \boldsymbol{B}(\mathfrak{H}_1, \mathfrak{H}_2)$ we define $D_T = (I_{\mathfrak{H}_1} - T^*T)^{1/2}$ and $\mathfrak{D}_T = \overline{\operatorname{ran}}\, D_T$.

Sectorial forms and operators Recall the following definitions and results from [Kato, 1995]. Let τ be a sesquilinear form in $\mathcal{H}$ defined on a linear subspace $\operatorname{dom} \tau$. The form τ is called symmetric if $\tau[u, v] = \overline{\tau[v, u]}$ for all $u, v \in \operatorname{dom} \tau$ and nonnegative if $\tau[u] := \tau[u, u] \geq 0$ for all $u \in$

$\operatorname{dom}\tau$. The form τ is called sectorial with the vertex at the point $\gamma \in \mathbb{C}$ and a semi-angle $\alpha \in [0, \pi/2)$ if its numerical range

$$W(\tau) = \{\tau[u],\ u \in \operatorname{dom}\tau,\ \|u\| = 1\}$$

is contained in the sector $\{z \in \mathbb{C} : |\arg(z-\gamma)| \le \alpha\}$, or equivalently

$$|\mathrm{Im}\,(\tau[u] - \gamma\|u\|^2)| \le \tan\alpha\,\mathrm{Re}\,(\tau[u] - \gamma\|u\|^2),\ u \in \operatorname{dom}\tau.$$

For a sesquilinear form τ, $\tau^*[u,v] := \overline{\tau[v,u]}$ defines the adjoint of τ; the forms

$$\tau_{\mathrm{R}} := \tfrac{1}{2}\,(\tau + \tau^*), \quad \tau_{\mathrm{I}} := \tfrac{1}{2\mathrm{i}}\,(\tau - \tau^*)$$

are called the real and the imaginary parts of τ, respectively.

A sequence $\{u_{\mathrm{n}}\}$ is called τ-convergent to the vector $u \in \mathfrak{H}$ if

$$\lim_{n\to\infty} u_{\mathrm{n}} = u \quad \text{and} \quad \lim_{n,m\to\infty} \tau[u_{\mathrm{n}} - u_m] = 0.$$

The form τ is called closed if for every sequence $\{u_{\mathrm{n}}\}$ τ-convergent to a vector u it follows that $u \in \operatorname{dom}\tau$ and $\lim_{n\to\infty}\tau[u - u_{\mathrm{n}}] = 0$. A sectorial form τ with vertex at the origin is closed if and only if the linear subspace $\operatorname{dom}\tau$ is a Hilbert space with the inner product $(u,v)_\tau - \tau_{\mathrm{R}}[u,v] + (u,v)$ [Kato, 1995]. The form τ is called closable if it has a closed extension; in this case the closure of τ is the smallest closed extension of τ. If T is a sectorial operator, then the form

$$T[u,v] := (Tu, v), \quad u, v \in \operatorname{dom}T$$

is sectorial and closable; the domain of the closure of the form $T[\cdot,-]$ is denoted by $\mathcal{D}[T]$.

If τ is a closed, densely defined sectorial form, then according to the First Representation Theorem [Kreĭn, 1947a; Kato, 1995] there exists a unique m-sectorial operator T in $\mathfrak{H}$, associated with τ, i.e.,

$$(Tu, v) = \tau[u,v], \quad u \in \operatorname{dom}T, \quad v \in \operatorname{dom}\tau.$$

Clearly, the adjoint operator T^* is associated with the adjoint form τ^*. In the sequel we shall consider sectorial forms with nonnegative real part, i.e., forms with the vertex at the origin. The nonnegative self-adjoint operator T_{R}, associated with the real part τ_{R} of τ is called the "real part" of T. According to the Second Representation Theorem [Kreĭn, 1947a; Kato, 1995]

$$\operatorname{dom}\tau = \operatorname{dom}T_{\mathrm{R}}^{\frac{1}{2}}$$

and the sectorial form τ with vertex at the origin has the representation

$$\tau[u,v] = ((I + \mathrm{i}M)T_{\mathrm{R}}^{\frac{1}{2}}u, T_{\mathrm{R}}^{\frac{1}{2}}v), \quad u, v \in \operatorname{dom}\tau,$$

where M is a bounded selfadjoint operator in the subspace $\overline{\operatorname{ran}}\, T_{\mathrm{R}}$ and $\|M\| \le \tan\alpha$. For the corresponding operator T one obtains

$$\begin{aligned}\operatorname{dom} T &= \{u \in \operatorname{dom} \tau : (I + \mathrm{i}M)T_{\mathrm{R}}^{\frac{1}{2}} u \in \operatorname{dom} \tau\},\\ Tu &= T_{\mathrm{R}}^{\frac{1}{2}}(I + \mathrm{i}M)T_{\mathrm{R}}^{\frac{1}{2}} u.\end{aligned}$$

If S is α-sectorial, then the open and connected set $\mathbb{C} \setminus \mathcal{S}(\alpha)$ consists of points of regular type, $\operatorname{ran}(S - zI)$ is closed and $\dim \mathfrak{N}_z$ does not depend on $z \in \mathbb{C} \setminus \mathcal{S}(\alpha)$, where

$$\mathfrak{N}_z := \ker(S^* - zI) = \mathfrak{H} \ominus \operatorname{ran}(S - \bar{z}I).$$

The operator S is called *coercive* if the quadratic form $\operatorname{Re}(Sf, f)$, $f \in \operatorname{dom} S$, is positive definite, i.e.,

$$\operatorname{Re}(Sf, f) \ge m\|f\|^2, \ f \in \operatorname{dom} S,$$

where $m > 0$. In the case of α-sectorial and coercive operator the dimension $\dim \mathfrak{N}_z$ does not depend on the choice of $z \in (\mathbb{C} \setminus \mathcal{S}(\alpha)) \cup \{\text{some neighborhood of the origin}\}$.

Linear relations By a *relation* in a Hilbert space $\mathfrak{H}$ we mean a closed linear manifold (subspace for short) in $\mathfrak{H} \oplus \mathfrak{H}$ equipped with the Hilbert space inner product $(u_1, v_1) + (u_2, v_2)$ for pairs $\langle u_1, u_2\rangle$ and $\langle v_1, v_2\rangle$. In particular the graph

$$\operatorname{Gr}(T) = \{\langle h, Th\rangle : h \in \operatorname{dom} T\}$$

of a linear operator T in $\mathfrak{H}$ provides an example of a relation. If $\boldsymbol{T}$ is a relation, then by definition

$$\begin{aligned}\operatorname{dom} \boldsymbol{T} &= \{x \in \mathfrak{H} : \langle x, x'\rangle \in \boldsymbol{T} \text{ for some } x' \in \mathfrak{H}\},\\ \operatorname{ran} \boldsymbol{T} &= \{x' \in \mathfrak{H} : \langle x, x'\rangle \in \boldsymbol{T} \text{ for some } x \in \mathfrak{H}\},\\ \ker \boldsymbol{T} &= \{x \in \operatorname{dom} \boldsymbol{T} : \langle x, 0\rangle \in \boldsymbol{T}\},\\ \lambda \boldsymbol{T} &= \{\langle x, \lambda x'\rangle, \ \langle x, x'\rangle \in \boldsymbol{T}\},\\ \boldsymbol{T}^{-1} &= \{\langle x', x\rangle : \langle x, x'\rangle \in \boldsymbol{T}\}.\end{aligned}$$

For $x \in \operatorname{dom} \boldsymbol{T}$ we set

$$\boldsymbol{T}x = \{x' \in \operatorname{ran} \boldsymbol{T} : \langle x, x'\rangle \in \boldsymbol{T}\}.$$

Since $\boldsymbol{T}$ is closed, the subspace $\boldsymbol{T} \ominus \langle 0, \boldsymbol{T}0\rangle$ is the graph of a closed linear operator T, $\operatorname{dom} T = \operatorname{dom} \boldsymbol{T}$, which is called the *operator part* of $\boldsymbol{T}$. Clearly, $\boldsymbol{T}x = Tx \oplus \boldsymbol{T}0$. The *adjoint* $\boldsymbol{T}^*$ of $\boldsymbol{T}$ is given by

$$\boldsymbol{T}^* = \mathfrak{H}^2 \ominus \{\langle -x', x\rangle : \langle x, x'\rangle \in \boldsymbol{T}\}.$$

The numerical range of a relation $\boldsymbol{T}$ is the set

$$W(\boldsymbol{T}) = \{(\boldsymbol{T}x, x) : x \in \operatorname{dom} \boldsymbol{T}, \ \|x\| = 1\}.$$

As has been shown in [Rofe-Beketov, 1985] if $W(\boldsymbol{T}) \neq \mathbb{C}$, then $\boldsymbol{T}0 = \mathfrak{H} \ominus \overline{\operatorname{dom} \boldsymbol{T}^*}$. A relation $\boldsymbol{T}$ is called

- *Hermitian* if $W(\boldsymbol{T}) \subseteq \mathbb{R} \iff \boldsymbol{T} \subseteq \boldsymbol{T}^*$;
- *selfadjoint* if $\boldsymbol{T} = \boldsymbol{T}^*$;
- *nonnegative* if $W(\boldsymbol{T}) \subseteq \mathbb{R}_+$;
- *accretive* if $W(\boldsymbol{T}) \subseteq \Pi_+$;
- *m-accretive* if $\boldsymbol{T}$ is accretive and has no accretive extensions in $\mathfrak{H} \oplus \mathfrak{H}$;
- *α-sectorial* if $W(\boldsymbol{T}) \subseteq \mathcal{S}(\alpha)$;
- *m-α-sectorial* if $\boldsymbol{T}$ is α-sectorial and m-accretive.

For a closed but non-densely defined sectorial form τ in the Hilbert space $\mathfrak{H}$, there is an associated m-sectorial relation $\boldsymbol{T}$ such that its operator part acts in $\overline{\operatorname{dom} \tau}$ and $\boldsymbol{T}0 = \mathfrak{H} \ominus \overline{\operatorname{dom} \tau}$ [Rofe-Beketov, 1985].

The Phillips approach Equip the Hilbert space $\mathfrak{H} \oplus \mathfrak{H}$ with an inner product

$$Q(\langle u_1, u_2 \rangle, \langle v_1, v_2 \rangle) = (u_2, v_1) + (u_1, v_2), \tag{3.1}$$

which makes it a Kreĭn space. Clearly, a relation $\boldsymbol{S}$ in $\mathfrak{H}$ is accretive (m-accretive) if and only if $\boldsymbol{S}$ is a nonnegative (maximal nonnegative) with respect to the Q-inner product. The subspace $\boldsymbol{N}_+ := \{\langle \psi, \psi \rangle : \psi \in \mathfrak{N}\}$, where $\mathfrak{N} = \ker(\boldsymbol{S}^* + I_{\mathfrak{H}}) = \{\psi : \langle \psi, -\psi \rangle \in \boldsymbol{S}^*\}$, is positive and complete with respect to the Q-norm. The relation $\boldsymbol{T} = \boldsymbol{S} \dotplus \boldsymbol{N}_+$ is an m-accretive extension of $\boldsymbol{S}$ and $-\boldsymbol{S}^* = -\boldsymbol{T}^* \oplus \boldsymbol{N}_+$, where the decomposition is both orthogonal and Q-orthogonal. The maximal nonpositive linear subspace $-\boldsymbol{T}^*$ need not be complete. The *Phillips abstract boundary space* is defined as follows

$$\widehat{\mathfrak{H}}_P := \boldsymbol{M}_- \oplus \boldsymbol{N}_+, \tag{3.2}$$

where $\boldsymbol{M}_-$ is the completion of $-\boldsymbol{T}^*$.

Theorem 3.1 [Phillips, 1969] *The formula*

$$-\boldsymbol{B} = \hat{\boldsymbol{B}} \cap -\boldsymbol{S}^*$$

gives a one-to-one correspondence between all m-accretive restrictions of $\boldsymbol{S}^*$ and all maximal nonpositive subspaces $\hat{\boldsymbol{B}}$ of $\widehat{\mathfrak{H}}_P$.

The above approach goes back to [Phillips, 1959a,b, 1969] for operators and [Arlinskiĭ, 2000b] for relations.

Fractional-linear transformations of sectorial relations On the set of all relations define the fractional linear transformation

$$\mathcal{K}(\boldsymbol{S}) = \boldsymbol{A} = \{\langle x + x', x - x' \rangle \,:\, \langle x, x' \rangle \in \boldsymbol{S}\}. \tag{3.3}$$

Clearly, $\mathcal{K}(\mathcal{K}(\boldsymbol{S})) = \boldsymbol{S}$. Let $\boldsymbol{S}$ be an accretive relation in $\mathfrak{H}$. Then $\boldsymbol{A}$ is the graph of a contraction A in $\mathfrak{H}$ with $\operatorname{dom} A = \operatorname{dom} \boldsymbol{A}$. Conversely, if A is a contraction defined on a subspace $\operatorname{dom} A \subseteq \mathfrak{H}$, then

$$\boldsymbol{S} = \{\langle (I + A)h, (I - A)h \rangle \,:\, h \in \operatorname{dom} A\} \tag{3.4}$$

is an accretive relation. Rewriting (3.3) and (3.4) gives

$$A = -I + 2(I + \boldsymbol{S})^{-1}, \quad \boldsymbol{S} = -I + 2(I + A)^{-1}.$$

The following statements are clear:

- $\boldsymbol{S}$ is the graph of an accretive operator if and only if $\ker (I + A) = \{0\}$;
- $\boldsymbol{S}$ is m-accretive if and only if $\operatorname{dom} A = \mathfrak{H}$;
- $\boldsymbol{S}$ is nonnegative selfadjoint if and only if A is a selfadjoint contraction.

Definition 3.2 [Arlinskiĭ, 1987, 1992; Arlinskiĭ and Tsekanovskiĭ, 1984a] Let $\alpha \in (0, \pi/2)$ and A be a closed linear operator in $\mathfrak{H}$. We say that a suboperator A is in $\mathcal{C}(\alpha)$ if

$$\|A \sin \alpha \pm \mathrm{i}\, I \cos \alpha\| \le 1,$$

or equivalently

$$2|\mathrm{Im}\,(Af, f)| \le (\|f\|^2 - \|Af\|^2) \tan \alpha, \quad f \in \operatorname{dom} A; \tag{3.5}$$

by a suboperator A we mean an operator with $\operatorname{dom} A \subset \mathfrak{H}$. Saying "an operator in $\mathcal{C}(\alpha)$" we emphasize the fact that $\operatorname{dom} A = \mathfrak{H}$.

Thus suboperators in $\mathcal{C}(\alpha)$ are contractions not necessarily defined everywhere.

Proposition 3.3 *The fractional-linear transformation* (3.3) *establishes a one-to-one correspondence between α-sectorial (m-α-sectorial) relations in $\mathfrak{H}$ and suboperators (operators) in $\mathcal{C}(\alpha)$.*

Due to (3.5), $\mathcal{C}(0) = \bigcap_{\alpha \in (0,\pi/2)} \mathcal{C}(\alpha)$. Therefore it is natural to consider Hermitian (selfadjoint) contractions in $\mathfrak{H}$ as suboperators (operators) in $\mathcal{C}(0)$. For later convenience we also introduce $\widetilde{\mathcal{C}} := \bigcup_{\alpha \in [0,\pi/2)} \mathcal{C}(\alpha)$.

Let A be a fractional-linear transformation (3.3) of an accretive relation $\boldsymbol{S}$ in $\mathfrak{H}$. Hence, there is a one-to-one correspondence between the set of all everywhere defined contractive extensions of A and the set of all m-accretive extensions of $\boldsymbol{S}$ and this correspondence is given by

$$\widetilde{\boldsymbol{S}} = \{\langle (I+\widetilde{A})h, (I-\widetilde{A})h\rangle \,:\, h \in \mathfrak{H}\}. \tag{3.6}$$

The description of the set of all contractive extensions is obtained in [Crandall, 1969]. Let us denote by $\mathfrak{N}$ the orthogonal complement in $\mathfrak{H}$ of the closed subspace $\operatorname{dom} A$ and by $P_{\mathfrak{N}}$ is the orthogonal projection in $\mathfrak{H}$ onto $\mathfrak{N}$. Let $A^* \in \boldsymbol{B}(\mathfrak{H}, \operatorname{dom} A)$ be the adjoint of $A \in \boldsymbol{B}(\operatorname{dom} A, \mathfrak{H})$.

Theorem 3.4 [Crandall, 1969] *The formula*

$$\widetilde{A} = AP_A + D_{A^*}\widetilde{K}P_{\mathfrak{N}}$$

establishes a one-to-one correspondence between all contractive extensions $\widetilde{A}$ of A and all contractions $\widetilde{K} \in \boldsymbol{B}(\mathfrak{N}, \mathfrak{D}_{A^*})$.

The next statement follows from Proposition 3.3.

Proposition 3.5 *Let the α-sectorial relation $\boldsymbol{S}$ and A in $C(\alpha)$ be connected by (3.3). For $\beta \in [\alpha, \pi/2)$ the following statements are equivalent:*

1. *the operator $\widetilde{A} \in C(\beta)$ is an extension of A;*
2. *$\widetilde{\boldsymbol{S}}$ in (3.6) is an m-β-sectorial extension of $\boldsymbol{S}$.*

3.3 Friedrichs and Kreĭn-von Neumann extensions

The Friedrichs extension For a densely defined α-sectorial operator S its *Friedrichs extension* S_{F} is the m-α-sectorial operator associated with the form $S[\cdot,-]$, cf. [Kato, 1995]. Then $\mathcal{D}[S] = \mathcal{D}[S_{\mathrm{F}}]$ and $\mathcal{D}[S] \cap \mathfrak{N}_z = \{0\}$, $z \in \rho(S_{\mathrm{F}})$, where $\mathfrak{N}_z$ is the defect subspace of S. Moreover, if S is also coercive then S_{F} is coercive as well and preserves the lower bound of the quadratic form $S[\cdot,-]$. It follows from the definition that

$$\inf\{\|f-\varphi\|^2 + \operatorname{Re}(S_{\mathrm{F}}(f-\varphi), f-\varphi) : \varphi \in \operatorname{dom} S\} = 0, \; f \in \operatorname{dom} S_{\mathrm{F}}.$$

If $\alpha = 0$, then it is well known that

$$\operatorname{dom} S_{\mathrm{F}} = \operatorname{dom} S^* \cap \mathcal{D}[S], \quad S_{\mathrm{F}} = S^* \upharpoonright \operatorname{dom} S_{\mathrm{F}}.$$

In fact, the Friedrichs extension S_{F} is a unique nonnegative self-adjoint extension having the domain in $\mathcal{D}[S]$.

The Kreĭn-von Neumann extension A densely defined nonnegative symmetric operator S has a minimal nonnegative self-adjoint extension, called in [Kreĭn, 1947a] the *soft extension* and it coincides with the von Neumann extension when S is positive definite. For this reason it is often called the Kreĭn or the Kreĭn-von Neumann extension. Kreĭn's construction of the soft extension is given by means of a fractional-linear transformation. Another definition of this extension is proposed in [Ando and Nishio, 1970] for operators and in [Coddington and de Snoo, 1978] for nonnegative relations. A different construction is suggested in [Arlinskiĭ and Tsekanovskiĭ, 2002, 2003, 2005] for densely defined operators.

Concerning α-sectorial operators with arbitrary α the analog of the Kreĭn-von Neumann extension can be defined in the way Ando and Nishio did for the nonnegative case, see [Arlinskiĭ, 1996, 1997]. Due to the inequality

$$|(\phi, Sh)|^2 \leq \frac{1}{\cos^2\alpha}\mathrm{Re}\,(S\phi,\phi)\,\mathrm{Re}\,(Sh,h),\ \ \phi, h\in \mathrm{dom}\, S,$$

the linear operator

$$T(S\phi) = P_{\overline{\mathrm{ran}}\,S}\phi,\ \ \phi\in \mathrm{dom}\, S,$$

with $P_{\overline{\mathrm{ran}}\,S}$ being the orthogonal projection onto $\overline{\mathrm{ran}}\,S$, is a well-defined α-sectorial operator with dense domain $\mathrm{dom}\,T = \mathrm{ran}\,S$ and dense range in $\overline{\mathrm{ran}}\,S$. Then the m-$\alpha$-sectorial operator

$$S_{\mathrm{N}} := T_{\mathrm{F}}^{-1} P_{\overline{\mathrm{ran}}\,S},$$

where T_{F} stands for the Friedrichs extensions of T, is called the Kreĭn-von Neumann extension of S. This definition can be rewritten as follows

$$S_{\mathrm{N}} = ((S^{-1})_{\mathrm{F}})^{-1},$$

where S^{-1} is the inverse relation of S. In the next theorem we collect the main properties of the operator S_{N}.

Theorem 3.6 [Arlinskiĭ, 1996, 1997] *Let S be a densely defined closed and sectorial operator. Then*

1. $\mathcal{D}[S_{\mathrm{N}}] = \{u\in H : \sup\{\left|(u,Sx)\right|^2/\mathrm{Re}\,(Sx,x) : x\in\mathrm{dom}\,S\}<\infty\}$*;*
2. $\mathcal{D}[S]\subset\mathcal{D}[\widetilde{S}]\subseteq\mathcal{D}[S_{\mathrm{N}}]$ *for any m-sectorial extension* $\widetilde{S}$ *of* S*;*
3. *for all* $u\in\mathrm{dom}\,S_{\mathrm{N}}$

$$\inf\{\|S_{\mathrm{N}}u - Sx\|^2 + \mathrm{Re}\,(S_{\mathrm{N}}(u-x), u-x) : x\in\mathrm{dom}\,S\} = 0;$$

4. $\mathcal{D}[S_{\mathrm{N}}] = \mathcal{D}[S] \dotplus (\mathfrak{N}_\lambda\cap\mathcal{D}[S_{\mathrm{N}}])$ *for each* $\lambda\in\rho(S_{\mathrm{F}}^*)$*;*

5. *S admits a unique m-sectorial extension if and only if for some (and hence for all) $\lambda \in \rho(S_{\mathrm{F}}^*)$*

$$\sup\{|(\varphi_\lambda, x)|^2 / \operatorname{Re}(Sx, x),\ x \in \operatorname{dom} S\} = \infty, \quad f_\lambda \in \mathfrak{N}_\lambda \setminus \{0\};$$

6. *if S is coercive, then*

$$S_{\mathrm{N}}[u, v] = S[\mathcal{P}u, \mathcal{P}v],\ u, v \in \mathcal{D}[S_{\mathrm{N}}],$$

where $\mathcal{P}$ is a skew projector onto $\mathcal{D}[S]$ corresponding to the decomposition $\mathcal{D}[S_{\mathrm{N}}] = \mathcal{D}[S] \dotplus \ker S^*$ and

$$\operatorname{dom} S_{\mathrm{N}} = \operatorname{dom} S \dotplus \ker S^*,\ S_{\mathrm{N}}(f + v) = Sf,$$

where $f \in \operatorname{dom} S$, $v \in \ker S^*$.

An m-accretive extension $\widetilde{S}$ of S is called *extremal* [Arlinskiĭ, 1996, 1997] if

$$\inf\{\operatorname{Re}(\widetilde{S}(u - x), u - x),\ x \in \operatorname{dom} S\} = 0, \quad u \in \operatorname{dom} \widetilde{S}.$$

The Friedrichs and Kreĭn-von Neumann extensions are extremal.

Proposition 3.7 [Arlinskiĭ, 1997] *The closed forms associated with extremal m-sectorial extensions are closed restrictions of the form $S_{\mathrm{N}}[\cdot, -]$ and the Kreĭn-von Neumann extension is the unique extremal m-sectorial extension having maximal domain of the associated closed form.*

Thus extremal m-sectorial extensions preserve the semi-angle.

Limit representations Let us define two families of operators $\widehat{S}_z$ and $\widetilde{S}_z$ for $z \in \Pi_-$:

$$\operatorname{dom} \widetilde{S}_z = \operatorname{dom} S \dotplus \mathfrak{N}_z,\ \widetilde{S}_z(f + \varphi_z) = Sf - z\varphi_z,\ f \in \operatorname{dom} S,\ \varphi_z \in \mathfrak{N}_z.$$

The operator $\widehat{S}_z$ is the Kreĭn-von Neumann extension of the coercive and sectorial operator $S - zI$ and the operator $\widetilde{S}_z$ is an m-accretive extension of S. Moreover, the following conditions are equivalent [Arlinskiĭ, 2000a, 1999a]

- $\widetilde{S}_z$ is m-sectorial for one (and then for all) $z \in \Pi_-$;
- $\operatorname{dom} S^* \subset \mathcal{D}[S_{\mathrm{N}}]$.

Theorem 3.8 [Arlinskiĭ, 1997, 1999a, 2006] *The Friedrichs and Kreĭn-von Neumann extensions satisfy*

$$S_{\mathrm{F}} = \lim_{z \to \infty} (\widehat{S}_z + zI) = \lim_{z \to \infty} \widetilde{S}_z, \quad S_{\mathrm{N}} = \lim_{z \to 0} (\widehat{S}_z + zI) = \lim_{z \to 0} \widetilde{S}_z,$$

where the limits are in the strong resolvent sense (non-tangential to the imaginary axis).

For a nonnegative operator S the equalities

$$S_{\mathrm{F}} = \lim_{a\to+\infty}((S + aI)_{\mathrm{N}} - aI), \quad S_{\mathrm{N}} = \lim_{a\to+0}((S + aI)_{\mathrm{N}} - aI)$$

were established in [Ando and Nishio, 1970] and in [Shtraus, 1973]. The case of nonnegative linear relations can be found in [Hassi, Malamud and de Snoo, 2004] and [Behrndt, Hassi, de Snoo and Wietsma, 2010].

3.4 Boundary pairs and closed forms associated with m-sectorial extensions

For a description of closed sectorial forms associated with m-sectorial extensions it is convenient to use boundary pairs.

Definition 3.9 [Arlinskiĭ, 1996] Let S be a closed densely defined α-sectorial operator in the Hilbert space $\mathfrak{H}$. A pair $\{\mathcal{H}, \Gamma\}$ is called a boundary pair for S if $\mathcal{H}$ is a Hilbert space and Γ is a continuous linear operator: $D[S_{\mathrm{N}}] \to \mathcal{H}$ such that

$$\ker\Gamma = \mathcal{D}[S], \quad \operatorname{ran}\Gamma = \mathcal{H}.$$

This notion is close to that of [Lyantse and Storozh, 1983] (see Section 3.10). The statement below gives a parametrization of all closed forms associated with m-sectorial extensions of S.

Theorem 3.10 [Arlinskiĭ, 1996] *Let S_{NR} be a real part of S_{N} and let $\{\mathcal{H}, \Gamma\}$ be a boundary pair for S. Then*

$$\begin{aligned} &\mathcal{D}[\widetilde{S}] = \Gamma^{-1}\operatorname{dom}\widetilde{w}, \\ &\widetilde{S}[u,v] = S_{\mathrm{N}}[u,v] + \widetilde{w}[\Gamma u, \Gamma v] + 2(\widetilde{X}\Gamma u, S_{\mathrm{NR}}^{1/2}v), \ u, v \in \mathcal{D}[\widetilde{S}] \end{aligned} \tag{3.7}$$

establishes a one-to-one correspondence between all closed forms associated with m-sectorial extensions $\widetilde{S}$ of S and all pairs $\langle\widetilde{w}, \widetilde{X}\rangle$ such that $\widetilde{w}$ is a closed sectorial form in $\mathcal{H}$, and the linear operator $\widetilde{X} : \operatorname{dom}\widetilde{w} \to \overline{\operatorname{ran}}\, S$ satisfies for some $\delta \in [0,1)$

$$\|\widetilde{X}e\|^2 \leqslant \delta^2 \operatorname{Re}\widetilde{w}[e], \quad e \in \operatorname{dom}\widetilde{w}.$$

If β is the semi-angle of the form w, then the semi-angle γ of the sectorial form $\widetilde{S}[\cdot,-]$ admits the following estimate

$$\alpha \leqslant \gamma \leqslant \arctan\{(1-\delta)^{-1}(\max\{\tan\alpha, \tan\beta\} + \delta)\}.$$

In the case of a coercive sectorial operator S one has

$$\operatorname{dom} S_{\mathrm{N}} = \operatorname{dom} S \dotplus \ker S^*, \quad \mathcal{D}[S_{\mathrm{N}}] = \mathcal{D}[S_{\mathrm{F}}] \dotplus \ker S^*.$$

Let $\{\mathcal{H}, \Gamma\}$ be a boundary pair for S. Then the operator

$$Z_0 := (\Gamma \restriction \ker S^*)^{-1}$$

maps $\mathcal{H}$ onto $\ker S^*$ and thus for each $e \in \mathcal{H}$ the boundary problem

$$S^* u = 0, \ \Gamma u = e$$

has the unique solution $Z_0 e$. In addition, the operator $Z_0 \Gamma$ is a skew projection onto $\ker S^*$ corresponding to the decomposition

$$\mathcal{D}[S_{\mathrm{N}}] = \mathcal{D}[S] \dotplus \ker S^*.$$

Define the quadratic functional

$$\mu[\varphi] = \sup\{\operatorname{Re} S[2\varphi - f, f] \ : \ f \in \mathcal{D}[S]\} \tag{3.8}$$

on $\mathcal{D}[S]$. For all $\varphi \in \mathcal{D}[S]$ one has

$$\mu[\varphi] = \sup\{|(\varphi, Sx)|^2 / \operatorname{Re}(Sx, x) \ : \ x \in \operatorname{dom} S\}.$$

Furthermore

$$\mu[\varphi] = 2\|(S_{\mathrm{F}}^{-1} + S_{\mathrm{F}}^{*-1})^{-1/2}\varphi\|^2$$

and in particular, if S is positive definite, then

$$\mu[\varphi] = S[\varphi].$$

The next theorem is a version of Theorem 3.10 for the case of coercive sectorial S.

Theorem 3.11 [Arlinskiĭ, 1996] *Let S be sectorial and coercive operator and let $\{\mathcal{H}, \Gamma\}$ be a boundary pair for S. Then*

$$\begin{aligned} &\mathcal{D}[\widetilde{S}] = \Gamma^{-1}\mathcal{D}[\widetilde{w}], \\ &\widetilde{S}[u, v] = S[u - Z_0\Gamma u + 2\widetilde{Y}\Gamma u, v - Z_0 \Gamma v] + \widetilde{w}[\Gamma u, \Gamma v] \end{aligned} \tag{3.9}$$

establishes a one-to-one correspondence between all closed forms associated with m-sectorial extensions $\widetilde{S}$ of S and all pairs $\langle \widetilde{w}, \widetilde{Y}\rangle$ such that $\widetilde{w}$ is a closed sectorial form in $\mathcal{H}$ and the linear operator $\widetilde{Y} : \mathcal{D}[\widetilde{w}] \to \mathcal{D}[S]$ satisfies for some $\delta \in [0, 1)$

$$\mu[\widetilde{Y} e] \leqslant \delta^2 \operatorname{Re} \widetilde{w}[e], \ \ e \in \mathcal{D}[\widetilde{w}].$$

An extension $\widetilde{S}$ is coercive if and only if the form $\widetilde{w}$ is coercive in $\mathcal{H}$.

Observe that for a nonnegative operator S in formulas (3.7) and (3.9) the pairs $\langle \widetilde{w}, 0\rangle$ correspond to quasi-self-adjoint m-sectorial extensions $\widetilde{S}$ of S ($S \subset \widetilde{S} \subset S^*$). Moreover, $\widetilde{S}$ is m-β-sectorial if and only if $\widetilde{w}$ is β-sectorial.

3.5 Boundary triplets and m-accretive extensions

The notion of a boundary triplet for densely defined symmetric operators with equal defect numbers and a description of all selfadjoint extensions in terms of boundary conditions appeared in [Kochubei, 1975; Bruk, 1976]; the idea goes back originally to [Calkin, 1939], see also Chapters 2 and 7. Special boundary spaces and boundary operators were constructed in [Vishik, 1952] for elliptic differential operators in a bounded domain of $\mathbb{R}^n$. Here we define specific boundary triplets in order to describe the domains and actions of m-accretive extensions and domains of m-accretive restrictions.

Definition 3.12 [Arlinskiĭ, 1998, 1999a, 2000a] Let S be a sectorial operator. A triplet $\{\mathcal{H}, G, \Gamma\}$ is called a boundary triplet for S^* if

1. $\{\mathcal{H}, \Gamma\}$ is a boundary pair for S,
2. $G : \mathcal{D}[S_{\rm N}] \cap \operatorname{dom} S^* \to \mathcal{H}$ is a linear operator such that

$$S_{\rm N}^*[u, v] = (S^*u, v) - (Gu, \Gamma v), \; u \in \mathcal{D}[S_{\rm N}] \cap \operatorname{dom} S^*, \; v \in \mathcal{D}[S_{\rm N}]. \quad (3.10)$$

As a consequence one has $\ker G = \operatorname{dom} S_{\rm N}^*$. The identity (3.10) can be viewed as an abstract analog of Green's identity for the Laplace operator. Note that a boundary triplet for S^* always exists, cf. [Arlinskiĭ, 1998, 1999a]. If S is nonnegative then the present boundary triplet coincides with the boundary triplet from [Arlinskiĭ, 1988; Arlinskiĭ, Hassi, Sebestyen and de Snoo, 2001].

With a boundary pair $\{\mathcal{H}, \Gamma\}$ of S is associated the γ-field

$$\gamma(\lambda) := (\Gamma \upharpoonright \mathfrak{N}_\lambda \cap \mathcal{D}[S_{\rm N}])^{-1}, \quad \lambda \in \rho(S_{\rm F}^*), \quad (3.11)$$

which clearly satisfies

$$\gamma(\lambda) = \gamma(z) + (\lambda - z)(S_{\rm F}^* - \lambda I)^{-1}, \quad \lambda, z \in \rho(S_{\rm F}^*).$$

The following operator-valued functions are associated with a boundary triplet $\{\mathcal{H}, G, \Gamma\}$, [Arlinskiĭ, 1998, 1999a]

$$\begin{aligned} &F(\lambda) := (S_{\rm NR}^{1/2}(S_{\rm F}^* - \bar{\lambda} I_{\mathfrak{H}})^{-1})^*, \; G(\lambda) := (S_{\rm NR}^{1/2}\gamma(\bar{\lambda}))^*, \\ &q(\lambda) := (G(S_{\rm F}^* - \bar{\lambda} I_{\mathfrak{H}})^{-1})^*, \; Q_{\rm F}(\lambda) := G\gamma(\lambda), \; \lambda \in \rho(S_{\rm F}). \end{aligned} \quad (3.12)$$

For $\lambda, \mu \in \rho(S_{\mathrm{F}})$ the relations

$$\begin{aligned}
F(\lambda) - F(\mu) &= (\lambda - \mu)(S_{\mathrm{F}} - \lambda I_{\mathfrak{H}})^{-1} F(\mu), \\
G(\lambda) - G(\mu) &= (\lambda - \mu)\gamma^*(\bar{\mu}) F(\lambda), \\
q(\lambda) - q(\mu) &= (\lambda - \mu)(S_{\mathrm{F}} - \lambda I_{\mathfrak{H}})^{-1}, \\
Q_{\mathrm{F}}(\lambda) - Q_{\mathrm{F}}(\mu) &= (\lambda - \mu) q^*(\bar{\lambda})\gamma(\mu).
\end{aligned}$$

hold. In addition, for $\lambda \in \rho(S_{\mathrm{N}}) \cap \rho(S_{\mathrm{F}})$

$$(S_{\mathrm{N}} - \lambda I)^{-1} = (S_{\mathrm{F}} - \lambda I)^{-1} - q(\lambda) Q_{\mathrm{F}}^{*-1}(\bar{\lambda})\gamma^*(\bar{\lambda}). \tag{3.13}$$

In the sequel we suppose that

$$\operatorname{dom} S^* \subset \mathcal{D}[S_{\mathrm{N}}] \tag{3.14}$$

in which case $\operatorname{dom} S_{\mathrm{F}} \cap \operatorname{dom} S_{\mathrm{N}} = \operatorname{dom} S$. Hence, the operator

$$\begin{aligned}
&\operatorname{dom} L = \operatorname{dom} S_{\mathrm{F}} + \operatorname{dom} S_{\mathrm{N}}, \\
&L(f_{\mathrm{F}} + f_{\mathrm{N}}) = S_{\mathrm{F}} f_{\mathrm{F}} + S_{\mathrm{N}} f_{\mathrm{N}},\ f_{\mathrm{F}} \in \operatorname{dom} S_{\mathrm{F}},\ f_{\mathrm{N}} \in \operatorname{dom} S_{\mathrm{N}}
\end{aligned} \tag{3.15}$$

is well defined. For $\lambda \in \rho(S_{\mathrm{N}}) \cap \rho(S_{\mathrm{F}})$ define a linear subspace

$$\mathfrak{L}_\lambda := ((S_{\mathrm{N}} - \lambda I_{\mathfrak{H}})^{-1} - (S_{\mathrm{F}} - \lambda I_{\mathfrak{H}})^{-1})\mathfrak{H}$$

and for other $\lambda \in \rho(S_{\mathrm{F}})$ define

$$\mathfrak{L}_\lambda = (I_{\mathfrak{H}} + (\lambda - \mu)(S_{\mathrm{F}} - \lambda I_{\mathfrak{H}})^{-1})\mathfrak{L}_\mu = (S_{\mathrm{F}} - \mu I_{\mathfrak{H}})(S_{\mathrm{F}} - \lambda I_{\mathfrak{H}})^{-1}\mathfrak{L}_\mu,$$

where $\mu \in \rho(S_{\mathrm{N}}) \cap \rho(S_{\mathrm{F}})$. Then $\ker(L - \lambda I_{\mathfrak{H}}) = \mathfrak{L}_\lambda$ and

$$\operatorname{dom} L = \operatorname{dom} S_{\mathrm{F}} \dotplus \mathfrak{L}_\lambda,\ \lambda \in \rho(S_{\mathrm{F}}).$$

If $\{\mathcal{H}, \Gamma\}$ is a boundary pair for S, then $\mathfrak{L}_\lambda = \operatorname{ran} q(\lambda)$,

$$Lu = S_{\mathrm{F}}(u - q(\lambda)\Gamma u) + \lambda q(\lambda)\Gamma u,\ u \in \operatorname{dom} L,\ \lambda \in \rho(S_{\mathrm{F}}),$$

and there exists a linear operator $G_* : \operatorname{dom} L \to \mathcal{H}$ such that

$$S_{\mathrm{N}}[u, v] = (Lu, v) - (G_* u, \Gamma v)_{\mathcal{H}},\ u \in \operatorname{dom} L,\ v \in \mathcal{D}[S_{\mathrm{N}}].$$

The triplet $\{\mathcal{H}, G_*, \Gamma\}$ is called the *boundary triplet* for L. The operators S_{F} and S_{N} are said to be *transversal* if

$$\operatorname{dom} S^* = \operatorname{dom} S_{\mathrm{F}}^* + \operatorname{dom} S_{\mathrm{N}}^*$$

If this happens then condition (3.14) implies [Arlinskiĭ, 1997, 2000a] that the sectorial operator $S_* := S^* \restriction (\operatorname{dom} S_{\mathrm{F}}^* \cap \operatorname{dom} S_{\mathrm{N}}^*)$ is densely defined and

$$S_{*\mathrm{F}} = S_{\mathrm{F}}^*,\ S_{*\mathrm{N}} = S_{\mathrm{N}}^*,\ L = S_*^*. \tag{3.16}$$

If S is nonnegative then condition (3.14) is equivalent [Malamud, 1992] to the transversality of $S_{\rm F}$ and $S_{\rm N}$ (in which case $L = S^*$). For non-symmetric sectorial operators transversality occurs if $\dim \mathfrak{N}_\lambda < \infty$ for $\operatorname{Re}\lambda < 0$, or if S is coercive. Notice Green's identity

$$(Lu, v) - (u, S^*v) = (G_*u, \Gamma v) - (\Gamma u, Gv), \quad u \in \operatorname{dom} L,\ v \in \operatorname{dom} S^*.$$

Theorem 3.13 [Arlinskiĭ, 1998, 2000b] *If condition* (3.14) *is satisfied then for each m-accretive extension* $\widetilde{S}$ *of* S *the inclusions*

$$\operatorname{dom}\widetilde{S} \subset \mathcal{D}[S_{\rm N}] = \operatorname{dom} S_{\rm NR}^{1/2}, \ \operatorname{ran} S \subset \operatorname{ran} S_{\rm FR}^{1/2}$$

hold. If $\{\mathcal{H}, \Gamma\}$ *is a boundary pair for* S, *then*

$$\begin{aligned}&(\widetilde{S}u, v) = S_{\rm N}[u, v] + (\widetilde{\boldsymbol{W}}\Gamma u, \Gamma v) + 2(\widetilde{X}\Gamma u, S_{\rm NR}^{1/2}v),\\ &u \in \operatorname{dom}\widetilde{S},\ v \in \mathcal{D}[S_{\rm N}]\end{aligned} \tag{3.17}$$

establishes a one-to-one correspondence between all m-accretive extensions $\widetilde{S}$ *of* S *and all pairs* $\langle \widetilde{\boldsymbol{W}}, \widetilde{X}\rangle$, *where* $\widetilde{\boldsymbol{W}}$ *is an m-accretive relation in* $\mathcal{H}$ *and* $\widetilde{X} : \operatorname{dom}\widetilde{\boldsymbol{W}} \to \overline{\operatorname{ran}}\, S$ *is a linear operator such that*

$$\|\widetilde{X}e\|^2 \le \operatorname{Re}(\widetilde{\boldsymbol{W}}e, e)_{\mathcal{H}}, \quad e \in \operatorname{dom}\widetilde{\boldsymbol{W}}.$$

The identity (3.17) gives the representation of m-accretive extensions in a weak form and is the basis for a subsequent determination of m-accretive extensions of S and m-accretive restrictions of S^* in terms of abstract boundary conditions.

Theorem 3.14 [Arlinskiĭ, 1998] *Let* $\{\mathcal{H}, \Gamma\}$ *be a boundary pair for* S *and let* $\{\mathcal{H}, G, \Gamma\}$, $\{\mathcal{H}, G_*, \Gamma\}$ *be boundary triplets for* S^* *and* L, *respectively. Then there is a one-to-one correspondence between all m-accretive extensions* $\widetilde{S}$ *of* S *and all pairs* $\langle \widetilde{\boldsymbol{W}}, \widetilde{X}\rangle$, *where* $\widetilde{W}$ *is an m-accretive relation in* $\mathcal{H}$, *and* $\widetilde{X} : \operatorname{dom}\widetilde{\boldsymbol{W}} \to \overline{\operatorname{ran}}\, S$ *is a linear operator such that*

$$\|\widetilde{X}e\|^2 \leqslant \operatorname{Re}(\widetilde{\boldsymbol{W}}e, e), \quad e \in \operatorname{dom}\widetilde{\boldsymbol{W}}.$$

For each $\lambda \in \rho(S_{\rm F})$ *this correspondence is given by*

$$\begin{aligned}&\operatorname{dom}\widetilde{S} = \left\{ u \in \mathcal{D}[S_{\rm N}] : \begin{array}{c} u - q(\lambda)\Gamma u + 2F(\lambda)\widetilde{X}\Gamma u \in \operatorname{dom} S_{\rm F} \\ G_*(u + 2F(\lambda)\widetilde{X}\Gamma u) \in (\widetilde{\boldsymbol{W}} + 2G(\lambda)\widetilde{X})\Gamma u \end{array} \right\},\\ &\widetilde{S}u = L(u + 2F(\lambda)\widetilde{X}\Gamma u) + 2F(\lambda)\widetilde{X}\Gamma u.\end{aligned}$$

The extension $\widetilde{S}$ *is m-sectorial if and only if* $\widetilde{\boldsymbol{W}}$ *is m-sectorial and for some* $\delta \in [0, 1)$

$$\|\widetilde{X}e\|^2 \leqslant \delta^2 \operatorname{Re}(\widetilde{\boldsymbol{W}}e, e), \ e \in \operatorname{dom}\widetilde{\boldsymbol{W}}.$$

Remark A number $\lambda \in \rho(S_{\mathrm{F}})$ belongs to $\rho(\widetilde{S})$ if and only if

$$(\widetilde{\boldsymbol{W}} - Q_{\mathrm{F}}^*(\overline{\lambda}) + 2G(\lambda)\widetilde{X})^{-1} \in \boldsymbol{B}(\mathcal{H}), \tag{3.18}$$

in which case the resolvent is of the form

$$\begin{aligned}(\widetilde{S} - \lambda I)^{-1} &= (S_{\mathrm{F}} - \lambda I)^{-1} \\ &+ \big(q(\lambda) - 2F(\lambda)\widetilde{X}\big)(\widetilde{\boldsymbol{W}} - Q_{\mathrm{F}}^*(\overline{\lambda}) + 2G(\lambda)\widetilde{X})^{-1}\gamma^*(\overline{\lambda}).\end{aligned}$$

The next theorem contains a description of all m-accretive restrictions of the adjoint operator S^*.

Theorem 3.15 [Arlinskiĭ, 1998] *Let S be a densely defined closed sectorial operator, satisfying condition* (3.14). *Assume that $\{\mathcal{H}, \Gamma\}$ is a boundary pair for S and $\{\mathcal{H}, G, \Gamma\}$ is the corresponding boundary triplet for S^*. Then*

$$\begin{aligned}&\operatorname{dom} T = \Big\{ v \in \operatorname{dom} S^* : (I + \mathcal{B})Gv - (I - \mathcal{B})\Gamma v = 2\Psi v \Big\}, \\ &T = S^* \upharpoonright \operatorname{dom} T\end{aligned}$$

establishes a one-to-one correspondence between all m-accretive restrictions T of S^* and all pairs $\langle \mathcal{B}, \Psi \rangle$, where $\mathcal{B} \in \boldsymbol{B}(\mathcal{H})$ is a contraction and $\Psi : \operatorname{dom} S^* \to \mathcal{H}$ is a linear operator such that

$$\operatorname{ran} \Psi \subseteq \operatorname{ran} D_{\mathcal{B}^*}, \quad \|D_{\mathcal{B}^*}^{-1} \Psi v\|^2 \leqslant \operatorname{Re} S_{\mathrm{N}}[v], \quad v \in \operatorname{dom} S^*.$$

The operator T is m-sectorial if and only if $\mathcal{B} \in \widetilde{\mathcal{C}}$ and for some $\delta \in [0, 1)$

$$\|D_{\mathcal{B}^*}^{-1} \Psi v\|^2 \leqslant \delta^2 \operatorname{Re} S_{\mathrm{N}}[v], \quad v \in \operatorname{dom} S^*.$$

In addition we assume S to be coercive. Recall that in this case $\operatorname{dom} S_{\mathrm{N}} = \operatorname{dom} S \dotplus \ker S^*$, $\mathcal{D}[S_{\mathrm{N}}] = \mathcal{D}[S_{\mathrm{F}}] \dotplus \ker S^*$. It follows that $\operatorname{dom} S_{\mathrm{F}}^* + \operatorname{dom} S_{\mathrm{N}}^* = \operatorname{dom} S^*$ and $L = (S^* \upharpoonright (\operatorname{dom} S_{\mathrm{F}}^* \cap \operatorname{dom} S_{\mathrm{N}}^*))^*$, see (3.16).

For coercive sectorial operators we have the following parametrization.

Theorem 3.16 [Arlinskiĭ, 1998, 1999a] *Let S be a coercive sectorial operator, let $\{\mathcal{H}, \Gamma\}$ be a boundary pair for S, and let $\{\mathcal{H}, G_*, \Gamma\}$ be a boundary triplet for L. Then*

$$\begin{aligned}&\operatorname{dom} \widetilde{S} = \left\{ u \in \mathcal{D}[S_{\mathrm{N}}] : \begin{array}{l} u - Z_0\Gamma u + 2\widetilde{Y}\Gamma u \in \operatorname{dom} S_{\mathrm{F}} \\ G_*(u + 2\widetilde{Y}\Gamma u) \in \widetilde{\boldsymbol{W}}\Gamma u \end{array} \right\}, \\ &\widetilde{S}u = S_{\mathrm{F}}(u - Z_0\Gamma u + 2\widetilde{Y}\Gamma u)\end{aligned}$$

establishes a bijective correspondence between all m-accretive extensions

$\widetilde{S}$ of S and all pairs $\langle \widetilde{\boldsymbol{W}}, \widetilde{Y}\rangle$, where $\widetilde{\boldsymbol{W}}$ is an m-accretive relation in $\mathcal{H}$ and $\widetilde{Y} : \operatorname{dom} \widetilde{\boldsymbol{W}} \to \mathcal{D}[S]$ is a linear operator such that

$$\mu\big[\widetilde{Y}e\big] \leqslant \operatorname{Re}\big(\widetilde{\boldsymbol{W}}e, e\big),\ e \in \operatorname{dom} \widetilde{\boldsymbol{W}}.$$

The operator $\widetilde{S}$ is m-sectorial if and only if $\widetilde{\boldsymbol{W}}$ is m-sectorial relation and for some $\delta \in [0, 1)$

$$\mu\big[\widetilde{Y}e\big] \leqslant \delta^2 \operatorname{Re}\big(\widetilde{\boldsymbol{W}}e, e\big),\ e \in \operatorname{dom} \widetilde{\boldsymbol{W}},$$

in which case for $u, v \in \mathcal{D}[\widetilde{S}] = \Gamma^{-1}\mathcal{D}[\widetilde{\boldsymbol{W}}]$

$$\widetilde{S}[u, v] = S_{\mathrm{N}}\big[u - Z_0\Gamma u + 2\overline{\widetilde{Y}}\Gamma u, v - Z_0\Gamma v\big] + \widetilde{\boldsymbol{W}}\big[\Gamma u, \Gamma v\big].$$

Remark The function $U(\lambda) := (I - \lambda S_{\mathrm{F}}^{-1})^{-1}$ can be used to describe some parts of the spectrum of $\widetilde{S}$. The number $\lambda \in \rho(S_{\mathrm{F}}) \cap \rho(\widetilde{S})$ if and only if

$$(\widetilde{\boldsymbol{W}} - \lambda Z_0^* U(\lambda)(Z_0 - 2\widetilde{Y}))^{-1} \in \boldsymbol{B}(\mathcal{H}),$$

in which case

$$\begin{aligned}\big(\widetilde{S} - \lambda I\big)^{-1} &= \big(S_{\mathrm{F}} - \lambda I\big)^{-1} \\ &\quad + U(\lambda)\big(Z_0 - 2\widetilde{Y}\big)(\widetilde{\boldsymbol{W}} - \lambda Z_0^* U(\lambda)\big(Z_0 - 2\widetilde{Y}\big))^{-1} Z_0^* U(\lambda);\end{aligned}$$

the number $\lambda \in \rho(S_{\mathrm{F}})$ is an eigenvalue of $\widetilde{S}$ if and only if

$$\ker(\widetilde{\boldsymbol{W}} - \lambda Z_0^* U(\lambda)(Z_0 - 2\widetilde{Y})) \neq \{0\}$$

in which case

$$\ker(\widetilde{S} - \lambda I) = U(\lambda)(Z_0 - 2\widetilde{Y})\ker(\widetilde{\boldsymbol{W}} - \lambda Z_0^* U(\lambda)(Z_0 - 2\widetilde{Y})).$$

In conclusion we mention that all results of Sections 3.3–3.5 remain valid for sectorial relations [Arlinskiĭ, 1997, 1999a, 2000b, 1999c, 2006].

3.6 W_{F}- and Q_{F}-functions

In [Derkach and Malamud, 1991] Weyl functions were introduced by means of boundary triplets of symmetric operators. Here we consider the case of sectorial operators.

Let $\{\mathcal{H}, \Gamma\}$ be a boundary pair for a closed densely defined sectorial operator S. Assume condition (3.14) and define the W_{F}*-function of* S, corresponding to the boundary pair $\{\mathcal{H}, \Gamma\}$, as follows [Arlinskiĭ, 1999a]:

$$(W_{\mathrm{F}}(\lambda,\, z)e,\, h) := S_{\mathrm{N}}^*[\gamma(\lambda)e,\, \gamma(z)h],\ e, h \in \mathcal{H},\ \lambda, z \in \rho(S_{\mathrm{F}}^*),$$

where $\gamma(\lambda) = (\Gamma\restriction\mathfrak{N}_\lambda)^{-1}$, $\lambda \in \rho(S_{\rm F}^*)$ (see (3.11)). The $W_{\rm F}$-function is holomorphic in λ and anti-holomorphic in z in the domain $\rho(S_{\rm F}^*)$. Moreover, the kernel appearing below is α-sectorial according to

$$|\mathrm{Im}\sum_{i,j=1}^{n}(W_{\rm F}(\lambda_i,\,\lambda_j)f_i,\,f_j)| \leqslant (\tan\alpha)\,\mathrm{Re}\sum_{i,j=1}^{n}(W_{\rm F}(\lambda_i,\,\lambda_j)f_i,\,f_j)$$

for every choice of numbers $\lambda_1,...,\lambda_n \in \rho(S_{\rm F}^*)$ and vectors $f_1,...,f_n \in \mathcal{H}$. In addition, for all $\lambda,\,z_1,\,z_2 \in \rho(S_{\rm F}^*)$ the equality

$$\lambda\gamma^*(z_1)\gamma(\lambda) - W_{\rm F}(\lambda,\,z_1) = \lambda\gamma^*(z_2)\gamma(\lambda) - W_{\rm F}(\lambda,\,z_2)$$

holds. It follows that the function

$$\lambda\gamma^*(z)\gamma(\lambda) - W_{\rm F}(\lambda,\,z)$$

does not depend on z and is holomorphic in $\lambda \in \rho(S_{\rm F}^*)$.

Theorem 3.17 [Arlinskiĭ, 1999a] *Let $\{\mathcal{H},\Gamma\}$ be a boundary pair for an α-sectorial operator S and let $\{\mathcal{H},G,\Gamma\}$ be a boundary triplet for S^*. Let $Q_{\rm F}(\lambda) = G\gamma(\lambda)$, see (3.12). Then*[1]

$$\begin{aligned}&Q_{\rm F}(\lambda) = \lambda\gamma^*(z)\gamma(\lambda) - W_{\rm F}(\lambda,\,z),\\ &Q_{\rm F}(\lambda) = -\lim\nolimits_{z\to\infty,\,z\in\mathbb{C}\setminus\mathcal{S}(\beta)} W_{\rm F}(\lambda,\,z) \textit{ for any } \beta \in (\alpha,\tfrac{\pi}{2}).\end{aligned}$$

When $\alpha = 0$ (S is nonnegative), then $Q_{\rm F}$ is the Weyl function corresponding to the boundary triplet $\{\mathcal{H},G,\Gamma\}$ and it coincides with the Q_μ-function in [Kreĭn and Ovčarenko, 1976, 1978], in which case

$$W_{\rm F}(\lambda,z) = \frac{\bar{z}Q_{\rm F}(\lambda) - \lambda Q_{\rm F}^*(z)}{\lambda - \bar{z}}.$$

The function $Q_{\rm F}$ is called the *$Q_{\rm F}$-function of S* corresponding to the boundary pair $\{\mathcal{H},\Gamma\}$. When S is coercive, then

$$\begin{aligned}&Q_{\rm F}(\lambda) = \lambda\gamma^*(0)(I_{\mathcal{H}} - \lambda S_{\rm F}^{*-1})^{-1}\gamma(0),\\ &W_{\rm F}(\lambda,z) = \lambda\bar{z}\gamma^*(0)(S_{\rm F} - \bar{z}I_{\mathfrak{H}})^{-1}(I_{\mathfrak{H}} - \lambda S_{\rm F}^{*-1})^{-1}\gamma(0).\end{aligned}$$

Observe that $Q_{\rm F}$-function is an essential part of the resolvent formulas (see (3.13), Theorems 3.14 and 3.16 and the discussion following them). A closed sectorial operator S is said to be *simple* if c.l.s. $\{\mathfrak{N}_\lambda,\ \lambda \in \Pi_-\} = \mathfrak{H}$.

[1] Whenever a limit for operators is considered we understand it in the strong operator topology.

Theorem 3.18 [Arlinskiĭ, 1999a] *Let S_1 and S_2 be simple closed sectorial, densely defined operators with disjoint Friedrichs and Kreĭn-von Neumann extensions and acting in the Hilbert spaces $\mathfrak{H}_1$ and $\mathfrak{H}_2$, respectively. Then S_1 and S_2 are unitary equivalent if and only if there is a bijective linear map X such that*

$$W_{1\mathrm{F}}(\lambda, z) = X^* W_{2\mathrm{F}}(\lambda, z) X, \quad \lambda, z \in \Pi_-.$$

The W_{F}-function characterizes a simple sectorial operator in the following sense.

Theorem 3.19 [Arlinskiĭ, 1999a] *Let W be a $\boldsymbol{B}(\mathcal{H})$-valued kernel, holomorphic in the left half-plane, with the following properties:*

1. *W is an α-sectorial kernel;*
2. *$Q(\lambda) = -\lim\limits_{x \to -\infty} W(\lambda, x)$ exists for every λ;*
3. *$\lim\limits_{\lambda \to 0} Q(\lambda) = 0$ and $\lim\limits_{\lambda \to -\infty} \lambda^{-1} Q(\lambda) = 0$;*
4. *$Q^{-1}(\lambda) \in \boldsymbol{B}(\mathcal{H})$, $\mathrm{Re}\,\lambda < 0$, and $\lim\limits_{\lambda \to -\infty} Q^{-1}(\lambda) = 0$;*
5. *$K(\lambda, z) = \lambda^{-1}(Q(\lambda) + W(\lambda, z))$ is positive definite and $K^{-1}(\lambda_0, \lambda_0) \in \boldsymbol{B}(\mathcal{H})$ for some λ_0.*

Then there exist a Hilbert space $\mathfrak{H}$ and a closed, densely defined and simple α–sectorial operator S in $\mathfrak{H}$ such that $\mathrm{dom}\, S^* \subset \mathcal{D}[S_{\mathrm{N}}]$ *and one of its W_{F}-functions coincides with W in the left half-plane.*

3.7 Realization of the Phillips boundary space

Let $\{\mathcal{H}, \Gamma\}$ be a boundary pair for S and let $\{\mathcal{H}, G, \Gamma\}$ be the corresponding boundary triplet for S^*. The linear space

$$\mathfrak{K} = \overline{\mathrm{ran}}\, S \times \mathcal{H} \times \mathcal{H}$$

equipped with the indefinite inner product,

$$Q(\langle f, e_1, e_2 \rangle, \langle g, h_1, h_2 \rangle) = -2(f, g) - (e_1, h_2) - (e_2, h_1) \tag{3.19}$$

becomes a Kreĭn space. Observe that the subspaces

$$\mathfrak{K}^+ = \{\langle 0, -h, h \rangle \,:\, h \in \mathcal{H}\},\ \mathfrak{K}^- = \{\langle f, g, g \rangle \,:\, f \in \overline{\mathrm{ran}}\, S,\ g \in \mathcal{H}\}$$

are maximal positive and maximal negative with respect to the Q-inner product, Q-orthogonal and

$$\mathfrak{K} = \mathfrak{K}^+ \oplus \mathfrak{K}_-. \tag{3.20}$$

Let $\boldsymbol{S}^* = \{\langle u, S^*u\rangle : u \in \operatorname{dom} S^*\}$ be the graph of the operator S^*. Define the linear operator $\Phi : -\boldsymbol{S}^* \to \mathfrak{K}$ by

$$\Phi\langle u, -S^*u\rangle = \langle S_{\mathrm{NR}}^{1/2}u, \Gamma u, Gu\rangle, \ u \in \operatorname{dom} S^*.$$

Due to (3.10) the operator Φ is isometric from $-\boldsymbol{S}^*$ with the indefinite inner product (3.1) into $\mathfrak{K}$ with the indefinite inner product (3.19). Hence, from the decomposition (3.2) it follows that Φ has an isometric extension $\overline{\Phi}$ to $\hat{\mathfrak{H}}_P$.

Proposition 3.20 [Arlinskiĭ, 2000b] *If the condition* (3.14) *is satisfied, then* $\operatorname{ran}\overline{\Phi} = \mathfrak{K}$.

Thus $\mathfrak{K}$ is a realization of the Phillips abstract boundary space. By (3.20) the maximal nonnegative subspaces in $\mathfrak{K}$ are of the form

$$\mathfrak{L}^+ = \{\langle Xh, -(I+Y)h, (I-Y)h\rangle : h \in \mathcal{H}\},$$

where $X, Y \in \boldsymbol{B}(\mathcal{H})$ and

$$\|Xh\|^2 + \|Yh\|^2 \le \|h\|^2, \ h \in \mathcal{H}.$$

Since a subspace in $\mathfrak{K}$ is maximal nonpositive if and only if it is Q-orthogonal to a maximal nonnegative subspace and Φ is isometric on $-\boldsymbol{S}^*$, Theorem 3.15 can be derived from Phillips' Theorem 3.1.

Related realizations and applications to positive definite ordinary differential operators are given in [Evans and Knowles, 1985, 1986; Mil'yo and Storozh, 1991, 1993; Wei and Xu, 2005; Xu et al., 2002].

3.8 Vishik-Birman-Grubb type formulas

In [Vishik, 1952] a dual pair $\{A, B\}$ of densely defined boundedly invertible operators is considered and a description of some classes of its extensions is given. In the particular case $A = B = S$, where S is a positive definite symmetric operator with dense domain the Vishik formula

$$\operatorname{dom}\widetilde{S} = \operatorname{dom} S \dotplus (S_{\mathrm{F}}^{-1} + \widetilde{C})\operatorname{dom}\widetilde{C} \dotplus (\ker S^* \ominus \operatorname{dom}\widetilde{C})$$

gives the parametrization of domains for all self-adjoint extensions of S. Here $\widetilde{C}$ is a selfadjoint operator with $N = \overline{\operatorname{dom}\widetilde{C}} \subseteq \ker S^*$. This approach was applied in [Birman, 1956] for nonnegative selfadjoint extensions. In [Grubb, 1968, 1970, 1971, 1973, 1974] the Vishik-Birman method has been developed applied to elliptic boundary problems; see in particular [Grubb, 1968, Theorems 1.1 and 2.1]. In this section an analog is presented in a sectorial setting. Recall that the quadratic functional μ in the following statements has been defined by (3.8).

Theorem 3.21 [Arlinskiĭ, 2000b] *Let S be a densely defined closed and coercive sectorial operator. Then*

$$\begin{aligned}
&\operatorname{dom}\widetilde{S} = \operatorname{dom} S \dotplus (S_{\mathrm{F}}^{-1} + (I-2Y)M)\operatorname{dom} M \dotplus \mathfrak{M},\\
&\widetilde{S}(x + S_{\mathrm{F}}^{-1}h + (I-2Y)Mh + e) = Sx + h,\\
&x \in \operatorname{dom} S,\ h \in \operatorname{dom} M,\ e \in \mathfrak{M}
\end{aligned}$$

establishes a bijective correspondence between all m-accretive extensions $\widetilde{S}$ of S and all triplets $\langle\mathfrak{M}, M, Y\rangle$, where $\mathfrak{M}$ is a closed subspace in $\ker S^*$, *M is an m-accretive operator in* $\ker S^* \ominus \mathfrak{M}$ *and the linear operator $Y : \operatorname{ran} M \to \mathcal{D}[S]$ is such that*

$$\mu[YMh] \leqslant \operatorname{Re}(Mh, h),\quad h \in \operatorname{dom} M.$$

Moreover, $\widetilde{S}$ is m-sectorial if and only if M is m-sectorial in $\ker S^* \ominus \mathfrak{M}$ *and for some $\delta \in [0,1)$*

$$\mu[YMh] \leqslant \delta^2 \operatorname{Re}(Mh,\, h),\quad h \in \operatorname{dom} M.$$

The closed form associated with $\widetilde{S}$ is given by

$$\begin{aligned}
&\mathcal{D}[\widetilde{S}] = \mathcal{D}[S] \dotplus \mathcal{R}[M] \dotplus \mathfrak{M},\\
&\widetilde{S}[\varphi + f + e, \phi + g + h] = S[\varphi + 2\overline{Y}f, \phi] + (M\restriction \overline{\operatorname{ran}}\, M)^{-1}[f, g],\\
&\varphi, \phi \in \mathcal{D}[S],\ f, g \in \operatorname{ran} M_R^{1/2},\ e, h \in \mathfrak{M}
\end{aligned}$$

where $\overline{Y}$ is an extension of Y to $\operatorname{ran} M_R^{1/2}$. *The operator $\widetilde{S}$ has bounded inverse if and only if* $\operatorname{dom} M = \ker S^*$, *in which case,*

$$\widetilde{S}^{-1} = S_{\mathrm{F}}^{-1} + (I-2Y)MP_0,$$

where P_0 is the orthogonal projection onto $\ker S^*$.

Theorem 3.22 [Arlinskiĭ, 1999c] *Let $\{S,T\}$ be a dual pair of densely defined closed and coercive sectorial operators. Suppose the equality $T_{\mathrm{F}} = S_{\mathrm{F}}^*$. Put*

$$\mathfrak{M}_0 = \ker S^* \cap \ker T^*$$

and denote by P_0^S, P_0^T the orthogonal projections in $\mathcal{D}[S_{\mathrm{N}}]$ and $\mathcal{D}[T_{\mathrm{N}}]$ onto $\ker S^*$ *and* $\ker T^*$ *with respect to the decompositions*

$$\mathcal{D}[S_{\mathrm{N}}] = \mathcal{D}[S] \dotplus \ker S^*,\quad \mathcal{D}[T_{\mathrm{N}}] = \mathcal{D}[S] \dotplus \ker T^*.$$

Then all m-accretive extensions of the dual pair $\{S,T\}$ are given by

$$\operatorname{dom}\widetilde{S} = \operatorname{dom} S \dotplus (S_{\mathrm{F}}^{-1} + P_0^T M)\operatorname{dom} M \dotplus \widehat{\mathfrak{M}}_0,\ \widetilde{S} = T^*\restriction \operatorname{dom}\widetilde{S}.$$

Here $\widehat{\mathfrak{M}}_0$ *is subspace of* $\mathfrak{M}_0$, M *is a m-accretive operator in the subspace* $\ker S^* \ominus \widehat{\mathfrak{M}}_0$ *with* $\operatorname{dom} M$ *and*

$$\operatorname{ran} M \subseteq P_0^S(\mathcal{D}[S_{\rm N}] \cap \mathcal{D}[T_{\rm N}]),\ \operatorname{Re}(Mh,h) \geqslant \tfrac{1}{4}\mu[P_{\rm F}^T Mh],\ h \in \operatorname{dom} M.$$

In terms of abstract boundary conditions the preceding theorem can be reformulated as follows.

Theorem 3.23 [Arlinskiĭ, 1999c] *Let the densely defined and coercive sectorial operators* S, T *form a dual pair. Let* $\{\mathcal{H}^S, \Gamma^S\}$ *and* $\{\mathcal{H}^T, \Gamma^T\}$ *be boundary pairs for* S *and* T, *respectively. Suppose that* $S_{\rm F} = T_{\rm F}^*$. *Let* $\{\mathcal{H}^S, G_*^S, \Gamma^S\}$ *be a boundary triplet for* L. *Define*

$$\begin{aligned} Z_0^S &= (\Gamma^S|\ker S^*)^{-1}, & E_0 &= \Gamma^S(\mathcal{D}[S_{\rm N}] \cap \mathcal{D}[T_{\rm N}]),\\ Z_0^T &= (\Gamma^T|\ker T^*)^{-1}, & Y_0 e &= \tfrac{1}{2}(Z_0^S - Z_0^T\Gamma^T Z_0^S)e,\ e \in E_0. \end{aligned}$$

Then

$$\begin{aligned} &\operatorname{dom}\widetilde{S} = \left\{u \in \mathcal{D}(T^*) \cap D[S_{\rm N}] : G_*^S(u - Z_0^T\Gamma^T u) \in \widetilde{\boldsymbol{W}}\Gamma^S u\right\},\\ &\widetilde{S}u = T^*u \end{aligned}$$

establishes a bijective correspondence between all m-accretive extensions $\widetilde{S}$ *of the dual pair* $\{S,T\}$ *and all m-accretive relations* $\widetilde{\boldsymbol{W}}$ *in* $\mathcal{H}^S$ *such that*

$$\operatorname{dom}\widetilde{\boldsymbol{W}} \subseteq E_0,\ \operatorname{Re}(\widetilde{\boldsymbol{W}}e,e)_{\mathcal{H}^S} \geqslant \mu[Y_0 e],\ \ e \in \operatorname{dom}\widetilde{\boldsymbol{W}}.$$

The operator $\widetilde{S}$ *is m-sectorial if and only if* $\widetilde{\boldsymbol{W}}$ *is m-sectorial and for some* $\delta \in [0,1)$

$$\delta^2\operatorname{Re}(\widetilde{\boldsymbol{W}}e,e)_{\mathcal{H}^S} \geqslant \mu[Y_0 e],\ \ e \in \operatorname{dom}\widetilde{\boldsymbol{W}}.$$

3.9 m-sectorial extensions via fractional-linear transformations

For $\alpha \in [0,\pi/2)$ let the suboperator A in $\mathcal{C}(\alpha)$ and the α-sectorial relation $\boldsymbol{S}$ be related by (3.3) and for some $\beta \in [\alpha,\pi/2)$ let the operator $\widetilde{A}$ in $\mathcal{C}(\beta)$ and the m-β-sectorial relation $\widetilde{\boldsymbol{S}}$ be related by (3.6). According to Proposition 3.5 $\widetilde{A}$ is an extension of A if and only if $\widetilde{\boldsymbol{S}}$ is an extension of $\boldsymbol{S}$.

Definition 3.24 [Arlinskiĭ, 1999b] Let A be a suboperator in $\mathcal{C}(\alpha)$ with $\operatorname{dom} A = \mathfrak{H}_0 \subset \mathfrak{H}$. The extension $\widetilde{A} \in \widetilde{\mathcal{C}}$ of A is called *hard* if

$$\inf\{\operatorname{Re}((I+\widetilde{A})(h-\phi),h-\phi),\ \phi \in \mathfrak{H}_0\} = 0 \quad \text{for all } h \in \mathfrak{H}$$

and *soft* if

$$\inf\{\operatorname{Re}((I-\widetilde{A})(h-\phi),h-\phi),\ \phi\in\mathfrak{H}_0\}=0\quad\text{for all } h\in\mathfrak{H}.$$

For a Hermitian contraction A these notions were introduced in [Kreĭn, 1947a]. Denote by A_μ and $A_{\rm M}$ the hard and soft extensions of A, respectively. Clearly, $A_{\rm M}=-(-A)_\mu$. The identities

$$2\operatorname{Re}\big((I\pm A)\phi,\phi\big)=\|(I\pm A)\phi\|^2+\|\phi\|-\|A\phi\|^2,\quad \phi\in\mathfrak{H}_0,$$

imply that if A is a suboperator in $\mathcal{C}(\alpha)$, then $I\pm A$ are α-sectorial. In [Arlinskiĭ, 1997] the following identities are shown:

$$A_\mu=(I+A)_{\rm N}-I,\quad A_{\rm M}=I-(I-A)_{\rm N},$$

where $(I\pm A)_{\rm N}$ are the Kreĭn-von Neumann extensions of $I\pm A$ and A_μ, $A_{\rm M}\in\mathcal{C}(\alpha)$. In terms of $\boldsymbol{S}$ one has [Arlinskiĭ, 1997]:

$$A_\mu=-I+2(I+\boldsymbol{S}_{\rm F})^{-1},\ A_{\rm M}=-I+2(I+\boldsymbol{S}_{\rm N})^{-1}.$$

Let $\mathfrak{N}=\mathfrak{H}\ominus\mathfrak{H}_0$ and let $P_{\mathfrak{H}_0}$, $P_{\mathfrak{N}}$ be the orthogonal projections in $\mathfrak{H}$ onto $\mathfrak{H}_0$ and $\mathfrak{N}$, respectively. The extensions A_μ and $A_{\rm M}$ can be written using Theorem 3.4 by means of special contractions $K_\mu, K_{\rm M}:\mathfrak{N}\to\mathfrak{D}_{A^*}$ [Arlinskiĭ, 1999b]. Here we give a representation of A_μ and $A_{\rm M}$ in block-operator matrix form with respect to the orthogonal decomposition $\mathfrak{H}=\mathfrak{H}_0\oplus\mathfrak{N}$. To do this we define the operators

$$A_0:=P_{\mathfrak{H}_0}A,\quad C:=P_{\mathfrak{N}}A,$$

so that $A_0\in\mathcal{C}(\alpha)$ in $\mathfrak{H}_0$. We can rewrite A in the block-matrix form

$$A=\begin{pmatrix}A_0\\ C\end{pmatrix}.$$

Since A is a contraction, we have $C=K_0D_{A_0}$, where $K_0\in\boldsymbol{B}(\mathfrak{D}_{A_0},\mathfrak{M})$ is a contraction. Let

$$\Theta_{A_0}(z)=\big[-A_0+zD_{A_0^*}(I-zA_0^*)^{-1}D_{A_0}\big]\upharpoonright\mathfrak{D}_{A_0}$$

be the Sz.-Nagy–Foias characteristic function [Sz.-Nagy and Foias, 1970] of the contraction A_0. Since $A_0\in\mathcal{C}(\alpha)$ in $\mathfrak{H}_0$, there exist strong non-tangential unitary limit values $\Theta_{A_0}(-1)$ and $\Theta_{A_0}(1)$ [Arlinskiĭ, 1987].

Then the operators A_μ and A_{M} take the form [Arlinskiĭ, 2006]

$$A_\mu = \begin{pmatrix} A_0 & -D_{A_0^*}\Theta_{A_0}(-1)K_0^* \\ K_0 D_{A_0} & K_0 A_0^*\Theta_{A_0}(-1)K_0^* - D^2_{K_0^*} \end{pmatrix},$$

$$A_{\mathrm{M}} = \begin{pmatrix} A_0 & D_{A_0^*}\Theta_{A_0}(1)K_0^* \\ K_0 D_{A_0} & -K_0 A_0^*\Theta_{A_0}(1)K_0^* + D^2_{K_0^*} \end{pmatrix}.$$

Define the operator

$$T_0 := \tfrac{1}{2}(A_\mu + A_{\mathrm{M}}),$$

so that $T_0 \in \mathcal{C}(\alpha)$ in $\mathfrak{H}$. According to the inequality

$$|((T_0 - T_0^*)f, f)| \le \tan\alpha \, \|D_{T_0} f\|^2, \; f \in \mathfrak{H}$$

there exists in $\mathfrak{D}_{T_0}$ a bounded selfadjoint operator Σ, such that

$$T_0 - T_0^* = 2\mathrm{i} D_{T_0}\Sigma D_{T_0}, \quad \|\Sigma\| \le \tfrac{1}{2}\tan\alpha.$$

The following characteristic function of T_0

$$\Theta_{T_0}(z) = \left[\, T_0 + z D_{T_0^*}(I - zT_0^*)^{-1} D_{T_0}\right] |\mathfrak{D}_{T_0}$$

has limit values $\Theta_{T_0}(\pm 1)$, so that [Arlinskiĭ, 1991]

$$2\mathrm{i}\Sigma = (I - \Theta_{T_0}^{-1}(-1)\Theta_{T_0}(1))^{-1}(I + \Theta_{T_0}^{-1}(-1)\Theta_{T_0}(1)).$$

Let P_{Ω_0} be the orthogonal projection from $\mathfrak{D}_{T_0}$ onto Ω_0, where

$$\Omega_0 = \{f \in \mathfrak{D}_{T_0} : D_{T_0} f \in \mathfrak{N}\}.$$

Theorem 3.25 [Arlinskiĭ, 1999b] *Let A be a suboperator in $\mathcal{C}(\alpha)$ and let $\beta \in [\alpha, \pi/2)$. The formula*

$$T = T_0 + D_{T_0^*}(I + \Theta_{T_0}(1)YT_0^*)^{-1}\Theta_{T_0} Y D_{T_0} \tag{3.21}$$

gives a one-to-one correspondence between the operators T in $\mathcal{C}(\beta)$, extending T_0, and the contractions $Y \in \boldsymbol{B}(\mathfrak{D}_{T_0})$ which satisfy:

$$\ker Y \supset \overline{D_{T_0}\mathfrak{H}_0}, \; (I + \Theta_{T_0}(1)YT_0^*)^{-1} \in \boldsymbol{B}(\mathfrak{D}_{T_0}),$$
$$|2\mathrm{i}((I - Y^*)\Sigma(I - Y) + Y - Y^*)h, h)| \le \tan\beta \|D_Y h\|^2, \; h \in \mathfrak{D}_{T_0}.$$

The extension A_{M} corresponds to $Y = P_{\Omega_0}$ and the extension A_μ corresponds to $Y = \Theta_{T_0}^{-1}(1)\Theta_{T_0}(-1)P_{\Omega_0}$.

If A is a Hermitian contraction ($\alpha = 0$), then A_μ and A_{M} are the extremal selfadjoint contractive extensions of A defined by Kreĭn and all selfadjoint contractive extensions of A form an operator interval $[A_\mu, A_{\rm M}]$ [Kreĭn, 1947a]. In this case $\Theta_{T_0}(\pm 1) = \pm I_{\mathfrak{D}_{T_0}}$ and (3.21) takes the form

$$T = T_0 + D_{T_0}(I + YT_0)^{-1} Y D_{T_0}, \tag{3.22}$$

where $Y \in \mathcal{C}(\beta)$, $(I + YT_0)^{-1} \in \boldsymbol{B}(\mathfrak{D}_{T_0})$ and $\ker Y \supset \overline{D_{T_0}\mathfrak{H}_0}$. If, in addition, $\ker Y^* \supset \overline{D_{T_0}\mathfrak{H}_0}$ then (3.22) can be transformed into

$$T = \tfrac{1}{2}(A_{\rm M} + A_\mu) + \tfrac{1}{2}(A_{\rm M} - A_\mu)^{1/2} Z (A_{\rm M} - A_\mu)^{1/2}, \tag{3.23}$$

where $Z \in \mathcal{C}(\beta)$ in $\mathfrak{L} = \overline{\operatorname{ran}}\,(A_{\rm M} - A_\mu)$. The latter gives a one-to-one correspondence between all quasi-selfadjoint extensions T of A ($T \supset A, T^* \supset A$) from the class $\mathcal{C}(\beta)$ [Arlinskiĭ and Tsekanovskiĭ, 1984a,b, 1988]. The equation (3.23) has been used for a description of accretive and sectorial boundary value problems for ordinary differential operators in [Arlinskiĭ and Tsekanovskiĭ, 1984b; Tsekanovskiĭ, 1985].

Theorem 3.26 [Malamud, 1989, 2002] *Let A be a suboperator in $\mathcal{C}(\alpha)$ in $\mathfrak{H}$. Then the operator T in $\mathcal{C}(\beta)$ is an extension of A if and only if*

$$T = A_\mu + KP_{\mathfrak{N}} \text{ and } R_\pm^{-1}(KP_{\mathfrak{N}} + Q_\pm) \text{ is a contraction,}$$

where

$$R_\pm = D_{A^*} \pm \mathrm{i}\cot\beta(AP_{\mathfrak{H}_0} - A^*) + \cot^2\beta P_{\mathfrak{N}}, \quad Q_\pm = (A_\mu \pm \mathrm{i}\cot\beta I)P_{\mathfrak{N}}.$$

In [Malamud, 2006] the existence and the description of $\widetilde{\mathcal{C}}$-extensions for some dual pairs of $\mathcal{C}(\alpha)$-operators have been considered.

3.10 Sectorial operators in divergence form

Now we consider a special case of sectorial operators, the so called operators in divergence form, [Arlinskiĭ, 2000a]. Assume that

(**A**) L_1 and L_2 are closed densely defined operators in the Hilbert space $\mathfrak{H}$ taking values in a Hilbert space H and such that $L_1 \subset L_2$;
(**B**) $Q \in \boldsymbol{B}(H)$ is a coercive operator.

Clearly, Q is m-sectorial. Consider the sesquilinear forms

$$S_k[u,v] = (QL_k u, L_k v)_H, \; u, v \in \operatorname{dom} L_k, \; k = 1, 2.$$

Due to (**A**) and (**B**) these forms are closed and sectorial, and

$$S_k = L_k^* Q L_k, \; k = 1, 2,$$

are the associated m-sectorial operators. We equip the subspaces dom L_k, $k = 1, 2$, with the graph norms. Assume in addition that

(C) $\operatorname{dom} L_1 \cap \operatorname{dom} S_2$ is dense in $\operatorname{dom} L_1$.

From **(A)**, **(B)**, **(C)** it follows that

$$S := L_2^* Q L_1 \tag{3.24}$$

is a closed densely defined sectorial operator and the operator S_1 is the Friedrichs extension of S. Such operators S are said to be in *in divergence form*; see [Hassi, Sandovici, de Snoo, and Winkler, 2007; Sandovici, 2006] for the case of relations. The representation (3.24) naturally arises in the theory of differential operators.

The case of finite codimension If, in addition to **(A)** and **(B)**, the condition

(D) $\dim(\operatorname{dom} L_2/\operatorname{dom} L_1) < \infty$

is satisfied then condition **(C)** is also satisfied [Arlinskiĭ, 2000a] and

- $S^* = L_1^* Q^* L_2$;
- $S_{\mathrm{N}} = L_2^* Q \mathcal{P} L_2$, where $\mathcal{P}$ is the projection in H onto $\overline{\operatorname{ran}} L_1$ with respect to the decomposition $\mathfrak{H} = \overline{\operatorname{ran}} L_1 \dotplus Q^{*-1}(\ker L_1^*)$;
- $\mathcal{D}[S_{\mathrm{N}}] = \operatorname{dom} L_2$, $S_{\mathrm{N}}[u, v] = (Q\mathcal{P} L_2 u, L_2 v)_{\mathfrak{H}}$, $u, v \in \operatorname{dom} L_2$.

Let L_1 and L_2 satisfy **(A)** and let $\{\mathcal{H}, \Gamma\}$ be a *boundary pair* for $L_1 \subset L_2$ in the sense of [Lyantse and Storozh, 1983], i.e.

- $\mathcal{H}$ is a Hilbert space;
- $\Gamma \in \boldsymbol{B}(\operatorname{dom} L_2, \mathcal{H})$;
- $\ker \Gamma = \operatorname{dom} L_1$, $\operatorname{ran} \Gamma = \mathcal{H}$.

Then there exists a linear operator $\Phi \in \boldsymbol{B}(\operatorname{dom} L_1^*, \mathcal{H})$ such that $\{\mathcal{H}, \Phi\}$ is a boundary pair for $L_2^* \subset L_1^*$ and Green's identity

$$(L_1^* f, u)_{\mathfrak{H}} - (f, L_2 u)_H = (\Phi f, \Gamma u)_{\mathcal{H}}, \; f \in \operatorname{dom} L_1^*, \; u \in \operatorname{dom} L_2 \tag{3.25}$$

holds. The set $\{\mathcal{H}, \Phi, \Gamma\}$ is called *the boundary triplet* for $L_1 \subset L_2$.

Let $\{\mathcal{H}, \Gamma\}$ be a boundary pair for $L_1 \subset L_2$. From **(A)**, **(B)**, **(D)** it follows that $\{\mathcal{H}, \Gamma\}$ is the boundary pair for $S = L_2^* Q L_1$. Moreover, if $\{\mathcal{H}, \Phi, \Gamma\}$ is a boundary triplet for $L_1 \subset L_2$ then $\{\mathcal{H}, \Phi Q^* L_2, \Gamma\}$ and $\{\mathcal{H}, \Phi Q L_2, \Gamma\}$ are the boundary triplets for S^* and $\mathcal{L} = L_1^* Q L_2$, respectively.

Theorem 3.27 [Arlinskiĭ, 1998, 2000a] *Assume the conditions* (**A**), (**B**), and (**D**). *Let* $\{\mathcal{H}, \Phi, \Gamma\}$ *be a boundary triplet for* $L_1 \subset L_2$. *Then*

$$\operatorname{dom}\widetilde{S} = \left\{ u \in \operatorname{dom} L_2 : \begin{array}{c} Q\mathcal{P}L_2u + 2Q_R^{1/2}\widetilde{X}\Gamma u \in \operatorname{dom} L_1^* \\ \Phi\left(Q\mathcal{P}L_2u + 2\mathcal{P}^*Q_R^{1/2}\widetilde{X}\Gamma u\right) \in \widetilde{\boldsymbol{W}}\Gamma u \end{array} \right\},$$
$$\widetilde{S}u = L_1^*\left(Q\mathcal{P}L_2u + 2Q_R^{1/2}\widetilde{X}\Gamma u\right)$$

establishes a one-to-one correspondence between all m-accretive extensions $\widetilde{S}$ *of* $S = L_2^*QL_1$ *and all pairs* $\langle\widetilde{\boldsymbol{W}}, \widetilde{X}\rangle$, *where* $\widetilde{\boldsymbol{W}}$ *is an m-accretive relation in* $\mathcal{H}$ *and* $\widetilde{X} : \operatorname{dom}\widetilde{\boldsymbol{W}} \to Q_R^{1/2}\,\overline{\operatorname{ran}}\,L_1$ *is a linear operator satisfying the condition*

$$\|\widetilde{X}e\|^2 \leqslant \operatorname{Re}(\widetilde{\boldsymbol{W}}e, e)_{\mathcal{H}}, \;\; e \in \operatorname{dom}\widetilde{\boldsymbol{W}}.$$

The extension $\widetilde{S}$ *is m-sectorial if and only if the relation* $\widetilde{\boldsymbol{W}}$ *is m-sectorial and for some* $\delta \in [0, 1)$

$$\|\widetilde{X}e\|^2 \leqslant \delta^2\operatorname{Re}(\widetilde{\boldsymbol{W}}e, e)_{\mathcal{H}}, \;\; e \in \operatorname{dom}\widetilde{\boldsymbol{W}}.$$

The adjoint operator $\widetilde{S}^*$ *is given by*

$$\operatorname{dom}\widetilde{S}^* = \{v \in \operatorname{dom} L_1^*Q^*L_2 : \; (\Phi\mathcal{P}^*Q^* - 2\widetilde{X}^*Q_R^{1/2}\mathcal{P})L_2v \in \widetilde{\boldsymbol{W}}^*\Gamma v\},$$
$$\widetilde{S}^*v = L_1^*Q^*L_2v.$$

An accretive extension $\widetilde{S}$ *satisfies the condition*

$$L_2^*QL_1 \subset \widetilde{S} \subset L_1^*QL_2$$

if and only if

$$\widetilde{X}\Gamma u = \frac{1}{2}P_0Q_R^{-1/2}Q(I - \mathcal{P})L_2u, \;\; u \in \operatorname{dom} L_2,$$

where P_0 *is the orthogonal projection onto* $Q_R^{1/2}\,\overline{\operatorname{ran}}\,L_1$.

Assume that, in addition to (**A**), (**B**) and (**D**) also the condition

(**E**) the operator L_1 has a bounded inverse

is satisfied. Then the operator S given in (3.24) is coercive. If L is defined by (3.15), then $\{\mathcal{H}, \Phi QL_1(I - Z_0\Gamma), \Gamma\}$ is the boundary triplet for L. This implies the following description of all m-accretive (m-sectorial) extensions $\widetilde{S}$ of S:

$$\operatorname{dom}\widetilde{S} = \left\{ u \in \operatorname{dom} L_2 : \begin{array}{c} u - (Z_0 - 2\widetilde{Y})\Gamma u \in \operatorname{dom} L_1^*QL_1 \\ \Phi QL_1(u - (Z_0 - 2\widetilde{Y})\Gamma u) \in \widetilde{\boldsymbol{W}}\Gamma u \end{array} \right\},$$
$$\widetilde{S}u = L_1^*QL_2(u - (Z_0 - 2\widetilde{Y})\Gamma u),$$

where $\widetilde{\boldsymbol{W}}$ is m-accretive (m-sectorial), $\widetilde{Y} : \operatorname{dom} \widetilde{\boldsymbol{W}} \to \operatorname{dom} L_1$ and

$$\mu[\widetilde{Y}e] \le \delta^2 \mathrm{Re}\,(\widetilde{\boldsymbol{W}}e, e)_{\mathcal{H}}\,, \quad e \in \operatorname{dom} \widetilde{\boldsymbol{W}},$$

with $\delta \in [0,1)$ (m-accretive) or $\delta = 1$ (m-sectorial).

The quadratic functional μ defined by (3.8) now takes the form

$$\mu[\varphi] = \mathrm{Re}\,(Q^{-1}\Pi Q L_1\varphi, \Pi Q L_1 \varphi)_H,$$

where Π is the projection in H onto $Q_R(\operatorname{ran} L_1)$ with respect to the decomposition $\ker L_1^* \dotplus Q_R(\operatorname{ran} L_1) = H$.

Let $Z_0' = (\Gamma\!\restriction \ker(L_1^* Q L_2))^{-1}$ and $Y_0 := \frac{1}{2}(Z_0 - Z_0')$. Then all m-accretive extensions $\widetilde{S}$ such that

$$L_2^* Q L_1 \subset \widetilde{S} \subset L_1^* Q L_2$$

can be parametrized by means of all pairs $\langle \widetilde{\boldsymbol{W}}, Y_0 \rangle$. For $\mu[Y_0 e]$ we have the relation

$$\mu[Y_0 e] = \tfrac{1}{4}\mathrm{Re}\,(Q^{-1}\Pi Q L_2 Z_0 e, \Pi Q L_2 Z_0 e)_H\,.$$

The case of infinite codimension Assume, in addition to (**A**), (**B**), (**C**) and (**E**), the condition

$$\dim(\operatorname{dom} L_2/\operatorname{dom} L_1) = \infty.$$

Then $S = L_2^* Q L_1$ is coercive and, see [Arlinskiĭ, 2000a],

- $\mathcal{D}[S_{\mathrm{N}}] \supseteq \operatorname{dom} L_2$;
- S^* is the closure of $L_1^* Q^* L_2$;
- $\operatorname{dom} L_2 \cap \ker S^* = \ker(L_1^* Q^* L_2)$ is dense in $\ker S^*$.

It is convenient to choose a boundary space as a rigged Hilbert space [Berezansky, 1968, Chapter 1, §1], [Lions and Magenes, 1968, Chapter 1, Section 2.4]:

$$\mathcal{H}^+ \subseteq \mathcal{H} \subseteq \mathcal{H}^-.$$

Let $\{\mathcal{H}^+, \Gamma\}$ be a boundary pair for $L_1 \subset L_2$. Then one can choose $\Phi : \operatorname{dom} L_1^* \to \mathcal{H}^-$ such that Green's identity (3.25) holds. Therefore,

$$\{\mathcal{H}^+ \subseteq \mathcal{H} \subseteq \mathcal{H}^-, \Phi, \Gamma\}$$

is a boundary triplet for $L_1 \subset L_2$.

Let $\mathcal{H}_-$ be the completion of $\mathcal{H}^+$ in the norm

$$\|\Gamma u\|_{\mathcal{H}_-} = \|u\|, \; u \in \operatorname{dom} L_2 \cap \ker S^*$$

and let $\overline{\Gamma}$ be an extension of Γ to $\mathcal{D}[S_{\mathrm{N}}] = \mathcal{D}[S] \dotplus \ker S^*$ with respect to the norm $\|\cdot\|_{\mathcal{H}_-}$. With $Z_0 = (\overline{\Gamma}|\ker S^*)^{-1}$ the operator L defined by (3.15) has the form

$$Lu = L_1^* Q L_1 (I - Z_0 \overline{\Gamma}) u, \quad u \in \operatorname{dom} L = \operatorname{dom} S_{\mathrm{F}} + \operatorname{dom} S_{\mathrm{N}}$$

and $\{\mathcal{H}_-, \overline{\Gamma}\}$ is a boundary pair for $S = L_2^* Q L_1$. Under the condition

(**F**) the Hilbert space $\mathcal{H}$ is dense and continuously embedded in $\mathcal{H}_-$

it can be proved that $\{\mathcal{H}_+ \subseteq \mathcal{H} \subseteq \mathcal{H}_-, \Phi Q L_1 (I - Z_0 \overline{\Gamma}), \overline{\Gamma}\}$ is a boundary triplet for L; here $\mathcal{H}_+ \subseteq \mathcal{H} \subseteq \mathcal{H}_-$ is the corresponding rigged Hilbert space. The operator $\Phi Q L_1 (I - Z_0 \overline{\Gamma})$ is a surjection from $\operatorname{dom} L$ onto $\mathcal{H}_+$.

Applications to differential operators We give some examples of the above results for sectorial ordinary or partial differential operators.

Example 3.28 [Arlinskiĭ, 1998, 2000a] Here $\mathfrak{H} = \mathcal{L}^2(\mathbb{R}_+)$; $\overset{\circ}{H}{}^1(\mathbb{R}_+)$, $H^1(\mathbb{R}_+)$, $\overset{\circ}{H}{}^2(\mathbb{R}_+)$, and $H^2(\mathbb{R}_+)$ are Sobolev spaces. Let

$$Su = -u'', \ u \in \operatorname{dom} S = \overset{\circ}{H}{}^2(\mathbb{R}_+).$$

Then $S = L_2^* L_1$, where $L_1 u = u'$, $L_2 u = u'$ and $\operatorname{dom} L_1 = \overset{\circ}{H}{}^1(\mathbb{R}_+)$, $\operatorname{dom} L_2 = H^1(\mathbb{R}_+)$, cf. (3.24). Clearly, $\dim(\operatorname{dom} L_2 / \operatorname{dom} L_1) = 1$, so that condition (**D**) is satisfied. Let

$$\mathcal{H} = \mathbb{C}, \ \Gamma u = \Phi u = u(0), \quad u \in H^1(\mathbb{R}_+).$$

Then $\{\mathcal{H}, \Gamma\}$ is the boundary pair and $\{\mathcal{H}, \Phi, \Gamma\}$ is the boundary triplet for $L_1 \subset L_2$ and Theorem 3.27 can be applied. The formula

$$\begin{aligned} &\operatorname{dom} \widetilde{S} = \left\{ u \in H^1(\mathbb{R}_+) \colon \begin{array}{c} u' + 2\sqrt{\operatorname{Re}\omega} u(0)\varphi \in H^1(\mathbb{R}_+) \\ (u' + 2\sqrt{\operatorname{Re}\omega} u(0)\varphi)\big|_{x=0} = \omega u(0) \end{array} \right\}, \\ &\widetilde{S} u = -(u' + 2\sqrt{\operatorname{Re}\omega}\, u(0)\varphi)' \end{aligned}$$

establishes a bijective correspondence between all m-accretive extensions $\widetilde{S} \neq S_{\mathrm{F}}$ of S and all pairs $\langle \omega, \varphi \rangle$, where $\omega \in \mathbb{C}$ with $\operatorname{Re}\omega \geq 0$, $\varphi \in \mathcal{L}^2(\mathbb{R}_+)$ with $\|\varphi\| \leqslant 1$. The extension $\widetilde{S}$ is m-sectorial if and only if $\operatorname{Re}\omega > 0$ and $\|\varphi\| < 1$. The associated closed form is given by

$$\begin{aligned} &\mathcal{D}[\widetilde{S}] = H^1(\mathbb{R}_+), \\ &\widetilde{S}[u, v] = \textstyle\int_0^\infty (u'(x) + 2\sqrt{\operatorname{Re}\omega}\, u(0)\varphi(x))\overline{v(x)} dx + \omega\, u(0)\overline{v(0)} \end{aligned}$$

and the linear subspace $\operatorname{dom}\widetilde{S}^*$ consists of all $u \in H^2(\mathbb{R}_+)$ such that

$$u'(0) = \overline{\omega}u(0) + 2\sqrt{\operatorname{Re}\omega}\int_0^\infty u'(x)\overline{\varphi(x)}dx.$$

The pair $\langle \infty, 0\rangle$ determines to the Friedrichs extension, while the pair $\langle 0, 0\rangle$ determines the Kreĭn-von Neumann extension. The pairs $\langle \omega, 0\rangle$ determine proper m-accretive extensions $\widetilde{S}$ $(S \subset \widetilde{S} \subset S^*)$.

Example 3.29 [Arlinskiĭ, 2000a, 1999c] Set $\mathfrak{H} = \mathcal{L}^2(\mathbb{R}_+)$ and

$$Su = -u'' + qu, \ \operatorname{dom} S = \overset{\circ}{H}{}^2(\mathbb{R}_+),$$

where $q \in \mathcal{L}^\infty(\mathbb{R}_+)$, $\operatorname{Re} q(x) = q_{\mathrm{R}}(x) \geqslant m > 0$. In this case (see (3.24))

$$\begin{aligned}
&\operatorname{dom} L_1 = \overset{\circ}{H}{}^1(\mathbb{R}_+), \ \operatorname{dom} L_2 = H^1(\mathbb{R}_+),\\
&L_k u = \begin{pmatrix} \sqrt{q_{\mathrm{R}}}u \\ u' \end{pmatrix}, \ u \in \operatorname{dom} L_k, \ k = 1, 2,\\
&Q\begin{pmatrix} f_1 \\ f_2 \end{pmatrix} = \begin{pmatrix} (1 + \mathrm{i}\dfrac{q_{\mathrm{I}}}{q_{\mathrm{R}}})f_1 \\ f_2 \end{pmatrix}, \ f_1, f_2 \in \mathcal{L}^2(\mathbb{R}_+).
\end{aligned}$$

Clearly condition (**D**) is satisfied. The Friedrichs extension is given by

$$S_{\mathrm{F}}u = -u'' + qu, \ u \in \operatorname{dom} S_{\mathrm{F}} = H^2(\mathbb{R}_+) \cap \overset{\circ}{H}{}^1(\mathbb{R}_+)$$

and $\mathcal{D}[S] = \overset{\circ}{H}{}^1(\mathbb{R}_+)$. The "real part" of S_{F} is given by

$$S_{\mathrm{FR}}u = -u'' + q_{\mathrm{R}}u, \ u \in \operatorname{dom} S_{\mathrm{FR}} = H^2(\mathbb{R}_+) \cap \overset{\circ}{H}{}^1(\mathbb{R}_+).$$

Let z_0 and ξ_0 be the solutions of the boundary value problems

$$\begin{aligned}
&-z_0'' + \overline{q}z_0 = 0, \ z_0 \in H^2(\mathbb{R}_+), \ z_0(0) = 1,\\
&-\xi_0'' + q\xi_0 = 0, \ \xi_0 \in H^2(\mathbb{R}_+), \ \xi_0(0) = 1,
\end{aligned}$$

and let $y_0 = (z_0 - \xi_0)/2$. The formula

$$\begin{aligned}
&\operatorname{dom}\widetilde{S} = \left\{ u \in H^1(\mathbb{R}_+) : \begin{array}{c} u' + 2y'u(0) \in H^1(\mathbb{R}_+) \\ [u' + 2y'u(0)]\Big|_{x=0} = (\omega + z_0'(0))u(0) \end{array} \right\},\\
&\widetilde{S}u = -(u' + 2y'\,u(0)\,)' + q(u + 2yu(0)) - 2\mathrm{i}q_I z_0(x)u(0)
\end{aligned}$$

establishes a bijective correspondence between all m-accretive extensions $\widetilde{S} \neq S_{\mathrm{F}}$ and all pairs $\langle \omega, y\rangle$, where

$$\operatorname{Re}\omega \geqslant 0, \ y \in \overset{\circ}{H}{}^1(\mathbb{R}_+) \text{ and } \mu[y] \leqslant \operatorname{Re}\omega$$

The extension $\widetilde{S}$ is m-sectorial if and only if $\mu[y] < \operatorname{Re}\omega$. For an explicit calculation of the functional μ, see [Arlinskiĭ, 1999c].

The linear manifold $\operatorname{dom}\widetilde{S}^*$ consists of all $v \in H^2(\mathbb{R}_+)$ such that

$$\overline{\omega} + z_0'(0)v(0) - v'(0) \\ = 2\int_0^\infty \left[v'(x)(\overline{y_0'(x)} - \overline{y'(x)}) + \overline{q(x)}v(x)(\overline{y_0(x)} - \overline{y(x)})\right]dx$$

and $\widetilde{S}^*v = -v'' + \bar{q}v$. The pair $\omega = 0$ and $y = 0$ corresponds to the Kreĭn-von Neumann extension and the pair $\omega = \infty$ and $y = 0$ corresponds to the Friedrichs extension.

With $y = y_0$ we obtain a description of all m-accretive and m-sectorial operators $\widetilde{S}$ such that $S \subset \widetilde{S} \subset S_{\max}$, where $S_{\max}$ is the maximal operator in $\mathcal{L}^2(\mathbb{R}_+)$ generated by the expression $l(u) = -u'' + qu$.

Example 3.30 [Arlinskiĭ, 2000a,b, 1999c] Let Ω be a bounded domain in $\mathbb{R}^n$ with $\mathcal{C}^\infty$ boundary $\partial\Omega$ and let $S = l_{\min}$ be the minimal operator generated by the second order partial differential expression

$$l(u) = -\sum_{j=1}^n \frac{\partial}{\partial x_j}\Big(\sum_{k=1}^n a_{jk}\frac{\partial u}{\partial x_k} + \alpha_j u\Big) + \sum_{k=1}^n a_k\frac{\partial u}{\partial x_k} + au$$

in $\mathcal{L}^2(\Omega)$. Suppose that the coefficients are in $\mathcal{C}^\infty(\bar{\Omega})$ and that the condition of uniform ellipticity is fulfilled, that is for some $c > 0$

$$\operatorname{Re}\Big(\sum_{j,k=0}^n q_{jk}(x)\xi_k\overline{\xi_j}\Big) \geqslant c\sum_{k=0}^n |\xi_k|^2,\ x \in \Omega,\ \xi = (\xi_0, ..., \xi_{\mathrm{N}})^{\mathsf{T}} \in \mathbb{C}^{n+1},$$

where $q_{j0} = \alpha_j, q_{0k} = a_k,\ q_{00} = a, q_{jk} = a_{jk},\ j,k = 1, ..., n$. Then S, with $\operatorname{dom} S = \mathring{H}^2(\Omega)$, has the divergence form (3.24) with the choice [Arlinskiĭ, 2000a]

$$L_k u = (u, \nabla u)^{\mathsf{T}},\ \operatorname{dom} L_1 = \mathring{H}^1(\Omega),\ \operatorname{dom} L_2 = H^1(\Omega),\ H = \mathcal{L}^2(\Omega)^{n+1}$$

and $Q = (q_{jk})_{j,k=0}^n$. Note that the operator L_1 has bounded inverse and $\dim(\operatorname{dom} L_2/\operatorname{dom} L_2) = \infty$.

The Friedrichs extension S_{F} of S is the operator $L_1^* Q L_1$ with

$$\operatorname{dom} S_{\mathrm{F}} = \mathring{H}^1(\Omega) \cap H^2(\Omega)$$

(Dirichlet boundary conditions) so that $\mathcal{D}[S] = \mathring{H}^1(\Omega)$. The boundary triplet for the pair $L_1 \subset L_2$ is

$$\{H^{1/2}(\partial\Omega) \subset L^2(\partial\Omega) \subset H^{-1/2}(\partial\Omega), \Phi, \Gamma\},$$

where the trace operator $\Gamma u = u\restriction\partial\Omega$ maps $H^1(\Omega)$ into $H^{1/2}(\partial\Omega)$, see [Lions and Magenes, 1968, Theorem 1.9.4], and Φ with $\operatorname{ran}\Phi = H^{-1/2}(\partial\Omega)$ extends

$$\Phi(f_0, \vec{f})^{\mathsf{T}} = \sum_{k=1}^{n} f_k \cos(\vec{n}, x_k)\restriction\partial\Omega,\ f_0 \in \mathcal{L}^2(\Omega),\ f_k \in H^1(\Omega),\ k=1,\dots,n,$$

to $\operatorname{dom} L_1^*$; here $\vec{f} = (f_1, \dots, f_n)$ and $\vec{n}$ is the continuous field of normal unit vectors of $\partial\Omega$ directed into the interior of Ω. Note that

$$\begin{aligned}
\operatorname{dom} L_1^* &= \left\{ (f_0, \vec{f})^{\mathsf{T}} : \begin{array}{l} f \in \mathcal{L}_2(\Omega),\ \vec{f} \in [\mathcal{L}_2(\Omega)]^n \\ \operatorname{div}\vec{f} \in \mathcal{L}_2(\Omega) \end{array} \right\}, \\
\operatorname{dom} L_2^* &= \left\{ (f_0, \vec{f})^{\mathsf{T}} \in \operatorname{dom} L_1^* : \vec{f}\cdot\vec{n}\restriction\partial\Omega = 0 \right\}, \\
L_1^*(f_0, \vec{f})^{\mathsf{T}} &= f_0 - \operatorname{div}\vec{f}.
\end{aligned}$$

Since the operator Γ has an extension $\overline{\Gamma}$ to $\mathcal{D}[S_{\mathrm{N}}] = \mathring{H}^1(\Omega) \dot{+} \ker S^*$ with values in $\mathcal{H}_- = H^{-1/2}(\partial\Omega)$, condition (**F**) is fulfilled.

Let $Z_0 e$ be a solution of the boundary value problem

$$S^* u = l^+_{\mathbf{max}} u = 0,\ \overline{\Gamma} u = u\restriction\partial\Omega = e,\ e \in H^{-1/2}(\partial\Omega),$$

where $l^+(u)$ is the adjoint expression:

$$l^+(u) = -\sum_{j=1}^{n} \frac{\partial}{\partial x_j}\Big(\sum_{k=1}^{n} \overline{a_{kj}} \frac{\partial u}{\partial x_k} + \overline{a_j} u\Big) + \sum_{k=1}^{n} \overline{\alpha_k} \frac{\partial u}{\partial x_k} + \overline{a} u.$$

The boundary operator

$$G_* : \operatorname{dom} L = \mathring{H}^1(\Omega) \cap H^2(\Omega) \dot{+} \ker S^* \to \mathcal{H}_+ = H^{1/2}(\partial\Omega)$$

is of the form

$$G_* u = \Phi Q L_1 (u - Z_0 \overline{\Gamma} u) = \frac{\partial(u - Z_0(u\restriction\partial\Omega))}{\partial\nu}\restriction\partial\Omega,$$

where

$$\frac{\partial u}{\partial\nu}\restriction\partial\Omega = \sum_{j=1}^{n}\Big(\sum_{k=1}^{n} a_{jk} \frac{\partial u}{\partial x_k} + \alpha_j u\Big)\cos(\vec{n}(x), x_j)\restriction\partial\Omega.$$

Then $\{H^{1/2}(\partial\Omega) \subset L^2(\partial\Omega) \subset H^{-1/2}(\partial\Omega), G_*, \overline{\Gamma}\}$ is the boundary triplet for L. The description of all m-accretive extensions $\widetilde{S}$ of S is given by

$$\begin{aligned}
\operatorname{dom}\widetilde{S} &= \left\{ u : \begin{array}{l} u - (Z_0 - 2Y)(u\restriction\partial\Omega)) \in \mathring{H}^1(\Omega) \cap H^2(\Omega) \\ \dfrac{\partial(u - (Z_0 - 2\widetilde{Y})(u\restriction\partial\Omega))}{\partial\nu}\restriction\partial\Omega \in \widetilde{\boldsymbol{W}}(u\restriction\partial\Omega) \end{array} \right\}, \\
\widetilde{S} u &= l(u - (Z_0 - 2\widetilde{Y})(u\restriction\partial\Omega)),
\end{aligned}$$

where the relation $\widetilde{\boldsymbol{W}} = \{\langle x, y\rangle : x \in H^{-1/2}(\partial\Omega),\ y \in H^{1/2}(\partial\Omega)\}$ is m-accretive in $H^{-1/2}(\partial\Omega) \times H^{1/2}(\partial\Omega)$ and $\widetilde{Y} : \operatorname{dom}\widetilde{\boldsymbol{W}} \to \overset{\circ}{H}{}^1(\Omega)$ satisfies the inequality

$$\mu[\widetilde{Y}e] \leq \operatorname{Re}(\widetilde{\boldsymbol{W}}e, e)_{\mathcal{L}_2(\partial\Omega)},\ e \in \operatorname{dom}\widetilde{\boldsymbol{W}}.$$

The operator $\widetilde{S}$ has a bounded inverse $\widetilde{S}^{-1}$ if and only if $\widetilde{\boldsymbol{W}}$ has a bounded inverse $\widetilde{\boldsymbol{W}}^{-1} : H^{1/2}(\partial\Omega) \to H^{-1/2}(\partial\Omega)$. Moreover, $\operatorname{dom}\widetilde{S} \subset H^2(\Omega)$ if and only if $\operatorname{dom}\widetilde{\boldsymbol{W}} \subset H^{3/2}(\partial\Omega)$, $\operatorname{ran}\widetilde{Y} \subset \overset{\circ}{H}{}^1(\Omega) \cap H^2(\Omega)$.

The minimal operator T generated by the adjoint expression l^+ takes the form $T = L_2^* Q^* L_1$, and $T_{\mathrm{F}} = L_1^* Q^* L_2 = S_{\mathrm{F}}^*$. Let $Z_0' = (\overline{\Gamma}{\restriction}\ker T^*)^{-1}$ so that $Z_0'e$ is the solution of the boundary value problem

$$T^*u = l_{\max}u = u,\ \overline{\Gamma}u = u{\restriction}\partial\Omega = e,\ \ e \in H^{-1/2}(\partial\Omega).$$

Then all m-accretive extensions $\widetilde{S}$ between $l_{\min}$ and $l_{\max}$ in $\mathcal{L}_2(\Omega)$, i.e., $S \subset \widetilde{S} \subset T^*$, can be parametrized by

$$\operatorname{dom}\widetilde{S} = \left\{u \in \operatorname{dom} l_{\max} \ :\ \frac{\partial(u - Z_0'\overline{\Gamma}u)}{\partial\nu}{\restriction}\partial\Omega \in \widetilde{\boldsymbol{W}}\overline{\Gamma}u\right\},$$

where $\widetilde{\boldsymbol{W}}$ is an m-accretive relation in $H^{-1/2}(\partial\Omega) \times H^{1/2}(\partial\Omega)$ such that $\operatorname{dom}\widetilde{\boldsymbol{W}} \subset \overline{\Gamma}(\mathcal{D}[S_{\mathrm{N}}] \cap \mathcal{D}[T_{\mathrm{N}}])$ and

$$\tfrac{1}{2}\mu[Z_0e - Z_0'\overline{\Gamma}Z_0e] \leq \operatorname{Re}(\widetilde{\boldsymbol{W}}e, e)_{\mathcal{L}_2(\partial\Omega)},\ \ e \in \operatorname{dom}\widetilde{\boldsymbol{W}}.$$

In addition, $\operatorname{dom}\widetilde{S} \subset H^2(\Omega)$ if and only if $\operatorname{dom}\widetilde{\boldsymbol{W}} \subset H^{3/2}(\partial\Omega)$.

References

Ando, T., and Nishio, K. 1970. Positive selfadjoint extensions of positive symmetric operators. *Tohóku Math. J.*, **22**, 65–75.

Arlinskiĭ, Yu. M. 1987. A class of contractions in Hilbert space, *Ukrain. Mat. Zh.*, **39**, 691-696 (Russian). English translation in *Ukrainian Math. J.*, **39**, 560-564.

Arlinskiĭ, Yu. M. 1988. Positive spaces of boundary values and sectorial extensions of nonnegative operator. *Ukrain. Mat. Zh.*, **40**, 8-14 (Russian). English translation in *Ukrainian Math. J.*, **40**, 5-10.

Arlinskiĭ, Yu. M. 1991. Characteristic functions of operators of the class $C(\alpha)$. *Izv. Vyssh. Uchebn. Zaved. Mat.*, No. 2, 13-21 (Russian).

Arlinskiĭ, Yu. 1992. On class of extensions of a $C(\alpha)$-suboperators. *Dokl. Akad. Nauk Ukr.*, No. 8 , 12–16 (Russian).

Arlinskiĭ, Yu. 1995. On proper accretive extensions of positive linear relations. *Ukrain. Mat. Zh.*, **47**, 723–730.

Arlinskiĭ, Yu. 1996. Maximal sectorial extensions and associated with them closed forms, *Ukrain. Mat. Zh.*, **48**, 723-738 (Russian). English translation in *Ukrainian Math. J.*, **48**, 809-827.

Arlinskiĭ, Yu. 1997. Extremal extensions of sectorial linear relations. *Mat. Stud.*, **7**, 81–96.

Arlinskiĭ, Yu. 1998, On m-accretive extensions and restrictions. *Methods Funct. Anal. Topology*, **4**, 1–26.

Arlinskiĭ, Yu. 1999a. On functions connected with sectorial operators and their extensions. *Integral Equations Operator Theory*, **33**, 125–152.

Arlinskiĭ, Yu. 1999b. On a class of nondensely defined contractions on a Hilbert space and their extensions. *J. Math. Sci. (N.Y.)*, **97**, 4390–4419. English translation from *Itogi Nauki i Tekhniki, Seriya Sovremennaya Matematika i Ee Prilozheniya. Tematicheskie Obzory. Vol. 50, Funktsional'nyi Analiz-5, 1997.*

Arlinskiĭ, Yu. 1999c. Maximal accretive extensions of sectorial operators. Doctor Sciences Dissertation, Institute of Mathematics Ukraine National Academy of Sciences, 288 p. (Russian).

Arlinskiĭ, Yu. 2000a. Abstract boundary conditions for maximal sectorial extensions of sectorial operators. *Math. Nachr.*, **209**, 5–36.

Arlinskiĭ, Yu. 2000b. M-accretive extensions of sectorial operators and Kreĭn spaces. *Oper. Theory Adv. Appl.*, **118**, 67–82.

Arlinskiĭ, Yu. 2006. Extremal extensions of a $C(\alpha)$-suboperator and their representations. *Oper. Theory Adv. Appl.*, **162**, 47–69.

Arlinskiĭ, Yu, Hassi, S., Sebestyen, Z., and de Snoo, H.S.V. 2001. On the class of extremal extensions of a nonnegative operators. *Oper. Theory Adv. Appl.*, **127**, 41–81.

Arlinskiĭ, Yu., Kovalev, Yu., and Tsekanovskiĭ, E. 2012. Accretive and sectorial extensions of nonnegative symmetric operators, *Complex Anal. Oper. Theory*, to appear.

Arlinskiĭ, Yu., and Tsekanovskiĭ, E. 1984a. On sectorial extensions of positive hermitian operators and their resolvents. *Dokl. Nats. Akad. Nauk Armen.*, **5**, 199–202 (Russian)

Arlinskiĭ, Yu., and Tsekanovskiĭ, E. 1984b. On resolvents of m-accretive extensions of symmetric differential operator. *Math. Phys. Nonlin. Mech.*, **1**, 11–16 (Russian).

Arlinskiĭ, Yu., and Tsekanovskiĭ, E. 1988. Quasi-self-adjoint contractive extensions of Hermitian contraction. *Theor. Functions, Funk. Anal. i Prilozhen.*, **50**, 9–16 (Russian). English translation in *J. Math. Sci. (N. Y.)*, **49** (1990), 1241–1247.

Arlinskiĭ, Yu., and Tsekanovskiĭ, E. 2002. On the theory of nonnegative extensions of a nonnegative symmetric operator. *Dopov. Nats. Akad. Nauk Ukr. Mat. Prirodozn. Tekh. Nauki*, No.11, 30–37.

Arlinskiĭ, Yu., and Tsekanovskiĭ, E. 2003. On von Neumann's problem in extension theory of nonnegative operators. *Proc. Amer. Math. Soc.*, **131**, 3143–3154.

Arlinskiĭ, Yu., and Tsekanovskiĭ, E. 2005. The von Neumann problem for nonnegative symmetric operators. *Integral Equations Operator Theory*, **51**, 319–356.

Arlinskiĭ, Yu., and Tsekanovskiĭ, E. 2009. Krein's research on semi-bounded operators, its contemporary developmentsand applications. *Oper. Theory Adv. Appl.*, **190**, 65–112.

Arsene, G., and Gheondea, A. 1982. Completing matrix contractions. *J. Operator Theory*, **7**, 179–189.

Behrndt, J., Hassi, S., de Snoo, H.S.V., and Wietsma, H.L. 2010 Monotone convergence theorems for semibounded operators and forms with applications. *Proc. Royal Soc. Edinburgh*, **140A**, 927–951.

Berezansky, Yu. 1968. *Expansions in eigenfunction of selfadjoint operators*, Transl. of Math. Monographs, **17**, Amer. Math. Soc., Providence.

Birman, M. S. 1956. On the selfadjoint extensions of positive definite operators. *Mat. Sb.*, **38**, 431–450 (Russian).

Bruk, V. M. 1976. On a class of boundary value problems with spectral parameters in the boundary conditions. *Mat. Sb.*, **100**, 186–192 (Russian).

Calkin, J. W. 1939. Symmetric boundary conditions. *Trans. Amer. Math. Soc.*, **45**, 369–442.

Crandall, M. 1969. Norm preserving extensions of linear transformations on Hilbert spaces. *Proc. Amer. Math. Soc.*, **21**, 335–340.

Coddington, E. A., and de Snoo, H. S. V. 1978. Positive selfadjoint extensions of positive symmetric subspaces. *Math. Z.*, **159**, 203–214.

Davis, C., Kahan, W. M., and Weinberger, H. F. 1982. Norm preserving dilations and their applications to optimal error bounds. *SIAM J. Numer. Anal.*, **19**, 445–469.

Derkach, V., and Malamud, M. 1991. Generalized resolvents and the boundary value problems for Hermitian operators with gaps. *J. Funct. Anal.*, **95**, 1–95.

Derkach, V., and Malamud, M. 1995. The extension theory of Hermitian operators and the moment problem. *J. Math. Sci. (N. Y.)*, **73**, 141–242.

Derkach, V., Malamud, M., and Tsekanovskiĭ, E. 1988. Sectorial extensions of a positive operator and characteristic function. *Soviet Math. Dokl.*, **37**, 106–110.

Derkach, V., Malamud, M., and Tsekanovskiĭ, E. 1989. Sectorial extensions of positive operator. *Ukrain. Mat. Zh.*, 41, 151–158 (Russian). English translation in *Ukrainian Math. J.*, **41**, 136-142.

Evans, W. D., and Knowles, I. 1985. On the extension problem for accretive differential operators. *J. Funct. Anal.*, **63**, 276–298.

Evans, W. D., and Knowles, I. 1986. On the extension problem for singular accretive differential operators. *J. Differential Equations*, **63**, 264–288.

Gorbachuk, M. L., and Gorbachuk, V. I. 1984. *Boundary value problems for differential-operator equations*, Naukova Dumka, Kiev (Russian).

Gorbachuk, M. L., Gorbachuk, V. I., and Kochubeĭ, A. N. 1989. Extension theory of symmetric operators and boundary value problems. *Ukrain. Mat. Zh.*, **41**, 1298–1313 (Russian).

Grubb, G. 1968. A characterization of the non-local boundary value problems associated with an elliptic operator. *Ann. Scuola Norm. Sup., Pisa (3)*, **22**, 425–513.

Grubb, G. 1970. Les problémes aux limites généraux d'un opérateur elliptique provenant de la théorie variationnelle. *Bull Sci. Math.*, **91**, 113–157.

Grubb, G. 1971. On coerciveness and semiboundedness of general boundary problems. *Israel J. Math.*, **10**, 32–95.

Grubb, G. 1973. Weakly semibounded boundary problems and sesquilinear forms. *Ann. Inst. Fourier (Grenoble)*, **23**, 145–194.

Grubb, G. 1974. Properties of normal boundary problems for elliptic even-order systems. *Ann. Scuola Norm. Sup., Pisa (4)*, **1**, 1–61.

Grubb, G. 1983. Spectral asymptotics for the "soft" self-adjoint extension of a symmetric elliptic differential operator. *J. Operator Theory*, **10**, 2–20.

Grubb, G. 2006. Known and unknown results on elliptic boundary problems. *Bull. Amer. Math. Soc. (N.S.)*, **43**, 227–230.

Hassi, S., Malamud, M. M., and de Snoo, H. S. V. 2004. On Kreĭn's extension theory of nonnegative operators. *Math. Nachr.*, **274/275**, 40–73.

Hassi, S., Sandovici, A., de Snoo, H. S. V., and Winkler, H. 2007 A general factorization approach to the extension theory of nonnegative operators and relations. *J. Operator Theory*, **58**, 351 - 386.

Kato, T. 1995. *Perturbation theory for linear operators*, Springer-Verlag, New York.

Kochubeĭ, A. N. (1975). Extensions of symmetric operators and symmetric binary relations. *Mat. Zametki*, **17**, 41-48 (Russian). English translation in *Math. Notes*, **17**, 25–28.

Kochubeĭ, A. N. 1979. On extensions of positive definite symmetric operator. *Dokl. Akad. Nauk Ukr. SSR, Ser. A*, 169–171 (Russian).

Kolmanovich, V., and Malamud M., 1985. Extensions of sectorial operators and dual pairs of contractions, Manuscript No. 4428-85, Deposited at VINITI, 1–57 (Russian).

Kreĭn, M. G. 1947a. The theory of selfadjoint extensions of semibounded Hermitian transformations and its applications, I. *Mat.Sb.*, **20**, 431–495 (Russian).

Kreĭn, M. G. 1947b. The theory of selfadjoint extensions of semibounded Hermitian transformations and its applications, II. *Mat.Sb.*, **21**, 365–404 (Russian).

Kreĭn, M. G., and Ovčarenko, I. E. 1976. On generalized resolvents and resolvent matrices of positive Hermitian operators. *Soviet Math. Dokl.*, **231**, 1063–1066 (Russian).

Kreĭn, M. G., and Ovčarenko, I. E. 1978. Inverse problems for Q-functions and resolvents matrices of positive Hermitian operators. *Soviet Math. Dokl.*, **242**, 521–524 (Russian).

Lions, J.-L., and Magenes, E. 1968. *Problèmes aux limites non homogènes et applications, vol. 1*, Editions Dunod, Paris. Russian translation in Mir, Moscow, 1971.

Lyantse, V. E. 1954. On a boundary problem for parabolic systems of differential equations with a strongly elliptic right-hand side. *Mat.Sb.*, **35**, 367–368 (Russian).

Lyantse, V. E., and Storozh, O. G. 1983. *Methods of the theory of unbounded operators*, Naukova Dumka, Kiev (Russian).

Malamud, M. 1989. On extensions of Hermitian and sectorial operators and dual pairs of contractions. *Dokl. Akad. Nauk SSSR*, **39**, 253–254 (Russian).

Malamud, M. 1992. On some classes of Hermitian operators with gaps. *Ukrain. Mat. Zh.*, **44**, 215–234 (Russian).

Malamud, M. 2001. On some classes of extensions of sectorial operators and dual pair of contractions. *Oper. Theory Adv. Appl.*, **124**, 401–448.

Malamud, M. 2006. Operator holes and extensions of sectorial operators and dual pair of contractions. *Math. Nachr.*, **279**, 625–655.

Mikhailets, V.A. 1980. Spectral analysis of diffential operators. *Sbornik Nauch. Trud., Kiev, Inst. of Math. of Ukrainian Acad. of Sci.*, 106–131 (Russian).

Mil'yo, O. Ya., and Storozh, O. G. 1991. On general form of maximal accretive extension of positive definite operator. *Dokl. Akad. Nauk Ukrain. SSR, Ser. A*, **6**, 19–22 (Russian).

Mil'yo, O. Ya., and Storozh, O. G. 1993. Maximal accretive extensions of positive definite operator with finite defect number, *Lviv University*, 31 pages, Deposited in GNTB of Ukraine 28.10.93, no 2139 Uk93 (Russian).

Mil'yo, O. Ya., and Storozh, O. G. 1999a. Maximal accretiveness and maximal nonnegativeness conditions for a class of finite-dimensional perturbations of a positive definite operator. *Mat. Stud.*, **12**, 90–100.

Mil'yo, O. Ya., and Storozh, O. G. 1999b. Sectoriality and solvability conditions for differential-boundary operators of Sturm-Liouville type with multipoint-integral boundary conditions. *J. Math. Sci. (N. Y.)*, **96**, 2961–2965.

Mil'yo, O. Ya., and Storozh, O. G. 2002. Sturm-Liouville type differential boundary operator on semiaxis with twopoint integral boundary conditions. *Ukrain. Mat. Zh.*, **54**, 1480–1485 Russian). English translation in *Ukrainian Math. J.*, **54**, 1793–1801.

Phillips, R. 1959. Dissipative operators and hyperbolic systems of partial differential equations. *Trans. Amer. Math. Soc.*, **90**, 192–254.

Phillips, R. 1959. Dissipative operators and parabolic partial diferential equations. *Comm. Pure Appl. Math.*, **12**, 249–276.

Phillips, R. 1969. On dissipative operators. *Lectures in Differential Equations*, **3**, 65–113.

Pipa, H. M., and Storozh, O. G. 2006. Accretive perturbations of some proper extensions of a positive definite operator. *Mat. Stud.*, **25**, 181–190.

Rofe-Beketov, F. S. 1985. Numerical range of a linear relation and maximal relations. *Theory of Functions, Funct. Anal. Appl.*, **44**, 103–112 (Russian).

Sandovici, A. 2006. *Contributions to the extension theory in Hilbert spaces*, PhD Dissertation, Groningen University, 193 p.

Shmul'yan, Yu., and Yanovskaya, R. 1981. On blocks of contractive operator matrix. *Izv. Vuzov Math.*, 72–75 (Russian).

Shtraus, A. V. 1973. On extensions of semibounded operators. *Dokl. Akad. Nauk SSSR*, **211**, 543–546 (Russian).

Storozh, O. G. 1990. Extremal extensions of nonnegative operator and accretive boundary problems,. *Ukrain. Mat. Zh.*, **42**, 857–860 (Russian). English translation in *Ukrainian Math. J.*, **42**, 758–760.

Sz.-Nagy, B., and Foias, C. 1970. *Harmonic analysis of operators on Hilbert space*, North-Holland, New York.

Tsekanovskiĭ, E. 1985. Characteristic function and description of accretive and sectorial boundary value problems for ordinary differential operators. *Dokl. Akad. Nauk Ukrain. SSR, Ser. A*, **6**, 21–24.

Vishik, M. I. 1952. On general boundary conditions for elliptic differential operators. *Trudy Moskov. Mat. Obsc.*, **1**, 187–246 (Russian). English translation in *Amer. Math. Soc. Transl.* **24** (1963), 107–172.

Wei, G., and Xu, Z. 2005. An expanded Phillips theory and its application to differential operators. *J. Funct. Anal.*, **222**, 29–60 .

Xu, Z., Wei, G., and Fergola, P. 2002. Maximal accretive realizations of regular Sturm-Liouville differential operators. *J. London Math. Soc.*, **66**, 175–197.

4
Boundary control state/signal systems and boundary triplets

Damir Z. Arov [a], Mikael Kurula and Olof J. Staffans

Abstract This chapter is an introduction to the basic theory of state/signal systems via boundary control theory. The $\mathcal{LC}$-transmission line illustrates the new concepts. It is shown that every boundary triplet can be interpreted as an impedance representation of a conservative boundary control state/signal system.

4.1 Introduction

We discuss the connection between some basic notions of boundary control state/signal systems on one hand, and classical boundary triplets on the other hand. Boundary triplets and their generalizations have been extensively utilized in the theory of self-adjoint extensions of symmetric operators in Hilbert spaces, see e.g. [Gorbachuk and Gorbachuk, 1991; Derkach and Malamud, 1995; Behrndt and Langer, 2007], and the references therein.

The notions related to standard input/state/output boundary control systems are discussed in Section 4.2, where we also introduce the boundary control state/signal system. In Section 4.3 we briefly discuss the concept of conservativity in the state/signal framework and in Section 4.4 we illustrate the abstract concepts we have introduced using the example of a finite-length conservative $\mathcal{LC}$-transmission line with distributed inductance and capacitance.

We conclude this chapter in Section 4.5, where we recall the definition of a boundary triplet for a symmetric operator and compare this object to a boundary control state/signal system. In particular, we show that

[a] Damir Z. Arov thanks Åbo Akademi for its hospitality and the Academy of Finland and the Magnus Ehrnrooth Foundation for their financial support during his visits to Åbo in 2003–2010.

every boundary triplet can be transformed into a conservative boundary control state/signal system in impedance form, but that the converse is not true. We make a few final remarks about common generalizations of boundary triplets, which leads over to Chapter 5, where we treat more general passive state/signal systems, not only conservative systems or systems of boundary-control type. There we show how general conservative state/signal systems are related to *boundary relations.*

4.2 Boundary control systems

In this section we introduce boundary control state/signal systems by first describing their predecessors, namely input/state/output systems of boundary-control type.

In boundary control one often investigates systems that can be abstractly written in the form

$$\Sigma_{i/s/o}: \quad \begin{cases} \dot{x}(t) = Lx(t), \\ u(t) = \Gamma_0 x(t), \qquad t \in \mathbb{R}^+,\ x(0) = x_0 \text{ given}, \\ y(t) = \Gamma_1 x(t), \end{cases} \tag{4.1}$$

where $\mathbb{R}^+ = [0,\infty)$ and $\dot{x} = \frac{\mathrm{d}x}{\mathrm{d}t}$. Here the *initial state* x_0 and the *current state* $x(t)$ belong to the Hilbert state space $\mathcal{X}$, the *input* $u(t)$ belongs to the Hilbert input space $\mathcal{U}$, and the *output* $y(t)$ belongs to the Hilbert output space $\mathcal{Y}$. The *main operator* L is an unbounded operator in $\mathcal{X}$ with domain $\operatorname{dom}(L)$, and the *boundary control operator* Γ_0 is an unbounded operator $\mathcal{X} \to \mathcal{U}$ with the same domain as L. The *observation operator* $\Gamma_1 \colon \mathcal{X} \to \mathcal{Y}$ may be bounded or unbounded, and it is defined at least on $\operatorname{dom}(L)$. All of these operators are linear. We denote the system (4.1) with these properties by $\Sigma_{i/s/o} = (L, \Gamma_0, \Gamma_1; \mathcal{X}, \mathcal{U}, \mathcal{Y})$.

In order for (4.1) to generate a dynamical system with good properties at least the properties listed in the following definition need to be assumed; see e.g. [Salamon, 1987], [Staffans, 2005], or [Malinen and Staffans, 2006] for details.

Definition 4.1 Assume that $\Sigma_{i/s/o} = (L, \Gamma_0, \Gamma_1; \mathcal{X}, \mathcal{U}, \mathcal{Y})$ is as described above. Then $\Sigma_{i/s/o}$ is a *boundary control input/state/output (i/s/o) node* if $\Sigma_{i/s/o}$ satisfies the following conditions:

1. The input operator Γ_0 is surjective and *strictly unbounded* in the sense that $\ker(\Gamma_0)$ is *dense* in $\mathcal{X}$.

2. The restriction $A := L|_{\ker(\Gamma_0)}$ of L to $\ker(\Gamma_0)$ *generates* a C_0-semigroup $t \mapsto \mathfrak{A}^t$, $t \in \mathbb{R}^+$.

A *boundary control state/signal system* is analogous to a boundary control i/s/o system, but we no longer specify which part of the "boundary signal" $w(t) := \begin{bmatrix} u(t) \\ y(t) \end{bmatrix}$ is the input, and which part is the output. Instead we combine the input and output spaces into one signal space $\mathcal{W} := \begin{bmatrix} \mathcal{U} \\ \mathcal{Y} \end{bmatrix} = \mathcal{U} \times \mathcal{Y}$, and denote $\Gamma := \begin{bmatrix} \Gamma_0 \\ \Gamma_1 \end{bmatrix}$. Then $\Gamma\colon \operatorname{dom}(L) \to \mathcal{W}$, and (4.1) can be rewritten in the form

$$\Sigma: \quad \begin{cases} \dot{x}(t) = Lx(t), \\ w(t) = \Gamma x(t), \end{cases} \quad t \in \mathbb{R}^+, \; x(0) = x_0 \text{ given.} \tag{4.2}$$

As before, the *initial state* x_0 and the *current state* $x(t)$ belong to the Hilbert state space $\mathcal{X}$. The (*interaction*) signal $w(t)$ belongs to the signal space $\mathcal{W}$, which we take to be an arbitrary Kreĭn space (the reason for this will be explained below). We thus no longer assume that $\mathcal{W}$ is of the form $\mathcal{W} = \begin{bmatrix} \mathcal{U} \\ \mathcal{Y} \end{bmatrix}$, where $\mathcal{U}$ and $\mathcal{Y}$ are the input and output spaces of a boundary control i/s/o node. The *main operator* L is still an unbounded operator $\mathcal{X} \to \mathcal{X}$ with domain $\operatorname{dom}(L)$, and the *boundary operator* Γ is an unbounded operator $\mathcal{X} \to \mathcal{W}$ with the same domain as L. We denote this system by $\Sigma = (L, \Gamma; \mathcal{X}, \mathcal{W})$.

Note that (4.2) can be written in the *graph form*:

$$\Sigma: \quad \begin{bmatrix} \dot{x}(t) \\ x(t) \\ w(t) \end{bmatrix} \in V, \quad t \in \mathbb{R}^+, \; x(0) = x_0,$$

where the *generating subspace* V is the graph of $\begin{bmatrix} L \\ \Gamma \end{bmatrix}$:

$$V := \left\{ \begin{bmatrix} Lx \\ x \\ \Gamma x \end{bmatrix} \,\middle|\, x \in \operatorname{dom}(L) \right\}. \tag{4.3}$$

The unbounded operator $\begin{bmatrix} L \\ \Gamma \end{bmatrix}$ is assumed to be closed, and this is equivalent to assuming that V is a closed subspace of the *node space* $\mathcal{X} \times \mathcal{X} \times \mathcal{W}$. The generating subspace is the key to generalizing the state/signal theory beyond boundary control, as we shall see in Chapter 5. We define the *dynamics* of a state/signal system using the generating subspace V.

Definition 4.2 Let V be a closed subspace of $\mathcal{X} \times \mathcal{X} \times \mathcal{W}$.

1. The pair $\begin{bmatrix} x \\ w \end{bmatrix}$ is a *classical trajectory* generated by V on $\mathbb{R}^+$ if $x \in C^1(\mathbb{R}^+; \mathcal{X})$, $w \in C(\mathbb{R}^+; \mathcal{W})$, and $\begin{bmatrix} \dot{x}(t) \\ x(t) \\ w(t) \end{bmatrix} \in V$ for all $t > 0$.

2. The pair $\left[\begin{smallmatrix} x \\ w \end{smallmatrix}\right]$ is a *generalized trajectory* generated by V on $\mathbb{R}^+$ if $x \in C(\mathbb{R}^+;\mathcal{X})$, $w \in L^2_{\mathrm{loc}}(\mathbb{R}^+;\mathcal{W})$, and there exists a sequence of classical trajectories $\left[\begin{smallmatrix} x_n \\ w_n \end{smallmatrix}\right]$ such that $x_n \to x$ uniformly on all bounded intervals $[0,T]$ and $w_n \to w$ in $L^2_{\mathrm{loc}}(\mathbb{R}^+;\mathcal{W})$.

Note that $\begin{bmatrix} \dot{x}(t) \\ x(t) \\ w(t) \end{bmatrix} \in V$ for all $t > 0$ in item 1 of Definition 4.2 if and only if $\begin{bmatrix} \dot{x}(t) \\ x(t) \\ w(t) \end{bmatrix} \in V$ for all $t \in \mathbb{R}^+$ when we interpret $\dot{x}(0)$ as the right-sided derivative of x at zero. We are now ready to define a boundary control s/s system.

Definition 4.3 A *boundary control state/signal (s/s) node* is a quadruple $\Sigma = (L,\Gamma;\mathcal{X},\mathcal{W})$ such that:

1. The space $\mathcal{X}$ is a Hilbert space and $\mathcal{W}$ is a Kreĭn space.
2. The operator $\left[\begin{smallmatrix} L \\ \Gamma \end{smallmatrix}\right] : \mathcal{X} \to \left[\begin{smallmatrix} \mathcal{X} \\ \mathcal{W} \end{smallmatrix}\right]$ is closed and densely defined.
3. The range of Γ is dense in $\mathcal{W}$.

By the *boundary control state/signal system* induced by a boundary control s/s node $(L,\Gamma;\mathcal{X},\mathcal{W})$ we mean this node together with the sets of classical and generalized trajectories generated by V in (4.3) on $\mathbb{R}^+$. We denote both the node and the system by $\Sigma = (L,\Gamma;\mathcal{X},\mathcal{W})$.

In Definition 4.4 below we will equip the node space $\mathcal{X}\times\mathcal{X}\times\mathcal{W}$ with an indefinite inner product which makes it a Kreĭn space.

4.3 Conservative state/signal systems in boundary control

In this section we shall focus our attention on s/s systems Σ whose classical trajectories on $\mathbb{R}^+$ satisfy the *power equality*

$$\frac{\mathrm{d}}{\mathrm{d}t}\|x(t)\|^2_{\mathcal{X}} = [w(t),w(t)]_{\mathcal{W}}, \quad t \in \mathbb{R}^+. \tag{4.4}$$

Here $\|x(t)\|^2_{\mathcal{X}}$ stands for (two times) the *internal energy* stored in the state x at time t and $[w(t),w(t)]_{\mathcal{W}}$ represents (two times) the power (energy flow per time unit) entering the system through the signal $w(t)$ at time t. This explains why we need to take $\mathcal{W}$ to be a Kreĭn space: we must allow the inner product $[\cdot,-]_{\mathcal{W}}$ in $\mathcal{W}$ to be indefinite. If the inner product in $\mathcal{W}$ is non-negative, then no energy can leave the system via the (interaction) signal, and if the inner product in $\mathcal{W}$ is non-positive, then no energy can enter the system via the signal.

The equality (4.4) says that the system has *no internal energy sources or sinks*. However, the equality is not enough to make the system Σ conservative: we need an additional *hypermaximality condition*. We give the full definition of a conservative boundary control s/s system in Definition 4.5 below.

After integration over the interval $[s,t] \subset \mathbb{R}^+$, one can rewrite (4.4) in the equivalent form

$$\|x(t)\|_{\mathcal{X}}^2 - \|x(s)\|_{\mathcal{X}}^2 = \int_s^t [w(v), w(v)]_{\mathcal{W}} \, \mathrm{d}v, \quad s, t \in \mathbb{R}^+, \ s \le t.$$

By the continuity of the inner product this inequality remains valid for generalized trajectories as well.

Carrying out the differentiation in (4.4), we get a third equivalent condition in terms of classical trajectories, namely

$$-(\dot{x}(t), x(t))_{\mathcal{X}} - (x(t), \dot{x}(t))_{\mathcal{X}} + [w(t), w(t)]_{\mathcal{W}} = 0, \quad t \in \mathbb{R}^+. \tag{4.5}$$

Using item 1 of Definition 4.2, we see that (4.5) always holds if

$$-(z, x)_{\mathcal{X}} - (x, z)_{\mathcal{X}} + [w, w]_{\mathcal{W}} = 0, \quad \begin{bmatrix} z \\ x \\ w \end{bmatrix} \in V. \tag{4.6}$$

It is now natural to make the following definition:

Definition 4.4 Let $\mathcal{X}$ be a Hilbert space and $\mathcal{W}$ a Kreĭn space. The corresponding *node space* is the product space $\mathfrak{K} = \mathcal{X} \times \mathcal{X} \times \mathcal{W}$ equipped with the indefinite inner product induced by the quadratic form in (4.6):

$$\left[\begin{bmatrix} z_1 \\ x_1 \\ w_1 \end{bmatrix}, \begin{bmatrix} z_2 \\ x_2 \\ w_2 \end{bmatrix}\right]_{\mathfrak{K}} = -(z_1, x_2)_{\mathcal{X}} - (x_1, z_2)_{\mathcal{X}} + [w_1, w_2]_{\mathcal{W}}. \tag{4.7}$$

Note that the the quadratic form in (4.6) is strictly indefinite, i.e., it takes both positive and negative values whenever $\mathcal{X} \ne \{0\}$. Furthermore, the inner product in (4.7) makes the node space $\mathfrak{K}$ a Kreĭn space.

The equality (4.6) says that V is a *neutral subspace* of $\mathfrak{K}$ with respect to the inner product (4.7), i.e., that $[v, v]_{\mathfrak{K}} = 0$ for all $v \in V$. The condition that a subspace V is a neutral subspace of $\mathfrak{K}$ can equivalently be written $V \subset V^{[\perp]}$, where

$$V^{[\perp]} := \left\{ k \in \mathfrak{K} \;\middle|\; [k, k']_{\mathfrak{K}} = 0 \text{ for all } k' \in V \right\}.$$

If instead $V^{[\perp]} \subset V$, then V is called co-neutral, and if $V^{[\perp]} = V$, then V is called *Lagrangian* or *hypermaximal neutral*.

Definition 4.5 A boundary control s/s system $\Sigma = (L, \Gamma; \mathcal{X}, \mathcal{W})$ is *conservative* if its generating subspace V in (4.3) is a Lagrangian subspace of the node space $\mathfrak{K}$, i.e., if $V = V^{[\perp]}$.

Since every orthogonal companion is closed, necessarily every Lagrangian subspace is closed. Moreover, in [Kurula et al., 2010, Th. 4.3] it was proved that if V in (4.3) is Lagrangian then $\ker(\Gamma)$ is dense in $\mathcal{X}$ and $\operatorname{ran}(\Gamma)$ is dense in $\mathcal{W}$. Since $\ker(\Gamma) \subset \operatorname{dom}(\Gamma) = \operatorname{dom}(L)$, the operator $\left[\begin{smallmatrix} L \\ \Gamma \end{smallmatrix}\right]$ is closed and automatically densely defined. Thus the conditions in Definition 4.3 are satisfied for every Lagrangian subspace V of the type (4.3). See also [Derkach et al., 2006, Cor. 2.4].

Remark In the boundary control case the neutrality condition $V \subset V^{[\perp]}$ means that

$$(Lx, x)_{\mathcal{X}} + (x, Lx)_{\mathcal{X}} = [\Gamma x, \Gamma x]_{\mathcal{W}}, \quad x \in \operatorname{dom}(L). \tag{4.8}$$

However, if V is only neutral, then V might for instance be the degenerate trivial system $\{0\}$. This case is excluded by the hypermaximality condition $V \supset V^{[\perp]}$, which in the case of boundary control means that

$$(z_1, x)_{\mathcal{X}} + (x_1, Lx)_{\mathcal{X}} = [w_1, \Gamma x]_{\mathcal{W}},\ x \in \operatorname{dom}(L) \Longrightarrow \begin{bmatrix} z_1 \\ x_1 \\ w_1 \end{bmatrix} \in V. \tag{4.9}$$

Letting $\mathcal{X}$ be a Hilbert space, $\mathcal{W}$ be a Kreĭn space, and $\left[\begin{smallmatrix} L \\ \Gamma \end{smallmatrix}\right] : \mathcal{X} \to \left[\begin{smallmatrix} \mathcal{X} \\ \mathcal{W} \end{smallmatrix}\right]$, we thus have that $\Sigma = (L, \Gamma; \mathcal{X}, \mathcal{W})$ is a conservative boundary control s/s system if and only if the conditions (4.8) and (4.9) are satisfied.

4.4 An example: the transmission line

An ideal transmission line of length ℓ can be modeled by the following equations, where $\xi \in [0, \ell]$ and $t \in \mathbb{R}^+$:

$$\begin{aligned} \frac{\partial}{\partial t} \begin{bmatrix} i(\xi,t) \\ v(\xi,t) \end{bmatrix} &= \begin{bmatrix} 0 & -\frac{1}{\mathcal{L}(\xi)} \frac{\partial}{\partial \xi} \\ -\frac{1}{\mathcal{C}(\xi)} \frac{\partial}{\partial \xi} & 0 \end{bmatrix} \begin{bmatrix} i(\xi,t) \\ v(\xi,t) \end{bmatrix}, \\ w(t) &= \begin{bmatrix} i(0,t) \\ v(0,t) \\ -i(\ell,t) \\ v(\ell,t) \end{bmatrix}, \quad \begin{bmatrix} i(\xi,0) \\ v(\xi,0) \end{bmatrix} = \begin{bmatrix} i_0(\xi) \\ v_0(\xi) \end{bmatrix}. \end{aligned} \tag{4.10}$$

Here $i(\xi, t)$ and $v(\xi, t)$ are the current and voltage, respectively, at the point $\xi \in [0, \ell]$ at time $t \in \mathbb{R}^+$. The functions $\mathcal{L}(\cdot) > 0$ and $\mathcal{C}(\cdot) > 0$ represent the *distributed inductance and capacitance*, respectively, of the line. For simplicity we assume that $\mathcal{C}(\cdot)$ and $\mathcal{L}(\cdot)$ are continuous on $[0, \ell]$, which implies that $\mathcal{C}$ and $\mathcal{L}$ are both bounded and bounded away from zero. The transmission line is illustrated in the following figure.

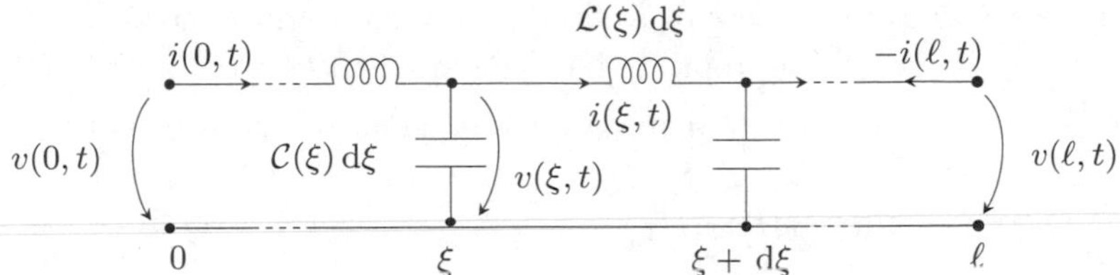

An ideal $\mathcal{LC}$-transmission line of length ℓ with *distributed* inductance $\mathcal{L}$ and capacitance $\mathcal{C}$. Here $i(\xi,t)$ and $v(\xi,t)$ denote the current and the voltage, respectively, at the point $\xi \in [0,\ell]$ at time $t \in \mathbb{R}^+$.

The natural state at time t of this transmission line is the current-voltage vector $x(t) = \left[\begin{smallmatrix} i(\cdot,t) \\ v(\cdot,t) \end{smallmatrix}\right]$, $t \in \mathbb{R}^+$, and the initial state is $x(0) = \left[\begin{smallmatrix} i(\cdot,0) \\ v(\cdot,0) \end{smallmatrix}\right] = \left[\begin{smallmatrix} i_0(\cdot) \\ v_0(\cdot) \end{smallmatrix}\right] =: x_0$. We take the state space $\mathcal{X}$ to be $L^2([0,\ell];\mathbb{C}^2)$ with inner product $(\cdot,-)_{\mathcal{X}}$ defined by

$$\left(\begin{bmatrix} i_1(\cdot) \\ v_1(\cdot) \end{bmatrix}, \begin{bmatrix} i_2(\cdot) \\ v_2(\cdot) \end{bmatrix}\right)_{\mathcal{X}} = \int_0^\ell \left(\mathcal{L}(\xi) i_1(\xi)\overline{i_2(\xi)} + \mathcal{C}(\xi) v_1(\xi)\overline{v_2(\xi)}\right) \mathrm{d}\xi. \quad (4.11)$$

In our setting the corresponding quadratic form $(x(t),x(t))_{\mathcal{X}}$ is equivalent to the standard inner product on $L^2([0,\ell];\mathbb{C}^2)$ and its value is twice the energy stored in the state $x(t)$ of the transmission line at time t.

The operator L is given by

$$L := \begin{bmatrix} 0 & -\frac{1}{\mathcal{L}(\xi)}\frac{\partial}{\partial\xi} \\ -\frac{1}{\mathcal{C}(\xi)}\frac{\partial}{\partial\xi} & 0 \end{bmatrix}, \quad \operatorname{dom}(L) := W^{1,2}([0,\ell];\mathbb{C}^2),$$

where $W^{1,2}([0,\ell];\mathbb{C}^2)$ is the Sobolev space of absolutely continuous functions in $L^2([0,\ell];\mathbb{C}^2)$ which have a distribution derivative in $L^2([0,\ell];\mathbb{C}^2)$. The signal space $\mathcal{W}$ is $\mathbb{C}^4$ equipped with the indefinite inner product

$$\left[\begin{bmatrix} i_{01} \\ v_{01} \\ i_{\ell 1} \\ v_{\ell 1} \end{bmatrix}, \begin{bmatrix} i_{02} \\ v_{02} \\ i_{\ell 2} \\ v_{\ell 2} \end{bmatrix}\right]_{\mathcal{W}} = \left(\begin{bmatrix} i_{01} \\ v_{01} \\ i_{\ell 1} \\ v_{\ell 1} \end{bmatrix}, J_{\mathcal{W}} \begin{bmatrix} i_{02} \\ v_{02} \\ i_{\ell 2} \\ v_{\ell 2} \end{bmatrix}\right)_{\mathbb{C}^4}, \quad J_{\mathcal{W}} = \begin{bmatrix} \left[\begin{smallmatrix} 0 & 1 \\ 1 & 0 \end{smallmatrix}\right] & 0 \\ 0 & \left[\begin{smallmatrix} 0 & 1 \\ 1 & 0 \end{smallmatrix}\right] \end{bmatrix}. \quad (4.12)$$

The boundary operator Γ has the same domain as L, and it is given by

$$\Gamma \begin{bmatrix} i(\cdot) \\ v(\cdot) \end{bmatrix} = \begin{bmatrix} i(0) \\ v(0) \\ -i(\ell) \\ v(\ell) \end{bmatrix}.$$

The operator $\left[\begin{smallmatrix} L \\ \Gamma \end{smallmatrix}\right]$ is closed as an operator from $\mathcal{X}$ to $\left[\begin{smallmatrix} \mathcal{X} \\ \mathcal{W} \end{smallmatrix}\right]$ with domain $\operatorname{dom}\left(\left[\begin{smallmatrix} L \\ \Gamma \end{smallmatrix}\right]\right) = \operatorname{dom}(L) = W^{1,2}([0,\ell];\mathbb{C}^2)$. With these definitions, the transmission line can be modeled as an example of a boundary control s/s system in the sense of Definition 4.3, as we now show.

We next derive the appropriate Lagrangian identity. Combining $x(t) = \left[\begin{smallmatrix} i(\cdot,t) \\ v(\cdot,t) \end{smallmatrix}\right]$, (4.10), and (4.11), we make the following computations for $t > 0$:

$$\begin{aligned}
\frac{\mathrm{d}}{\mathrm{d}t}\|x(t)\|_{\mathcal{X}}^2 &= 2\mathrm{Re}\ (x(t), \dot{x}(t))_{\mathcal{X}} \\
&= 2\mathrm{Re} \int_0^\ell \left(\mathcal{L}(\xi) i(\xi,t) \overline{\frac{\partial}{\partial t} i(\xi,t)} + \mathcal{C}(\xi) v(\xi,t) \overline{\frac{\partial}{\partial t} v(\xi,t)} \right) \mathrm{d}\xi \\
&= -2 \int_0^\ell \mathrm{Re} \left(i(\xi,t) \overline{\frac{\partial}{\partial \xi} v(\xi,t)} + \overline{\frac{\partial}{\partial \xi} i(\xi,t)} v(\xi,t) \right) \mathrm{d}\xi \\
&= -2 \int_0^\ell \mathrm{Re}\ \frac{\partial}{\partial \xi} \left(i(\xi,t)\overline{v(\xi,t)} \right) \mathrm{d}\xi \\
&= -2\mathrm{Re} \left[i(\xi,t)\overline{v(\xi,t)} \right]_{\xi=0}^{\ell} \\
&= 2\mathrm{Re}\, i(0,t)\overline{v(0,t)} - 2\mathrm{Re}\, i(\ell,t)\overline{v(\ell,t)} \\
&= \left(\begin{bmatrix} i(0,t) \\ v(0,t) \\ -i(\ell,t) \\ v(\ell,t) \end{bmatrix}, \begin{bmatrix} \left[\begin{smallmatrix} 0 & 1 \\ 1 & 0 \end{smallmatrix}\right] & 0 \\ 0 & \left[\begin{smallmatrix} 0 & 1 \\ 1 & 0 \end{smallmatrix}\right] \end{bmatrix} \begin{bmatrix} i(0,t) \\ v(0,t) \\ -i(\ell,t) \\ v(\ell,t) \end{bmatrix} \right)_{\mathbb{C}^4} \\
&= [\Gamma x(t), \Gamma x(t)]_{\mathcal{W}},
\end{aligned}$$

where we have used that ($'$ denotes spatial derivative)

$$2\mathrm{Re}\,(i\overline{v'} + \overline{i'}v) = i\overline{v'} + \overline{i'}v + \overline{i}v' + i'\overline{v} = 2\mathrm{Re}\,(i\overline{v})'$$

in the fourth equality. Thus, $[w(t), w(t)]_{\mathcal{W}} = [\Gamma x(t), \Gamma x(t)]_{\mathcal{W}}$ is two times the power entering the transmission line through the terminals at the ends $\xi = 0$ and $\xi = \ell$ of the line at time $t \geq 0$.

These computations tell us that the generating subspace V is a neutral subspace of the node space $\mathfrak{K}$, i.e., that (4.8) holds. It is not difficult to show that this subspace is not only neutral, but even Lagrangian, so that (4.9) also holds; see Example 4.9 below for the proof idea. Thus, *the transmission line gives rise to a conservative boundary control s/s system.*

Remark Set $\mathcal{U} := \mathbb{C}^2$, $R := iL|_{\ker \Gamma}$, and

$$\Gamma_0 \begin{bmatrix} i(\cdot) \\ v(\cdot) \end{bmatrix} := \begin{bmatrix} i(0) \\ -i(\ell) \end{bmatrix} \quad \text{and} \quad \Gamma_1 \begin{bmatrix} i(\cdot) \\ v(\cdot) \end{bmatrix} := \begin{bmatrix} v(0) \\ v(\ell) \end{bmatrix}. \tag{4.13}$$

Then R is a closed, densely defined and symmetric operator in the

Hilbert space $\mathcal{X}$, and the triple $(\Gamma_0, -\mathrm{i}\Gamma_1; \mathcal{U})$ is a *boundary triplet* for $R^* = \mathrm{i}L$ in the standard sense; see below. The boundary triplet and its connection to boundary-control state/signal systems is the topic of the last section of this chapter.

Recall that $[w(t), w(t)]_{\mathcal{W}}$ is two times the power entering the transmission line through the terminals at the ends $\xi = 0$ and $\xi = \ell$ of the line at time $t \geq 0$. The decomposition in (4.13) of Γ into an input map Γ_0 and an output map Γ_1 corresponds to choosing the current entering the system at $\xi = 0$ and $\xi = \ell$ as input and the voltages at both ends as output, cf. (4.1). We refer to this as an *impedance decomposition* of the external signal w.

Several other choices of input and output would have been possible, such as for example

$$\begin{aligned} \widetilde{\Gamma}_0 \begin{bmatrix} i(\cdot) \\ v(\cdot) \end{bmatrix} &:= \frac{1}{\sqrt{2}}(\Gamma_1 + \Gamma_0) \begin{bmatrix} i(\cdot) \\ v(\cdot) \end{bmatrix} = \frac{1}{\sqrt{2}} \begin{bmatrix} v(0)+i(0) \\ v(\ell)-i(\ell) \end{bmatrix} \quad \text{and} \\ \widetilde{\Gamma}_1 \begin{bmatrix} i(\cdot) \\ v(\cdot) \end{bmatrix} &:= \frac{1}{\sqrt{2}}(\Gamma_1 - \Gamma_0) \begin{bmatrix} i(\cdot) \\ v(\cdot) \end{bmatrix} = \frac{1}{\sqrt{2}} \begin{bmatrix} v(0)-i(0) \\ v(\ell)+i(\ell) \end{bmatrix}, \quad \text{or} \end{aligned} \tag{4.14}$$

$$\widehat{\Gamma}_0 \begin{bmatrix} i(\cdot) \\ v(\cdot) \end{bmatrix} := \begin{bmatrix} i(0) \\ v(0) \end{bmatrix} \quad \text{and} \quad \widehat{\Gamma}_1 \begin{bmatrix} i(\cdot) \\ v(\cdot) \end{bmatrix} := \begin{bmatrix} -i(\ell) \\ v(\ell) \end{bmatrix}. \tag{4.15}$$

In (4.14) we have

$$\left\| \widetilde{\Gamma}_0 \begin{bmatrix} i(\cdot) \\ v(\cdot) \end{bmatrix} \right\|_{\mathbb{C}^2}^2 - \left\| \widetilde{\Gamma}_1 \begin{bmatrix} i(\cdot) \\ v(\cdot) \end{bmatrix} \right\|_{\mathbb{C}^2}^2 = \left[\Gamma \begin{bmatrix} i(\cdot) \\ v(\cdot) \end{bmatrix}, \Gamma \begin{bmatrix} i(\cdot) \\ v(\cdot) \end{bmatrix} \right]_{\mathcal{W}},$$

where $[\cdot, -]_{\mathcal{W}}$ still denotes the inner product (4.12). This decomposition is an example of a *scattering decomposition.* In (4.15) we choose voltage and current at $\xi = 0$ as input and the voltage and current at $\xi = \ell$ as output, and this is an example of a *transmission decomposition.*

Remark 4.6 Making a different choice of input and output signals results in a different map from the input to the output, i.e., a different input/state/output representation, with possibly widely different properties. However, the physical system, i.e., the $\mathcal{LC}$-transmission line with length ℓ, is still the same. This “input/output-free” paradigm is inherent in the state/signal philosophy.

4.5 The connection to boundary triplets

Boundary triplets originate from the extension theory of symmetric operators on Hilbert spaces. The following definition is adapted from [Gorbachuk and Gorbachuk, 1991, pp. 154–155], using the more recent terminology and notations from [Derkach et al., 2006, Def. 5.1]; see also Chapters 3, 6 and 7.

Definition 4.7 Let R be a closed densely defined symmetric operator on the Hilbert space $\mathcal{X}$ with equal (finite or infinite) defect numbers $n_\pm := \dim\ker(R \mp i)$. Let $\mathcal{U}$ be another Hilbert space, the "external Hilbert space", and let Γ_j, $j = 0, 1$, be linear operators mapping $\mathrm{dom}\,(R^*)$ into $\mathcal{U}$.

The triplet $(\Gamma_0, \Gamma_1; \mathcal{U})$ is called a *boundary triplet* for the operator R^* if the following two conditions hold:

1. For all $x_1, x_2 \in \mathrm{dom}\,(R^*)$ we have
$$(R^*x_1, x_2)_{\mathcal{X}} - (x_1, R^*x_2)_{\mathcal{X}} = (\Gamma_0 x_1, \Gamma_1 x_2)_{\mathcal{U}} - (\Gamma_1 x_1, \Gamma_0 x_2)_{\mathcal{U}}.$$
2. The range of the combined operator $\Gamma := \left[\begin{smallmatrix} \Gamma_0 \\ \Gamma_1 \end{smallmatrix}\right]$ is $\left[\begin{smallmatrix} \mathcal{U} \\ \mathcal{U} \end{smallmatrix}\right]$.

Here condition 1 is the *Lagrangian identity* and condition 2 can be interpreted as a *regularity condition* or a *(hyper)maximality condition.*

By a *direct-sum decomposition* $\mathcal{W} = \mathcal{U} \dotplus \mathcal{Y}$ of a Kreĭn space we mean that $\mathcal{U}$ and $\mathcal{Y}$ are closed subspaces of $\mathcal{W}$, such that $\mathcal{U} + \mathcal{Y} = \mathcal{W}$ and $\mathcal{U} \cap \mathcal{Y} = \{0\}$. This decomposition is *Lagrangian* if $\mathcal{U}$ and $\mathcal{Y}$ are both Lagrangian subspaces: $\mathcal{U} = \mathcal{U}^{[\perp]}$ and $\mathcal{Y} = \mathcal{Y}^{[\perp]}$. For every Hilbert space $\mathcal{U}$, the direct-sum decomposition

$$\mathcal{W} = \widetilde{\mathcal{U}} \dotplus \widetilde{\mathcal{Y}} := \left[\begin{smallmatrix} \mathcal{U} \\ \{0\} \end{smallmatrix}\right] \dotplus \left[\begin{smallmatrix} \{0\} \\ \mathcal{U} \end{smallmatrix}\right] \tag{4.16}$$

of $\mathcal{W} = \mathcal{U}^2$ is *Lagrangian* if $\mathcal{W}$ has the inner product

$$\left[\left[\begin{smallmatrix} u_1 \\ y_1 \end{smallmatrix}\right], \left[\begin{smallmatrix} u_2 \\ y_2 \end{smallmatrix}\right]\right]_{\mathcal{W}} = (u_1, y_2)_{\mathcal{U}} + (y_1, u_2)_{\mathcal{U}}. \tag{4.17}$$

For instance, the impedance decomposition in the transmission line example, where we take the currents as input and voltages as outputs, is a Lagrangian decomposition.

For a proof of the following result, see [Malinen and Staffans, 2007, Sec. 5]:

Theorem 4.8 *Let R be a closed and densely defined symmetric operator on $\mathcal{X}$ with equal defect numbers, and let $(\Gamma_0, \Gamma_1; \mathcal{U})$ be a boundary triplet for R^*. Take $\mathcal{W} := \left[\begin{smallmatrix} \mathcal{U} \\ \mathcal{U} \end{smallmatrix}\right]$ with the indefinite inner product* (4.17) *and define $\Gamma := \left[\begin{smallmatrix} \Gamma_0 \\ \mathrm{i}\Gamma_1 \end{smallmatrix}\right]$ with* $\mathrm{dom}\,(\Gamma) = \mathrm{dom}\,(R^*)$.

Then $\Sigma = (\mathrm{i}R^*, \Gamma; \mathcal{X}, \mathcal{W})$ is a boundary control s/s system *in the sense of Definition 4.3. The system is moreover* conservative*:* $V = V^{[\perp]}$*, where V is given by* (4.3).

Consider the conservative boundary control s/s system Σ in Th. 4.8. The *input/state/output representation*

$$\Sigma_{i/s/o} = \left(\mathrm{i}R^*, \left[\begin{smallmatrix}\Gamma_0\\ \{0\}\end{smallmatrix}\right], \left[\begin{smallmatrix}\{0\}\\ \mathrm{i}\Gamma_1\end{smallmatrix}\right]; \mathcal{X}, \left[\begin{smallmatrix}\mathcal{U}\\ \{0\}\end{smallmatrix}\right], \left[\begin{smallmatrix}\{0\}\\ \mathcal{U}\end{smallmatrix}\right]\right)$$

corresponding to the Lagrangian decomposition (4.16) is an example of an *impedance representation* of Σ. We investigate these concepts in more detail in Section 5.3.

The converse of Theorem 4.8 is not true: *there do exist conservative boundary control s/s systems which are not induced by any boundary triplet of the type in Definition 4.7.* These examples are of two types:

1. The signal space $\mathcal{W}$ *need not have a Lagrangian decomposition.* A necessary and sufficient condition for the existence of a Lagrangian decomposition is that $\mathrm{ind}_+\mathcal{W} = \mathrm{ind}_-\mathcal{W}$ ($\leq \infty$); see Example 4.9 below. In the case of a boundary triplet we always have at least the Lagrangian decomposition (4.16).
2. Even if the signal space $\mathcal{W}$ has a Lagrangian decomposition *the main operator L need not be closed*, and we can thus not have $L = \mathrm{i}R^*$. Moreover, *the operator $\Gamma := \left[\begin{smallmatrix}\Gamma_0\\ \Gamma_1\end{smallmatrix}\right]$ need not be surjective.* See [Malinen and Staffans, 2007] for an example.

   More precisely, let $\Sigma = (L, \Gamma; \mathcal{X}, \mathcal{W})$ be a conservative boundary control s/s system. According to [Kurula et al., 2010, Prop. 4.5], *L is closed if and only if the range of Γ is closed.* Combining this with the condition that Γ has dense range, we obtain that L is closed if and only if Γ is surjective. The same conclusion can be made based on [Derkach et al., 2006, Prop. 2.3 and Cor. 2.4].

We now give an example of a conservative boundary control s/s system that is not induced by a boundary triplet. In a scattering setting this system has no input and a one-dimensional output, and the C_0-semigroup describing the system dynamics is the left shift in $L^2(\mathbb{R}^+;\mathbb{C})$.

Example 4.9 Choose $\mathcal{X} := L^2(\mathbb{R}^+;\mathbb{C})$ with its standard Hilbert-space inner product, set $\mathcal{W} := -\mathbb{C}$, and define

$$V := \left\{ \begin{bmatrix} \frac{\mathrm{d}x}{\mathrm{d}\xi} \\ x \\ x(0) \end{bmatrix} \,\middle|\, x \in W^{1,2}(\mathbb{R}^+;\mathbb{C}) \right\} \subset \mathcal{X} \times \mathcal{X} \times \mathcal{W}.$$

It is clear that $\left[\begin{smallmatrix} z\\ 0\\ 0\end{smallmatrix}\right] \in V$ implies that $z = 0$, and we will now show that $V = V^{[\perp]}$, i.e., that $(V; \mathcal{X}, \mathcal{W})$ is a conservative boundary control s/s system. Note that the signal space $\mathcal{W}$ has no Lagrangian decompositions.

We first prove that $V^{[\perp]} \subset V$. By definition $\left[\begin{smallmatrix}\widetilde{z}\\ \widetilde{x}\\ \widetilde{w}\end{smallmatrix}\right] \in V^{[\perp]}$ if and only if $\left[\begin{smallmatrix}\widetilde{z}\\ \widetilde{x}\\ \widetilde{w}\end{smallmatrix}\right] \in \mathfrak{K} = L^2(\mathbb{R}^+;\mathbb{C}) \times L^2(\mathbb{R}^+;\mathbb{C}) \times \mathbb{C}$ and for all $x \in W^{1,2}(\mathbb{R}^+;\mathbb{C})$:

$$\left[\left[\begin{bmatrix}\widetilde{z}\\ \widetilde{x}\\ \widetilde{w}\end{bmatrix}, \begin{bmatrix}\frac{\mathrm{d}x}{\mathrm{d}\xi}\\ x\\ x(0)\end{bmatrix}\right]\right]_{\mathfrak{K}} = -\widetilde{w}\,\overline{x(0)} - \int_0^\infty \left(\widetilde{x}(\xi)\overline{\frac{\mathrm{d}x}{\mathrm{d}\xi}(\xi)} + \widetilde{z}(\xi)\overline{x(\xi)}\right)\mathrm{d}\xi = 0. \quad (4.18)$$

In particular, if we let x vary over the set of test functions in C^∞ with support contained in $(0,\infty)$, then $x(0) = 0$, and we find that $\frac{\mathrm{d}\widetilde{x}}{\mathrm{d}\xi} = \widetilde{z}$ in the distribution sense. Since both $\widetilde{x}$ and $\widetilde{z}$ belong to $L^2(\mathbb{R}^+;\mathbb{C})$, this implies that $\widetilde{x} \in W^{1,2}(\mathbb{R}^+;\mathbb{C})$. This makes it is possible to integrate by parts in (4.18), using that $\widetilde{z}(\xi) = \frac{\mathrm{d}\widetilde{x}}{\mathrm{d}\xi}(\xi)$, in order to get that

$$\widetilde{w}\,\overline{x(0)} - \widetilde{x}(0)\,\overline{x(0)}, \quad x \in W^{1,2}(\mathbb{R}^+;\mathbb{C}).$$

Thus $\widetilde{w} = \widetilde{x}(0)$, and this proves that $V^{[\perp]} \subset V$.

In order to show that $V \subset V^{[\perp]}$, we choose $\widetilde{x} \in W^{1,2}(\mathbb{R}^+;\mathbb{C})$ arbitrarily, and we set $\widetilde{z} := \frac{\mathrm{d}\widetilde{x}}{\mathrm{d}\xi}$ and $\widetilde{w} := \widetilde{x}(0)$. Then (4.18) holds for all $x, \widetilde{x} \in W^{1,2}(\mathbb{R}^+;\mathbb{C})$, i.e., $V \subset V^{[\perp]}$. We are done proving that $V = V^{[\perp]}$, and therefore, that $(V;\mathcal{X},\mathcal{W})$ is a conservative boundary control s/s system whose signal space $\mathcal{W} = -\mathbb{C}$ has no Lagrangian decompositions.

The i/s/o case where $\Gamma = \left[\begin{smallmatrix}\Gamma_0\\ \Gamma_1\end{smallmatrix}\right] : \mathcal{X} \to \mathcal{U}^2$ has dense but non-closed range has been treated using *generalized boundary triplets* in [Derkach and Malamud, 1995] and using *quasi boundary triplets* in [Behrndt and Langer, 2007]. Interconnection of conservative boundary control i/s/o systems with surjective $\left[\begin{smallmatrix}\Gamma_0\\ \Gamma_1\end{smallmatrix}\right]$ was worked out in detail in [Kurula et al., 2010].

A considerably more general notion than that of a boundary triplet is that of a *boundary relation* which was extensively studied in e.g. [Derkach et al., 2006]. The topic of Chapter 5, which is more detailed than the present one, is to show how boundary relations are connected to general (non-boundary control) s/s systems. There the main point is to show that the notion of a boundary relation is connected to the notion of a conservative state/signal system in the same way as the boundary triplet is related to the boundary control s/s system: the former arises as a particular i/s/o impedance representation of the latter.

A preliminary version of the notion of a boundary control s/s system can be found in an implicit form in the early work by Calkin presented in Chapter 2. Our Kreĭn signal space can be identified with the Kreĭn space $\mathfrak{K}$ described in Section 2.4, and the operator Γ in (4.2) can be identified with a Calkin's *reduction operator* U in Definition 2.1. As noticed in the text before Proposition 2.19, the surjectivity condition on Γ in Definition 4.7 is equivalent to the boundedness of the reduction operator U, and consequently the reduction operator induced by a boundary control s/s system need not be bounded. Also note that in Calkin's setting the signal space $\mathcal{W}$ need not have equal positive and negative indices, i.e., $\mathcal{W}$ need not have a Lagrangian decomposition.

References

Behrndt, J., and Langer, M. 2007. Boundary value problems for elliptic partial differential operators on bounded domains. *J. Funct. Anal.*, **243**, 536–565.

Derkach, V. A., and Malamud, M. M. 1995. The extension theory of Hermitian operators and the moment problem. *J. Math. Sci.*, **73**, 141–242. Analysis. 3.

Derkach, V. A., Hassi, S., Malamud, M. M., and de Snoo, H. S. V. 2006. Boundary relations and their Weyl families. *Trans. Amer. Math. Soc.*, **358**, 5351–5400.

Gorbachuk, V. I., and Gorbachuk, M. L. 1991. *Boundary value problems for operator differential equations*. Mathematics and its Applications (Soviet Series), vol. 48. Dordrecht: Kluwer Academic Publishers Group, translation and revised from the 1984 Russian original.

Kurula, M. M., Zwart, H. J., van der Schaft, A. J., and Behrndt, J. 2010. Dirac structures and their composition on Hilbert spaces. *J. Math. Anal. Appl.*, **372**, 402–422.

Malinen, J., and Staffans, O. J. 2006. Conservative boundary control systems. *J. Differential Equ.*, **231**, 290–312.

Malinen, J., and Staffans, O. J. 2007. Impedance passive and conservative boundary control systems. *Complex Anal. Oper. Theory*, **1**, 279–30.

Salamon, D. 1987. Infinite dimensional linear systems with unbounded control and observation: a functional analytic approach. *Trans. Amer. Math. Soc.*, **300**, 383–431.

Staffans, O. J. 2005. *Well-Posed Linear Systems*. Cambridge and New York: Cambridge University Press.

5

Passive state/signal systems and conservative boundary relations

Damir Z. Arov [a], Mikael Kurula and Olof J. Staffans

Abstract This chapter is a continuation and deepening of Chapter 4. In the present chapter the state/signal theory is extended beyond boundary control and beyond conservative systems. The main aim is to clarify the basic connections between the state/signal theory and that of (conservative) boundary relations. It is described how one can represent a state/signal system using input/state/output systems in different ways by making different choices of input signal and output signal. There is an "almost one-to-one" relationship between conservative state/signal systems and boundary relations, and this connection is used in order to introduce dynamics to a boundary relation. Consequently, a boundary relation is such a general object that it mathematically has rather little to do with boundary control. The Weyl family and γ-field of a boundary relation are connected to the frequency-domain characteristics of a state/signal system.

5.1 Introduction

The theory of boundary relations has been developed by a number of authors in the framework of the theory of self-adjoint extensions of symmetric operators and relations in Hilbert spaces; see e.g. the recent articles [Derkach et al., 2006; Derkach, 2009; Derkach et al., 2009; Behrndt et al., 2009]. Boundary relations are described in detail in Chapter 7.

One way of introducing the notion of a state/signal (s/s) system is to start from an input/state/output (i/s/o) system. By a standard i/s/o system we mean a system of equations of the type

$$\Sigma_{i/s/o}: \quad \begin{cases} \dot{x}(t) = Ax(t) + Bu(t) \\ y(t) = Cx(t) + Du(t), \end{cases} \quad t \in \mathbb{R}^+, \ x(0) = x_0 \text{ given,} \tag{5.1}$$

[a] Damir Z. Arov thanks Åbo Akademi for its hospitality and the Academy of Finland and the Magnus Ehrnrooth Foundation for their financial support during his visits to Åbo in 2003–2010.

where $\dot{x}$ stands for the time derivative of x. Here x, u and y take values in the Hilbert spaces $\mathcal{X}$, $\mathcal{U}$ and $\mathcal{Y}$, that are called the "state space", the "input space" and the "output space", respectively. For now the linear operators A, B, C, and D are assumed to be bounded, but we soon drop this restrictive assumption.

The system (5.1) can be viewed as an i/s/o representation of a s/s system by setting $\mathcal{W} := \left[\begin{smallmatrix}\mathcal{U}\\ \mathcal{Y}\end{smallmatrix}\right]$ and using the graph V of the operator $\left[\begin{smallmatrix}A & B\\ C & D\end{smallmatrix}\right]$:

$$\begin{aligned}\Sigma: \quad & \begin{bmatrix}\dot{x}(t)\\ x(t)\\ \left[\begin{smallmatrix}u(t)\\ y(t)\end{smallmatrix}\right]\end{bmatrix} \in V, \quad t \in \mathbb{R}^+, \ x(0) = x_0, \quad \text{where}\\ V &:= \left\{ \begin{bmatrix} z\\ x\\ \left[\begin{smallmatrix}u\\ y\end{smallmatrix}\right]\end{bmatrix} \in \begin{bmatrix}\mathcal{X}\\ \mathcal{X}\\ \mathcal{W}\end{bmatrix} \;\middle|\; \begin{matrix} z = Ax + Bu\\ y = Cx + Du\end{matrix} \right\}.\end{aligned} \tag{5.2}$$

This reformulation might seem trivial, but many concepts, such as that of a passive or a conservative system, are much simpler to formulate in the s/s framework than in the i/s/o counterpart; see Remark 5.17 below. Moreover, the input/output-free approach of the s/s theory permits the study of a physical system *as such* by looking at the geometric properties of V instead of merely studying *a particular representation* $\left[\begin{smallmatrix}A & B\\ C & D\end{smallmatrix}\right]$ of the system, cf. Remark 4.6 in Chapter 4.

We give the general definition of a s/s system, which does not a priori assume a representation (5.2), and we discuss well-posed i/s/o representations in Section 5.2. In Section 5.3 we study passive and conservative systems in more detail. The topic of Section 5.4 is frequency domain theory, and here we introduce the characteristic node bundle of a s/s system, which extends the notions of the γ-field and the Weyl family of a boundary relation. We make the precise connection between s/s systems and boundary relations in Section 5.5, where we also describe exactly how to transform a conservative s/s system into a boundary relation and vice versa.

5.2 Continuous-time state/signal systems

In this section we extend the ideas in Section 4.2 to more general generating subspaces V than those arising from either an i/s/o representation of the type (5.2) or from a boundary control s/s system.

General definitions We first introduce the s/s node and the s/s system that it induces.

Definition 5.1 Let $\mathcal{X}$ be a Hilbert space, let $\mathcal{W}$ be a Kreĭn space, and let V be a *closed* subspace of the node space $\mathfrak{K} = \mathcal{X} \times \mathcal{X} \times \mathcal{W}$ equipped with the indefinite inner product induced by the quadratic form

$$\left[\left[\begin{smallmatrix} z_1 \\ x_1 \\ w_1 \end{smallmatrix}\right], \left[\begin{smallmatrix} z_2 \\ x_2 \\ w_2 \end{smallmatrix}\right]\right]_{\mathfrak{K}} = -(z_1, x_2)_{\mathcal{X}} - (x_1, z_2)_{\mathcal{X}} + [w_1, w_2]_{\mathcal{W}}.$$

The pair $\left[\begin{smallmatrix} x \\ w \end{smallmatrix}\right]$ is a *classical trajectory* generated by V on $\mathbb{R}^+$ if $x \in C^1(\mathbb{R}^+; \mathcal{X})$, $w \in C(\mathbb{R}^+; \mathcal{W})$, and

$$\begin{bmatrix} \dot{x}(t) \\ x(t) \\ w(t) \end{bmatrix} \in V, \quad t \in \mathbb{R}^+, \tag{5.3}$$

where $\dot{x}$ stands for the time derivative of x (at $t = 0$ this is the right-sided derivative of x at zero). The closure of the set of classical trajectories on $\mathbb{R}^+$ in $\left[\begin{smallmatrix} C(\mathbb{R}^+;\mathcal{X}) \\ L^2_{\mathrm{loc}}(\mathbb{R}^+;\mathcal{W}) \end{smallmatrix}\right]$ is the set of *generalized trajectories* on $\mathbb{R}^+$ generated by V.

Moreover, $\Sigma = (V; \mathcal{X}, \mathcal{W})$ is a *state/signal node (s/s node)* if V has the following properties in addition to being closed:

1. The generating subspace V satisfies the condition
$$\left[\begin{smallmatrix} z \\ 0 \\ 0 \end{smallmatrix}\right] \in V \quad \Longrightarrow \quad z = 0. \tag{5.4}$$
2. For every $\left[\begin{smallmatrix} z_0 \\ x_0 \\ w_0 \end{smallmatrix}\right] \in V$ there exists a classical trajectory $\left[\begin{smallmatrix} x \\ w \end{smallmatrix}\right]$ of Σ on $\mathbb{R}^+ := [0, \infty)$ that satisfies $\begin{bmatrix} \dot{x}(0) \\ x(0) \\ w(0) \end{bmatrix} = \left[\begin{smallmatrix} z_0 \\ x_0 \\ w_0 \end{smallmatrix}\right]$.

By the *state/signal system (s/s system)* induced by a s/s node $(V; \mathcal{X}, \mathcal{W})$ we mean the s/s node itself together with its sets of classical and generalized trajectories on $\mathbb{R}^+$ generated by V.

It follows immediately from part 2 of Definition 5.1 that a space of classical trajectories determines its generating subspace uniquely through

$$V = \left\{ \begin{bmatrix} \dot{x}(0) \\ x(0) \\ w(0) \end{bmatrix} \,\middle|\, \left[\begin{smallmatrix} x \\ w \end{smallmatrix}\right] \text{ is a classical trajectory} \right\}.$$

It is less obvious, but still true, that a space of generalized trajectories determines its generating s/s node uniquely. This is because the space of generalized trajectories of a s/s node determines the space of classical trajectories uniquely. Indeed, a generalized trajectory $\left[\begin{smallmatrix} x \\ w \end{smallmatrix}\right]$ is in fact a classical trajectory if and only if $x \in C^1(\mathbb{R}^+; \mathcal{X})$ and $w \in C(\mathbb{R}^+; \mathcal{W})$. For proof, see [Kurula and Staffans, 2011, Cor. 3.2].

Example 5.2 Let $(L,\Gamma;\mathcal{X},\mathcal{U},\mathcal{Y})$ be a boundary control s/s node as given in Definition 4.3. This does in general not imply that $\Sigma := (V;\mathcal{X},\mathcal{W})$ is a s/s system, where V is given by

$$V = \left\{ \begin{bmatrix} Lx \\ x \\ \Gamma x \end{bmatrix} \,\middle|\, x \in \operatorname{dom}(L) \right\},$$

because V might not have property 2 of Definition 5.1. We prove in Example 5.7 below that Σ is indeed a s/s system when L and Γ arise from a boundary control i/s/o system of the type described in Definition 4.1.

The fact that the generating subspace V is independent of the time variable t means that the state/signal system is *time invariant.* Moreover, condition (5.4) means that V is the graph of some linear operator $G\colon \left[\begin{smallmatrix}\mathcal{X}\\ \mathcal{W}\end{smallmatrix}\right] \to \mathcal{X}$ with domain $\operatorname{dom}(G) \subset \left[\begin{smallmatrix}\mathcal{X}\\ \mathcal{W}\end{smallmatrix}\right]$, i.e., that

$$V = \left\{ \left[\begin{smallmatrix} z \\ x \\ w \end{smallmatrix}\right] \mid z = G\left[\begin{smallmatrix} x \\ w \end{smallmatrix}\right],\ \left[\begin{smallmatrix} x \\ w \end{smallmatrix}\right] \in \operatorname{dom}(G) \right\}.$$

The assumption that V is closed means that G is a closed operator. Now (5.3) can alternatively be written in the form

$$\begin{bmatrix} x(t) \\ w(t) \end{bmatrix} \in \operatorname{dom}(G) \quad \text{and} \quad \dot{x}(t) = G\begin{bmatrix} x(t) \\ w(t) \end{bmatrix}, \quad t \in \mathbb{R}^+, \tag{5.5}$$

and all classical trajectories generated by V satisfy this condition.

Example 5.3 If V is given by (5.2) then the operator G defined above is given by

$$\begin{aligned} G &= \begin{bmatrix} A & \begin{bmatrix} B & 0 \end{bmatrix} \end{bmatrix}\Big|_{\operatorname{dom}(G)} \quad \text{with} \\ \operatorname{dom}(G) &= \left\{ \begin{bmatrix} \left[\begin{smallmatrix} x \\ u \\ Cx+Du \end{smallmatrix}\right] \end{bmatrix} \,\middle|\, \left[\begin{smallmatrix} x \\ u \end{smallmatrix}\right] \in \left[\begin{smallmatrix} \mathcal{X} \\ \mathcal{U} \end{smallmatrix}\right] \right\}. \end{aligned} \tag{5.6}$$

Note, however, that the operators A, B, C, and D in (5.2), and therefore also in (5.6), by construction depend on a particular choice of i/o (input/output) decomposition $\mathcal{W} = \left[\begin{smallmatrix}\mathcal{U}\\ \mathcal{Y}\end{smallmatrix}\right]$, whereas (5.5) does not. In this sense (5.5) is a truly coordinate-free differential-equation representation of a s/s system.

Let us now go back to the general case. Condition 2 in Definition 5.1 means that there for all $\left[\begin{smallmatrix} x_0 \\ w_0 \end{smallmatrix}\right] \in \operatorname{dom}(G)$ exists a classical trajectory $\left[\begin{smallmatrix} x \\ w \end{smallmatrix}\right] \in \begin{bmatrix} C^1(\mathbb{R}^+;\mathcal{X}) \\ C(\mathbb{R}^+;\mathcal{W}) \end{bmatrix}$ such that $\begin{bmatrix} x(0) \\ w(0) \end{bmatrix} = \left[\begin{smallmatrix} x_0 \\ w_0 \end{smallmatrix}\right]$. From the condition $\begin{bmatrix} \dot{x}(0) \\ x(0) \\ w(0) \end{bmatrix} \in V$ we immediately obtain that this trajectory also satisfies $\dot{x}(0) = G\left[\begin{smallmatrix} x_0 \\ w_0 \end{smallmatrix}\right]$.

It is an interesting observation that we can represent an arbitrary s/s system by a closed operator G, and it helps to build intuition, but we shall not make any significant use of this operator in this exposition. For our present purposes it is more convenient to use (5.3).

As is well-known, an arbitrary Kreĭn space $\mathcal{W}$ can be interpreted as a Hilbert space consisting of the same vectors as $\mathcal{W}$. This is done by equipping $\mathcal{W}$ with an *admissible Hilbert-space inner product*; see Remark 5.10 below. An important consequence is that, from a topological point of view, every closed subspace of a Kreĭn space can be regarded as a Hilbert space, and we make frequent use of this.

Definition 5.4 A direct-sum decomposition $\mathcal{W} = \mathcal{U} \dotplus \mathcal{Y}$ of a Kreĭn space is *i/s/o well-posed* for the s/s system $\Sigma = (V; \mathcal{X}, \mathcal{W})$ if the following two conditions hold:

1. For every $x_0 \in \mathcal{X}$ and $u \in L^2_{\text{loc}}(\mathbb{R}^+; \mathcal{U})$ there exists a generalized trajectory $\left[\begin{smallmatrix} x \\ w \end{smallmatrix}\right] \in \left[\begin{smallmatrix} C(\mathbb{R}^+;\mathcal{X}) \\ L^2_{\text{loc}}(\mathbb{R}^+;\mathcal{W}) \end{smallmatrix}\right]$ of Σ on $\mathbb{R}^+$ with $x(0) = x_0$ and $P^{\mathcal{Y}}_{\mathcal{U}} w = u$.
2. There exists a positive nondecreasing function K on $\mathbb{R}^+$ such that every generalized trajectory $\left[\begin{smallmatrix} x \\ w \end{smallmatrix}\right] \in \left[\begin{smallmatrix} C(\mathbb{R}^+;\mathcal{X}) \\ L^2_{\text{loc}}(\mathbb{R}^+;\mathcal{W}) \end{smallmatrix}\right]$ of Σ on $\mathbb{R}^+$ satisfies

$$\begin{aligned} &\|x(t)\|^2_{\mathcal{X}} + \int_0^t \|P^{\mathcal{U}}_{\mathcal{Y}} w(s)\|^2_{\mathcal{W}}\, \mathrm{d}s \\ &\quad \le K(t) \left(\|x(0)\|^2_{\mathcal{X}} + \int_0^t \|P^{\mathcal{Y}}_{\mathcal{U}} w(s)\|^2_{\mathcal{W}}\, \mathrm{d}s \right), \quad t \in \mathbb{R}^+. \end{aligned} \tag{5.7}$$

Here $\|\cdot\|_{\mathcal{W}}$ stands for an arbitrary admissible norm in $\mathcal{W}$.

The s/s system $\Sigma = (V; \mathcal{X}, \mathcal{W})$ is *well-posed* if there exists at least one i/s/o well-posed decomposition $\mathcal{W} = \mathcal{U} \dotplus \mathcal{Y}$ of the signal space $\mathcal{W}$.

For more details on well-posed s/s systems, see [Kurula and Staffans, 2009]. In the next section we elaborate on the topic of representing state/signal systems by i/s/o systems.

Input/state/output representations The simplest example of a s/s system may be constructed by starting from a bounded classical linear i/s/o continuous-time system $\left[\begin{smallmatrix} A & B \\ C & D \end{smallmatrix}\right]$ as we did in the introduction.

However, applications often require that the operator $\left[\begin{smallmatrix} A & B \\ C & D \end{smallmatrix}\right]$ is unbounded. In the unbounded case the operator $\left[\begin{smallmatrix} A & B \\ C & D \end{smallmatrix}\right]$ in (5.1) can be replaced by an *i/s/o system node* operator $S = \left[\begin{smallmatrix} A\&B \\ C\&D \end{smallmatrix}\right]$. Here the top and bottom rows are denoted by $A\&B$ and $C\&D$ in order to indicate the connection to (5.1), but this notation is purely symbolic. In general it is possible to extend $A\&B$ into an operator $\begin{bmatrix} A_{-1} & B \end{bmatrix}$ which maps $\left[\begin{smallmatrix} \mathcal{X} \\ \mathcal{U} \end{smallmatrix}\right]$ continuously into a larger extrapolation space $\mathcal{X}_{-1}$. The operator A_{-1} is the continuous extension to $\mathcal{X}$ of the generator A of a C_0-semigroup on $\mathcal{X}$. Unfortunately, $C\&D$ does not split correspondingly. One can define an operator C, whose domain is a subspace of $\mathcal{X}$ containing the domain of A, but there is no uniquely defined operator corresponding to D in the general unbounded case. See [Staffans, 2005, Chapter 5] for details.

We now give a definition of an abstract system node, which is based on [Staffans, 2005, Lem. 4.7.7].

Definition 5.5 By an *i/s/o-system node* $(S;\mathcal{X},\mathcal{U},\mathcal{Y})$ we mean a triple of Hilbert spaces $\mathcal{X}$ (the state space), $\mathcal{U}$ (the input space), and $\mathcal{Y}$ (the output space), together with a linear operator

$$S = \left[\begin{smallmatrix} A\&B \\ C\&D \end{smallmatrix}\right] : \mathrm{dom}\,(S) \to \left[\begin{smallmatrix} \mathcal{X} \\ \mathcal{Y} \end{smallmatrix}\right], \quad \mathrm{dom}\,(S) \subset \left[\begin{smallmatrix} \mathcal{X} \\ \mathcal{U} \end{smallmatrix}\right],$$

with the following properties:

1. The operator $\left[\begin{smallmatrix} A\&B \\ C\&D \end{smallmatrix}\right] : \left[\begin{smallmatrix} \mathcal{X} \\ \mathcal{U} \end{smallmatrix}\right] \to \left[\begin{smallmatrix} \mathcal{X} \\ \mathcal{Y} \end{smallmatrix}\right]$ is closed as an operator mapping $\left[\begin{smallmatrix} \mathcal{X} \\ \mathcal{U} \end{smallmatrix}\right]$ into $\left[\begin{smallmatrix} \mathcal{X} \\ \mathcal{Y} \end{smallmatrix}\right]$ with domain $\mathrm{dom}\,(S)$.
2. The operator $A\&B : \left[\begin{smallmatrix} \mathcal{X} \\ \mathcal{U} \end{smallmatrix}\right] \to \mathcal{X}$ is closed with domain $\mathrm{dom}\,(S)$.
3. The *main operator* A of $\left[\begin{smallmatrix} A\&B \\ C\&D \end{smallmatrix}\right]$, defined by

$$Ax = A\&B \left[\begin{smallmatrix} x \\ 0 \end{smallmatrix}\right] \quad \text{on} \quad \mathrm{dom}\,(A) = \left\{x \in \mathcal{X} \mid \left[\begin{smallmatrix} x \\ 0 \end{smallmatrix}\right] \in \mathrm{dom}\,(S)\right\},$$

   generates a strongly continuous semigroup $t \mapsto \mathfrak{A}^t$ on $\mathcal{X}$.
4. For all $u \in \mathcal{U}$ there exists an $x \in \mathcal{X}$ such that $\left[\begin{smallmatrix} x \\ u \end{smallmatrix}\right] \in \mathrm{dom}\,(S)$.

The triple (u, x, y) is said to be a *classical i/s/o trajectory* of the i/s/o system node $(S;\mathcal{X},\mathcal{U},\mathcal{Y})$ if $u \in C(\mathbb{R}^+;\mathcal{U})$, $x \in C^1(\mathbb{R}^+;\mathcal{X})$, $y \in C(\mathbb{R}^+;\mathcal{Y})$, and

$$\begin{bmatrix} \dot{x}(t) \\ y(t) \end{bmatrix} = S \begin{bmatrix} x(t) \\ u(t) \end{bmatrix}, \quad t \in \mathbb{R}^+. \tag{5.8}$$

As we can see from the above definition, in the unbounded case (5.1) is replaced by (5.8). The i/s/o system (5.8) can again be interpreted as an i/s/o representation of a s/s system $\Sigma = (V;\mathcal{X},\mathcal{W})$ by taking $\mathcal{W} := \left[\begin{smallmatrix} \mathcal{U} \\ \mathcal{Y} \end{smallmatrix}\right]$ and defining

$$V := \left\{ \left[\begin{smallmatrix} z \\ x \\ \left[\begin{smallmatrix} u \\ y \end{smallmatrix}\right] \end{smallmatrix}\right] \subset \left[\begin{smallmatrix} \mathcal{X} \\ \mathcal{X} \\ \mathcal{W} \end{smallmatrix}\right] \;\middle|\; \left[\begin{smallmatrix} z \\ y \end{smallmatrix}\right] = S \left[\begin{smallmatrix} x \\ u \end{smallmatrix}\right],\ \left[\begin{smallmatrix} x \\ u \end{smallmatrix}\right] \in \mathrm{dom}\,(S) \right\}. \tag{5.9}$$

Here $\left[\begin{smallmatrix} \mathcal{U} \\ \mathcal{Y} \end{smallmatrix}\right]$ stands for the product of $\mathcal{U}$ and $\mathcal{Y}$ which can be turned into a Kreĭn space by equipping it with any of several indefinite inner products. A few important choices of inner product will be described later.

Recall that the component spaces $\mathcal{U}$ and $\mathcal{Y}$ of every direct-sum decomposition $\mathcal{W} = \mathcal{U} \dotplus \mathcal{Y}$ of a Kreĭn space can be interpreted as Hilbert spaces with inner products inherited from some admissible inner product in $\mathcal{W}$.

Remark In the sequel we call $\Sigma_{i/s/o} = (S; \mathcal{X}, \mathcal{U}, \mathcal{Y})$, where $\mathcal{U}$ and $\mathcal{Y}$ are arbitrary closed subspaces of some Kreĭn spaces, an i/s/o system node if $\Sigma_{i/s/o}$ is an i/s/o system node in the sense of Definition 5.5 with $\mathcal{U}$ and $\mathcal{Y}$ equipped with admissible inner products.

We define an i/s/o representation of a general s/s system $(V; \mathcal{X}, \mathcal{W})$.

Definition 5.6 Let $\Sigma = (V; \mathcal{X}, \mathcal{W})$ be a s/s system and let $\mathcal{W} = \mathcal{U} \dotplus \mathcal{Y}$ be an arbitrary direct-sum decomposition of the signal space.

Assume that V can be written on the form (5.9), where $\Sigma_{i/s/o} := (S; \mathcal{X}, \mathcal{U}, \mathcal{Y})$ is an i/s/o system node. Then we call $\Sigma_{i/s/o}$ the *i/s/o representation* of Σ corresponding to the i/o (input/output) decomposition $\mathcal{W} = \left[\begin{smallmatrix} \mathcal{U} \\ \{0\} \end{smallmatrix}\right] \dotplus \left[\begin{smallmatrix} \{0\} \\ \mathcal{Y} \end{smallmatrix}\right]$, and we call the i/o decomposition *system-node admissible*, or shortly just *admissible*.

The i/s/o representation $\Sigma_{i/s/o}$ is uniquely determined by the s/s system $\Sigma = (V; \mathcal{X}, \mathcal{W})$ and the decomposition $\mathcal{W} = \mathcal{U} \dotplus \mathcal{Y}$ (except for the fact that the norms and inner products in $\mathcal{U}$ and $\mathcal{Y}$ are determined only up to equivalence), since V is the graph of S in the sense of (5.9). In general, a s/s system Σ has several i/s/o representations, one induced by every admissible i/o decomposition of $\mathcal{W}$.

Example 5.7 Let $\Sigma_{i/s/o} = (L, \Gamma_0, \Gamma_1; \mathcal{X}, \mathcal{U}, \mathcal{Y})$ be an i/s/o boundary control system of the type in Definition 4.1. We let $\mathcal{W} := \left[\begin{smallmatrix} \mathcal{U} \\ \mathcal{Y} \end{smallmatrix}\right]$, equipped with an arbitrary Kreĭn-space inner product, e.g. the standard Hilbert-space inner product, and we define

$$V := \left\{ \begin{bmatrix} Lx \\ x \\ \begin{bmatrix} \Gamma_0 x \\ \Gamma_1 x \end{bmatrix} \end{bmatrix} \middle| \, x \in \operatorname{dom}(L) \right\}. \tag{5.10}$$

We now prove that $\Sigma = (V; \mathcal{X}, \mathcal{W})$ is a s/s system with admissible i/o decomposition $\mathcal{W} = \left[\begin{smallmatrix} \mathcal{U} \\ \{0\} \end{smallmatrix}\right] \dotplus \left[\begin{smallmatrix} \{0\} \\ \mathcal{Y} \end{smallmatrix}\right]$. We find the corresponding i/s/o representation $(S; \mathcal{X}, \mathcal{U}, \mathcal{Y})$ by identifying $\mathcal{U} = \left[\begin{smallmatrix} \mathcal{U} \\ \{0\} \end{smallmatrix}\right]$ and $\mathcal{Y} = \left[\begin{smallmatrix} \{0\} \\ \mathcal{Y} \end{smallmatrix}\right]$, and by noting that the map from $\left[\begin{smallmatrix} x \\ u \end{smallmatrix}\right]$ to $\left[\begin{smallmatrix} z \\ y \end{smallmatrix}\right]$, where $\begin{bmatrix} z \\ x \\ \left[\begin{smallmatrix} u \\ y \end{smallmatrix}\right] \end{bmatrix} \in V$, is given by

$$S = \left[\begin{smallmatrix} L \\ \Gamma_1 \end{smallmatrix}\right] \left[\begin{smallmatrix} 1 \\ \Gamma_0 \end{smallmatrix}\right]^{-1}, \quad \operatorname{dom}(S) = \left\{ \left[\begin{smallmatrix} x \\ \Gamma_0 x \end{smallmatrix}\right] \middle| \, x \in \operatorname{dom}(L) \right\}.$$

A detailed investigation of the connections between $\begin{bmatrix} L \\ \Gamma_0 \\ \Gamma_1 \end{bmatrix}$ and S can be found in Section 2 of [Malinen and Staffans, 2006]. In particular, S is a system node by [Malinen and Staffans, 2006, Th. 2.3], and from [Kurula, 2010a, Prop. 2.7] it then follows that Σ is a s/s node, as we claimed above. This shows how nicely boundary control can be incorporated into the general s/s framework.

If $\Sigma_{i/s/o} = (S; \mathcal{X}, \mathcal{U}, \mathcal{Y})$ is an i/s/o representation of $\Sigma = (V; \mathcal{X}, \mathcal{W})$, so that V is given by (5.9), then the well-posedness condition (5.7) is equivalent to the condition that every classical trajectory (u, x, y) of $\Sigma_{i/s/o}$ satisfies the following inequality for all $t \in \mathbb{R}^+$:

$$\|x(t)\|_{\mathcal{X}}^2 + \int_0^t \|y(s)\|_{\mathcal{Y}}^2 \,\mathrm{d}s \le K(t) \left(\|x(0)\|_{\mathcal{X}}^2 + \int_0^t \|u(s)\|_{\mathcal{U}}^2 \,\mathrm{d}s \right). \tag{5.11}$$

Definition 5.8 Input/state/output systems whose classical trajectories (u, x, y) satisfy (5.11) for a positive nondecreasing function K, which does not depend on the trajectory, are called *well-posed.*

It follows directly from Definitions 5.4 and 5.6 that if an i/o decomposition is both admissible and well-posed for a s/s system then the corresponding i/s/o representation is i/s/o-well-posed. In fact, every well-posed i/o decomposition is admissible by [Kurula and Staffans, 2009, Ths 4.9 and 6.4]. For more detailed information about i/s/o system nodes and well-posed i/s/o systems we refer the reader to [Staffans, 2005].

5.3 Passive and conservative state/signal systems

In this section we describe the concepts of passivity and conservativity within the state/signal system framework.

We need the notion of an *anti-Hilbert space.* A Kreĭn space $\mathcal{Y}$ is an anti-Hilbert space if $-\mathcal{Y}$, i.e., the space of all vectors in $\mathcal{Y}$ equipped with the inner product $-[\cdot, -]_{\mathcal{Y}}$, is a Hilbert space.

Definition 5.9 A direct-sum decomposition $\mathcal{W} = \mathcal{W}_1 \dotplus \mathcal{W}_2$ of a Kreĭn space is called:

1. *orthogonal* if every vector $w_1 \in \mathcal{W}_1$ is orthogonal to every vector in $w_2 \in \mathcal{W}_2$: $[w_1, w_2]_{\mathcal{W}} = 0$, and we write this as $\mathcal{W} = \mathcal{W}_1 \boxplus \mathcal{W}_2$.
2. *fundamental* if $\mathcal{W} = \mathcal{W}_1 \boxplus \mathcal{W}_2$, where $\mathcal{W}_1$ is a Hilbert space and $\mathcal{W}_2$ is an anti-Hilbert space in the inner product inherited from $\mathcal{W}$. In

this case we denote $\mathcal{W}_+ := \mathcal{W}_1$ and $\mathcal{W}_- := -\mathcal{W}_2$, so that $\mathcal{W}_+$ always is the Hilbert space component and $-\mathcal{W}_-$ is the anti-Hilbert space component, and we write $\mathcal{W} = \mathcal{W}_+ \boxplus -\mathcal{W}_-$.

3. *Lagrangian* if $\mathcal{W}_1$ and $\mathcal{W}_2$ are both Lagrangian: $\mathcal{W}_j = \mathcal{W}_j^{[\perp]}$. We introduce and explain the special notation $\mathcal{W} = \mathcal{W}_1 \stackrel{\Psi}{+} \mathcal{W}_2$ for Lagrangian decompositions in Definition 5.20 below.

When $\mathcal{W}$ is the signal space of a s/s system we typically use $\mathcal{W}_1$ as input space and $\mathcal{W}_2$ as output space in i/s/o representations. In this connection we do not always use the inner products inherited from $\mathcal{W}$ in $\mathcal{W}_1$ and $\mathcal{W}_2$. In a Lagrangian decomposition the subspaces do not even inherit a unique inner product from $\mathcal{W}$. In the case of orthogonal (and fundamental) decompositions we throughout take the input space to be $\mathcal{U} := \mathcal{W}_1$ with the inner product inherited from $\mathcal{W}$ and the output space to be $\mathcal{Y} := -\mathcal{W}_2$. Thus in the case of a fundamental decomposition both $\mathcal{U}$ and $\mathcal{Y}$ are Hilbert spaces.

If $\mathcal{W} = \mathcal{U} \boxplus \mathcal{Y}$, then in fact $\mathcal{Y} = \mathcal{U}^{[\perp]}$ and both $\mathcal{U}$ and $\mathcal{Y}$ are themselves Kreĭn spaces. Every Kreĭn space, which is neither a Hilbert space nor an anti-Hilbert space, has an uncountable number of fundamental decompositions $\mathcal{W} = \mathcal{W}_+ \boxplus -\mathcal{W}_-$. For every fundamental decomposition it holds that

$$
\begin{aligned}
[w_+, w_+]_{\mathcal{W}} &= (w_+, w_+)_{\mathcal{W}_+} > 0, \qquad & w_+ \in \mathcal{W}_+,\ w_+ \neq 0, \\
[w_-, w_-]_{\mathcal{W}} &= -(w_-, w_-)_{\mathcal{W}} < 0, \qquad & w_- \in -\mathcal{W}_-,\ w_- \neq 0.
\end{aligned}
$$

Remark 5.10 Let $\mathcal{W} = \mathcal{W}_+ \boxplus -\mathcal{W}_-$ be a fundamental decomposition of a Kreĭn space. Then $\mathcal{W}$ can be viewed as a Hilbert space with the inner product

$$
\begin{aligned}
(w_{1,+} + w_{1,-}, w_{2,+} + w_{2,-})_{\mathcal{W}} &= (w_{1,+}, w_{2,+})_{\mathcal{W}_+} + (w_{1,-}, w_{2,-})_{\mathcal{W}_-}, \\
& w_{1,+}, w_{2,+} \in \mathcal{W}_+, \quad w_{1,-}, w_{2,-} \in -\mathcal{W}_-.
\end{aligned}
$$

This inner product is called an *admissible inner product* and the norm induced by this inner product is called an *admissible norm.*

If $\mathcal{W}$ is either a Hilbert space or an anti-Hilbert space, then $\mathcal{W}$ has one unique fundamental decomposition, but in all other case $\mathcal{W}$ has infinitely many fundamental decompositions, and consequently also infinitely many admissible norms. However, all of these norms are equivalent.

Thus, once a fundamental decomposition $\mathcal{W} = \mathcal{W}_+ \boxplus -\mathcal{W}_-$ has been fixed, each $w \in \mathcal{W}$ has a unique decomposition $w = w_+ + w_-$ with

$w_\pm \in \mathcal{W}_\pm$, and

$$[w,w]_\mathcal{W} = (w_+,w_+)_{\mathcal{W}_+} - (w_-,w_-)_{\mathcal{W}_-} = \|w_+\|^2_{\mathcal{W}_+} - \|w_-\|^2_{\mathcal{W}_-}. \quad (5.12)$$

The dimensions of $\mathcal{W}_\pm$ do not depend on the choice of fundamental decomposition $\mathcal{W} = \mathcal{W}_+ \boxplus -\mathcal{W}_-$. They are called the positive and negative indices of $\mathcal{W}$ and are denoted by $\mathrm{ind}_\pm\mathcal{W}$. A Lagrangian decomposition of $\mathcal{W}$ exists if and only if $\mathrm{ind}_+\mathcal{W} = \mathrm{ind}_-\mathcal{W}$.

Passive s/s systems and scattering representations We first recall that a subspace V of a Kreĭn space $\mathfrak{K}$ is called *non-negative, non-positive*, or *neutral* if every vector $v \in V$ satisfies

$$[v,v]_\mathfrak{K} \geq 0, \quad [v,v]_\mathfrak{K} \leq 0, \quad \text{or} \quad [v,v]_\mathfrak{K} = 0,$$

respectively. A non-negative (or non-positive) subspace is called maximal non-negative (or maximal non-positive) if it is not strictly contained in any other non-negative (or non-positive) subspace. Such a subspace is automatically closed. A subspace V is *Lagrangian* if $V = V^{[\perp]}$, where $V^{[\perp]}$ is given by

$$V^{[\perp]} := \left\{k \in \mathfrak{K} \mid [k,k']_\mathfrak{K} = 0 \text{ for all } k' \in V\right\}.$$

Since many physical systems lack internal energy sources, it is natural to require the generating subspace V to be non-negative in the node space $\mathfrak{K} := \mathcal{X} \times \mathcal{X} \times \mathcal{W}$ which is equipped with the inner product

$$\left[\left[\begin{smallmatrix} z_1 \\ x_1 \\ w_1 \end{smallmatrix}\right], \left[\begin{smallmatrix} z_2 \\ x_2 \\ w_2 \end{smallmatrix}\right]\right]_\mathfrak{K} = -(z_1,x_2)_\mathcal{X} - (x_1,z_2)_\mathcal{X} + [w_1,w_2]_\mathcal{W}; \quad (5.13)$$

cf. Definition 5.1.

The node space $\mathfrak{K}$ is a Kreĭn space with the fundamental decomposition $\mathfrak{K} = \mathfrak{K}_+ \boxplus -\mathfrak{K}_-$, where

$$\mathfrak{K}_\pm = \left\{\left[\begin{smallmatrix} \mp x \\ x \\ w_\pm \end{smallmatrix}\right] \middle|\, x \in \mathcal{X},\ w_\pm \in \mathcal{W}_\pm\right\}$$

and $\mathcal{W} = \mathcal{W}_+ \boxplus -\mathcal{W}_-$ is an arbitrary fundamental decomposition of $\mathcal{W}$. As an immediate consequence, we have that $\mathrm{ind}_\pm\mathfrak{K} = \dim\mathcal{X} + \mathrm{ind}_\pm\mathcal{W}$.

Just as in the case of boundary control, it is immediate that all classical trajectories on $\mathbb{R}^+$ generated by a non-negative V satisfy

$$\frac{\mathrm{d}}{\mathrm{d}t}\|x(t)\|^2_\mathcal{X} \leq [w(t),w(t)]_\mathcal{W}, \quad t \in \mathbb{R}^+, \quad \text{and} \quad (5.14)$$

$$\|x(t)\|^2_\mathcal{X} - \|x(s)\|^2_\mathcal{X} \leq \int_s^t [w(v),w(v)]_\mathcal{W}\,\mathrm{d}v, \quad s,t \in \mathbb{R}^+,\ s \leq t, \quad (5.15)$$

where the second inequality holds also for the generalized trajectories.

However, non-negativity of V does not yet imply that $(V;\mathcal{X},\mathcal{W})$ is a s/s node. The situation is analogous to the situation in semigroup theory: the generator of a contraction semigroup is not just dissipative, but even *maximal* dissipative; see the Lumer-Phillips Theorem [Staffans, 2005, Th. 3.4.8].

Definition 5.11 A s/s system $\Sigma = (V;\mathcal{X};\mathcal{W})$ is said to be *passive* if V is a maximal non-negative subspace of the node space $\mathfrak{K}$, i.e., with respect to the inner product (5.13). The system Σ is *conservative* if $V = V^{[\perp]}$.

I/s/o representations corresponding to fundamental decompositions of the signal space of a passive s/s system are exceptionally well-behaved, and we now investigate these in more detail.

Definition 5.12 Let $\Sigma = (V;\mathcal{X},\mathcal{W})$ be a s/s system and let $\Sigma_{i/s/o} = (S;\mathcal{X},\mathcal{U},\mathcal{Y})$ be an i/s/o representation of Σ in the sense of Definition 5.6. Then $\Sigma_{i/s/o}$ is called a *scattering representation* of Σ if $\mathcal{U} = \mathcal{W}_+$ and $\mathcal{Y} = \mathcal{W}_-$, where $\mathcal{W} = \mathcal{W}_+ \boxplus -\mathcal{W}_-$ is a fundamental decomposition.

Let $\mathcal{W} = \mathcal{W}_+ \boxplus -\mathcal{W}_-$ be a fundamental decomposition, and set $\mathcal{U} := \mathcal{W}_+$ and $\mathcal{Y} := \mathcal{W}_-$. Combining (5.15) and (5.12) we obtain that every classical trajectory of a passive s/s system satisfies (with $u(v) \in \mathcal{U}$ and $y(v) \in \mathcal{Y}$):

$$\|x(t)\|_{\mathcal{X}}^2 - \|x(s)\|_{\mathcal{X}}^2 \le \int_s^t \|u(v)\|_{\mathcal{U}}^2 - \|y(v)\|_{\mathcal{Y}}^2 \, dv \qquad (5.16)$$

for every $s,t \in \mathbb{R}^+$ such that $s \le t$. This is the well-known *scattering-passivity inequality.* Note that (5.16) implies (5.7) with $K(t) = 1, t \in \mathbb{R}^+$.

The first part of the following further development of the above ideas was proved as [Kurula, 2010a, Th. 4.5 and Prop. 5.8]. The second part follows from the first part and Definition 5.4.

Theorem 5.13 *Assume that V is a maximal non-negative subspace of $\mathfrak{K}$ satisfying (5.4): $\left[\begin{smallmatrix} z \\ 0 \\ 0 \end{smallmatrix}\right] \in V$ only if $z = 0$. Then $(V;\mathcal{X},\mathcal{W})$ is a passive well-posed s/s node for which every fundamental decomposition $\mathcal{W} = \mathcal{W}_+ \boxplus -\mathcal{W}_-$ is (admissible and) well-posed and the corresponding scattering representation with input space $\mathcal{U} = \mathcal{W}_+$ and output space $\mathcal{Y} = \mathcal{W}_-$ is well-posed.*

In particular, for every $x_0 \in \mathcal{X}$ and $u \in L^2_{\mathrm{loc}}(\mathbb{R}^+;\mathcal{U})$ there exists *a unique generalized trajectory* $\left[\begin{smallmatrix} x \\ w \end{smallmatrix}\right] \in \left[\begin{smallmatrix} C(\mathbb{R}^+;\mathcal{X}) \\ L^2_{\mathrm{loc}}(\mathbb{R}^+;\mathcal{W}) \end{smallmatrix}\right]$ *of* Σ *on* $\mathbb{R}^+$ *with* $x(0) = x_0$ *and* $P^{\mathcal{Y}}_{\mathcal{U}} w = u$.

Thus a triple $(V; \mathcal{X}, \mathcal{W})$ is a passive s/s system if and only if V is a maximal non-negative subspace of $\mathfrak{K}$ with the property (5.4).

Let $\Sigma = (V; \mathcal{X}, \mathcal{W})$ be a passive s/s system. Each different fundamental decomposition $\mathcal{W} = \mathcal{W}_+ \boxplus -\mathcal{W}_-$ gives rise to a different scattering representation, so there always exist uncountably many scattering representations of a given *passive* s/s system (except for the degenerate cases where the energy exchange through the external signal is unidirectional).

Now suppose that $\Sigma = (V; \mathcal{X}, \mathcal{W})$ has the property that V is *maximal non-positive*. Then (5.14) is replaced by

$$\frac{\mathrm{d}}{\mathrm{d}t}\|x(t)\|^2_{\mathcal{X}} \geq [w(t), w(t)]_{\mathcal{W}}, \quad t \in \mathbb{R}^+,$$

and an analogue of Theorem 5.13 can be formulated for Σ, which says that Σ is *well-posed in the backward time direction*, and that every fundamental decomposition $\mathcal{W} = \mathcal{W}_+ \boxplus -\mathcal{W}_-$ yields a well-posed i/s/o representation if we take the *output* space to be $\mathcal{Y} = \mathcal{W}_+$ and the *input* space to be $\mathcal{U} = \mathcal{W}_-$.

Definition 5.14 We call a triple $\Sigma = (V; \mathcal{X}, \mathcal{W})$ with a maximal non-positive generating subspace V satisfying (5.4) an *anti-passive s/s node* (in the backward time direction), i.e., it has properties 1 and 2 in Definition 5.1 with $\mathbb{R}^+$ replaced by $\mathbb{R}^-$.

It is well-known that $V = V^{[\perp]}$ if and only if V is both maximal non-negative and maximal non-positive. A conservative s/s system is thus one that is at the same time both passive and anti-passive. We conclude that conservative s/s systems are *i/s/o well-posed both in the forward and in the backward time directions.* This does *not* imply that the signal space $\mathcal{W}$ has a direct sum decomposition $\mathcal{W} = \mathcal{U} \dotplus \mathcal{Y}$ which is i/s/o well-posed both in the forward and backward time direction. We provided a conservative system for which no decomposition of the signal space is admissible both in the forward and backward time directions in Example 4.9. Indeed, when that system is solved in forward time, $x(0) \in \mathbb{C}$ is the unique output and there is no input, and when the system is solved in backward time, $x(0)$ is the unique input and there is no output. See [Kurula, 2010a, Th. 4.11] for more details on conservative s/s systems.

Remark 5.15 The *maximal* non-negativity of V in a passive s/s system

$\Sigma = (V; \mathcal{X}, \mathcal{W})$ intuitively means that it has "enough" trajectories to make sense as a system.

More precisely, the maximal non-negativity of V implies that the *state/signal dual* $(V^{[\perp]}; \mathcal{X}, \mathcal{W})$ of Σ is anti-passive and thus very well-structured. If V is replaced by a smaller space then the dual becomes larger, and in particular, if $V = \{0\}$, then $V^{[\perp]} = \mathfrak{K}$ which has no meaning as a s/s system at all.

We also note that a s/s system is conservative if and only if it coincides with its own s/s dual.

See [Kurula, 2010a, Sec. 3] for more details on the dual s/s system and its i/s/o representations.

We now return to i/s/o representations of passive s/s systems. It is well known that the adjoint S^* of an i/s/o system node operator S is also an i/s/o system node operator which represents the adjoint system; see [Staffans, 2005, Lemma 6.2.14]. If $\mathcal{U}$ is the input space and $\mathcal{Y}$ is the output space of S then $\mathcal{Y}$ is the input space and $\mathcal{U}$ is the output space of S^*.

Definition 5.16 An i/s/o system node $\Sigma_{i/s/o} = (S; \mathcal{X}, \mathcal{U}, \mathcal{Y})$ is *scattering passive* if all its classical trajectories (u, x, y) satisfy (5.16) for every $s, t \in \mathbb{R}^+$ such that $s \le t$.

The i/s/o system node $\Sigma_{i/s/o}$ is *scattering conservative* if all classical trajectories (u, x, y) and (y^d, x^d, u^d) on $\mathbb{R}^+$ of S and S^*, respectively, satisfy

$$\|x(t)\|^2_{\mathcal{X}} - \|x(s)\|^2_{\mathcal{X}} = \int_s^t \|u(v)\|^2_{\mathcal{U}} - \|y(v)\|^2_{\mathcal{Y}} \,\mathrm{d}v \quad \text{and}$$

$$\|x^d(t)\|^2_{\mathcal{X}} - \|x^d(s)\|^2_{\mathcal{X}} = \int_s^t \|y^d(v)\|^2_{\mathcal{Y}} - \|u^d(v)\|^2_{\mathcal{U}} \,\mathrm{d}v$$

for every $s, t \in \mathbb{R}^+$ such that $s \le t$. (Compare this to (5.16).)

Remark 5.17 Recall that a triple $(V; \mathcal{X}, \mathcal{W})$ is a passive s/s system if and only if V is a maximal non-negative subspace of $\mathfrak{K}$ with the property (5.4). Comparing this to Definitions 5.5 and 5.16, which are necessary for defining only a special class of passive i/s/o systems, we see that the s/s definition is both more general and considerably simpler. Moreover, the definitions of conservative i/s/o systems are even more complicated, since we need to formulate conditions on the *dual system* but a conservative s/s system is very elegantly characterized by the properties $V = V^{[\perp]}$ and (5.4).

The following proposition was proved in [Kurula, 2010a, Prop. 5.6].

Proposition 5.18 *All scattering representations of a* passive (conservative) *s/s system are scattering passive (conservative) i/s/o systems.*

Conversely, let $\Sigma_{i/s/o} = (S; \mathcal{X}, \mathcal{U}, \mathcal{Y})$ be a scattering passive (conservative) i/s/o system node, so that $\mathcal{U}$ and $\mathcal{Y}$ are both Hilbert spaces. Define $\mathcal{W} := \left[\begin{smallmatrix} \mathcal{U} \\ -\mathcal{Y} \end{smallmatrix}\right]$ with inner product $[[\begin{smallmatrix} u_1 \\ y_1 \end{smallmatrix}], [\begin{smallmatrix} u_2 \\ y_2 \end{smallmatrix}]] := (u_1, u_2)_{\mathcal{U}} - (y_1, y_2)_{\mathcal{Y}}$. Then $\mathcal{W}$ is a Kreĭn space with fundamental decomposition $\Sigma = \left[\begin{smallmatrix} \mathcal{U} \\ \{0\} \end{smallmatrix}\right] \boxplus \left[\begin{smallmatrix} \{0\} \\ -\mathcal{Y} \end{smallmatrix}\right]$. Moreover, $(V; \mathcal{X}, \mathcal{W})$ with V given by (5.9), is the unique passive (conservative) s/s system whose scattering representation induced by the above fundamental decomposition is $\Sigma_{i/s/o}$.

Scattering passive i/s/o systems are discussed in, e.g., [Arov and Nudelman, 1996] and [Staffans, 2005, Chapter 11]. The connection between different well-posed i/s/o representations of a s/s system, and thus in particular, between different scattering representations of a passive s/s system, is described in [Kurula and Staffans, 2009, Section 4].

Impedance and transmission representations In the context of boundary relations, another type of i/s/o representation is in fact more important than the scattering representation, namely the *impedance* representation.

Definition 5.19 An *impedance representation* of a s/s system $\Sigma = (V; \mathcal{X}, \mathcal{W})$ is an i/s/o representation corresponding to a system-node admissible *Lagrangian* decomposition $\mathcal{W} = \mathcal{U} \dotplus \mathcal{Y}$ of the signal space $\mathcal{W}$.

Since not all Kreĭn spaces have Lagrangian decompositions, there exist passive s/s systems which have no impedance representations. However, assume that $\mathcal{W} = \mathcal{U} \dotplus \mathcal{Y}$ indeed is a Lagrangian decomposition of $\mathcal{W}$, i.e., that $\mathcal{U}$ and $\mathcal{Y}$ are both Lagrangian subspaces of $\mathcal{W}$. By [Arov and Staffans, 2007a, Lemma 2.3] there exist admissible Hilbert-space inner products on $\mathcal{U}$ and $\mathcal{Y}$ and a unitary operator $\Psi : \mathcal{Y} \to \mathcal{U}$, such that the Kreĭn-space inner product on $\mathcal{W}$ is given by the following (where $u_1, u_2 \in \mathcal{U},\ y_1, y_2 \in \mathcal{Y}$):

$$[y_1 + u_1, y_2 + u_2]_{\mathcal{W}} = (\Psi y_1, u_2)_{\mathcal{U}} + (u_1, \Psi y_2)_{\mathcal{U}}. \tag{5.17}$$

Definition 5.20 By writing $\mathcal{W} = \mathcal{U} \overset{\Psi}{\dotplus} \mathcal{Y}$ we mean that the Kreĭn space $\mathcal{W}$ is decomposed into the direct sum of $\mathcal{U}$ and $\mathcal{Y}$, and that the inner product $[\cdot, \cdot]_{\mathcal{W}}$ in $\mathcal{W}$ may be written in the form (5.17), where Ψ is a unitary operator from $\mathcal{Y}$ to $\mathcal{U}$.

It follows from (5.17) that both $\mathcal{U}$ and $\mathcal{Y}$ are Lagrangian subspaces of $\mathcal{W}$, i.e., that the decomposition in Definition 5.20 is always Lagrangian. See Section 2 of [Arov and Staffans, 2007a] for more details on Lagrangian decompositions of $\mathcal{W}$. If $\mathcal{W} = \mathcal{U} \overset{\Psi}{\dotplus} \mathcal{Y}$, then the inequality (5.14) becomes

$$\frac{\mathrm{d}}{\mathrm{d}t}\|x(t)\|^2_{\mathcal{X}} \leq 2\mathrm{Re}\,(u(t), \Psi y(t))_{\mathcal{U}}, \quad t \in \mathbb{R}^+.$$

Moreover, the inequality (5.15) takes the form

$$\|x(t)\|^2_{\mathcal{X}} - \|x(s)\|^2_{\mathcal{X}} \leq 2\mathrm{Re} \int_s^t (u(v), \Psi y(v))_{\mathcal{U}}\,\mathrm{d}v, \;\; s,t \in \mathbb{R}^+,\; t \geq s, \tag{5.18}$$

and this is the *impedance-passivity inequality.*

Definition 5.21 An i/s/o system node $\Sigma_{i/s/o} = (S; \mathcal{X}, \mathcal{U}, \mathcal{Y})$ is *impedance passive* if all its classical trajectories (u, x, y) satisfy (5.18) for some unitary operator $\Psi : \mathcal{Y} \to \mathcal{U}$. (One commonly has $\mathcal{Y} = \mathcal{U}$ and $\Psi = 1_{\mathcal{U}}$.)

The i/s/o system node $\Sigma_{i/s/o}$ is *impedance conservative* if all classical trajectories (u, x, y) and (y, x, u) on $\mathbb{R}^+$ of S and S^*, respectively, satisfy (5.18) with equality instead of inequality.

An analogue of Proposition 5.18 relating impedance representations and impedance passive i/s/o systems can be formulated simply by replacing "scattering" by "impedance" and the fundamental decomposition $\mathcal{W} = \left[\begin{smallmatrix} \mathcal{U} \\ \{0\} \end{smallmatrix}\right] \boxplus \left[\begin{smallmatrix} \{0\} \\ -\mathcal{Y} \end{smallmatrix}\right]$ by a Lagrangian decomposition $\mathcal{W} = \mathcal{U} \overset{\Psi}{\dotplus} \mathcal{Y}$, where $\Psi : \mathcal{Y} \to \mathcal{U}$ is an arbitrary unitary operator.

Theorem 5.22 *The following claims are true for an impedance conservative i/s/o system node* $\Sigma_{i/s/o} = (S; \mathcal{X}, \mathcal{U}, \mathcal{Y})$.

1. *The main operator* A *of* S*, see item 3 of Definition 5.5, is skew-adjoint and* A *generates a unitary group* $t \mapsto \mathfrak{A}^t$, $t \in \mathbb{R}$, *on* $\mathcal{X}$.
2. *For every* $u \in W^{2,1}_{\mathrm{loc}}(\mathbb{R}^+; \mathcal{U})$ *and initial state* $x_0 \in \mathcal{X}$*, such that* $\left[\begin{smallmatrix} x_0 \\ u(0) \end{smallmatrix}\right] \in \mathrm{dom}\,(S)$*, the system*

$$\begin{bmatrix} \dot{x}(t) \\ y(t) \end{bmatrix} = S \begin{bmatrix} x(t) \\ u(t) \end{bmatrix}, \quad t \in \mathbb{R}^+, \tag{5.19}$$

has a unique classical *trajectory* (u, x, y) *with* $x(0) = x_0$*;* $W^{2,1}_{\mathrm{loc}}(\mathbb{R}^+; \mathcal{U})$ *denotes the space of functions that together with their first and second distribution derivatives lie in* $L^1_{\mathrm{loc}}(\mathbb{R}^+; \mathcal{U})$.

3. *For every $x_0 \in \mathcal{X}$ there exists a* generalized *trajectory (u, x, y) of $\Sigma_{i/s/o}$, such that $x(0) = x_0$. This trajectory is uniquely determined by the initial state x_0 and the input u.*
4. *The system* (5.19) *can also be solved in backwards time, i.e., for $t \in \mathbb{R}^- = (-\infty, 0]$, with the initial state $x_0 \in \mathcal{X}$ given at $t = 0$. In particular, every trajectory (u, x, y) of $\Sigma_{i/s/o}$ with $x(0) = x_0$ and $u = 0$ satisfies $x(t) = \mathfrak{A}^t x_0$ for all $t \in \mathbb{R}$, and if $x_0 \in \mathrm{dom}\,(A)$, then this trajectory is classical. This trajectory is the unique trajectory of $\Sigma_{i/s/o}$ with the given state x_0 at time* 0 *and input $u(t) = 0$, $t \in \mathbb{R}$.*

Proof One can verify that the i/s/o system node $\Sigma_{i/s/o}$ is impedance conservative with some given Ψ if and only if V defined in (5.9) is a Lagrangian subspace of $\mathfrak{K}$: $V = V^{[\perp]}$, where $\mathcal{W} := \mathcal{U} \overset{\Psi}{\dotplus} \mathcal{Y}$.

1. According to [Staffans, 2002a, Th. 4.7(4)] we have $A = -A^*$, and thus A generates a unitary group by Stone's theorem [Pazy, 1983, Th. 10.8].

2. This follows from [Staffans, 2005, Lem. 4.7.8].

3. The s/s system induced by an impedance conservative i/s/o system node is conservative, and therefore in particular passive. By Theorem 5.13, $(V; \mathcal{X}, \mathcal{W})$ is well-posed, and according to condition 1 of Definition 5.4, every $x_0 \in \mathcal{X}$ can be taken as the initial state of some generalized trajectory. Moreover, if $x_0 = 0$ and $u(t) = 0$ for all $t \in \mathbb{R}^+$, then $x(t) = 0$ and $y(t) = 0$ for all ≥ 0 by claim 2.

This is a consequence of Remark 5.15, [Staffans, 2005, Th. 3.8.2], and the previous claims in this theorem. □

There are several ways to add dynamics to a boundary relation. Using Theorem 5.22 is one way, as we will show at the end of Section 5.5.

Remark 5.23 Note that the input u in Theorem 5.22 corresponds to a system-node admissible Lagrangian decomposition of the signal space of a conservative s/s system, and that this decomposition need not be well-posed in general. Indeed, the corresponding impedance representation $(S; \mathcal{X}, \mathcal{U}, \mathcal{Y})$ need not be well-posed, i.e., the i/s/o system node S in (5.9) need not satisfy (5.11).

If the decomposition $\mathcal{W} = \mathcal{U} \dotplus \mathcal{Y}$ happens to be well-posed then we have from Definition 5.4 that the set

$$\{u \mid (u, x, y) \text{ is a generalized trajectory of } \Sigma_{i/s/o} \text{ with } x(0) = x_0\}$$

equals all of $L^2_{\mathrm{loc}}(\mathbb{R}^+; \mathcal{U})$ for all $x_0 \in \mathcal{X}$, but in the ill-posed case we can make no such conclusion.

On the contrary, every scattering representation of a passive s/s system is well-posed, cf. Theorem 5.13. This explains why the scattering formalism is sometimes useful for solving technical difficulties in the boundary relations theory, cf. [Behrndt et al., 2009], where this technique is used extensively.

There exist conservative s/s systems for which no Lagrangian decompositions are system-node admissible, see [Arov and Staffans, 2007a, Ex. 5.13], which can also be formulated for continuous time with trivial modifications. It follows from Theorem 5.35 and Proposition 5.36 below that the following two conditions together are sufficient and necessary for a Lagrangian decomposition $\mathcal{W} = \mathcal{U} \dot{+} \mathcal{Y}$ to be admissible for a conservative s/s system $(V; \mathcal{X}, \mathcal{W})$:

1. $\left[\begin{smallmatrix} z \\ 0 \\ \left[\begin{smallmatrix} 0 \\ y \end{smallmatrix}\right] \end{smallmatrix}\right] \in V \implies \left[\begin{smallmatrix} z \\ y \end{smallmatrix}\right] = 0.$

2. for each $u \in \mathcal{U}$ there exist $z, x \in \mathcal{X}$ and $y \in \mathcal{Y}$ such that $\left[\begin{smallmatrix} z \\ x \\ \left[\begin{smallmatrix} u \\ y \end{smallmatrix}\right] \end{smallmatrix}\right] \in V$.

Well-posed impedance passive i/s/o systems were studied in [Staffans, 2002a]; the ill-posed impedance case was considered in [Staffans, 2002b].

Remark The energy inequalities (5.16) and (5.18) correspond to fundamental and Lagrangian decompositions of $\mathcal{W}$, respectively, but the *property of passivity* is characterized by the maximal non-negativity of V. Thus passivity is a *state/signal characteristic*, i.e., passivity *does not depend on any particular decomposition of the signal space into an input space and an output space.*

A third, fairly common, type of representation is the *transmission representation.*

Definition 5.24 An i/s/o representation of a passive s/s system corresponding to an admissible *orthogonal* decomposition $\mathcal{W} = \mathcal{W}_1 \boxplus \mathcal{W}_2$ of the signal space, with input space $\mathcal{U} = \mathcal{W}_1$ and output space $\mathcal{Y} = -\mathcal{W}_2$ is called a *transmission representation.*

Every scattering representation can also be interpreted as a transmission representation.

Example 5.25 We continue the transmission line example in Section 4.4. As we saw there, this is a conservative boundary control system. The choice of input and output maps Γ_0, Γ_1 in (4.13) corresponds to the Lagrangian decomposition $\mathcal{W} = \begin{bmatrix} \mathbb{C} \\ \{0\} \\ \mathbb{C} \\ \{0\} \end{bmatrix} \overset{\Psi}{\dotplus} \begin{bmatrix} \{0\} \\ \mathbb{C} \\ \{0\} \\ \mathbb{C} \end{bmatrix}$, where $\Psi = \left[\begin{smallmatrix} 1 & 0 \\ 0 & 1 \end{smallmatrix}\right]$, and according to Example 5.7, this decomposition is admissible. The choice $\widetilde{\Gamma}_0, \widetilde{\Gamma}_1$ made in (4.14) corresponds to the fundamental decomposition $\mathcal{W} = \left\{ \left[\begin{smallmatrix} a \\ a \\ b \\ b \end{smallmatrix}\right] \middle| \, a, b \in \mathbb{C} \right\} \boxplus \left\{ \left[\begin{smallmatrix} -a \\ a \\ -b \\ b \end{smallmatrix}\right] \middle| \, a, b \in \mathbb{C} \right\}$, and according to Theorem 5.13, also this decomposition is admissible. The choice $\widehat{\Gamma}_0, \widehat{\Gamma}_1$ in (4.15) corresponds to the orthogonal (but non-fundamental) decomposition $\mathcal{W} = \begin{bmatrix} \mathbb{C} \\ \mathbb{C} \\ \{0\} \\ \{0\} \end{bmatrix} \boxplus \begin{bmatrix} \{0\} \\ \{0\} \\ \mathbb{C} \\ \mathbb{C} \end{bmatrix}$, which is *not* admissible.

The non-admissible orthogonal and Lagrangian decompositions which do not yield i/s/o representations can be treated using continuous-time analogues of the *affine representations* developed in [Arov and Staffans, 2007b].

5.4 The frequency domain characteristics of a state/signal system

The input-state/state-output resolvent matrix Suppose that x, $\dot{x}$, y, and u are all Laplace transformable, with the Laplace transforms converging in $\mathbb{C}^+ = \{\lambda \in \mathbb{C} \mid \operatorname{Re} \lambda > 0\}$, the right half-plane. Take Laplace transforms in the i/s/o equation

$$\Sigma_{i/s/o}: \quad \begin{bmatrix} \dot{x}(t) \\ y(t) \end{bmatrix} = \begin{bmatrix} A & B \\ C & D \end{bmatrix} \begin{bmatrix} x(t) \\ u(t) \end{bmatrix}, \quad t \in \mathbb{R}^+, \quad x(0) = x_0,$$

in order to get

$$\begin{bmatrix} \lambda \hat{x}(\lambda) - x_0 \\ \hat{y}(\lambda) \end{bmatrix} = \begin{bmatrix} A & B \\ C & D \end{bmatrix} \begin{bmatrix} \hat{x}(\lambda) \\ \hat{u}(\lambda) \end{bmatrix}, \quad \lambda \in \mathbb{C}^+. \tag{5.20}$$

At least in the case where $\left[\begin{smallmatrix} A & B \\ C & D \end{smallmatrix}\right]$ is a bounded operator in a scattering representation of a passive s/s system it is possible to solve $\left[\begin{smallmatrix} \hat{x}(\lambda) \\ \hat{y}(\lambda) \end{smallmatrix}\right]$ in terms of $\left[\begin{smallmatrix} x_0 \\ \hat{u}(\lambda) \end{smallmatrix}\right]$ from the identity (5.20) for all $\lambda \in \mathbb{C}^+$. The map $\left[\begin{smallmatrix} x_0 \\ \hat{u}(\lambda) \end{smallmatrix}\right] \mapsto \left[\begin{smallmatrix} \hat{x}(\lambda) \\ \hat{y}(\lambda) \end{smallmatrix}\right]$ turns out to be a bounded linear operator that we denote by $\widehat{\mathfrak{S}}(\lambda) = \left[\begin{smallmatrix} \widehat{\mathfrak{A}}(\lambda) & \widehat{\mathfrak{B}}(\lambda) \\ \widehat{\mathfrak{C}}(\lambda) & \widehat{\mathfrak{D}}(\lambda) \end{smallmatrix}\right]$. More explicitly,

$$\begin{aligned} \begin{bmatrix} \hat{x}(\lambda) \\ \hat{y}(\lambda) \end{bmatrix} &= \begin{bmatrix} \widehat{\mathfrak{A}}(\lambda) & \widehat{\mathfrak{B}}(\lambda) \\ \widehat{\mathfrak{C}}(\lambda) & \widehat{\mathfrak{D}}(\lambda) \end{bmatrix} \begin{bmatrix} x_0 \\ \hat{u}(\lambda) \end{bmatrix}, \quad \lambda \in \mathbb{C}^+, \quad \text{where} \\ \begin{bmatrix} \widehat{\mathfrak{A}}(\lambda) & \widehat{\mathfrak{B}}(\lambda) \\ \widehat{\mathfrak{C}}(\lambda) & \widehat{\mathfrak{D}}(\lambda) \end{bmatrix} &= \begin{bmatrix} (\lambda - A)^{-1} & (\lambda - A)^{-1}B \\ C(\lambda - A)^{-1} & C(\lambda - A)^{-1}B + D \end{bmatrix}. \end{aligned} \tag{5.21}$$

Definition 5.26 The operator $\widehat{\mathfrak{S}} := \left[\begin{smallmatrix} \widehat{\mathfrak{A}} & \widehat{\mathfrak{B}} \\ \widehat{\mathfrak{C}} & \widehat{\mathfrak{D}} \end{smallmatrix}\right]$ is called the *is/so (input-state/state-output) resolvent matrix of* $\Sigma_{i/s/o}$. The different components of this resolvent matrix are named as follows:

1. $\widehat{\mathfrak{A}}$ is the *state/state* resolvent function,
2. $\widehat{\mathfrak{B}}$ is the *input/state* resolvent function,
3. $\widehat{\mathfrak{C}}$ is the *state/output* resolvent function, and
4. $\widehat{\mathfrak{D}}$ is the *input/output* resolvent function.

Of course, the state/state resolvent function is the familiar *resolvent* of the main operator A. The other components of $\widehat{\mathfrak{S}}$ has different names in different parts of the literature, and we make the connections to the corresponding notions in the theory of boundary relations in Theorem 5.34 below. In the i/s/o tradition the input/output resolvent function is usually called the *transfer function* of $\Sigma_{i/s/o}$.

Remark A significant part of formula (5.21) remains valid with the appropriate interpretation of the operators A, B, and C if we replace $\left[\begin{smallmatrix} A & B \\ C & D \end{smallmatrix}\right]$ by a *system node* operator S of the type described in Definition 5.5; see [Staffans, 2002b, Sec. 2].

In the case of a scattering passive i/s/o system $\Sigma_{i/s/o}$, the function $\widehat{\mathfrak{D}}$ is often called the *scattering matrix* of $\Sigma_{i/s/o}$. If in addition, $\Sigma_{i/s/o}$ is *conservative*, then $\widehat{\mathfrak{D}}$ is also called the *characteristic function* of the corresponding i/s/o system node, or of its main operator A; see Definition 5.5. In this case A is a maximal dissipative operator in $\mathcal{X}$.

In the case where $\Sigma_{i/s/o}$ is transmission passive, $\widehat{\mathfrak{D}}$ is called the *transmission matrix* of $\Sigma_{i/s/o}$. Also here $\widehat{\mathfrak{D}}$ is called the *characteristic function* if $\Sigma_{i/s/o}$ is conservative; see e.g. [Tsekanovskiĭ and Šmuljan, 1977]. In a transmission passive i/s/o system, the main operator A is often not dissipative, and this lack of dissipativity causes many of the technical problems associated with transmission passive systems.

Finally, in the case where $\Sigma_{i/s/o}$ is impedance passive, $\widehat{\mathfrak{D}}$ is called the *impedance matrix* of $\Sigma_{i/s/o}$. If $\Sigma_{i/s/o}$ is conservative then the main operator A is skew-adjoint, cf. Theorem 5.22.

See [Šmuljan, 1986; Salamon, 1987; Curtain and Weiss, 1989; Arov and Nudelman, 1996], or [Staffans, 2005] for more information on transfer functions (input/output resolvent functions).

The characteristic node bundle In order to derive the analogue of an i/s/o resolvent matrix for a s/s system, we rewrite the identity (5.20) so that it uses the generating subspace V instead of the system node operator S.

Suppose therefore that $\left[\begin{smallmatrix} x \\ w \end{smallmatrix}\right]$ is a classical trajectory of a s/s node, and that x, $\dot{x}$, and w are all Laplace transformable with the Laplace transforms converging in the whole right half-plane $\mathbb{C}^+$. Taking Laplace transforms in $\begin{bmatrix} \dot{x}(t) \\ x(t) \\ w(t) \end{bmatrix} \in V$, $t \in \mathbb{R}^+$, we get

$$\begin{bmatrix} \lambda\hat{x}(\lambda) - x_0 \\ \hat{x}(\lambda) \\ \widehat{w}(\lambda) \end{bmatrix} \in V, \quad \lambda \in \mathbb{C}^+.$$

Definition 5.27 Let $\mathcal{W} = \mathcal{U} \dotplus \mathcal{Y}$ be a direct sum decomposition of $\mathcal{W}$. The domain of the *generalized i/s/o resolvent matrix* with respect to this decomposition and the generalized i/s/o resolvent matrix itself are defined by

$$\operatorname{dom}(\widehat{\mathfrak{S}}) = \left\{ \lambda \in \mathbb{C} \;\middle|\; \begin{array}{l} \text{for all } \left[\begin{smallmatrix} x_0 \\ u \end{smallmatrix}\right] \in \left[\begin{smallmatrix} \mathcal{X} \\ \mathcal{U} \end{smallmatrix}\right] \text{ there exists} \\ \text{a unique pair } \left[\begin{smallmatrix} x \\ y \end{smallmatrix}\right] \in \left[\begin{smallmatrix} \mathcal{X} \\ \mathcal{Y} \end{smallmatrix}\right] \\ \text{such that } \begin{bmatrix} \lambda x - x_0 \\ x \\ \left[\begin{smallmatrix} u \\ y \end{smallmatrix}\right] \end{bmatrix} \in V \end{array} \right\},$$

$$\widehat{\mathfrak{S}}(\lambda) \left[\begin{smallmatrix} x_0 \\ u \end{smallmatrix}\right] := \begin{bmatrix} \widehat{\mathfrak{A}}(\lambda) & \widehat{\mathfrak{B}}(\lambda) \\ \widehat{\mathfrak{C}}(\lambda) & \widehat{\mathfrak{D}}(\lambda) \end{bmatrix} \left[\begin{smallmatrix} x_0 \\ u \end{smallmatrix}\right] := \left[\begin{smallmatrix} x \\ y \end{smallmatrix}\right], \; \lambda \in \operatorname{dom}(\widehat{\mathfrak{S}}),$$

$$\text{where } \left[\begin{smallmatrix} x \\ y \end{smallmatrix}\right] \text{ is the unique pair for which } \begin{bmatrix} \lambda x - x_0 \\ x \\ \left[\begin{smallmatrix} u \\ y \end{smallmatrix}\right] \end{bmatrix} \in V.$$

Of course, in this definition only those decompositions $\mathcal{W} = \mathcal{U} \dotplus \mathcal{Y}$ of the signal space for which the domain of the generalized resolvent matrix $\widehat{\mathfrak{S}}$ is nonempty are interesting.

Example 5.28 We continue Example 5.7 by computing the i/s/o resolvent matrix $\widehat{\mathfrak{S}} = \left[\begin{smallmatrix} \widehat{\mathfrak{A}} & \widehat{\mathfrak{B}} \\ \widehat{\mathfrak{C}} & \widehat{\mathfrak{D}} \end{smallmatrix}\right]$ of the boundary control i/s/o system $\Sigma_{i/s/o} = (L, \Gamma_0, \Gamma_1; \mathcal{X}, \mathcal{U}, \mathcal{Y})$ in Definition 4.1. Therefore we again let

V be given by (5.10) and we carry out the following computations:

$$\begin{bmatrix} \lambda x - x_0 \\ x \\ \left[\begin{smallmatrix} u \\ y \end{smallmatrix}\right] \end{bmatrix} \in V = \left\{ \begin{bmatrix} Lx \\ x \\ \left[\begin{smallmatrix} \Gamma_0 x \\ \Gamma_1 x \end{smallmatrix}\right] \end{bmatrix} \;\middle|\; x \in \operatorname{dom}(L) \right\} \iff$$

$$x_0 = (\lambda - L)x, \quad y = \Gamma_1 x, \quad \text{and} \quad u = \Gamma_0 x,$$

$$\text{so that} \quad \widehat{\mathfrak{S}}(\lambda) : \begin{bmatrix} (\lambda - L)x \\ \Gamma_0 x \end{bmatrix} \mapsto \begin{bmatrix} x \\ \Gamma_1 x \end{bmatrix}, \quad x \in \operatorname{dom}(L).$$

One can show that $\mathbb{C}^+ \subset \operatorname{dom}(\widehat{\mathfrak{S}})$ if the system $\Sigma_{i/s/o}$ is *passive*, i.e., if V is maximal non-negative, and if $\mathcal{U} = \mathcal{W}_+$ for some fundamental decomposition $\mathcal{W} = \mathcal{W}_+ \boxplus -\mathcal{W}_-$.

It is possible to further extend the notion of a generalized i/s/o resolvent matrix by allowing $\widehat{\mathfrak{S}}(\lambda)$ to be a *relation* instead of a function. This extension is implemented by the following notion:

Definition 5.29 The *characteristic node bundle* of the (not necessarily passive) s/s system $\Sigma = (V; \mathcal{X}, \mathcal{W})$ is the family $\{\widehat{\mathfrak{E}}(\lambda)\}_{\lambda \in \mathbb{C}}$ of subspaces of the node space $\mathfrak{K}$, where each $\widehat{\mathfrak{E}}(\lambda)$ is given by

$$\widehat{\mathfrak{E}}(\lambda) = \left\{ \begin{bmatrix} x_0 \\ x \\ w \end{bmatrix} \middle| \begin{bmatrix} \lambda x - x_0 \\ x \\ w \end{bmatrix} \in V \right\}.$$

The subspace $\widehat{\mathfrak{E}}(\lambda)$ is called the *fiber of* $\widehat{\mathfrak{E}}$ *at* $\lambda \in \mathbb{C}$.

By using the above state/signal characteristic node bundle we can reformulate the definition of the generalized i/s/o resolvent matrix $\left[\begin{smallmatrix} \widehat{\mathfrak{A}} & \widehat{\mathfrak{B}} \\ \widehat{\mathfrak{C}} & \widehat{\mathfrak{D}} \end{smallmatrix}\right]$ as follows.

Remark Let $\mathcal{W} = \mathcal{U} \dotplus \mathcal{Y}$ be a direct sum decomposition of $\mathcal{W}$. The domain of the *generalized i/s/o resolvent matrix* $\widehat{\mathfrak{S}}$ of the passive s/s system $\Sigma = (V; \mathcal{X}, \mathcal{W})$ with respect to this decomposition consists of those points $\lambda \in \mathbb{C}$ for which *the fiber* $\widehat{\mathfrak{E}}(\lambda)$ *of the characteristic node bundle is the graph of a bounded linear operator* $\begin{bmatrix} \mathcal{X} \\ \{0\} \\ \mathcal{U} \end{bmatrix} \to \begin{bmatrix} \{0\} \\ \mathcal{X} \\ \mathcal{Y} \end{bmatrix}$, and $\widehat{\mathfrak{S}}(\lambda) = \left[\begin{smallmatrix} \widehat{\mathfrak{A}}(\lambda) & \widehat{\mathfrak{B}}(\lambda) \\ \widehat{\mathfrak{C}}(\lambda) & \widehat{\mathfrak{D}}(\lambda) \end{smallmatrix}\right]$ is defined to be this operator. Note that we require that $\operatorname{dom}(\widehat{\mathfrak{S}}(\lambda)) = \left[\begin{smallmatrix} \mathcal{X} \\ \mathcal{U} \end{smallmatrix}\right]$ for all $\lambda \in \operatorname{dom}(\widehat{\mathfrak{S}})$.

However, even if $\widehat{\mathfrak{E}}(\lambda)$ is not the graph of an operator, it can always be interpreted *as the graph of a closed relation* $\left[\begin{smallmatrix} \mathcal{X} \\ \mathcal{U} \end{smallmatrix}\right] \to \left[\begin{smallmatrix} \mathcal{X} \\ \mathcal{Y} \end{smallmatrix}\right]$. With this interpretation it makes sense to call this relation the *is/so resolvent relation at the point* $\lambda \in \mathbb{C}$. This resolvent relation is *defined for all* $\lambda \in \mathbb{C}$ but now $\operatorname{dom}(\widehat{\mathfrak{E}}(\lambda))$ may depend on λ.

Observe that unlike the above mentioned resolvent matrices and resolvent relations, *the fiber* $\widehat{\mathfrak{E}}(\lambda)$ *is a state/signal characteristic*, i.e., it does not depend on any particular decomposition $\mathcal{W} = \mathcal{U} \dotplus \mathcal{Y}$ of the signal space. Thus, although the s/s system Σ has many different resolvent relations, each corresponding to a different decomposition $\mathcal{W} = \mathcal{U} \dotplus \mathcal{Y}$, *all resolvent relations have the same graph.* The different resolvent relations are simply different *representations* of the characteristic node bundle corresponding to different input/output decompositions.

We refer the reader to [Arov and Staffans, 2012] for more information on characteristic node bundles.

5.5 Conservative boundary relations

As we showed in Section 4.5, boundary triplets can be obtained as the i/s/o representations of conservative boundary control systems in case the boundary mapping Γ is surjective and the external signal space $\mathcal{W}$ has equal positive and negative indices. Here we show that a Lagrangian decomposition of the signal space of a conservative s/s system gives rise to a *boundary relation*, even if the decomposition of the signal space does not induce an i/s/o representation. We also prove the converse: every conservative boundary relation can be interpreted as a conservative state/signal system.

Definitions The following definition of a boundary relation has been adapted from [Derkach et al., 2009, Def. 3.1], with some minor change of notation; see also Definition 7.34.

Definition 5.30 Let R be a closed symmetric linear relation in a Hilbert space $\mathcal{X}$ (with arbitrary defect numbers), and let $\mathcal{U}$ be an auxiliary Hilbert space. A linear relation $\Gamma\colon \mathcal{X}^2 \to \mathcal{U}^2$ is called a *conservative boundary relation* for R^* if

1. dom (Γ) is dense in R^*,
2. the identity

$$(z_1, x_2)_{\mathcal{X}} - (x_1, z_2)_{\mathcal{X}} = (y_1, u_2)_{\mathcal{U}} - (u_1, y_2)_{\mathcal{U}} \tag{5.22}$$

   holds for every $\{[\begin{smallmatrix} x_1 \\ z_1 \end{smallmatrix}], [\begin{smallmatrix} u_1 \\ y_1 \end{smallmatrix}]\}$, $\{[\begin{smallmatrix} x_2 \\ z_2 \end{smallmatrix}], [\begin{smallmatrix} u_2 \\ y_2 \end{smallmatrix}]\} \in \Gamma$, and
3. Γ is maximal in the sense that if $\{[\begin{smallmatrix} x_1 \\ z_1 \end{smallmatrix}], [\begin{smallmatrix} u_1 \\ y_1 \end{smallmatrix}]\} \in \mathcal{X}^2 \times \mathcal{U}^2$ satisfies (5.22) for every $\{[\begin{smallmatrix} x_2 \\ z_2 \end{smallmatrix}], [\begin{smallmatrix} u_2 \\ y_2 \end{smallmatrix}]\} \in \Gamma$, then $\{[\begin{smallmatrix} x_1 \\ z_1 \end{smallmatrix}], [\begin{smallmatrix} u_1 \\ y_1 \end{smallmatrix}]\} \in \Gamma$.

We remark that what we here call "conservative boundary relation" is simply called "boundary relation" in [Derkach et al., 2009] and it is called a "unitary boundary relation" in Chapter 7. We have added the word "conservative" because of the close resemblance to conservative s/s systems. As was shown in [Derkach et al., 2009, Prop. 3.1], $\ker(\Gamma) = R$ for the relation Γ and the operator R in Definition 5.30.

The following definition is an adaptation of [Derkach et al., 2009, Defs 3.4 and 3.5]; see also Definition 7.35.

Definition 5.31 Let R be a closed symmetric linear relation in the Hilbert space $\mathcal{X}$ and let $\Gamma : \left[\begin{smallmatrix}\mathcal{X}\\ \mathcal{X}\end{smallmatrix}\right] \to \left[\begin{smallmatrix}\mathcal{U}\\ \mathcal{U}\end{smallmatrix}\right]$ be a conservative boundary relation for R^*.

The *Weyl family* (of $R = \ker(\Gamma)$) corresponding to Γ is the family

$$M(\lambda) := \left\{\{u, y\} \mid \{\left[\begin{smallmatrix}x\\ \lambda x\end{smallmatrix}\right], \left[\begin{smallmatrix}u\\ y\end{smallmatrix}\right]\} \in \Gamma\right\}, \quad \lambda \in \mathbb{C} \setminus \mathbb{R}.$$

The γ*-field* (of $R = \ker(\Gamma)$) corresponding to Γ is the relation

$$\gamma(\lambda) := \left\{\{u, x\} \mid \{\left[\begin{smallmatrix}x\\ \lambda x\end{smallmatrix}\right], \left[\begin{smallmatrix}u\\ y\end{smallmatrix}\right]\} \in \Gamma\right\}, \quad \lambda \in \mathbb{C} \setminus \mathbb{R}.$$

By [Derkach et al., 2006, Sec. 4.2], the γ field of a boundary relation is in fact single-valued for $\lambda \in \mathbb{C} \setminus \mathbb{R}$. Note that

$$\operatorname{dom}(M(\lambda)) = \operatorname{dom}(\gamma(\lambda)) = \left\{u \mid \{\left[\begin{smallmatrix}x\\ \lambda x\end{smallmatrix}\right], \left[\begin{smallmatrix}u\\ y\end{smallmatrix}\right]\} \in \Gamma\right\}$$

in general depends on λ. This is analogous to the dependence of the domain of the i/s/o resolvent relation of a s/s system on λ in the general case. In [Derkach et al., 2006, Sect. 4.3] it is studied in which cases $\operatorname{dom}(M(\lambda))$ is independent of λ.

Connections to conservative state/signal systems We now proceed essentially in the same way as we did in Section 4.5 in order to explain the connection between a conservative boundary relation and a conservative s/s system.

Let R be a closed symmetric linear relation in $\mathcal{X}$ and let $\Gamma : \mathcal{X}^2 \to \mathcal{U}^2$ be a conservative boundary relation for R^*. We construct a s/s system by taking the signal space $\mathcal{W}$ to be $\mathcal{W} := \left[\begin{smallmatrix}\mathcal{U}\\ \mathcal{U}\end{smallmatrix}\right]$ with the indefinite inner product

$$[\left[\begin{smallmatrix}u_1\\ y_1\end{smallmatrix}\right], \left[\begin{smallmatrix}u_2\\ y_2\end{smallmatrix}\right]]_{\mathcal{W}} := (u_1, y_2)_{\mathcal{U}} + (y_1, u_2)_{\mathcal{U}}, \tag{5.23}$$

corresponding to the Lagrangian decomposition $\mathcal{W} = \left[\begin{smallmatrix}\mathcal{U}\\ \{0\}\end{smallmatrix}\right] \stackrel{\Psi}{+} \left[\begin{smallmatrix}\{0\}\\ \mathcal{U}\end{smallmatrix}\right]$ with

$\Psi = 1_{\mathcal{U}}$, and defining

$$V := \left\{ \begin{bmatrix} \mathrm{i}z \\ x \\ \left[\begin{smallmatrix} u \\ \mathrm{i}y \end{smallmatrix}\right] \end{bmatrix} \in \begin{bmatrix} \mathcal{X} \\ \mathcal{X} \\ \mathcal{W} \end{bmatrix} \middle| \{ [\begin{smallmatrix} x \\ z \end{smallmatrix}], [\begin{smallmatrix} u \\ y \end{smallmatrix}] \} \in \Gamma \right\}. \tag{5.24}$$

We will prove in Lemma 5.33 below that V is a Lagrangian subspace of the Kreĭn space $\mathfrak{K} := \begin{bmatrix} \mathcal{X} \\ \mathcal{X} \\ \mathcal{W} \end{bmatrix}$ equipped with the inner product (5.13).

Thus, if we knew that also (5.4) holds, then $\Sigma = (V; \mathcal{X}, \mathcal{W})$ would be a conservative s/s system. However, conditions 2 and 3 of Definition 5.30 alone do not yet imply that V satisfies (5.4). Indeed, let $\mathcal{X}$ be an arbitrary nontrivial Hilbert space, and set $\mathcal{W} = \{0\}$ and $\Gamma := \{ \{ [\begin{smallmatrix} 0 \\ z \end{smallmatrix}], [\begin{smallmatrix} 0 \\ 0 \end{smallmatrix}] \} \mid z \in \mathcal{X} \}$. Then $V = \left\{ \begin{bmatrix} z \\ 0 \\ \left[\begin{smallmatrix} 0 \\ 0 \end{smallmatrix}\right] \end{bmatrix} \middle| z \in \mathcal{X} \right\} = V^{[\perp]}$ in $\mathfrak{K}$.

Fortunately, it is possible to meet condition (5.4) by replacing the state space $\mathcal{X}$ by a smaller space, and this can be done without essential loss of generality. The following proposition follows from [Kurula, 2010a, Prop. 4.7].

Proposition 5.32 *Let V be a maximal non-negative subspace of $\mathfrak{K}$. Denote*

$$\widetilde{\mathcal{X}} := \mathcal{X} \ominus \left\{ z \middle| \left[\begin{smallmatrix} z \\ 0 \\ 0 \end{smallmatrix}\right] \in V \right\} \quad \text{and} \quad \widetilde{V} = V \cap \begin{bmatrix} \widetilde{\mathcal{X}} \\ \widetilde{\mathcal{X}} \\ \mathcal{W} \end{bmatrix}. \tag{5.25}$$

Then $\widetilde{\Sigma} := (\widetilde{V}; \widetilde{\mathcal{X}}, \mathcal{W})$ is a passive s/s system and the sets of classical and generalized trajectories generated by V and $\widetilde{V}$ are the same.

The s/s system $\widetilde{\Sigma}$ is conservative if and only if $V = V^{[\perp]}$.

In this way every conservative boundary relation induces a unique conservative s/s system. See Theorem 5.34 below for the exact statement.

Conversely, let $\Sigma = (V; \mathcal{X}, \mathcal{W})$ be a conservative s/s node, such that the signal space has a Lagrangian decomposition $\mathcal{W} = \mathcal{U} \stackrel{\Psi}{+} \mathcal{Y}$ with the inner product (5.17). Define a linear relation on $\mathcal{X}^2 \times \mathcal{U}^2$ by

$$\Gamma := \left\{ \left\{ \begin{bmatrix} x \\ -\mathrm{i}z \end{bmatrix}, \begin{bmatrix} P_{\mathcal{U}}^{\mathcal{Y}} w \\ -\mathrm{i}\Psi P_{\mathcal{Y}}^{\mathcal{U}} w \end{bmatrix} \right\} \mid \left[\begin{smallmatrix} z \\ x \\ w \end{smallmatrix}\right] \in V \right\}. \tag{5.26}$$

In order to prove that Γ is a conservative boundary relation, we need to recall the *main transform* $\mathcal{J}(\Gamma)$ of Γ defined in [Derkach et al., 2006, Sect. 2.4] by

$$\mathcal{J}(\Gamma) := \{ \{ [\begin{smallmatrix} x \\ u \end{smallmatrix}], [\begin{smallmatrix} z \\ -y \end{smallmatrix}] \} \mid \{ [\begin{smallmatrix} x \\ z \end{smallmatrix}], [\begin{smallmatrix} u \\ y \end{smallmatrix}] \} \in \Gamma \}, \tag{5.27}$$

and to state the following lemma:

Lemma 5.33 *The space $\mathcal{U}^2$ with the indefinite inner product* (5.23) *is a Kreĭn space. Moreover, the following claims are equivalent for an arbitrary Lagrangian decomposition $\mathcal{W} = \mathcal{U} \overset{\Psi}{\dotplus} \mathcal{Y}$:*

1. *The subspace $V \subset \mathfrak{K}$ satisfies $V = V^{[\perp]}$.*
2. *The relation $\Gamma : \mathcal{X}^2 \to \mathcal{U}^2$ given by* (5.26) *satisfies conditions 2 and 3 of Definition 5.30.*
3. *The relation $\mathcal{J}(\Gamma)$ in $\mathcal{X} \times \mathcal{U}$ is self-adjoint.*

Proof The reader may verify that that $\mathcal{U}^2 = \left[\begin{smallmatrix} 1_{\mathcal{U}} \\ 1_{\mathcal{U}} \end{smallmatrix}\right] \mathcal{U} \boxplus - \left[\begin{smallmatrix} -1_{\mathcal{U}} \\ 1_{\mathcal{U}} \end{smallmatrix}\right] \mathcal{U}$ is a fundamental decomposition, and therefore $\mathcal{U}^2$ is a Kreĭn space with the given indefinite inner product.

In order to prove the equivalence of the three listed claims, first note that

$$\begin{aligned} \left[\begin{smallmatrix} z \\ x \\ \left[\begin{smallmatrix} u \\ y \end{smallmatrix}\right] \end{smallmatrix}\right] \in V \quad &\iff \quad \left\{\left[\begin{smallmatrix} x \\ -\mathrm{i}z \end{smallmatrix}\right], \left[\begin{smallmatrix} u \\ -\mathrm{i}\Psi y \end{smallmatrix}\right]\right\} \subset \Gamma \\ &\iff \quad \left\{\left[\begin{smallmatrix} x \\ u \end{smallmatrix}\right], \left[\begin{smallmatrix} \mathrm{i}z \\ \mathrm{i}\Psi y \end{smallmatrix}\right]\right\} \in \mathcal{J}(\Gamma). \end{aligned}$$

Moreover, $\left[\begin{smallmatrix} z \\ x \\ \left[\begin{smallmatrix} u \\ y \end{smallmatrix}\right] \end{smallmatrix}\right] \in V^{[\perp]}$ if and only if

$$\begin{aligned} (u, \Psi\widetilde{y}) + (\Psi y, \widetilde{u}) - (z, \widetilde{x}) - (x, \widetilde{z}) = 0, \quad & \left[\begin{smallmatrix} \widetilde{z} \\ \widetilde{x} \\ \left[\begin{smallmatrix} \widetilde{u} \\ \widetilde{y} \end{smallmatrix}\right] \end{smallmatrix}\right] \in V \quad \iff \\ (-\mathrm{i}z, \widetilde{x}) - (x, -\mathrm{i}\widetilde{z}) = (-\mathrm{i}\Psi y, \widetilde{u}) - (u, -\mathrm{i}\Psi\widetilde{y}), \quad & \\ & \left\{\left[\begin{smallmatrix} \widetilde{x} \\ -\mathrm{i}\widetilde{z} \end{smallmatrix}\right], \left[\begin{smallmatrix} \widetilde{u} \\ -\mathrm{i}\Psi\widetilde{y} \end{smallmatrix}\right]\right\} \in \Gamma \quad \iff \\ \left(\left[\begin{smallmatrix} -\mathrm{i}z \\ \mathrm{i}\Psi y \end{smallmatrix}\right], \left[\begin{smallmatrix} \widetilde{x} \\ \widetilde{u} \end{smallmatrix}\right]\right) = \left(\left[\begin{smallmatrix} x \\ u \end{smallmatrix}\right], \left[\begin{smallmatrix} -\mathrm{i}\widetilde{z} \\ \mathrm{i}\Psi\widetilde{y} \end{smallmatrix}\right]\right), \quad & \left\{\left[\begin{smallmatrix} \widetilde{x} \\ \widetilde{u} \end{smallmatrix}\right], \left[\begin{smallmatrix} -\mathrm{i}\widetilde{z} \\ \mathrm{i}\Psi\widetilde{y} \end{smallmatrix}\right]\right\} \in \mathcal{J}(\Gamma), \end{aligned}$$

where the last line is equivalent to $\left\{\left[\begin{smallmatrix} x \\ u \end{smallmatrix}\right], \left[\begin{smallmatrix} -\mathrm{i}z \\ \mathrm{i}\Psi y \end{smallmatrix}\right]\right\} \in \mathcal{J}(\Gamma)^*$.

Thus $V \subset V^{[\perp]}$ if and only if condition 2 of Definition 5.30 holds, which in turn is true if and only if $\mathcal{J}(\Gamma) \subset \mathcal{J}(\Gamma)^*$. Analogously, $V^{[\perp]} \subset V$ if and only if condition 3 of Definition 5.30 holds, which in turn is true if and only if $\mathcal{J}(\Gamma)^* \subset \mathcal{J}(\Gamma)$. □

If the signal space $\mathcal{W}$ has no Lagrangian decomposition, which is the case, e.g., when the dimension of $\mathcal{W}$ is finite and odd, then Σ is not induced by any conservative boundary relation, cf. Example 4.9. We collect our observations in the following theorem:

Theorem 5.34 *The following claims are true:*

1. *Let $(V; \mathcal{X}, \mathcal{W})$ be a* conservative *s/s node and assume that there exists*

a Lagrangian decomposition $\mathcal{W} = \mathcal{U} \overset{\Psi}{\dotplus} \mathcal{Y}$. *Define* Γ *by* (5.26) *and set* $R := \ker(\Gamma)$.

Then R *is a closed symmetric* operator *in* $\mathcal{X}$, R^* *is the closure of* $\operatorname{dom}(\Gamma)$ *in* $\mathcal{X}^2$, Γ *is a conservative boundary relation for* R^*, *and* V *can be recovered using the following expression, which reduces to* (5.24) *when* $\mathcal{Y} = \mathcal{U}$ *and* $\Psi = 1_{\mathcal{U}}$*:*

$$V = \left\{ \begin{bmatrix} \mathrm{i}z \\ x \\ \begin{bmatrix} u \\ \mathrm{i}\Psi^* y \end{bmatrix} \end{bmatrix} \in \begin{bmatrix} \mathcal{X} \\ \mathcal{X} \\ \mathcal{W} \end{bmatrix} \middle| \left\{ \left[\begin{smallmatrix} x \\ z \end{smallmatrix}\right], \left[\begin{smallmatrix} u \\ y \end{smallmatrix}\right] \right\} \in \Gamma \right\}. \tag{5.28}$$

2. *Conversely, let* R *be a closed symmetric linear relation in the Hilbert space* $\mathcal{X}$ *and let* $\Gamma : \mathcal{X}^2 \to \mathcal{U}^2$ *be a conservative boundary relation for* R^*. *Let* $\mathcal{W} := \mathcal{U}^2$ *be the Kreĭn space with the indefinite inner product* (5.23) *(corresponding to* $\Psi = 1_{\mathcal{U}}$*). Define* V *by* (5.24), *and* $\widetilde{\mathcal{X}}$ *and* $\widetilde{V}$ *by* (5.25).

   Then $\widetilde{\Sigma} = (\widetilde{V}; \widetilde{\mathcal{X}}, \mathcal{W})$ *is a conservative s/s node with state space* $\widetilde{\mathcal{X}} = \mathcal{X} \ominus \operatorname{mul}(R)$, *where* $\operatorname{mul}(R) = \{z \mid \{0, z\} \in R\}$ *is the* multi-valued part *of* R. *Moreover, if we define* $\widetilde{\Gamma}$ *by the right-hand side of* (5.26) *with* V *replaced by* $\widetilde{V}$ *and* $\Psi = 1_{\mathcal{U}}$, *then*

$$\widetilde{\Gamma} = \Gamma\big|_{\operatorname{dom}(\Gamma) \cap \widetilde{\mathcal{X}}^2} = \Gamma\big|_{\operatorname{dom}(\Gamma) \cap \left[\begin{smallmatrix} \widetilde{\mathcal{X}} \\ \mathcal{X} \end{smallmatrix}\right]}. \tag{5.29}$$

3. *Let the conservative boundary relation* Γ *and the conservative s/s node* $\Sigma = (\widetilde{V}; \widetilde{\mathcal{X}}, \mathcal{W})$ *be related as in* (5.28) *and* (5.25). *Denote the Weyl family and* γ*-field of* Γ *by* M *and* γ, *respectively, and let* $\widehat{\mathfrak{E}}$ *be the characteristic node bundle of* Σ. *Then*

$$\begin{aligned} M(\lambda) &= \left\{ \{u, -\mathrm{i}\Psi y\} \middle| \begin{bmatrix} 0 \\ x \\ \left[\begin{smallmatrix} u \\ y \end{smallmatrix}\right] \end{bmatrix} \in \widehat{\mathfrak{E}}(\mathrm{i}\lambda) \right\} \quad \textit{and} \\ \gamma(\lambda) &= \left\{ \{u, x\} \middle| \begin{bmatrix} 0 \\ x \\ \left[\begin{smallmatrix} u \\ y \end{smallmatrix}\right] \end{bmatrix} \in \widehat{\mathfrak{E}}(\mathrm{i}\lambda) \right\}, \quad \lambda \in \mathbb{C} \setminus \mathbb{R}. \end{aligned} \tag{5.30}$$

Proof First note that if $R = \ker(\Gamma)$ then (5.26) implies that

$$z \in \operatorname{mul}(R) \iff \left\{ \left[\begin{smallmatrix} 0 \\ z \end{smallmatrix}\right], \left[\begin{smallmatrix} 0 \\ 0 \end{smallmatrix}\right] \right\} \in \Gamma \iff \begin{bmatrix} z \\ 0 \\ 0 \end{bmatrix} \in V. \tag{5.31}$$

1. Since every Lagrangian V is closed and neutral, Γ and its kernel R are also closed, and from (5.22) it follows that R is symmetric. By (5.31), R is single-valued. Lemma 5.33 yields that $\mathcal{J}(\Gamma)$ is self-adjoint, and applying [Derkach et al., 2006, Prop. 3.5], we obtain that Γ is a conservative boundary relation for R^*. Condition 1 of Definition 5.30 says that R^* is the closure of $\operatorname{dom}(\Gamma)$ in $\mathcal{X}^2$. It is easy to verify that (5.26) and (5.28) are equivalent.

2. Setting $\mathcal{Y} = \mathcal{U}$ and $\Psi = 1_{\mathcal{U}}$ in Lemma 5.33, we obtain that $\mathcal{W}$ is a

Kreĭn space and that $V = V^{[\perp]}$, and according to Proposition 5.32, $\widetilde{\Sigma}$ is then a conservative s/s system. By [Derkach et al., 2006, Prop. 3.2], $R = \ker(\Gamma)$, and therefore (5.31) and (5.25) imply that $\widetilde{\mathcal{X}} = \mathcal{X} \ominus \operatorname{mul}(R)$.

The first equality in (5.29) follows by noting that

$$\{[\begin{smallmatrix} x \\ z \end{smallmatrix}], [\begin{smallmatrix} u \\ y \end{smallmatrix}]\} \in \widetilde{\Gamma} \iff \begin{bmatrix} \mathrm{i}z \\ x \\ [\begin{smallmatrix} u \\ \mathrm{i}y \end{smallmatrix}] \end{bmatrix} \in \widetilde{V} \iff$$

$$\begin{bmatrix} z \\ x \\ [\begin{smallmatrix} u \\ \mathrm{i}y \end{smallmatrix}] \end{bmatrix} \in V,\ z, x \in \widetilde{\mathcal{X}} \iff \{[\begin{smallmatrix} x \\ z \end{smallmatrix}], [\begin{smallmatrix} u \\ y \end{smallmatrix}]\} \in \Gamma,\ z, x \in \widetilde{\mathcal{X}}.$$

The second equality holds, since we by (5.13) always have

$$\begin{bmatrix} \mathrm{i}z \\ x \\ [\begin{smallmatrix} u \\ \mathrm{i}y \end{smallmatrix}] \end{bmatrix} \in V = V^{[\perp]} \implies (x, \widetilde{z})_{\mathcal{X}} = 0,\ \begin{bmatrix} \widetilde{z} \\ 0 \\ 0 \end{bmatrix} \in V, \tag{5.32}$$

i.e., $x \in \widetilde{\mathcal{X}}$ automatically when $\begin{bmatrix} \mathrm{i}z \\ x \\ [\begin{smallmatrix} u \\ \mathrm{i}y \end{smallmatrix}] \end{bmatrix} \in V$ for a Lagrangian V.

3. The equalities (5.30) now follow from Definition 5.31 once we observe that

$$\begin{bmatrix} 0 \\ x \\ [\begin{smallmatrix} u \\ y \end{smallmatrix}] \end{bmatrix} \in \widehat{\mathfrak{C}}(\mathrm{i}\lambda) \iff \begin{bmatrix} \mathrm{i}\lambda x \\ x \\ [\begin{smallmatrix} u \\ y \end{smallmatrix}] \end{bmatrix} \in V,\ x \in \widetilde{\mathcal{X}} \iff \begin{bmatrix} \mathrm{i}\lambda x \\ x \\ [\begin{smallmatrix} u \\ y \end{smallmatrix}] \end{bmatrix} \in V$$

$$\iff \{[\begin{smallmatrix} x \\ \lambda x \end{smallmatrix}], [\begin{smallmatrix} u \\ -\mathrm{i}\Psi y \end{smallmatrix}]\} \in \Gamma$$

where we used Definition 5.29, (5.32), and (5.28), respectively. □

Claim 3 of Theorem 5.34 shows that $\widehat{\mathfrak{C}} \cap \begin{bmatrix} \{0\} \\ \mathcal{X} \\ \mathcal{W} \end{bmatrix}$ can be identified with the product of the γ-field and the Weyl family of the conservative boundary relation Γ in (5.26). Note, however, that there is an extra rotation of the complex plane in (5.30), due to the fact that in the boundary relation theory one works with self-adjoint operators that have $\mathbb{C} \setminus \mathbb{R}$ in their resolvent set, whereas in the s/s theory the convention is to use skew-adjoint operators whose resolvent sets contain $\mathbb{C} \setminus \mathrm{i}\mathbb{R}$. Also note that the ordering of the two internal variables z and x is different on the left-hand and the right-hand sides of (5.28), which is due to different conventions in different fields of mathematics.

A systems theory interpretation We now introduce dynamics to a conservative boundary relation by giving a systems and control theory interpretation. At the same time, the following results show that boundary relations, in spite of their name, are much more closely related to the general i/s/o systems in Section 5.2 than to the boundary control systems in Chapter 4.

Theorem 5.35 *Assume that $\Gamma \subset \mathcal{X}^2 \times \mathcal{U}^2$ is a conservative boundary relation with the following properties:*

1. *If $\left\{\left[\begin{smallmatrix}0\\z\end{smallmatrix}\right], \left[\begin{smallmatrix}0\\y\end{smallmatrix}\right]\right\} \in \Gamma$ then $z = 0$ and $y = 0$.*
2. *The set $V_u := \left\{u \mid \left\{\left[\begin{smallmatrix}x\\z\end{smallmatrix}\right], \left[\begin{smallmatrix}u\\y\end{smallmatrix}\right]\right\} \in \Gamma\right\}$ equals $\mathcal{U}$.*

Then Γ has the representation

$$\Gamma = \left\{\left\{\left[\begin{smallmatrix}x\\-\mathrm{i}z\end{smallmatrix}\right], \left[\begin{smallmatrix}u\\-\mathrm{i}y\end{smallmatrix}\right]\right\} \mid \left[\begin{smallmatrix}x\\u\end{smallmatrix}\right] \in \mathrm{dom}\,(S),\ \left[\begin{smallmatrix}z\\y\end{smallmatrix}\right] = S\left[\begin{smallmatrix}x\\u\end{smallmatrix}\right]\right\}, \tag{5.33}$$

where $(S; \mathcal{X}, \mathcal{U}, \mathcal{U})$ is an impedance conservative i/s/o system node.

Moreover, $\Sigma := (V; \mathcal{X}, \mathcal{W})$ is a conservative s/s node, where V is defined by (5.24) and $\mathcal{W} = \mathcal{U}^2$ with the inner product (5.23). The Lagrangian decomposition $\mathcal{W} = \left[\begin{smallmatrix}\mathcal{U}\\\{0\}\end{smallmatrix}\right] \overset{\Psi}{\dotplus} \left[\begin{smallmatrix}\{0\}\\\mathcal{U}\end{smallmatrix}\right]$, $\Psi = 1_{\mathcal{U}}$, is admissible, and $(S; \mathcal{X}, \mathcal{U}, \mathcal{U})$ is the corresponding impedance representation.

Proof From claim 2 of Theorem 5.34 it follows that V defined in (5.24) generates a conservative s/s system. The representation (5.33) for some (single-valued) operator S follows from assumption 1. Then $V = V^{[\perp]}$, assumption 2, and [Ball and Staffans, 2006, Prop. 4.11] imply that S is an impedance conservative i/s/o system node operator, which is an impedance representation of V:

$$\left[\begin{smallmatrix}z\\x\\\left[\begin{smallmatrix}u\\y\end{smallmatrix}\right]\end{smallmatrix}\right] \in V \iff \left\{\left[\begin{smallmatrix}x\\-\mathrm{i}z\end{smallmatrix}\right], \left[\begin{smallmatrix}u\\-\mathrm{i}y\end{smallmatrix}\right]\right\} \in \Gamma \iff \left[\begin{smallmatrix}z\\y\end{smallmatrix}\right] = S\left[\begin{smallmatrix}x\\u\end{smallmatrix}\right],$$

where we have used Definition 5.6, (5.24) and (5.33). □

It follows from Theorems 5.22 and 5.35 that for every $u \in W^{2,1}_{\mathrm{loc}}(\mathbb{R}^+; \mathcal{U})$ and every initial state $x_0 \in \mathcal{X}$ with $\left\{\left[\begin{smallmatrix}x_0\\z\end{smallmatrix}\right], \left[\begin{smallmatrix}u(0)\\y\end{smallmatrix}\right]\right\} \in \Gamma$, the system

$$\left\{\begin{bmatrix}x(t)\\-\mathrm{i}\dot{x}(t)\end{bmatrix}, \begin{bmatrix}u(t)\\-\mathrm{i}y(t)\end{bmatrix}\right\} \in \Gamma, \quad t \in \mathbb{R}^+, \tag{5.34}$$

has a unique *classical* solution (u, x, y) with $x(0) = x_0$.

We have the following converse to Theorem 5.35:

Proposition 5.36 *If $(S; \mathcal{X}, \mathcal{U}, \mathcal{U})$ is an impedance-conservative i/s/o system node then Γ in (5.33) is a conservative boundary relation for R^*, where $R := \ker(\Gamma)$. Moreover, Γ has properties 1 and 2 in Theorem 5.35.*

Proof If $S = \left[\begin{smallmatrix}A\&B\\C\&D\end{smallmatrix}\right]$ is an impedance conservative i/s/o system node operator then $S = \left[\begin{smallmatrix}A\&B\\-C\&D\end{smallmatrix}\right]$ is skew-adjoint by [Staffans, 2002b, Th. 4.3], and this is equivalent to $S = \left[\begin{smallmatrix}-\mathrm{i}A\&B\\\mathrm{i}C\&D\end{smallmatrix}\right]$ being self-adjoint. By (5.27) and (5.33),

$$\mathcal{J}(\Gamma) = \left\{\left\{\left[\begin{smallmatrix}x\\u\end{smallmatrix}\right], \left[\begin{smallmatrix}-\mathrm{i}A\&B\\\mathrm{i}C\&D\end{smallmatrix}\right]\left[\begin{smallmatrix}x\\u\end{smallmatrix}\right]\right\} \mid \left[\begin{smallmatrix}x\\u\end{smallmatrix}\right] \in \mathrm{dom}\left(\left[\begin{smallmatrix}-\mathrm{i}A\&B\\\mathrm{i}C\&D\end{smallmatrix}\right]\right)\right\},$$

and we obtain from [Derkach et al., 2006, Prop. 3.5] that Γ is a conservative boundary relation for R^*.

Moreover, by condition 4 of Definition 5.5, for all $u \in \mathcal{U}$ there exists some $x \in \mathcal{X}$ such that $\left[\begin{smallmatrix} x \\ u \end{smallmatrix}\right] \in \operatorname{dom}\left(\left[\begin{smallmatrix} A\&B \\ C\&D \end{smallmatrix}\right]\right)$. From (5.33) it now follows that condition 2 in Theorem 5.35 is met, and also that $\left\{\left[\begin{smallmatrix} 0 \\ z \end{smallmatrix}\right], \left[\begin{smallmatrix} 0 \\ y \end{smallmatrix}\right]\right\} \in \Gamma$ implies $\left[\begin{smallmatrix} z \\ y \end{smallmatrix}\right] = 0$. □

Using Proposition 5.32, one can reformulate Theorem 5.35 slightly in such a way that condition 1 is replaced by the weaker condition that

$$\left\{\left[\begin{smallmatrix} 0 \\ z \end{smallmatrix}\right], \left[\begin{smallmatrix} 0 \\ y \end{smallmatrix}\right]\right\} \in \Gamma \quad \Longrightarrow \quad y = 0. \tag{5.35}$$

Moreover, condition 1 implies that $\operatorname{dom}(S)$ is dense in $\left[\begin{smallmatrix} \mathcal{X} \\ \mathcal{U} \end{smallmatrix}\right]$ when Γ is a boundary relation, and therefore, condition 2 can be weakened to the condition that V_u is closed. We formulate the result but we leave the proof to the reader.

Corollary 5.37 *Assume that Γ is a conservative boundary relation such that* (5.35) *holds and the set V_u in Theorem 5.35 is closed. Let $\mathcal{W} := \left[\begin{smallmatrix} \mathcal{U} \\ \mathcal{U} \end{smallmatrix}\right]$ with the indefinite inner product* (5.23), *let V be given by* (5.24), *and let $\widetilde{\mathcal{X}}, \widetilde{V}$ be given by* (5.25).

Then Γ has the representation

$$\Gamma = \left\{\left\{\left[\begin{smallmatrix} x \\ -\mathrm{i}z \end{smallmatrix}\right], \left[\begin{smallmatrix} u \\ -\mathrm{i}y \end{smallmatrix}\right]\right\} \;\middle|\; \left[\begin{smallmatrix} x \\ u \end{smallmatrix}\right] \in \operatorname{dom}(S),\ z \in \mathcal{X},\ \left[\begin{smallmatrix} Pz \\ y \end{smallmatrix}\right] = S\left[\begin{smallmatrix} x \\ u \end{smallmatrix}\right]\right\},$$

where P is the orthogonal projection of $\mathcal{X}$ onto $\widetilde{\mathcal{X}}$, and $(S; \mathcal{X}, \mathcal{U}, \mathcal{U})$ is the impedance representation corresponding to the admissible Lagrangian decomposition $\mathcal{W} = \left[\begin{smallmatrix} \mathcal{U} \\ \{0\} \end{smallmatrix}\right] \stackrel{\Psi}{+} \left[\begin{smallmatrix} \{0\} \\ \mathcal{U} \end{smallmatrix}\right]$, $\Psi = 1_{\mathcal{U}}$, of the conservative s/s system $\Sigma = (\widetilde{V}; \widetilde{\mathcal{X}}, \mathcal{W})$.

We could also have introduced dynamics to the boundary relation simply by considering the classical and generalized solutions of (5.34). However, without using claim 2 of Theorem 5.22, or changing to a scattering representation and using Theorem 5.13, we would not know that the sets of classical and generalized trajectories in fact are large.

Dirac structures The *Dirac structures* described in Section 9.4 are also related to conservative s/s systems. For background on Dirac structures (and port-Hamiltonian systems), we recommend the references at the beginning of Chapter 9 and [Kurula, 2010b, Chap. 4]. We have the following result:

Proposition 5.38 *Let $(V;\mathcal{X},\mathcal{W})$ be a conservative s/s node with a Lagrangian decomposition $\mathcal{W}=\mathcal{U}\overset{\Psi}{\dotplus}\mathcal{Y}$ and define*

$$\mathcal{D}:=\left\{\left\{\begin{bmatrix} z \\ P_{\mathcal{Y}}^{\mathcal{U}}w\end{bmatrix},\begin{bmatrix} x \\ P_{\mathcal{U}}^{\mathcal{Y}}w\end{bmatrix}\right\}\;\middle|\;\begin{bmatrix} z \\ x \\ w\end{bmatrix}\in V\right\}. \tag{5.36}$$

Then $\mathcal{D}$ is a Dirac structure in the sense of Section 9.4 with $\mathfrak{H}=\mathcal{X}$, $\mathcal{H}_0=\mathcal{U}$, $\mathcal{H}_1=\mathcal{Y}$, and $\mathfrak{w}=\Psi^*$. The generating subspace V can be recovered from $\mathcal{D}$ as

$$V=\left\{\begin{bmatrix} z \\ x \\ e+f\end{bmatrix}\in\begin{bmatrix}\mathcal{X} \\ \mathcal{X} \\ \mathcal{W}\end{bmatrix}\;\middle|\;\{\left[\begin{smallmatrix} z \\ f\end{smallmatrix}\right],\left[\begin{smallmatrix} x \\ e\end{smallmatrix}\right]\}\in\mathcal{D}\right\}. \tag{5.37}$$

Conversely, let $\mathcal{D}$ be a Dirac structure as described after Definition 9.8. Equip $\mathcal{W}:=\left[\begin{smallmatrix}\mathcal{H}_0 \\ \mathcal{H}_1\end{smallmatrix}\right]$ with the Kreĭn-space inner product

$$[\left[\begin{smallmatrix} e_1 \\ f_1\end{smallmatrix}\right],\left[\begin{smallmatrix} e_2 \\ f_2\end{smallmatrix}\right]]_{\mathcal{W}}:=(e_1,\mathfrak{w}^*f_2)_{\mathcal{H}_0}+(\mathfrak{w}^*f_1,e_2)_{\mathcal{H}_0},$$

so that $\mathcal{W}=\mathcal{H}_0\overset{\Psi}{\dotplus}\mathcal{H}_1$ with $\Psi=\mathfrak{w}^*$. Define V by (5.37), identifying $e+f=\left[\begin{smallmatrix} e \\ f\end{smallmatrix}\right]$. Letting $\widetilde{\mathcal{X}}$ and $\widetilde{V}$ be given by (5.25), we obtain that $(\widetilde{V};\widetilde{\mathcal{X}},\mathcal{W})$ is a conservative s/s node.

Proof Assume that $(V;\mathcal{X},\mathcal{W})$ is a conservative s/s node with a Lagrangian decomposition $\mathcal{W}=\mathcal{U}\overset{\Psi}{\dotplus}\mathcal{Y}$. From the following simple calculation, which uses (5.13), Definition 5.11, and (5.17), it follows that $\mathcal{D}$ is neutral, i.e., that $\mathcal{D}\subset\mathcal{D}^{[\perp]}$:

$$\left(\begin{bmatrix} 0 & 0 & 1 & 0 \\ 0 & 0 & 0 & -\mathfrak{w} \\ 1 & 0 & 0 & 0 \\ 0 & -\mathfrak{w}^* & 0 & 0\end{bmatrix}\begin{bmatrix} z \\ f \\ x \\ e\end{bmatrix},\begin{bmatrix} z \\ f \\ x \\ e\end{bmatrix}\right)=2\mathrm{Re}\,(x,z)-2\mathrm{Re}\,(\Psi^*e,f)=0.$$

It is not difficult to show that $\mathcal{D}$ is even *hypermaximal* neutral, i.e., that also $\mathcal{D}^{[\perp]}\subset\mathcal{D}$, and hence $\mathcal{D}$ in (5.36) is a Dirac structure if $(V;\mathcal{X},\mathcal{W})$ is a conservative s/s node.

The formula (5.37) follows trivially from (5.36). Along the lines of the proof of the first assertion one easily shows that $\mathcal{D}$ being a Dirac structure implies that V in (5.37) is a Lagrangian subspace of the node space $\mathfrak{K}:=\mathcal{X}\times\mathcal{X}\times\mathcal{W}$, equipped with the inner product in (5.13). Proposition 5.32 now completes the proof of the second assertion.

We remark that one also can prove the first assertion by combining Theorem 5.34 with Proposition 9.9. □

If $(V;\mathcal{X},\mathcal{W})$ is a conservative s/s node with a Lagrangian decomposition $\mathcal{W} = \mathcal{U} \overset{\Psi}{+} \mathcal{Y}$ then the graph of $\mathcal{D}$:

$$\operatorname{graph}(\mathcal{D}) := \left\{ \begin{bmatrix} z \\ P_{\mathcal{Y}}^{\mathcal{U}} w \\ x \\ P_{\mathcal{U}}^{\mathcal{Y}} w \end{bmatrix} \middle| \begin{bmatrix} z \\ x \\ w \end{bmatrix} \in V \right\}$$

is a *split Dirac structure* in the sense of [Kurula et al., 2010, Def. 3.1], with $\mathcal{E}_1 = \mathcal{F}_1 = \mathcal{X}$, $\mathcal{E}_2 = \mathcal{U}$, $\mathcal{F}_2 = \mathcal{Y}$, $r_{\mathcal{E}_1,\mathcal{F}_1} = -1_{\mathcal{X}}$, and $r_{\mathcal{E}_2,\mathcal{F}_2} = \Psi^*$. The proof is trivial.

5.6 Conclusions

We have presented the fundamentals of the state/signal approach to systems theory and we have made the basic connections between this theory and that of conservative boundary relations. We can conclude that the main objects of the two fields, namely the s/s system and the (conservative) boundary relation, are very closely related.

Sometimes technical complications arise from the way a s/s system is represented by an i/s/o system and not from the s/s system itself. For instance, the characteristic node bundle of a s/s system is much cleaner and more general than an i/s/o resolvent matrix. Moreover, in many cases it is useful to change from a impedance representation to a scattering representations in order to obtain a well-posed system which describes the dynamics of the system in a clear way; see Remark 5.23. The s/s formalism provides a firm basis for doing this. In particular, the families of all classical and generalized trajectories of a passive s/s system are in general more easily characterized by means of a scattering representation than by means of an impedance representation.

Passivity is a good example of a property which refers to a physical system, and not to any one of its input/output representations. Indeed, the property of passivity of a s/s system $(V; \mathcal{X}, \mathcal{W})$ simply means that the generating subspace V is maximal non-negative, whereas different i/s/o *representations* of the s/s system *are passive in different senses*, cf. Definitions 5.16 and 5.21.

For instance the flexibility in choosing i/s/o representations, the introduction of dynamics, the connection to control theory made in Theorem 5.35, and the work done on passive nonconservative s/s systems could potentially turn out to be useful for future research in the theory of boundary relations.

Conversely, it is interesting to look for new directions for the future development of the state/signal theory by studying the theory of boundary relations. In particular, the realization results [Derkach et al., 2006, Th. 3.9], [Behrndt et al., 2009, Th. 6.1], and [Derkach, 2009, Th. 2.5] can be utilized directly for conservative s/s systems in the case where a Lagrangian decomposition of the external signal space exists, i.e., when the signal space has equal positive and negative indices. An intriguing question is exactly how these realizations are related to those developed in [Arov et al., 2011] and their frequency-domain counterparts. A related question is to what extent the available results on Weyl families and their connections to the associated boundary relation can be employed in order to explore the properties of the characteristic node bundles of s/s systems.

It is our sincere hope that this exposition will increase the interaction between researchers of boundary relations and state/signal systems, thus preventing overlapping research, and that it gives rise to future cooperation on common research interests.

References

Arov, D. Z., Kurula, M., and Staffans, O. J. 2011. Canonical State/Signal Shift Realizations of Passive Continuous Time Behaviors. *Complex Anal. Oper. Theory*, **5**, 331–402.

Arov, D. Z., and Nudelman, M. A. 1996. Passive linear stationary dynamical scattering systems with continuous time. *Integral Equations Operator Theory*, **24**, 1–45.

Arov, D. Z., and Staffans, O. J. 2007a. State/signal linear time-invariant systems theory. Part III: Transmission and impedance representations of discrete time systems. Pages 101–140 of: *Operator Theory, Structured Matrices, and Dilations, Tiberiu Constantinescu Memorial Volume*. Bucharest Romania: Theta Foundation. available from American Mathematical Society.

Arov, D. Z., and Staffans, O. J. 2007b. State/signal linear time-invariant systems theory. Part IV: Affine representations of discrete time systems. *Complex Anal. Oper. Theory*, **1**, 457–521.

Arov, D. Z., and Staffans, O. J. 2012. Symmetries in special classes of passive state/signal systems. *J. Funct. Anal.*, to appear.

Ball, J. A., and Staffans, O. J. 2006. Conservative state-space realizations of dissipative system behaviors. *Integral Equations Operator Theory*, **54**, 151–213.

Behrndt, J., Hassi, S., and de Snoo, H. S. V. 2009. Boundary relations, unitary colligations, and functional models. *Complex Anal. Oper. Theory*, **3**, 57–98.

Curtain, R. F., and Weiss, G. 1989. Well posedness of triples of operators (in the sense of linear systems theory). Pages 41–59 of: *Control and Optimization of Distributed Parameter Systems*. International Series of Numerical Mathematics, vol. 91. Basel Boston Berlin: Birkhäuser-Verlag.

Derkach, V. 2009. Abstract interpolation problem in Nevanlinna classes. Pages 197–236 of: *Modern analysis and applications. The Mark Krein Centenary Confer-*

ence. Vol. 1: Operator theory and related topics. Oper. Theory Adv. Appl., vol. 190. Basel: Birkhäuser Verlag.

Derkach, V. A., Hassi, S., Malamud, M. M., and de Snoo, H. S. V. 2006. Boundary relations and their Weyl families. *Trans. Amer. Math. Soc.*, **358**, 5351–5400.

Derkach, V. A., Hassi, S., Malamud, M. M., and de Snoo, H. S. V. 2009. Boundary relations and generalized resolvents of symmetric operators. *Russ. J. Math. Phys.*, **16**, 17–60.

Kurula, M. 2010. On passive and conservative state/signal systems in continuous time. *Integral Equations Operator Theory*, **67**, 377–424, 449.

Kurula, M. 2010b. *Towards input/output-free modelling of linear infinite-dimensional systems in continuous time.* Ph.D. thesis, ISBN 978-952-12-2410-2, 230 pages, electronic summary `http://urn.fi/URN:ISBN:978-952-12-2418-8`.

Kurula, M., Zwart, H., van der Schaft, A., and Behrndt, J. 2010. Dirac structures and their composition on Hilbert spaces. *J. Math. Anal. Appl.*, **372**, 402–422.

Kurula, M., and Staffans, O. J. 2009. Well-posed state/signal systems in continuous time. *Complex Anal. Oper. Theory*, **4**, 319–390.

Kurula, M., and Staffans, O. J. 2011. Connections between smooth and generalized trajectories of a state/signal system. *Complex Anal. Oper. Theory*, **5**, 403–422.

Malinen, J., and Staffans, O. J. 2006. Conservative boundary control systems. *J. Differential Equations*, **231**, 290–312.

Pazy, A. 1983. *Semi-Groups of Linear Operators and Applications to Partial Differential Equations.* Berlin: Springer-Verlag.

Salamon, D. 1987. Infinite dimensional linear systems with unbounded control and observation: a functional analytic approach. *Trans. Amer. Math. Soc.*, **300**, 383–431.

Šmuljan, Yu. L. 1986. *Invariant subspaces of semigroups and the Lax–Phillips scheme.* Deposited in VINITI, No. 8009-B86, Odessa, 49 pages.

Staffans, O. J. 2002a. Passive and conservative continuous-time impedance and scattering systems. Part I: Well-posed systems. *Math. Control Signals Systems*, **15**, 291–315.

Staffans, O. J. 2002b. Passive and conservative infinite-dimensional impedance and scattering systems (from a personal point of view). Pages 375–414 of: *Mathematical Systems Theory in Biology, Communication, Computation, and Finance.* IMA Volumes in Mathematics and its Applications, vol. 134. New York: Springer-Verlag.

Staffans, O. J. 2005. *Well-Posed Linear Systems.* Cambridge and New York: Cambridge University Press.

Tsekanovskiĭ, E. R., and Šmuljan, Yu. L. 1977. The theory of biextensions of operators in rigged Hilbert spaces. Unbounded operator colligations and characteristic functions. *Uspehi Mathem. Nauk SSSR*, **32**, 69–124.

6

Elliptic operators, Dirichlet-to-Neumann maps and quasi boundary triples

Jussi Behrndt and Matthias Langer

Abstract The notion of quasi boundary triples and their Weyl functions is reviewed and applied to self-adjointness and spectral problems for a class of elliptic, formally symmetric, second order partial differential expressions with variable coefficients on bounded domains.

6.1 Introduction

Boundary triples and associated Weyl functions are a powerful and efficient tool to parameterize the self-adjoint extensions of a symmetric operator and to describe their spectral properties. There are numerous papers applying boundary triple techniques to spectral problems for various types of ordinary differential operators in Hilbert spaces; see, e.g. [Behrndt and Langer, 2010; Behrndt, Malamud and Neidhardt, 2008; Behrndt and Trunk, 2007; Brasche, Malamud and Neidhardt, 2002; Brüning, Geyler and Pankrashkin, 2008; Derkach, Hassi and de Snoo, 2003; Gorbachuk and Gorbachuk, 1991; Derkach and Malamud, 1995; Karabash, Kostenko and Malamud, 2009; Kostenko and Malamud, 2010; Posilicano, 2008] and the references therein.

The abstract notion of boundary triples and Weyl functions is strongly inspired by Sturm–Liouville operators on a half-line and their Titchmarsh–Weyl coefficients. To make this more precise, let us consider the ordinary differential expression $\ell = -D^2 + q$ on the positive half-line $\mathbb{R}^+ = (0, \infty)$, where D denotes the derivative, and suppose that q is a real-valued L^∞-function. The maximal operator associated with ℓ in $L^2(\mathbb{R}^+)$ is defined on the Sobolev space $H^2(\mathbb{R}^+)$ and turns out to be the adjoint of the minimal operator $Sf = \ell(f)$, $\operatorname{dom} S = H_0^2(\mathbb{R}^+)$, where $H_0^2(\mathbb{R}^+)$ is the subspace of $H^2(\mathbb{R}^+)$ consisting of functions f that satisfy the boundary conditions $f(0) = f'(0) = 0$. Here S is a densely defined closed symmetric operator in $L^2(\mathbb{R}^+)$ with deficiency numbers $(1,1)$. In this situation it is natural to define boundary mappings Γ_0 and Γ_1 on the domain $H^2(\mathbb{R}^+)$ of the maximal operator S^* (the adjoint of S) by

$$\Gamma_0, \Gamma_1 \colon \operatorname{dom} S^* \to \mathbb{C}, \qquad \Gamma_0 f := f(0) \quad \text{and} \quad \Gamma_1 f := f'(0).$$

The mapping $(\Gamma_0; \Gamma_1)^\top : \operatorname{dom} S^* \to \mathbb{C}\times\mathbb{C}$ is surjective, and the Lagrange identity reads as

$$(S^* f, g) - (f, S^* g) = (\Gamma_1 f, \Gamma_0 g) - (\Gamma_0 f, \Gamma_1 g), \qquad f, g \in \operatorname{dom} S^*,$$

where on the right-hand side the standard inner product of the boundary space $\mathbb{C}$ is used. The abstract Lagrange identity and the surjectivity of $(\Gamma_0; \Gamma_1)^\top$ are the defining relations for a boundary triple, in this case the triple $\{\mathbb{C}, \Gamma_0, \Gamma_1\}$. In a general situation, one has a triple $\{\mathcal{G}, \Gamma_0, \Gamma_1\}$ where $\mathcal{G}$ is an auxiliary Hilbert space (the space of boundary values) and Γ_0, Γ_1 are linear mappings from $\operatorname{dom} S^*$ to $\mathcal{G}$; see Definition 6.1 below for details. The self-adjoint extensions of the symmetric Sturm–Liouville operator S can be parameterized in the form

$$A_\alpha f = \ell(f), \qquad \operatorname{dom} A_\alpha = \bigl\{ f \in \operatorname{dom} S^* : \Gamma_1 f = \alpha \Gamma_0 f \bigr\},$$

where $\alpha \in \mathbb{R} \cup \{\infty\}$. For an arbitrary closed symmetric operator with equal deficiency indices and a boundary triple $\{\mathcal{G}, \Gamma_0, \Gamma_1\}$ for its adjoint, the domains of the self-adjoint extensions A_Θ are characterized formally in the same way, namely by the boundary conditions $\Gamma_1 f = \Theta \Gamma_0 f$, where Θ is a self-adjoint operator (or relation) in $\mathcal{G}$; cf. Proposition 6.2.

If S is an arbitrary closed symmetric operator with equal deficiency indices in some Hilbert space and $\{\mathcal{G}, \Gamma_0, \Gamma_1\}$ is a boundary triple for the adjoint S^*, then the corresponding Weyl function M is defined as the map $\Gamma_0 f_\lambda \mapsto \Gamma_1 f_\lambda$, where f_λ belongs to $\ker(S^* - \lambda)$ and $\lambda \in \mathbb{C}\backslash\mathbb{R}$. In the Sturm–Liouville case the Weyl function corresponding to the boundary triple $\{\mathbb{C}, \Gamma_0, \Gamma_1\}$ is a scalar analytic function defined on $\mathbb{C}\backslash\mathbb{R}$, which maps Dirichlet boundary values $f_\lambda(0)$ of H^2-solutions f_λ of the differential equation $\ell(u) = \lambda u$ onto their Neumann boundary values $f'_\lambda(0)$. We note that the Weyl function M coincides with the Titchmarsh–Weyl coefficient associated with ℓ. In Sturm–Liouville theory it is well known that the complete spectral information of the self-adjoint realizations is encoded in the Titchmarsh–Weyl coefficient, that is, in the Weyl function of the boundary triple $\{\mathbb{C}, \Gamma_0, \Gamma_1\}$. For example, the spectrum of the Dirichlet operator equals the set of real numbers to which M cannot be continued analytically; isolated eigenvalues coincide with poles of M.

Similar considerations can be made for a regular Sturm–Liouville expression $\ell = -D^2 + q$ on the interval $[0,1]$; here a usual choice of a boundary triple is

$$\mathcal{G} = \mathbb{C}^2, \qquad \Gamma_0 f = \begin{pmatrix} f(0) \\ f(1) \end{pmatrix} \quad \text{and} \quad \Gamma_1 f = \begin{pmatrix} f'(0) \\ -f'(1) \end{pmatrix},$$

and the poles of the corresponding Weyl function (which is a 2×2-matrix function in this case) coincide with the eigenvalues of the Dirichlet operator. For more details, see Section 6.2.

Motivated by the above considerations for the case of second order ordinary differential operators it seems very desirable and natural to adapt the boundary triple concept in such a form that it can be applied to elliptic, formally symmetric, second order differential operators of the form

$$\mathcal{L} = -\sum_{j,k=1}^{n} \partial_j \, a_{jk} \, \partial_k + a$$

with variable coefficients a_{jk} and a on a bounded domain $\Omega \subset \mathbb{R}^n$ by choosing the boundary mappings

$$\Gamma_0 f := f|_{\partial\Omega} \qquad \text{and} \qquad \Gamma_1 f := -\frac{\partial f}{\partial \nu_{\mathcal{L}}}\Big|_{\partial\Omega} = -\sum_{j,k=1}^{n} a_{jk} \mathfrak{n}_j \partial_k f\big|_{\partial\Omega} \quad (6.1)$$

as the Dirichlet and (oblique) Neumann trace map, respectively. Here $\mathfrak{n}$ denotes the outward normal vector on $\partial\Omega$. One of the main motivations to choose the boundary maps in (6.1) is that in applications usually Dirichlet and Neumann data are used and that (formally) the corresponding Weyl function M coincides (up to a minus sign) with the Dirichlet-to-Neumann map. For $f, g \in H^2(\Omega)$, Green's identity takes the form

$$(\mathcal{L}f, g) - (f, \mathcal{L}g) = (\Gamma_1 f, \Gamma_0 g) - (\Gamma_0 f, \Gamma_1 g),$$

where the $L^2(\Omega)$ and $L^2(\partial\Omega)$ inner products appear on the left-hand and right-hand sides, respectively. However, $H^2(\Omega)$ is only a core for the maximal operator associated with $\mathcal{L}$, which is defined on the set

$$\mathfrak{D}_{\max} = \{ f \in L^2(\Omega) : \mathcal{L}(f) \in L^2(\Omega) \};$$

moreover, the mapping $(\Gamma_0; \Gamma_1)^\top : H^2(\Omega) \to L^2(\partial\Omega) \times L^2(\partial\Omega)$ is not surjective, but its range is only dense. Green's identity in the above form cannot be extended to functions f, g in $\mathfrak{D}_{\max}$. Therefore the triple $\{L^2(\partial\Omega), \Gamma_0, \Gamma_1\}$ with Γ_0 and Γ_1 as in (6.1) is not a boundary triple in the usual sense and cannot be turned into one by enlarging the domain of the boundary mappings. These simple observations led to a generalization of the notion of boundary triples in [Behrndt and Langer, 2007]. There the concept of *quasi boundary triples* was introduced in an abstract setting and applied to second order elliptic differential operators on bounded domains. The $H^2(\Omega)$ setting, as well as the case of the larger domain

$$H^{3/2}_{\mathcal{L}}(\Omega) := \{ f \in H^{3/2}(\Omega) \colon \mathcal{L}(f) \in L^2(\Omega) \}$$

for the boundary mappings was discussed there in detail. In contrast to (ordinary) boundary triples there is no bijective correspondence of self-adjoint extensions A_Θ of the underlying symmetric operator S and self-adjoint parameters Θ in the boundary space via the formula

$$\Theta \mapsto A_\Theta = S^* \restriction \{ f \in \operatorname{dom} S^* : \Gamma_1 f = \Theta \Gamma_0 f \}.$$

However, sufficient conditions for self-adjointness can be given with the help of a version of Krein's formula (see Theorems 6.20, 6.21 and 6.24). Many papers have been written on Krein's formula; see, e.g. [Derkach and Malamud, 1991; Gesztesy, Makarov and Tsekanovskii, 1998; Kreĭn, 1946; Langer and Textorius, 1977; Pankrashkin, 2006; Saakjan, 1965] for the general case and, e.g. [Gesztesy and Mitrea, 2008, 2009, 2011; Grubb, 2008; Posilicano and Raimondi, 2009] for applications to PDEs. Here we present a version of Krein's formula that is slightly more general than the one in [Behrndt and Langer, 2007] (but compare also [Behrndt, Langer and Lotoreichik, 2011]):

$$(A_\Theta - \lambda)^{-1} = (A_{\mathrm{D}} - \lambda)^{-1} - \gamma(\lambda)\big(\Theta - M(\lambda)\big)^{-1}\gamma(\bar{\lambda})^*,$$

which holds for any Θ such that $\rho(A_\Theta) \cap \rho(A_{\mathrm{D}}) \neq \varnothing$ (see Theorem 6.26, and Theorem 6.16, and their corollaries for the abstract setting see also Theorem 7.26 and Proposition 7.27); here $\gamma(\lambda)$ is the so-called γ-field, which maps boundary values y onto solutions f of the equation $\mathcal{L}(f) = \lambda f$ with $\Gamma_0 f = y$.

Let us point out that a concept that is similar to quasi boundary triples was introduced and studied by V. Ryzhov independently in [Ryzhov, 2007, 2009]. In [Malinen and Staffans, 2007, Section 6.2] the idea to restrict boundary mappings in connection with colligations and boundary nodes is also used. A boundary triple concept for first order operators was introduced in [Post, 2007]. Other generalizations of boundary triples and abstract concepts of boundary mappings were studied, e.g. in [Arlinskiĭ, 2000; Derkach et al., 2006, 2009; Kopachevskiĭ and Kreĭn, 2004;

Posilicano, 2004, 2008; Mogilevskiĭ, 2006, 2009], see also Chapters 3, 4, 5, 7 and 9 for other related concepts.

The aim of the present chapter is to give an introduction to and an overview of the properties of quasi boundary triples, associated γ-fields and Weyl functions, and to demonstrate how conveniently this technique can be applied to boundary value and spectral problems for elliptic operators. For simplicity, we restrict ourselves to the case that the underlying symmetric operator S is densely defined, so that the adjoint S^* is an operator and not a multi-valued linear relation. For the general case we refer the reader to [Behrndt and Langer, 2007]. In Section 6.2 we start by recalling the notion of boundary triples and Weyl functions and collect some well-known properties of these objects. Furthermore, we show in examples how boundary triples can be applied to ordinary, as well as elliptic differential operators. The boundary triple for an elliptic operator from the last subsection of Section 6.2 can already be found in a slightly different context in [Grubb, 1968] and was studied in slight variations in [Brown, Grubb and Wood, 2009; Brown et al., 2009, 2008; Malamud, 2010; Posilicano, 2008; Posilicano and Raimondi, 2009] as well as in Chapter 8. We emphasize that there the boundary mapping Γ_1 is a regularized variant of the Neumann trace in (6.1) and that the corresponding Weyl function is not the Dirichlet-to-Neumann map.

The notion of quasi boundary triples, their γ-fields and Weyl functions is reviewed in Section 6.3. We also provide a full proof of Krein's formula, which is difficult to find in the literature in this form; cf. [Behrndt and Langer, 2007; Behrndt, Langer and Lotoreichik, 2011]. Furthermore, we give some sufficient criteria for self-adjointness of the extensions of the underlying symmetric operator. In Section 6.4 the quasi boundary triple concept is then applied to the elliptic differential expression $\mathcal{L}$. Here we have decided to work on the scale of spaces $H^s_{\mathcal{L}}(\Omega)$, $s \in [\frac{3}{2}, 2]$, the largest possible range of values of s for our purposes. We stress again that the essential idea here is to use the Dirichlet and Neumann boundary mappings Γ_0 and Γ_1 from (6.1) and to identify the corresponding Weyl function with the Dirichlet-to-Neumann map from the theory of elliptic differential equations. Furthermore, we compare and connect the quasi boundary triple $\{L^2(\partial\Omega), \Gamma_0, \Gamma_1\}$ and its Weyl function with the regularized (ordinary) boundary triple from Section 6.2 and the associated Weyl function.

Let us finish this introduction by fixing some notation. For Hilbert spaces $\mathcal{H}_1$, $\mathcal{H}_2$, denote by $\boldsymbol{B}(\mathcal{H}_1, \mathcal{H}_2)$ the space of everywhere defined bounded linear operators from $\mathcal{H}_1$ to $\mathcal{H}_2$; moreover, we set $\boldsymbol{B}(\mathcal{H}_1) :=$

$\boldsymbol{B}(\mathcal{H}_1, \mathcal{H}_1)$. For abstract boundary conditions we need linear relations in the boundary space; so let us recall a couple of definitions. A linear relation (or, in short, relation) T in a Hilbert space $\mathcal{H}$ is a subspace of the cartesian product $\mathcal{H} \times \mathcal{H}$; operators are identified as linear relations via their graphs. The elements in a linear relation T are denoted in the form $(f;g)^\top$ or as column vectors. For a linear relation T we define its domain, range, kernel, multi-valued part, inverse and adjoint as

$$\begin{aligned}
\operatorname{dom} T &:= \{f : \exists\, g \text{ such that } (f;g)^\top \in T\},\\
\operatorname{ran} T &:= \{g : \exists\, f \text{ such that } (f;g)^\top \in T\},\\
\ker T &:= \{f : \text{ such that } (f;0)^\top \in T\},\\
\operatorname{mul} T &:= \{g : \text{ such that } (0;g)^\top \in T\},\\
T^{-1} &:= \{(g;f)^\top : (f;g)^\top \in T\},\\
T^* &:= \{(f;g)^\top : (v,f) = (u,g) \text{ for all } (u;v)^\top \in T\}.
\end{aligned}$$

A linear relation T in a Hilbert space $\mathcal{H}$ is called *symmetric* if $T \subset T^*$, and *self-adjoint* if $T = T^*$. The relation T is called *dissipative* if $\operatorname{Im}(g,f) \geq 0$ for all $(f;g)^\top \in T$ and *accumulative* if $\operatorname{Im}(g,f) \leq 0$ for all $(f;g)^\top \in T$; T is called *maximal dissipative* (*maximal accumulative*) if T is dissipative (or accumulative, respectively) and $\operatorname{ran}(T - \lambda) = \mathcal{H}$ for $\lambda \in \mathbb{C}^-$ ($\lambda \in \mathbb{C}^+$, respectively), where $\mathbb{C}^+$ and $\mathbb{C}^-$ denote the upper and lower half-planes. The sum of two linear relations T_1, T_2 is defined as

$$T_1 + T_2 := \{(f;g+h)^\top : (f;g)^\top \in T_1,\ (f;h)^\top \in T_2\}.$$

The real and imaginary parts of a linear relation are defined as

$$\operatorname{Re} T = \frac{1}{2}(T + T^*), \qquad \operatorname{Im} T = \frac{1}{2\mathrm{i}}(T - T^*).$$

6.2 Boundary triples and Weyl functions for ordinary and partial differential operators

In this section we first review the concept of boundary triples and associated Weyl functions from abstract extension theory of symmetric operators in Hilbert spaces. As a standard example we discuss the case of a regular Sturm–Liouville operator. Furthermore, as a second example a class of second order elliptic differential operators is studied.

Boundary triples and Weyl functions Let $(\mathcal{H}, (\cdot,-))$ be a Hilbert space and let in the following S be a densely defined closed symmetric operator in $\mathcal{H}$. Everything what follows can be done also for non-densely defined operators S, in which case the adjoint S^* is a proper relation, but for simplicity we restrict ourselves to the case that S is densely defined. This is also sufficient for the applications we have in mind here. First we recall the notion of a boundary triple (originally also called boundary value space), which nowadays is very popular in extension theory of symmetric operators; cf. [Brown et al., 2008; Bruk, 1976; Brüning, Geyler and Pankrashkin, 2008; Derkach and Malamud, 1991, 1995; Gorbachuk and Gorbachuk, 1991; Kochubei, 1975; Malamud, 1992].

Definition 6.1 A triple $\{\mathcal{G}, \Gamma_0, \Gamma_1\}$ is said to be a *boundary triple* for the adjoint operator S^* if $(\mathcal{G}, (\cdot, -))$ is a Hilbert space and $\Gamma_0, \Gamma_1 : \operatorname{dom} S^* \to \mathcal{G}$ are linear mappings such that

$$(S^* f, g) - (f, S^* g) = (\Gamma_1 f, \Gamma_0 g) - (\Gamma_0 f, \Gamma_1 g) \tag{6.2}$$

holds for all $f, g \in \operatorname{dom} S^*$ and the map $\Gamma := (\Gamma_0; \Gamma_1)^\top : \operatorname{dom} S^* \to \mathcal{G} \times \mathcal{G}$ is surjective.

A boundary triple $\{\mathcal{G}, \Gamma_0, \Gamma_1\}$ for S^* exists if and only if the deficiency numbers $n_\pm(S) = \dim \ker (S^* \pm i)$ are equal, that is, if and only if S admits self-adjoint extensions in $\mathcal{H}$. It follows that $\dim \mathcal{G} = n_\pm(S)$, and we point out that $\dim \mathcal{G}$ may be infinite. Moreover, if $S \neq S^*$ then a boundary triple for S^* (if it exists) is not unique.

In the following let $\{\mathcal{G}, \Gamma_0, \Gamma_1\}$ be a boundary triple for S^*. Then the map $\Gamma = (\Gamma_0; \Gamma_1)^\top : \operatorname{dom} S^* \to \mathcal{G} \times \mathcal{G}$ is closed, continuous with respect to the graph norm of S^*, and

$$\operatorname{dom} S = \ker \Gamma = \ker \Gamma_0 \cap \ker \Gamma_1$$

holds. Furthermore, the restrictions of S^* to the dense subspaces $\ker \Gamma_0$ and $\ker \Gamma_1$,

$$A_0 := S^* \upharpoonright \ker \Gamma_0 \qquad \text{and} \qquad A_1 := S^* \upharpoonright \ker \Gamma_1,$$

are self-adjoint extensions of S in $\mathcal{H}$ which are transversal, that is, $\operatorname{dom} A_0 \cap \operatorname{dom} A_1 = \operatorname{dom} S$ and $\operatorname{dom} A_0 + \operatorname{dom} A_1 = \operatorname{dom} S^*$ hold. With the help of the boundary triple $\{\mathcal{G}, \Gamma_0, \Gamma_1\}$ all closed extensions of S which are restrictions of S^* can be parameterized in a convenient way; see, e.g. [Derkach and Malamud, 1995, Proposition 1.4].

Proposition 6.2 *Let $\{\mathcal{G}, \Gamma_0, \Gamma_1\}$ be a boundary triple for S^*. Then*

$$\Theta \mapsto A_\Theta := S^* \upharpoonright \ker (\Gamma_1 - \Theta \Gamma_0) \tag{6.3}$$

establishes a bijective correspondence between the closed linear relations Θ in $\mathcal{G}$ and the closed extensions $A_\Theta \subset S^*$ of S. Furthermore, for every closed linear relation Θ in $\mathcal{G}$ the identity

$$(A_\Theta)^* = A_{\Theta^*}$$

holds, and A_Θ is a self-adjoint (symmetric, (maximal) dissipative, (maximal) accumulative) operator in $\mathcal{H}$ if and only if Θ is a self-adjoint (symmetric, (maximal) dissipative, (maximal) accumulative, respectively) relation in $\mathcal{G}$.

We mention that the dense subspace $\ker(\Gamma_1 - \Theta\Gamma_0)$ on the right-hand side of (6.3) coincides with

$$\{f \in \operatorname{dom} S^* : \Gamma f = (\Gamma_0 f; \Gamma_1 f)^\top \in \Theta\} = \Gamma^{-1}(\Theta) \tag{6.4}$$

and that the expression $\Gamma_1 - \Theta\Gamma_0$ has to be interpreted in the sense of linear relations if $\operatorname{mul}\Theta \neq \{0\}$. Observe that a linear relation Θ in $\mathcal{G}$ is self-adjoint if and only if there exists a pair $\{\Phi, \Psi\}$ of operators in $\mathcal{G}$ with the properties

$$\Phi, \Psi \in \boldsymbol{B}(\mathcal{G}), \quad \Psi^*\Phi = \Phi^*\Psi \quad \text{and} \quad 0 \in \rho(\Psi \pm i\Phi) \tag{6.5}$$

such that

$$\Theta = \{(\Phi k; \Psi k)^\top : k \in \mathcal{G}\} = \{(h; h')^\top \in \mathcal{G} \times \mathcal{G} : \Psi^* h = \Phi^* h'\}. \tag{6.6}$$

With the help of this representation the condition (6.4) can also be written in the form

$$\{f \in \operatorname{dom} S^* : \Psi^*\Gamma_0 f = \Phi^*\Gamma_1 f\},$$

and hence the corresponding self-adjoint operator A_Θ in (6.3) is given by

$$A_\Theta = S^* \restriction \ker(\Psi^*\Gamma_0 - \Phi^*\Gamma_1).$$

The following theorem from [Behrndt and Langer, 2010] is of a certain inverse nature and can be used to determine the adjoint of a given symmetric operator with the help of boundary mappings that satisfy (6.2) and a maximality condition. Very roughly speaking the problem of determining the adjoint is reduced to the much easier problem of checking self-adjointness. The method is inspired by the theory of isometric and unitary operators between indefinite inner product spaces; see, e.g. [Azizov and Iokhvidov, 1989; Derkach et al., 2006; Šmuljan, 1976].

Theorem 6.3 *Let T be a linear operator in $\mathcal{H}$ and let $\mathcal{G}$ be a Hilbert space. Assume that $\Gamma_0, \Gamma_1 \colon \operatorname{dom} T \to \mathcal{G}$ are linear mappings which satisfy the following conditions:*

(i) *there exists a symmetric operator or relation Θ in $\mathcal{G}$ such that*

$$T \upharpoonright \ker(\Gamma_1 - \Theta\Gamma_0)$$

is the extension of a self-adjoint operator A in $\mathcal{H}$;

(ii) $\operatorname{ran}(\Gamma_0; \Gamma_1)^\top = \mathcal{G} \times \mathcal{G}$*;*

(iii) $(Tf, g) - (f, Tg) = (\Gamma_1 f, \Gamma_0 g) - (\Gamma_0 f, \Gamma_1 g)$ *for all $f, g \in \operatorname{dom} T$.*

Then the operator

$$S := T \upharpoonright \ker\Gamma_0 \cap \ker\Gamma_1$$

is a densely defined closed symmetric operator in $\mathcal{H}$ such that $S^* = T$ and $\{\mathcal{G}, \Gamma_0, \Gamma_1\}$ is a boundary triple for S^*. Furthermore, Θ is a self-adjoint operator or relation in $\mathcal{G}$, and $A = S^* \upharpoonright \ker(\Gamma_1 - \Theta\Gamma_0) = A_\Theta$ holds.

Next the notion and essential properties of the γ-field and Weyl function corresponding to a boundary triple are recalled. Let again S be a densely defined closed symmetric operator in $\mathcal{H}$ and let $\{\mathcal{G}, \Gamma_0, \Gamma_1\}$ be a boundary triple for S^* with $A_0 = S^* \upharpoonright \ker\Gamma_0$. We first define $\mathcal{N}_\lambda(S^*) := \ker(S^* - \lambda)$ for $\lambda \in \mathbb{C}$. It follows from $A_0 = A_0^*$ that for all $\lambda \in \rho(A_0)$ the domain of S^* can be decomposed into a direct sum:

$$\operatorname{dom} S^* = \operatorname{dom} A_0 \dotplus \mathcal{N}_\lambda(S^*) = \ker\Gamma_0 \dotplus \mathcal{N}_\lambda(S^*). \tag{6.7}$$

In particular, the restriction of the map Γ_0 to $\mathcal{N}_\lambda(S^*)$, $\lambda \in \rho(A_0)$, is injective, and as a consequence of $\operatorname{ran}\Gamma_0 = \mathcal{G}$ it follows that $\Gamma_0 \upharpoonright \mathcal{N}_\lambda(S^*) \to \mathcal{G}$ is bijective.

Definition 6.4 The *γ-field* γ and *Weyl function* M corresponding to $\{\mathcal{G}, \Gamma_0, \Gamma_1\}$ for S^* are defined by

$$\left.\begin{aligned} \gamma(\lambda) &:= \big(\Gamma_0 \upharpoonright \mathcal{N}_\lambda(S^*)\big)^{-1}, \\ M(\lambda) &:= \Gamma_1\gamma(\lambda) = \Gamma_1\big(\Gamma_0 \upharpoonright \mathcal{N}_\lambda(S^*)\big)^{-1}, \end{aligned}\right. \qquad \lambda \in \rho(A_0).$$

In the next two propositions we collect the basic properties of the γ-field and Weyl function of a boundary triple; see [Derkach and Malamud, 1991, Lemma 1 and Theorem 1] and Proposition 6.14 (iv) for the particular form of M in (6.9).

Proposition 6.5 *Let $\{\mathcal{G},\Gamma_0,\Gamma_1\}$ be a boundary triple for S^* with $A_0 = S^* \upharpoonright \ker\Gamma_0$. Then the corresponding γ-field $\lambda \mapsto \gamma(\lambda)$ is a holomorphic $\boldsymbol{B}(\mathcal{G},\mathcal{H})$-valued function on $\rho(A_0)$, and the identities*

$$\gamma(\lambda) = \big(I + (\lambda-\mu)(A_0-\lambda)^{-1}\big)\gamma(\mu) \quad \text{and} \quad \gamma(\bar{\lambda})^* = \Gamma_1(A_0-\lambda)^{-1}$$

hold for all $\lambda,\mu \in \rho(A_0)$.

Proposition 6.6 *Let $\{\mathcal{G},\Gamma_0,\Gamma_1\}$ be a boundary triple for S^* with $A_0 = S^* \upharpoonright \ker\Gamma_0$. Then the corresponding Weyl function M is a holomorphic $\boldsymbol{B}(\mathcal{G})$-valued function on $\rho(A_0)$, and the identities*

$$M(\lambda) - M(\mu)^* = (\lambda-\bar{\mu})\gamma(\mu)^*\gamma(\lambda) \tag{6.8}$$

and for all $\lambda,\mu \in \rho(A_0)$ and any fixed $\lambda_0 \in \rho(A_0)$

$$\begin{aligned} M(\lambda) = \operatorname{Re} M(\lambda_0) + \gamma(\lambda_0)^*\big((\lambda - \operatorname{Re}\lambda_0) \\ + (\lambda-\lambda_0)(\lambda-\bar{\lambda}_0)(A_0-\lambda)^{-1}\big)\gamma(\lambda_0). \end{aligned} \tag{6.9}$$

The identity (6.8) yields that the Weyl function M is a so-called *Nevanlinna* (or *Herglotz*) *function*, that is, M is holomorphic on $\mathbb{C}\backslash\mathbb{R}$, $M(\lambda) = M(\bar{\lambda})^*$ for all $\lambda \in \mathbb{C}\backslash\mathbb{R}$ and $\operatorname{Im} M(\lambda)$ is a non-negative operator for all λ in the upper half-plane $\mathbb{C}^+$; see, e.g. [Gesztesy and Tsekanovskii, 2000; Kac and Kreĭn, 1974]. Moreover, it follows from (6.8) that $0 \in \rho(\operatorname{Im} M(\lambda))$ if $\lambda \in \mathbb{C}\backslash\mathbb{R}$, i.e. M is a *uniformly strict* Nevanlinna function; cf. [Derkach et al., 2006, p. 5354]. Conversely, every uniformly strict Nevanlinna function is the Weyl function of some boundary triple (where S may be non-densely defined); see [Derkach and Malamud, 1995, Section 5] and [Langer and Textorius, 1977]. We also mention, that a $\boldsymbol{B}(\mathcal{G})$-valued function N is a Nevanlinna function if and only if there exist self-adjoint operators $\alpha,\beta \in \boldsymbol{B}(\mathcal{G})$, $\beta \geq 0$, and a non-decreasing self-adjoint operator function $t \mapsto \Sigma(t) \in \boldsymbol{B}(\mathcal{G})$ on $\mathbb{R}$ such that $\int_{\mathbb{R}} \frac{1}{1+t^2} d\Sigma(t) \in \boldsymbol{B}(\mathcal{G})$ and

$$N(\lambda) = \alpha + \lambda\beta + \int_{-\infty}^{\infty} \Big(\frac{1}{t-\lambda} - \frac{t}{1+t^2}\Big)\, d\Sigma(t), \qquad \lambda \in \mathbb{C}\backslash\mathbb{R}.$$

Let again $\{\mathcal{G},\Gamma_0,\Gamma_1\}$ be a boundary triple for S^*. With the help of the corresponding Weyl function M the spectral properties of the closed extensions of S can be described. Roughly speaking the spectrum of A_Θ can be described by means of the singularities of the function $\lambda \mapsto (\Theta - M(\lambda))^{-1}$. The following theorem, see, e.g. [Derkach and Malamud, 1991, Propositions 1 and 2], illustrates this and provides a variant of

Krein's formula for canonical extensions (which are not necessarily self-adjoint).

Theorem 6.7 *Let $\{\mathcal{G}, \Gamma_0, \Gamma_1\}$ be a boundary triple for S^* with γ-field γ and Weyl function M. Let $A_0 = S^* \upharpoonright \ker \Gamma_0$ and let A_Θ be a closed extension of S corresponding to some Θ via* (6.3)–(6.4). *Then the following statements hold for all $\lambda \in \rho(A_0)$:*

(i) $\lambda \in \rho(A_\Theta)$ *if and only if* $0 \in \rho(\Theta - M(\lambda))$;

(ii) $\lambda \in \sigma_i(A_\Theta)$ *if and only if* $0 \in \sigma_i(\Theta - M(\lambda))$, $i = \mathrm{p}, \mathrm{c}, \mathrm{r}$, *where* σ_p, σ_c, σ_r *denote the point, continuous and residual spectrum, respectively;*

(iii) *for all* $\lambda \in \rho(A_0) \cap \rho(A_\Theta)$,

$$(A_\Theta - \lambda)^{-1} = (A_0 - \lambda)^{-1} + \gamma(\lambda)\big(\Theta - M(\lambda)\big)^{-1}\gamma(\bar{\lambda})^*.$$

We mention that Krein's formula in Theorem 6.7 (iii) above can also be written in the form

$$(A_\Theta - \lambda)^{-1} = (A_0 - \lambda)^{-1} + \gamma(\lambda)\Phi\big(\Psi - M(\lambda)\Phi\big)^{-1}\gamma(\bar{\lambda})^*$$

if Θ is a self-adjoint relation in $\mathcal{G}$ which is represented by a pair $\{\Phi, \Psi\}$ as in (6.6).

We point out that the spectral characterization with the help of the Weyl function in Theorem 6.7 (i)–(ii) is only valid on $\rho(A_0)$; see [Brüning, Geyler and Pankrashkin, 2008, Section 4] for a certain extension to points which are isolated eigenvalues of A_0. We also want to point out that, if the symmetric operator S is simple, i.e. there exist no non-trivial reducing subspaces on which S is self-adjoint, it is well known that in the case $\Theta \in \boldsymbol{B}(\mathcal{G})$ the function

$$\lambda \mapsto (\Theta - M(\lambda))^{-1}$$

can be minimally represented by the extension A_Θ. In particular, this implies that the complete spectrum of A_Θ can be characterized with an analytic extension of the function $(\Theta - M(\cdot))^{-1}$. Moreover, again under the condition that S is simple, the spectrum of A_0 can be characterized with the singularities of the Weyl function M; cf. [Brasche, Malamud and Neidhardt, 2002].

Boundary triples for Sturm–Liouville operators Let (a, b) be a bounded interval and let p, q, w be real-valued functions on (a, b) such that $p \neq 0$, $w > 0$ almost everywhere and $1/p$, $q, w \in L^1(a, b)$. In the following we consider the regular Sturm–Liouville differential expression

$$\ell = \frac{1}{w}\left(-\frac{d}{dx}p\frac{d}{dx} + q\right)$$

and differential operators associated with ℓ. For simplicity, only the regular case is discussed here, for singular problems in the limit circle case; see, e.g. [Allakhverdiev, 1991; Behrndt and Langer, 2010]. The limit point case is very well known and is also briefly discussed in the introduction of the present paper. For more general ordinary differential operators and canonical systems see [Behrndt, Hassi, de Snoo and Wietsma, 2011].

Let $L^2_w(a, b)$ denote the space of (equivalence classes) of complex-valued measurable functions on (a, b) such that $|f|^2 w \in L^1(a, b)$ and equip $L^2_w(a, b)$ with the inner product

$$(f, g) = \int_a^b f(x)\overline{g(x)}w(x)dx, \qquad f, g \in L^2_w(a, b).$$

The differential operators associated with ℓ act in the Hilbert space $(L^2_w(a, b), (\cdot, -))$ and are defined as follows: let

$$\begin{aligned}\mathfrak{D}_{\max} = \big\{f \in L^2_w(a, b) : f,\, pf' \text{ absolutely continuous on } (a, b)\\ \text{and } \ell(f) \in L^2_w(a, b)\big\},\\ \mathfrak{D}_{\min} = \big\{f \in \mathfrak{D}_{\max} : f(a) = (pf)'(a) = f(b) = (pf)'(b) = 0\big\},\end{aligned}$$

and let $Sf = \ell(f)$ with $\operatorname{dom} S = \mathfrak{D}_{\min}$ be the minimal operator associated with ℓ. Then S is a densely defined closed symmetric operator in $L^2_w(a, b)$ with deficiency numbers $n_\pm(S) = 2$. The maximal realization of ℓ coincides with the adjoint of the minimal operator:

$$S^*f = \ell(f), \qquad \operatorname{dom} S^* = \mathfrak{D}_{\max}, \qquad f, g \in L^2_w(a, b).$$

As a basis for the two-dimensional space $\mathcal{N}_\lambda(S^*)$, $\lambda \in \mathbb{C}$, we choose the unique solutions φ_λ and ψ_λ of $\ell(f) = \lambda f$ fixed by the initial conditions

$$\begin{aligned}\varphi_\lambda(a) = 1, \qquad (p\varphi'_\lambda)(a) = 0,\\ \psi_\lambda(a) = 0, \qquad (p\psi'_\lambda)(a) = 1.\end{aligned}$$

The proof of the next proposition is straightforward and left to the reader.

Proposition 6.8 *The triple* $\{\mathbb{C}^2, \Gamma_0, \Gamma_1\}$, *where*

$$\Gamma_0 f := \begin{pmatrix} f(a) \\ f(b) \end{pmatrix} \qquad \textit{and} \qquad \Gamma_1 f := \begin{pmatrix} (pf')(a) \\ -(pf')(b) \end{pmatrix},$$

is a boundary triple for the maximal operator

$$S^* f = \ell(f), \quad \operatorname{dom} S^* = \mathfrak{D}_{\max}$$

such that $A_0 = S^* \upharpoonright \ker \Gamma_0$ *and* $A_1 = S^* \upharpoonright \ker \Gamma_1$ *are the Dirichlet realization and the Neumann realization of* ℓ. *For* $\lambda \in \rho(A_0)$ *the corresponding* γ*-field and Weyl function are given by*

$$\gamma(\lambda)\eta = \eta_1 \left(\varphi_\lambda - \frac{\varphi_\lambda(b)}{\psi_\lambda(b)} \psi_\lambda \right) + \eta_2 \frac{1}{\psi_\lambda(b)} \psi_\lambda, \qquad \eta = \begin{pmatrix} \eta_1 \\ \eta_2 \end{pmatrix} \in \mathbb{C}^2,$$

and

$$M(\lambda) = \frac{1}{\psi_\lambda(b)} \begin{pmatrix} -\varphi_\lambda(b) & 1 \\ 1 & -(p\psi_\lambda')(b) \end{pmatrix}.$$

Note that the poles of M are exactly the eigenvalues of A_0, i.e. the Dirichlet eigenvalues.

Boundary triples for second order elliptic differential operators Let Ω be a bounded domain in $\mathbb{R}^n$, $n > 1$, with C^∞-boundary $\partial\Omega$ and consider the second order differential expression

$$\mathcal{L} = -\sum_{j,k=1}^{n} \partial_j \, a_{jk} \, \partial_k + a \tag{6.10}$$

on Ω with real-valued coefficients $a_{jk} \in C^\infty(\overline{\Omega})$, $a \in L^\infty(\Omega)$ such that $a_{jk} = a_{kj}$ for all $j, k = 1, \dots, n$. In addition, it is assumed that the ellipticity condition

$$\sum_{j,k=1}^{n} a_{jk}(x)\xi_j\xi_k \geq C \sum_{k=1}^{n} \xi_k^2, \qquad \xi = (\xi_1, \dots, \xi_n)^\top \in \mathbb{R}^n, \ x \in \overline{\Omega},$$

holds for some constant $C > 0$.

The Sobolev space of kth order on Ω is denoted by $H^k(\Omega)$, $k \in \mathbb{N}$, and the closure of $C_0^\infty(\Omega)$ in $H^k(\Omega)$ by $H_0^k(\Omega)$. Sobolev spaces on the boundary are denoted by $H^s(\partial\Omega)$, $s \in \mathbb{R}$. Let $(\cdot, -)_{-1/2 \times 1/2}$ and $(\cdot, \cdot)_{-3/2 \times 3/2}$ be the extensions of the $L^2(\partial\Omega)$ inner product to $H^{-1/2}(\partial\Omega) \times H^{1/2}(\partial\Omega)$ and $H^{-3/2}(\partial\Omega) \times H^{3/2}(\partial\Omega)$, respectively, and let

$$\iota_\pm : H^{\pm 1/2}(\partial\Omega) \to L^2(\partial\Omega)$$

be isomorphisms such that $(x,y)_{-1/2\times 1/2} = (\iota_- x, \iota_+ y)$ holds for every $x \in H^{-1/2}(\partial\Omega)$ and $y \in H^{1/2}(\partial\Omega)$.

Recall that the *Dirichlet operator*

$$A_{\mathrm{D}} f = \mathcal{L}(f), \qquad \operatorname{dom} A_{\mathrm{D}} = H^2(\Omega) \cap H_0^1(\Omega),$$

associated with the elliptic differential expression $\mathcal{L}$ in (6.10) is self-adjoint in $L^2(\Omega)$ and the resolvent of A_{D} is compact, cf. [Edmunds and Evans, 1987, Theorem VI.1.4] and, e.g. [Lions and Magenes, 1972; Wloka, 1987]. Furthermore, the *minimal operator*

$$Sf = \mathcal{L}(f), \qquad \operatorname{dom} S = H_0^2(\Omega),$$

is a densely defined closed symmetric operator in $L^2(\Omega)$ with equal infinite deficiency numbers, and the adjoint operator $S^* f = \mathcal{L}(f)$ is defined on the maximal domain

$$\operatorname{dom} S^* = \mathfrak{D}_{\max} = \big\{ f \in L^2(\Omega) \colon \mathcal{L}(f) \in L^2(\Omega) \big\}. \tag{6.11}$$

Let us fix some $\eta \in \mathbb{R} \cap \rho(A_{\mathrm{D}})$. Then for each function $f \in \mathfrak{D}_{\max}$ there is a unique decomposition $f = f_{\mathrm{D}} + f_\eta$, where $f_{\mathrm{D}} \in \operatorname{dom} A_{\mathrm{D}}$ and $f_\eta \in \mathcal{N}_\eta(S^*) = \ker(S^* - \eta)$; cf. (6.7).

Next we recall the definition and some properties of the Dirichlet and (oblique) Neumann trace operators. Let $\mathfrak{n} = (\mathfrak{n}_1, \dots, \mathfrak{n}_n)^\top$ be the unit outward normal of Ω. It is well known that the map

$$C^\infty(\overline{\Omega}) \ni f \mapsto \left\{ f|_{\partial\Omega}, \frac{\partial f}{\partial \nu_{\mathcal{L}}}\Big|_{\partial\Omega} \right\}, \text{ where } \frac{\partial f}{\partial \nu_{\mathcal{L}}} := \sum_{j,k=1}^n a_{jk} \mathfrak{n}_j \partial_k f, \tag{6.12}$$

can be extended to a linear operator from $\mathfrak{D}_{\max}$ into the product space $H^{-1/2}(\partial\Omega) \times H^{-3/2}(\partial\Omega)$ and that for $f \in \mathfrak{D}_{\max}$ and $g \in H^2(\Omega)$, Green's identity

$$(S^* f, g) - (f, S^* g) = \left(f|_{\partial\Omega}, \frac{\partial g}{\partial \nu_{\mathcal{L}}}\Big|_{\partial\Omega} \right)_{-\frac12 \times \frac12} - \left(\frac{\partial f}{\partial \nu_{\mathcal{L}}}\Big|_{\partial\Omega}, g|_{\partial\Omega} \right)_{-\frac32 \times \frac32} \tag{6.13}$$

holds; see [Grubb, 1968; Lions and Magenes, 1972; Wloka, 1987].

The boundary triple in the following proposition can also be found in, e.g. [Brown, Grubb and Wood, 2009; Brown et al., 2008; Grubb, 2009; Malamud, 2010; Posilicano and Raimondi, 2009] and is already essentially contained in the classical paper [Grubb, 1968], see also Sections 8.5 and 8.6 of Chapter 8. In [Posilicano, 2008] also a more abstract version of this construction was considered. For the convenience of the reader we repeat the short proof of the next proposition from [Behrndt, 2010] which is based on the general observations in [Grubb, 1968, 1971].

Proposition 6.9 *Let $\eta \in \mathbb{R} \cap \rho(A_{\mathrm{D}})$. The triple $\{L^2(\partial\Omega), \Gamma_0, \Gamma_1\}$, where*

$$\Gamma_0 f := \iota_- f_\eta|_{\partial\Omega} = \iota_- f|_{\partial\Omega} \qquad \text{and} \qquad \Gamma_1 f := -\iota_+ \frac{\partial f_{\mathrm{D}}}{\partial \nu_{\mathcal{L}}}\Big|_{\partial\Omega}$$

$$\text{with } f = f_{\mathrm{D}} + f_\eta \in \mathfrak{D}_{\max}, \quad f_{\mathrm{D}} \in \operatorname{dom} A_{\mathrm{D}}, \quad f_\eta \in \ker(S^* - \eta),$$

is a boundary triple for the maximal operator

$$S^* f = \mathcal{L}(f), \quad \operatorname{dom} S^* = \mathfrak{D}_{\max},$$

such that $A_{\mathrm{D}} = S^* \restriction \ker \Gamma_0$. The corresponding γ-field and Weyl function are, for $\lambda \in \rho(A_{\mathrm{D}})$ and $y \in L^2(\partial\Omega)$, given by

$$\gamma(\lambda) y = \big(I + (\lambda - \eta)(A_{\mathrm{D}} - \lambda)^{-1}\big) f_\eta(y),$$

$$M(\lambda) y = (\eta - \lambda)\iota_+ \frac{\partial (A_{\mathrm{D}} - \lambda)^{-1} f_\eta(y)}{\partial \nu_{\mathcal{L}}}\Big|_{\partial\Omega}$$

respectively, where $f_\eta(y)$ is the unique function in $\ker(S^* - \eta)$ satisfying $\iota_- f_\eta(y)|_{\partial\Omega} = y$.

Proof Let $f, g \in \mathfrak{D}_{\max}$ be decomposed in the form $f = f_{\mathrm{D}} + f_\eta$ and $g = g_{\mathrm{D}} + g_\eta$. Since A_{D} is self-adjoint and $\eta \in \mathbb{R}$, we find

$$(S^* f, g) - (f, S^* g) = (A_{\mathrm{D}} f_{\mathrm{D}}, g_\eta) - (f_{\mathrm{D}}, S^* g_\eta) + (S^* f_\eta, g_{\mathrm{D}}) - (f_\eta, A_{\mathrm{D}} g_{\mathrm{D}}).$$

Then $f_{\mathrm{D}}|_{\partial\Omega} = g_{\mathrm{D}}|_{\partial\Omega} = 0$ together with Green's identity (6.13) implies that

$$(S^* f, g) - (f, S^* g) = -\left(\frac{\partial f_{\mathrm{D}}}{\partial \nu_{\mathcal{L}}}\Big|_{\partial\Omega}, g_\eta|_{\partial\Omega}\right)_{\frac{1}{2}\times -\frac{1}{2}} + \left(f_\eta|_{\partial\Omega}, \frac{\partial g_{\mathrm{D}}}{\partial \nu_{\mathcal{L}}}\Big|_{\partial\Omega}\right)_{-\frac{1}{2}\times\frac{1}{2}}$$

$$= (\Gamma_1 f, \Gamma_0 g) - (\Gamma_0 f, \Gamma_1 g).$$

Hence (6.2) in Definition 6.1 holds. By the classical trace theorem the map $H^2(\Omega) \cap H^1_0(\Omega) \ni f_{\mathrm{D}} \mapsto \frac{\partial f_{\mathrm{D}}}{\partial \nu_{\mathcal{L}}}|_{\partial\Omega} \in H^{1/2}(\partial\Omega)$ is onto and the same holds for the map $\ker(S^* - \eta) \ni f_\eta \mapsto f_\eta|_{\partial\Omega} \in H^{-1/2}(\partial\Omega)$, which is an isomorphism according to [Grubb, 1971, Theorem 2.1]. Hence $(\Gamma_0; \Gamma_1)^\top$ maps $\operatorname{dom} S^*$ onto $L^2(\partial\Omega) \times L^2(\partial\Omega)$ and therefore $\{L^2(\partial\Omega), \Gamma_0, \Gamma_1\}$ is a boundary triple for S^* with $A_{\mathrm{D}} = \ker \Gamma_0$.

It remains to show that the corresponding γ-field and Weyl function have the asserted form. For this let $y \in L^2(\partial\Omega)$, choose the unique function $f_\eta(y)$ in $\ker(S^* - \eta)$ such that $y = \iota_- f_\eta(y)|_{\partial\Omega}$, and set

$$f_\lambda := (\lambda - \eta)(A_{\mathrm{D}} - \lambda)^{-1} f_\eta(y) + f_\eta(y) \tag{6.14}$$

for $\lambda \in \rho(A_D)$. Then $(S^*-\lambda)f_\lambda = 0$ and since $(A_D-\lambda)^{-1}f_\eta(y) \in \operatorname{dom} A_D$ and $f_\eta(y) \in \ker(S^* - \eta)$, we obtain

$$\Gamma_0 f_\lambda = \iota_- f_\eta(y)|_{\partial\Omega} = y,$$

i.e. $\gamma(\lambda)y = f_\lambda = (I+(\lambda-\eta)(A_D-\lambda)^{-1})f_\eta(y)$. Finally, by the definition of the Weyl function and (6.14) we have

$$M(\lambda)y = \Gamma_1 f_\lambda = (\eta - \lambda)\,\iota_+ \frac{\partial(A_D-\lambda)^{-1}f_\eta(y)}{\partial\nu_{\mathcal{L}}}\Big|_{\partial\Omega}. \qquad \square$$

Note that $A_1 = S^* \upharpoonright \ker\Gamma_1$ is not the Neumann operator but the restriction of S^* to the domain

$$\operatorname{dom} S \dotplus \ker(S^* - \eta) = H_0^2(\Omega) \dotplus \ker(S^* - \eta).$$

If $\eta = 0$, then the operator A_1 is the Krein–von Neumann or “soft” extension of S, which was studied, e.g. in [Ashbaugh et al., 2010; Behrndt and Langer, 2007; Everitt and Markus, 2003; Everitt, Markus and Plum, 2005; Grubb, 1968, 1983, 2006].

6.3 Quasi boundary triples and their Weyl functions

The notion of quasi boundary triples was introduced by the authors in [Behrndt and Langer, 2007] with a particular focus on the applicability to elliptic boundary value problems and elliptic differential operators. The concept is a natural generalization of the concept of boundary triples from the previous section and so-called generalized boundary triples from [Derkach and Malamud, 1995, Section 6] and [Derkach et al., 2006, Section 5.2]. In this section again $(\mathcal{H}, (\cdot,-))$ is assumed to be a Hilbert space and S a densely defined closed symmetric operator in $\mathcal{H}$. The idea is that boundary mappings are defined not on the domain of S^* but only on a core of S^*, and that the abstract Green identity is supposed to be valid on this core. The restriction of S^* to this core is called T in the following.

Definition 6.10 A triple $\{\mathcal{G}, \Gamma_0, \Gamma_1\}$ is said to be a *quasi boundary triple* for the operator S^* if $(\mathcal{G}, (\cdot,-))$ is a Hilbert space and there exists an operator T such that $\overline{T} = S^*$, and $\Gamma_0, \Gamma_1 \colon \operatorname{dom} T \to \mathcal{G}$ are linear mappings satisfying

$$(Tf, g) - (f, Tg) = (\Gamma_1 f, \Gamma_0 g) - (\Gamma_0 f, \Gamma_1 g) \qquad (6.15)$$

for all $f, g \in \operatorname{dom} T$, the range of the map $\Gamma := (\Gamma_0; \Gamma_1)^\top : \operatorname{dom} T \to \mathcal{G} \times \mathcal{G}$ is dense and $A_0 = T \upharpoonright \ker \Gamma_0$ is self-adjoint in $\mathcal{H}$.

A quasi boundary triple $\{\mathcal{G}, \Gamma_0, \Gamma_1\}$ for S^* exists if and only if the deficiency numbers of S are equal, and it follows that $\dim \mathcal{G} = n_\pm(S)$, just as for (ordinary) boundary triples. Clearly, every boundary triple is also a quasi boundary triple, and we point out that for the case of finite deficiency numbers also the converse holds. We also remark that a quasi boundary triple with the additional property $\operatorname{ran} \Gamma_0 = \mathcal{G}$ is a generalized boundary triple in the sense of [Derkach and Malamud, 1995, Definition 6.1] (cf. [Behrndt and Langer, 2007, Corollary 3.7]) and that quasi boundary triples are not necessarily boundary relations as studied in [Derkach et al., 2006, 2009], see also Section 7.8 of Chapter 7.

In the following, let $\{\mathcal{G}, \Gamma_0, \Gamma_1\}$ be a quasi boundary triple for S^*. Then the map $\Gamma = (\Gamma_0; \Gamma_1)^\top$ as a mapping from $\operatorname{dom} T$ endowed with the graph norm of T to $\mathcal{G} \times \mathcal{G}$ is closable (which follows from (6.15) and the assumption that $\operatorname{ran} \Gamma$ is dense), and

$$\operatorname{dom} S = \ker \Gamma = \ker \Gamma_0 \cap \ker \Gamma_1$$

holds; see [Behrndt and Langer, 2007, Proposition 2.2].

For a linear operator or relation Θ in $\mathcal{G}$ (not necessarily closed) we define the extension A_Θ of S in analogy to (6.3)–(6.4) by

$$A_\Theta := T \upharpoonright \ker (\Gamma_1 - \Theta \Gamma_0) = T \upharpoonright \{f \in \operatorname{dom} T : \Gamma f \in \Theta\}. \tag{6.16}$$

In contrast to boundary triples, relation (6.16) in general does not induce a bijective correspondence between the self-adjoint extensions of S and the self-adjoint operators and relations Θ in $\mathcal{G}$; cf. Proposition 6.2. However, if Θ is symmetric (dissipative, accumulative) in $\mathcal{G}$, then the corresponding extension A_Θ in (6.16) is also symmetric (dissipative, accumulative) in $\mathcal{H}$, but simple counterexamples show that self-adjointness of Θ does not even imply essential self-adjointness of A_Θ; see [Behrndt and Langer, 2007, Proposition 4.11].

The following result is a variant of Theorem 6.3 and will turn out to be useful when defining quasi boundary triples for elliptic operators in the next section. The advantage of this theorem is that one starts with some operator T and then constructs S and one does not have to show that $\operatorname{dom} T$ is a core of S^*; this follows from the theorem. Moreover, one only has to show that $T \upharpoonright \ker \Gamma_0$ is an extension of a self-adjoint operator and not that it is equal to one. For the proof see [Behrndt and Langer, 2007, Theorem 2.3].

Theorem 6.11 *Let T be a linear operator in $\mathcal{H}$ and let $\mathcal{G}$ be a Hilbert space. Assume that $\Gamma_0, \Gamma_1 \colon \operatorname{dom} T \to \mathcal{G}$ are linear mappings which satisfy the following conditions:*

(i) $T \upharpoonright \ker \Gamma_0$ *is the extension of a self-adjoint operator A in $\mathcal{H}$;*

(ii) $\operatorname{ran}(\Gamma_0; \Gamma_1)^\top$ *is dense in $\mathcal{G} \times \mathcal{G}$ and $\ker \Gamma_0 \cap \ker \Gamma_1$ is dense in $\mathcal{H}$;*

(iii) $(Tf, g) - (f, Tg) = (\Gamma_1 f, \Gamma_0 g) - (\Gamma_0 f, \Gamma_1 g)$ *for all $f, g \in \operatorname{dom} T$.*

Then the operator

$$S := T \upharpoonright \ker \Gamma_0 \cap \ker \Gamma_1$$

is a densely defined closed symmetric operator in $\mathcal{H}$ such that $S^* = \overline{T}$, and $\{\mathcal{G}, \Gamma_0, \Gamma_1\}$ is a quasi boundary triple for S^* with $A = T^* \upharpoonright \ker \Gamma_0 = A_0$.

Let $\{\mathcal{G}, \Gamma_0, \Gamma_1\}$ be a quasi boundary triple for $\overline{T} = S^*$. In the following we set $\mathcal{G}_0 := \operatorname{ran} \Gamma_0$ and $\mathcal{G}_1 := \operatorname{ran} \Gamma_1$. Because $\operatorname{ran} \Gamma$ is dense in $\mathcal{G} \times \mathcal{G}$, it follows that $\mathcal{G}_0$ and $\mathcal{G}_1$ are dense subspaces of $\mathcal{G}$. Since $A_0 = T \upharpoonright \ker \Gamma_0$ is a self-adjoint extension of S in $\mathcal{H}$, the decomposition

$$\operatorname{dom} T = \operatorname{dom} A_0 \dotplus \mathcal{N}_\lambda(T), \qquad \mathcal{N}_\lambda(T) := \ker(T - \lambda),$$

holds for all $\lambda \in \rho(A_0)$; cf. (6.7). The γ-field and Weyl function of a quasi boundary triple are defined in analogy to Definition 6.4.

Definition 6.12 Let S be a densely defined closed symmetric operator in $\mathcal{H}$ and let $\{\mathcal{G}, \Gamma_0, \Gamma_1\}$ be a quasi boundary triple for $\overline{T} = S^*$ with $A_0 = T \upharpoonright \ker \Gamma_0$. Then the functions γ and M defined by

$$\begin{aligned} \gamma(\lambda) &:= \big(\Gamma_0 \upharpoonright \mathcal{N}_\lambda(T)\big)^{-1}, \\ M(\lambda) &:= \Gamma_1 \gamma(\lambda) = \Gamma_1 \big(\Gamma_0 \upharpoonright \mathcal{N}_\lambda(T)\big)^{-1}, \end{aligned} \qquad \lambda \in \rho(A_0), \tag{6.17}$$

are called the *γ-field* and *Weyl function* corresponding to the quasi boundary triple $\{\mathcal{G}, \Gamma_0, \Gamma_1\}$.

Note that $\gamma(\lambda)$ is a mapping from $\mathcal{G}_0$ to $\mathcal{H}$, and $M(\lambda)$ is a mapping from $\mathcal{G}_0$ to $\mathcal{G}_1 \subset \mathcal{G}$ for $\lambda \in \rho(A_0)$. In the next propositions we collect some properties of the γ-field and the Weyl function of a quasi boundary triple, which are extensions of well-known properties of the γ-field and Weyl function of an ordinary boundary triple; cf. Propositions 6.5 and 6.6. For the convenience of the reader we repeat the proofs from [Behrndt and Langer, 2007, Proposition 2.6] which are similar to the ones for γ-fields and Weyl functions of ordinary boundary triples; cf. [Derkach and Malamud, 1991, 1995].

Proposition 6.13 *Let $\{\mathcal{G}, \Gamma_0, \Gamma_1\}$ be a quasi boundary triple with $A_0 = T \upharpoonright \ker \Gamma_0$ and γ-field γ. Then the following assertions hold for all $\lambda, \mu \in \rho(A_0)$:*

(i) *the mapping $\gamma(\lambda)$ is a bounded operator from $\mathcal{G}$ to $\mathcal{H}$ with dense domain $\operatorname{dom}\gamma(\lambda) = \mathcal{G}_0$ and range $\operatorname{ran}\gamma(\lambda) = \mathcal{N}_\lambda(T)$, and hence $\overline{\gamma(\lambda)} \in \boldsymbol{B}(\mathcal{G}, \mathcal{H})$;*

(ii) *the function $\lambda \mapsto \gamma(\lambda)g$ is holomorphic on $\rho(A_0)$ for every $g \in \mathcal{G}_0$, and the relation*

$$\gamma(\lambda) = \big(I + (\lambda - \mu)(A_0 - \lambda)^{-1}\big)\gamma(\mu) \tag{6.18}$$

holds;

(iii) *$\gamma(\bar{\lambda})^* \in \boldsymbol{B}(\mathcal{H}, \mathcal{G})$, $\operatorname{ran}\gamma(\bar{\lambda})^* \subset \mathcal{G}_1$ and for all $h \in \mathcal{H}$ we have*

$$\gamma(\bar{\lambda})^* h = \Gamma_1(A_0 - \lambda)^{-1}h. \tag{6.19}$$

Proof Let $\lambda \in \rho(A_0)$. Since Γ is closable from $\operatorname{dom} T$ (with the graph norm) to $\mathcal{G} \times \mathcal{G}$, it follows that $\Gamma(A_0 - \lambda)^{-1}$ is closable and hence bounded from $\mathcal{H}$ to $\mathcal{G} \times \mathcal{G}$ by the closed graph theorem, which implies that the mapping $\Gamma_1(A_0 - \lambda)^{-1} : \mathcal{H} \to \mathcal{G}$ is bounded. For $h \in \mathcal{H}$ and $x \in \operatorname{dom}\gamma(\bar{\lambda}) = \mathcal{G}_0$ we have (where we use (6.15), (6.17), the relation $T\gamma(\bar{\lambda})x = \bar{\lambda}\gamma(\bar{\lambda})x$ and the fact that Γ_0 vanishes on $\operatorname{dom} A_0$)

$$\begin{aligned}
(h, \gamma(\bar{\lambda})x) &= \big((T - \lambda)(A_0 - \lambda)^{-1}h, \gamma(\bar{\lambda})x\big) \\
&= \big(T(A_0 - \lambda)^{-1}h, \gamma(\bar{\lambda})x\big) - \lambda\big((A_0 - \lambda)^{-1}h, \gamma(\bar{\lambda})x\big) \\
&= \big(T(A_0 - \lambda)^{-1}h, \gamma(\bar{\lambda})x\big) - \big((A_0 - \lambda)^{-1}h, T\gamma(\bar{\lambda})x\big) \\
&= \big(\Gamma_1(A_0 - \lambda)^{-1}h, \Gamma_0\gamma(\bar{\lambda})x\big) - \big(\Gamma_0(A_0 - \lambda)^{-1}h, \Gamma_1\gamma(\bar{\lambda})x\big) \\
&= \big(\Gamma_1(A_0 - \lambda)^{-1}h, x\big),
\end{aligned}$$

which shows relation (6.19). The latter also yields $\operatorname{ran}\gamma(\bar{\lambda})^* \subset \mathcal{G}_1$, and the boundedness of $\Gamma_1(A_0 - \lambda)^{-1}$ implies (i). The resolvent identity and (6.19) show that the following equality is true for $\lambda, \mu \in \rho(A_0)$:

$$\gamma(\lambda)^* - \gamma(\mu)^* = (\bar{\lambda} - \bar{\mu})\gamma(\mu)^*(A_0 - \bar{\lambda})^{-1}.$$

Taking the adjoint and rearranging we obtain (6.18), which also implies the analyticity of $\gamma(\cdot)g$, $g \in \mathcal{G}_0$. □

The first five items of the next proposition are taken from [Behrndt and Langer, 2007, Proposition 2.6]; for the last item see [Behrndt, Langer and Lotoreichik, 2011].

Proposition 6.14 *Let* $\{\mathcal{G}, \Gamma_0, \Gamma_1\}$ *be a quasi boundary triple with* $A_0 = T \upharpoonright \ker \Gamma_0$ *and Weyl function* M. *Then the following assertions hold for all* $\lambda, \mu \in \rho(A_0)$:

(i) $M(\lambda)$ *maps* $\mathcal{G}_0$ *into* $\mathcal{G}_1$. *If also* $A_1 := T \upharpoonright \ker \Gamma_1$ *is a self-adjoint operator in* $\mathcal{H}$ *and* $\lambda \in \rho(A_1)$, *then* $M(\lambda)$ *maps* $\mathcal{G}_0$ *onto* $\mathcal{G}_1$.

(ii) $M(\lambda)\Gamma_0 f_\lambda = \Gamma_1 f_\lambda$ *for all* $f_\lambda \in \mathcal{N}_\lambda(T)$.

(iii) $M(\lambda) \subset M(\bar{\lambda})^*$ *and*

$$M(\lambda) - M(\mu)^* = (\lambda - \bar{\mu})\gamma(\mu)^*\gamma(\lambda). \tag{6.20}$$

(iv) *The function* $\lambda \mapsto M(\lambda)$ *is holomorphic in the sense that it can be written as the sum of the possibly unbounded symmetric operator* $\operatorname{Re} M(\mu)$, *where* μ *is fixed, and a bounded holomorphic operator function:*

$$\begin{aligned} M(\lambda) &= \operatorname{Re} M(\mu) \\ &\quad + \gamma(\mu)^*\Big((\lambda - \operatorname{Re}\mu) + (\lambda - \mu)(\lambda - \bar{\mu})(A_0 - \lambda)^{-1}\Big)\gamma(\mu). \end{aligned} \tag{6.21}$$

(v) $\operatorname{Im} M(\lambda)$ *is a densely defined bounded operator in* $\mathcal{G}$. *Moreover, for* $\lambda \in \mathbb{C}^+ (\mathbb{C}^-)$ *the operator* $\operatorname{Im} M(\lambda)$ *is positive (negative, respectively).*

(vi) *If* $M(\lambda_0)$ *is bounded for some* $\lambda_0 \in \rho(A_0)$, *then* $M(\lambda)$ *is bounded for all* $\lambda \in \rho(A_0)$. *In this case,*

$$\frac{1}{\operatorname{Im}\lambda} \operatorname{Im} \overline{M(\lambda)} > 0, \qquad \lambda \in \mathbb{C}\backslash\mathbb{R}, \tag{6.22}$$

and, in particular, $\ker \overline{M(\lambda)} = \{0\}$ *for* $\lambda \in \mathbb{C}\backslash\mathbb{R}$.

Proof The first assertion in (i) and the statement in (ii) follow immediately from the definition of $M(\lambda)$. The second assertion in (i) follows from the relation $\operatorname{dom} T = \operatorname{dom} A_1 \dotplus \mathcal{N}_\lambda(T)$ for $\lambda \in \rho(A_1)$.

(iii) Let $x, y \in \mathcal{G}_0$ and $\lambda, \mu \in \rho(A_0)$. Then

$$\begin{aligned}
\big((M(\lambda)x, y) - (x, M(\mu)y) &= (\Gamma_1\gamma(\lambda)x, \Gamma_0\gamma(\mu)y) - (\Gamma_0\gamma(\lambda)x, \Gamma_1\gamma(\mu)y) \\
&= (T\gamma(\lambda)x, \gamma(\mu)y) - (\gamma(\lambda)x, T\gamma(\mu)y) \\
&= (\lambda\gamma(\lambda)x, \gamma(\mu)y) - (\gamma(\lambda)x, \mu\gamma(\mu)y) \\
&= (\lambda - \bar{\mu})\big(\gamma(\lambda)x, \gamma(\mu)y\big). \qquad (6.23)
\end{aligned}$$

For $\mu = \bar{\lambda}$ one obtains $(M(\lambda)x, y) = (x, M(\bar{\lambda})y)$, which shows that $\mathcal{G}_0 \subset \operatorname{dom} M(\bar{\lambda})^*$ and that $M(\lambda)$ is a restriction of $M(\bar{\lambda})^*$. Now it follows from (6.23) that

$$\big(M(\lambda)x, y\big) - \big(M(\mu)^*x, y\big) = (\lambda - \bar{\mu})\big(\gamma(\mu)^*\gamma(\lambda)x, y\big),$$

for all $x, y \in \mathcal{G}_0$, which yields (6.20) since $\mathcal{G}_0$ is dense in $\mathcal{G}$ and the operators on both sides of (6.20) are defined on $\mathcal{G}_0$.

(iv) Using (6.20) and (6.18) we obtain the following relations, which are valid on $\mathcal{G}_0$:

$$\begin{aligned}
&M(\lambda) - \operatorname{Re} M(\mu) = M(\lambda) - \tfrac{1}{2}\big(M(\mu) + M(\mu)^*\big) \\
&= M(\lambda) - M(\mu)^* - \tfrac{1}{2}\big(M(\mu) - M(\mu)^*\big) \\
&= (\lambda - \bar{\mu})\gamma(\mu)^*\gamma(\lambda) - \tfrac{1}{2}(\mu - \bar{\mu})\gamma(\mu)^*\gamma(\mu) \\
&= \gamma(\mu)^*\Big[(\lambda - \bar{\mu})\big(I + (\lambda - \mu)(A_0 - \lambda)^{-1}\big)\gamma(\mu) - \tfrac{1}{2}(\mu - \bar{\mu})\gamma(\mu)\Big] \\
&= \gamma(\mu)^*\Big[\lambda - \bar{\mu} + (\lambda - \bar{\mu})(\lambda - \mu)(A_0 - \lambda)^{-1} - \tfrac{1}{2}(\mu - \bar{\mu})\Big]\gamma(\mu) \\
&= \gamma(\mu)^*\Big[\lambda - \operatorname{Re}\mu + (\lambda - \bar{\mu})(\lambda - \mu)(A_0 - \lambda)^{-1}\Big]\gamma(\mu).
\end{aligned}$$

This shows (6.21) and the analyticity as claimed.

(v) Let $\lambda \in \mathbb{C}^+$; the case $\lambda \in \mathbb{C}^-$ is analogous. From (6.20) we obtain

$$\begin{aligned}
\operatorname{Im} M(\lambda) &= \tfrac{1}{2\mathrm{i}}\big(M(\lambda) - M(\lambda)^*\big) = \tfrac{1}{2\mathrm{i}}(\lambda - \bar{\lambda})\gamma(\lambda)^*\gamma(\lambda) \\
&= (\operatorname{Im}\lambda)\gamma(\lambda)^*\gamma(\lambda),
\end{aligned} \qquad (6.24)$$

which is a bounded, positive operator since $\gamma(\lambda)$ is bounded and injective; it is defined on the dense subspace $\mathcal{G}_0$.

(vi) The first assertion follows immediately from (6.21). For the inequality (6.22), assume without loss of generality that $\operatorname{Im}\lambda > 0$. Observe that $\operatorname{Im}\overline{M(\lambda)} = \overline{\operatorname{Im} M(\lambda)}$ since $M(\lambda)$ is bounded. It follows from (v) that $\operatorname{Im} M(\lambda) > 0$. Hence it is sufficient to show that

$$\ker\big(\operatorname{Im}\overline{M(\lambda)}\big) = \{0\}.$$

Let $x \in \ker(\operatorname{Im}\overline{M(\lambda)}) = \ker(\overline{\operatorname{Im} M(\lambda)})$. Then there exist $x_n \in \operatorname{dom} M(\lambda)$

so that $x_n \to x$ and $(\operatorname{Im} M(\lambda))x_n \to 0$ when $n \to \infty$. By (6.24) we have

$$\big((\operatorname{Im} M(\lambda))x_n, x_n\big) = \big((\operatorname{Im}\lambda)\gamma(\lambda)^*\gamma(\lambda)x_n, x_n\big) = (\operatorname{Im}\lambda)\|\gamma(\lambda)x_n\|^2,$$

and and since $\operatorname{Im}\lambda \neq 0$, this implies that $\gamma(\lambda)x_n \to 0$. Then $\Gamma_1\gamma(\lambda)x_n = M(\lambda)\Gamma_0\gamma(\lambda)x_n$, $\Gamma_0\gamma(\lambda)x_n = x_n$, and the boundedness of $M(\lambda)$ imply

$$\Gamma\gamma(\lambda)x_n = \begin{pmatrix}\Gamma_0\gamma(\lambda)x_n\\ \Gamma_1\gamma(\lambda)x_n\end{pmatrix} = \begin{pmatrix}x_n\\ M(\lambda)x_n\end{pmatrix} \to \begin{pmatrix}x\\ \overline{M(\lambda)}x\end{pmatrix}.$$

Since $\gamma(\lambda)x_n$ converges to 0 in the graph norm of T, the closability of Γ implies that $x = 0$, which shows (6.22).

If $x \neq 0$, then $\operatorname{Im}\big(\overline{M(\lambda)}x, x\big) = \big((\operatorname{Im}\overline{M(\lambda)})x, x\big) \neq 0$,which implies that $x \notin \ker\overline{M(\lambda)}$. Hence $\ker\overline{M(\lambda)} = \{0\}$. □

The next theorem gives a characterization of the class of Weyl functions corresponding to quasi boundary triples. It is a reformulation of [Alpay and Behrndt, 2009, Theorem 2.6] and can be regarded as a generalization of [Langer and Textorius, 1977, Theorem 2.2 and Theorem 2.4], [Derkach and Malamud, 1991, Corollary 2] and [Derkach and Malamud, 1995, Theorem 6.1]; see also [Derkach et al., 2006, Section 5].

Theorem 6.15 *Let $\mathcal{G}_0$ be a dense subspace of $\mathcal{G}$, $\lambda_0 \in \mathbb{C}\backslash\mathbb{R}$, and let M be a function defined on $\mathbb{C}\backslash\mathbb{R}$ whose values $M(\lambda)$ are linear operators in $\mathcal{G}$ with* $\operatorname{dom} M(\lambda) = \mathcal{G}_0$, $\lambda \in \mathbb{C}\backslash\mathbb{R}$. *Then the following two statements are equivalent.*

(i) *There exists a separable Hilbert space $\mathcal{H}$, a densely defined closed symmetric operator S and a quasi boundary triple $\{\mathcal{G}, \Gamma_0, \Gamma_1\}$ for $\overline{T} = S^*$ such that M is the corresponding Weyl function.*

(ii) *There exists a unique $\boldsymbol{B}(\mathcal{G})$-valued Nevanlinna function N with the properties (α), (β) and (γ):*

(α) *the relations*

$$M(\lambda)h - \operatorname{Re} M(\lambda_0)h = N(\lambda)h,$$
$$M(\lambda)^*h - \operatorname{Re} M(\lambda_0)h = N(\lambda)^*h$$

hold for all $h \in \mathcal{G}_0$ and $\lambda \in \mathbb{C}\backslash\mathbb{R}$;

(β) $\operatorname{Im} N(\lambda)h = 0$ *for some $h \in \mathcal{G}_0$ and $\lambda \in \mathbb{C}\backslash\mathbb{R}$ implies $h = 0$;*

(γ) *the conditions*

$$\lim_{\eta\to+\infty}\frac{1}{\eta}\big(N(\mathrm{i}\eta)k, k\big) = 0 \quad \textit{and} \quad \lim_{\eta\to+\infty}\eta\operatorname{Im}\big(N(\mathrm{i}\eta)k, k\big) = \infty$$

are valid for all $k \in \mathcal{G}$, $k \neq 0$.

The following theorem and corollary contain a variant of Krein's formula for the resolvents of canonical extensions parameterized by quasi boundary triples via (6.16). The theorem generalizes [Derkach and Malamud, 1991, Proposition 2] and can be found in a similar form in [Behrndt and Langer, 2007] and [Behrndt, Langer and Lotoreichik, 2011], see also Theorem 7.26 and Proposition 7.27. For completeness, the full proof is given after the corollaries below.

Theorem 6.16 *Let S be a densely defined closed symmetric operator in $\mathcal{H}$ and let $\{\mathcal{G}, \Gamma_0, \Gamma_1\}$ be a quasi boundary triple for $\overline{T} = S^*$ with $A_0 = T \restriction \ker \Gamma_0$, γ-field γ and Weyl function M. Further, let Θ be a relation in $\mathcal{G}$ and assume that $\lambda \in \rho(A_0)$ is not an eigenvalue of A_Θ, or, equivalently, that $\ker(\Theta - M(\lambda)) = \{0\}$. Then the following assertions are true.*

(i) *$g \in \operatorname{ran}(A_\Theta - \lambda)$ if and only if $\gamma(\bar\lambda)^* g \in \operatorname{dom}(\Theta - M(\lambda))^{-1}$.*

(ii) *For all $g \in \operatorname{ran}(A_\Theta - \lambda)$ we have*

$$(A_\Theta - \lambda)^{-1} g = (A_0 - \lambda)^{-1} g + \gamma(\lambda)\bigl(\Theta - M(\lambda)\bigr)^{-1}\gamma(\bar\lambda)^* g. \quad (6.25)$$

If $\rho(A_\Theta) \cap \rho(A_0) \neq \varnothing$ or $\rho(\overline{A_\Theta}) \cap \rho(A_0) \neq \varnothing$, e.g. if A_Θ is self-adjoint or essentially self-adjoint, respectively, then for $\lambda \in \rho(\overline{A_\Theta}) \cap \rho(A_0)$, relation (6.25) is valid on $\mathcal{H}$ or a dense subset of $\mathcal{H}$, respectively. This, together with the fact that $\gamma(\lambda)^*$ is an everywhere defined bounded operator and

$$\gamma(\lambda)\bigl(\Theta - M(\lambda)\bigr)^{-1}\gamma(\lambda)^* \subset \overline{\gamma(\lambda)\bigl(\Theta - M(\lambda)\bigr)^{-1}}\gamma(\bar\lambda)^*$$

implies the following corollary.

Corollary 6.17 *Let the assumptions be as in Theorem 6.16. Then the following assertions hold.*

(i) *If $\lambda \in \rho(A_\Theta) \cap \rho(A_0)$, then*

$$(A_\Theta - \lambda)^{-1} = (A_0 - \lambda)^{-1} + \gamma(\lambda)\bigl(\Theta - M(\lambda)\bigr)^{-1}\gamma(\bar\lambda)^*. \quad (6.26)$$

(ii) *If $\lambda \in \rho(\overline{A_\Theta}) \cap \rho(A_0)$, then*

$$(\overline{A_\Theta} - \lambda)^{-1} = (A_0 - \lambda)^{-1} + \overline{\gamma(\lambda)\bigl(\Theta - M(\lambda)\bigr)^{-1}}\gamma(\bar\lambda)^*.$$

In particular, if A_Θ is self-adjoint, then Krein's formula (6.26) holds at least for all non-real λ.

If the relation Θ in Theorem 6.16 is self-adjoint, then Krein's formula can be rewritten as follows: let $\{\Phi,\Psi\}$ be a pair of bounded operators in $\mathcal{G}$ such that

$$\Theta = \{(\Phi k;\Psi k)^\top : k \in \mathcal{G}\} \tag{6.27}$$

holds; cf. (6.6) and note that (6.5) has to be satisfied. It follows that $\ker(\Theta - M(\lambda)) = \{0\}$ if and only if $\ker(\Psi - M(\lambda)\Phi) = \{0\}$. Then

$$(\Theta - M(\lambda))^{-1} = \{(\Psi k - M(\lambda)\Phi k;\Phi k)^\top : k \in \mathcal{G},\ \Phi k \in \operatorname{dom} M(\lambda)\}$$

together with Theorem 6.16 yield the following corollary.

Corollary 6.18 *Let S be a densely defined closed symmetric operator in $\mathcal{H}$ and let $\{\mathcal{G},\Gamma_0,\Gamma_1\}$ be a quasi boundary triple for $\overline{T} = S^*$ with $A_0 = T \upharpoonright \ker\Gamma_0$, γ-field γ and Weyl function M. Further, let Θ be a self-adjoint relation in $\mathcal{G}$ represented with a pair $\{\Phi,\Psi\}$ in the form* (6.27) *and assume that $\lambda \in \rho(A_0)$ is not an eigenvalue of A_Θ, or, equivalently, that* $\ker(\Psi - M(\lambda)\Phi) = \{0\}$. *Then the following assertions are true.*

(i) *$g \in \operatorname{ran}(A_\Theta - \lambda)$ if and only if $\gamma(\bar\lambda)^* g \in \operatorname{dom}(\Psi - M(\lambda)\Phi)^{-1}$.*

(ii) *For all $g \in \operatorname{ran}(A_\Theta - \lambda)$ we have*

$$(A_\Theta - \lambda)^{-1} g = (A_0 - \lambda)^{-1} g + \gamma(\lambda)\Phi\big(\Psi - M(\lambda)\Phi\big)^{-1}\gamma(\bar\lambda)^* g.$$

Let us now turn to the proof of Theorem 6.16.

Proof of Theorem 6.16 Let us first show that $\lambda \in \rho(A_0)$ is not an eigenvalue of A_Θ if and only if $\ker(\Theta - M(\lambda)) = \{0\}$. Assume, e.g. that $f \in \ker(A_\Theta - \lambda)$ and $f \neq 0$. Then $f \in \mathcal{N}_\lambda(T)$ and as $\Gamma f \in \Theta$, we obtain

$$\begin{pmatrix}\Gamma_0 f\\ 0\end{pmatrix} = \begin{pmatrix}\Gamma_0 f\\ \Gamma_1 f - M(\lambda)\Gamma_0 f\end{pmatrix} \in \Theta - M(\lambda).$$

Moreover, $\Gamma_0 f \neq 0$ because otherwise $f \in \operatorname{dom} A_0 \cap \mathcal{N}_\lambda(T)$, which would imply $f = 0$. Conversely, if $y \in \ker(\Theta - M(\lambda))$ and $y \neq 0$, then

$$\begin{pmatrix} y\\ M(\lambda)y\end{pmatrix} \in \Theta,$$

and for $f := \gamma(\lambda)y \in \mathcal{N}_\lambda(T)$ we obtain

$$\begin{pmatrix}\Gamma_0 f\\ \Gamma_1 f\end{pmatrix} = \begin{pmatrix} y\\ M(\lambda)y\end{pmatrix} \in \Theta.$$

Therefore $f \in \operatorname{dom} A_\Theta$, i.e. $\gamma(\lambda)y \in \ker(A_\Theta - \lambda)$. Thus $\lambda \in \rho(A_0)$ is not an eigenvalue of A_Θ if and only if $\ker(\Theta - M(\lambda)) = \{0\}$.

Now fix some point $\lambda \in \rho(A_0)$ which is not an eigenvalue of A_Θ. Then $(A_\Theta - \lambda)^{-1}$ and $(\Theta - M(\lambda))^{-1}$ are operators in $\mathcal{H}$ and $\mathcal{G}$, respectively. Let $g \in \operatorname{ran}(A_\Theta - \lambda)$. We show that $\gamma(\bar{\lambda})^* g \in \operatorname{dom}(\Theta - M(\lambda))^{-1}$ and that formula (6.25) holds. Set

$$f := (A_\Theta - \lambda)^{-1} g - (A_0 - \lambda)^{-1} g \quad \text{and} \quad h := (A_\Theta - \lambda)^{-1} g.$$

Then we have $f \in \mathcal{N}_\lambda(T)$ and $h \in \operatorname{dom} A_\Theta$. Moreover,

$$\Gamma_0 f = \Gamma_0 h - \Gamma_0 (A_0 - \lambda)^{-1} g = \Gamma_0 h$$

since $(A_0 - \lambda)^{-1} g \in \operatorname{dom} A_0 = \ker \Gamma_0$, and

$$\Gamma_1 f = \Gamma_1 h - \Gamma_1 (A_0 - \lambda)^{-1} g = \Gamma_1 h - \gamma(\bar{\lambda})^* g$$

by Proposition 6.13 (iii). These equalities together with Proposition 6.14 (ii) yield

$$\gamma(\bar{\lambda})^* g = \Gamma_1 h - \Gamma_1 f = \Gamma_1 h - M(\lambda)\Gamma_0 f = \Gamma_1 h - M(\lambda)\Gamma_0 h.$$

Since $h \in \operatorname{dom} A_\Theta$, we have $(\Gamma_0 h; \Gamma_1 h)^\top \in \Theta$ by (6.16) and hence

$$\begin{pmatrix} \Gamma_0 h \\ \gamma(\bar{\lambda})^* g \end{pmatrix} = \begin{pmatrix} \Gamma_0 h \\ \Gamma_1 h - M(\lambda)\Gamma_0 h \end{pmatrix} \in \Theta - M(\lambda), \qquad (6.28)$$

which implies $\gamma(\lambda)^* g \in \operatorname{dom}(\Theta - M(\lambda))^{-1}$, i.e. $\Rightarrow$ in (i) is proved. Furthermore, it follows from (6.28) that $\Gamma_0 h = (\Theta - M(\lambda))^{-1}\gamma(\bar{\lambda})^* g$ since $(\Theta - M(\lambda))^{-1}$ is an operator. Therefore

$$\begin{aligned} \gamma(\lambda)\big(\Theta - M(\lambda)\big)^{-1}\gamma(\bar{\lambda})^* g &= \gamma(\lambda)\Gamma_0 h = \gamma(\lambda)\Gamma_0 f \\ &= f = (A_\Theta - \lambda)^{-1} g - (A_0 - \lambda)^{-1} g, \end{aligned}$$

which shows relation (6.25).

For $\Leftarrow$ in (i) assume that $\Theta - M(\lambda)$ is injective and let $\gamma(\bar{\lambda})^* g \in \operatorname{ran}(\Theta - M(\lambda))$ for some $g \in \mathcal{H}$. Then $(\Theta - M(\lambda))^{-1}\gamma(\bar{\lambda})^* g$ belongs to $\operatorname{dom}(\Theta - M(\lambda)) \subset \mathcal{G}_0$, and we claim that

$$f := \gamma(\lambda)\big(\Theta - M(\lambda)\big)^{-1}\gamma(\bar{\lambda})^* g + (A_0 - \lambda)^{-1} g \in \operatorname{dom} A_\Theta.$$

Clearly, $f \in \operatorname{dom} T$. Moreover, the relations

$$\begin{aligned} \Gamma_0 f &= \big(\Theta - M(\lambda)\big)^{-1}\gamma(\bar{\lambda})^* g, \\ \Gamma_1 f &= M(\lambda)\big(\Theta - M(\lambda)\big)^{-1}\gamma(\bar{\lambda})^* g + \gamma(\bar{\lambda})^* g \end{aligned}$$

and

$$\begin{pmatrix} \big(\Theta - M(\lambda)\big)^{-1}\gamma(\bar{\lambda})^* g \\ \gamma(\bar{\lambda})^* g \end{pmatrix} \in \Theta - M(\lambda),$$

imply that

$$\begin{pmatrix}\Gamma_0 f\\ \Gamma_1 f\end{pmatrix} = \begin{pmatrix}(\Theta - M(\lambda))^{-1}\gamma(\bar{\lambda})^* g\\ M(\lambda)(\Theta - M(\lambda))^{-1}\gamma(\bar{\lambda})^* g + \gamma(\bar{\lambda})^* g\end{pmatrix}$$
$$\in M(\lambda) + (\Theta - M(\lambda)) \subset \Theta,$$

that is, $f \in \operatorname{dom} A_\Theta$. Since $\gamma(\lambda)$ maps into $\ker(T - \lambda)$, we have

$$(A_\Theta - \lambda)f = (T-\lambda)\gamma(\lambda)(\Theta - M(\lambda))^{-1}\gamma(\bar{\lambda})^* g + (T-\lambda)(A_0 - \lambda)^{-1} g = g,$$

which shows that $g \in \operatorname{ran}(A_\Theta - \lambda)$. □

If Θ is a self-adjoint (maximal dissipative, maximal accumulative) relation in $\mathcal{G}$, then we can decompose Θ as follows. Let $\mathcal{G}_\infty := \operatorname{mul}\Theta$, $\mathcal{G}_{\rm op} := \mathcal{G}_\infty^\perp$ and denote by $P_{\rm op}$, P_∞ the orthogonal projections onto $\mathcal{G}_{\rm op}$ and $\mathcal{G}_\infty$, respectively. Then the relation Θ can be written as

$$\Theta = \Theta_{\rm op} \oplus \Theta_\infty,$$

where $\Theta_{\rm op}$ is a self-adjoint (maximal dissipative, maximal accumulative, respectively) operator in $\mathcal{G}_{\rm op}$ and $\Theta_\infty = \{(0; y)^\top : y \in \mathcal{G}_\infty\}$. In the next corollary Krein's formula is rewritten in terms of this decomposition. The canonical embedding of $\mathcal{G}_{\rm op}$ in $\mathcal{G}$ is denoted by $\iota_{\rm op}$.

Corollary 6.19 *Let S, T, $\{\mathcal{G}, \Gamma_0, \Gamma_1\}$, γ and M be as in Theorem 6.16. Further, let Θ be a self-adjoint, maximal dissipative or maximal accumulative relation in $\mathcal{G}$ and assume that $\lambda \in \rho(A_0)$ is not an eigenvalue of A_Θ. Then*

$$\Theta_{\rm op} - P_{\rm op} M(\lambda)|_{\mathcal{G}_{\rm op}}$$

defined on $\operatorname{dom}\Theta_{\rm op} \cap \operatorname{dom} M(\lambda)$ *is an injective operator in* $\mathcal{G}_{\rm op}$ *and*

$$\begin{aligned}&(A_\Theta - \lambda)^{-1} g\\ &\quad = (A_0 - \lambda)^{-1} g + \gamma(\lambda)\,\iota_{\rm op}\big(\Theta_{\rm op} - P_{\rm op} M(\lambda)|_{\mathcal{G}_{\rm op}}\big)^{-1} P_{\rm op}\gamma(\bar{\lambda})^* g \qquad (6.29)\end{aligned}$$

holds for all $g \in \operatorname{ran}(A_\Theta - \lambda)$.

Proof First we show that $\Theta_{\rm op} - P_{\rm op}M(\lambda)|_{\mathcal{G}_{\rm op}}$ is an injective operator in $\mathcal{G}_{\rm op}$. Let $x \in \operatorname{dom}\Theta_{\rm op} \cap \operatorname{dom} M(\lambda)$ be such that $(\Theta_{\rm op} - P_{\rm op}M(\lambda))x = 0$. Then we have

$$\big(x; (\Theta_{\rm op} - P_{\rm op}M(\lambda))x \oplus (y - (I - P_{\rm op})M(\lambda)x)\big) \in \Theta - M(\lambda), \quad y \in \mathcal{G}_\infty,$$

and according to Theorem 6.16, $\Theta - M(\lambda)$ is injective; thus $x = 0$. It remains to show the equality

$$\big(\Theta - M(\lambda)\big)^{-1} = \iota_{\mathrm{op}}\big(\Theta_{\mathrm{op}} - P_{\mathrm{op}}M(\lambda)|_{\mathcal{G}_{\mathrm{op}}}\big)^{-1}P_{\mathrm{op}}, \tag{6.30}$$

which was proved in [Langer and Textorius, 1977, (1.3)] for the case when $M(\lambda) \in \boldsymbol{B}(\mathcal{G})$. We have the following chain of equivalences (note that $\mathrm{dom}\,\Theta = \mathrm{dom}\,\Theta_{\mathrm{op}} \subset \mathcal{G}$):

$$\begin{aligned}
&(x;y)^{\top} \in \big(\Theta - M(\lambda)\big)^{-1} \\
&\iff y \in \mathrm{dom}\,\Theta \cap \mathrm{dom}\,M(\lambda),\ \exists u \in \mathcal{G}\colon (y;u)^{\top} \in \Theta,\ x = u - M(\lambda)y \\
&\iff y \in \mathrm{dom}\,\Theta \cap \mathrm{dom}\,M(\lambda),\ \exists u \in \mathcal{G}\colon P_{\mathrm{op}}u = \Theta_{\mathrm{op}}y,\ x = u - M(\lambda)y \\
&\iff y \in \mathrm{dom}\,\Theta \cap \mathrm{dom}\,M(\lambda),\ P_{\mathrm{op}}x = \big(\Theta_{\mathrm{op}} - P_{\mathrm{op}}M(\lambda)\big)y \\
&\iff (P_{\mathrm{op}}x;y)^{\top} \in \big(\Theta_{\mathrm{op}} - P_{\mathrm{op}}M(\lambda)|_{\mathcal{G}_{\mathrm{op}}}\big)^{-1},
\end{aligned}$$

which shows (6.30). Now formula (6.29) follows from Theorem 6.16. □

With the help of Krein's formula one can show the following theorem, which provides a sufficient condition for self-adjointness of the extension A_Θ and which was proved in [Behrndt, Langer and Lotoreichik, 2011]. We make use of the notation

$$\Theta^{-1}(X) := \left\{ x \in \mathcal{G}\colon \exists y \in X \text{ so that } \begin{pmatrix} x \\ y \end{pmatrix} \in \Theta \right\}$$

for a linear relation Θ in $\mathcal{G}$ and a subspace $X \subset \mathcal{G}$.

Theorem 6.20 *Let S be a densely defined closed symmetric operator in $\mathcal{H}$ and let $\{\mathcal{G}, \Gamma_0, \Gamma_1\}$ be a quasi boundary triple for $\overline{T} = S^*$ with $A_i = T \restriction \ker \Gamma_i$, $i = 0, 1$, and Weyl function M. Assume that A_1 is self-adjoint and that $\overline{M(\lambda_0)}$ is a compact operator in $\mathcal{G}$ for some $\lambda_0 \in \mathbb{C}\backslash\mathbb{R}$. If Θ is a self-adjoint relation in $\mathcal{G}$ such that*

$$0 \notin \sigma_{\mathrm{ess}}(\Theta) \qquad \textit{and} \qquad \Theta^{-1}\big(\mathrm{ran}\,\overline{M(\lambda_\pm)}\big) \subset \mathcal{G}_0 \tag{6.31}$$

hold for some $\lambda_+ \in \mathbb{C}^+$ and some $\lambda_- \in \mathbb{C}^-$, then A_Θ as defined in (6.16) is a self-adjoint operator in $\mathcal{H}$. In particular, the second condition in (6.31) is satisfied if $\mathrm{dom}\,\Theta \subset \mathcal{G}_0$.

Remark We also mention that if in the above theorem Θ is assumed to be maximal dissipative (maximal accumulative) and the second condition in (6.31) is replaced by the condition

$$\Theta^{-1}\big(\mathrm{ran}\,\overline{M(\lambda)}\big) \subset \mathcal{G}_0$$

for some $\lambda \in \mathbb{C}^-$ ($\lambda \in \mathbb{C}^+$, respectively), then the operator A_Θ in (6.16) is maximal dissipative (maximal accumulative) in $\mathcal{H}$.

We formulate another variant of Theorem 6.20 below, which will be used later on. Observe that if $\{\mathcal{G}, \Gamma_0, \Gamma_1\}$ is a quasi boundary triple for $\overline{T} = S^*$ with corresponding Weyl function M and if, in addition, $A_1 = T \upharpoonright \ker \Gamma_1$ is self-adjoint, then $\{\mathcal{G}, \widetilde{\Gamma}_0, \widetilde{\Gamma}_1\}$, where $\widetilde{\Gamma}_0 := -\Gamma_1$ and $\widetilde{\Gamma}_1 := \Gamma_0$, is also a quasi boundary triple for $\overline{T} = S^*$ with corresponding Weyl function $\widetilde{M} = -M^{-1}$ and self-adjoint operator $\widetilde{A}_1 = T \upharpoonright \ker \widetilde{\Gamma}_1 = A_0$. Moreover,

$$A_\Theta = T \upharpoonright \{f \in \operatorname{dom} T : \Gamma f \in \Theta\} = T \upharpoonright \{f \in \operatorname{dom} T : \widetilde{\Gamma} f \in \widetilde{\Theta}\} = \widetilde{A}_{\widetilde{\Theta}}$$

holds with $\widetilde{\Theta} = -\Theta^{-1}$. This transformation of quasi boundary triples leads to the following theorem.

Theorem 6.21 *Let S be a densely defined closed symmetric operator in $\mathcal{H}$ and let $\{\mathcal{G}, \Gamma_0, \Gamma_1\}$ be a quasi boundary triple for $\overline{T} = S^*$ with $A_i = T \upharpoonright \ker \Gamma_i$, $i = 0, 1$, and Weyl function M. Assume that A_1 is self-adjoint and that $\overline{M(\lambda_0)^{-1}}$ is a compact operator in $\mathcal{G}$ for some $\lambda_0 \in \mathbb{C} \backslash \mathbb{R}$. If Θ is a bounded self-adjoint operator in $\mathcal{G}$ such that*

$$\Theta\big(\operatorname{dom} \overline{M(\lambda_\pm)}\big) \subset \mathcal{G}_1 \tag{6.32}$$

holds for some $\lambda_+ \in \mathbb{C}^+$ and some $\lambda_- \in \mathbb{C}^-$, then

$$A_\Theta = T \upharpoonright \ker (\Gamma_1 - \Theta \Gamma_0) \tag{6.33}$$

is a self-adjoint operator in $\mathcal{H}$. In particular, the condition (6.32) *is satisfied if* $\operatorname{ran} \Theta \subset \mathcal{G}_1$.

Remark For completeness we also remark that for a dissipative (accumulative) bounded operator Θ with the property

$$\Theta\big(\operatorname{dom} \overline{M(\lambda)}\big) \subset \mathcal{G}_1$$

for some $\lambda \in \mathbb{C}^-$ ($\lambda \in \mathbb{C}^+$, respectively) it follows that the operator A_Θ in (6.33) is maximal dissipative (maximal accumulative) in $\mathcal{H}$.

6.4 Quasi boundary triples for elliptic operators and Dirichlet-to-Neumann maps

In this section we consider the same type of elliptic operators as in Section 6.2 and we define and study a family of quasi boundary triples for the maximal operator. As boundary mappings we choose the Dirichlet and (oblique) Neumann trace so that the associated Weyl function turns out to be the Dirichlet-to-Neumann map. Only the case of a bounded domain Ω is treated here, although the considerations for unbounded domains with compact boundaries (so-called exterior domains) are very similar; cf. [Behrndt, Langer and Lotoreichik, 2011].

Let again Ω be a bounded domain in $\mathbb{R}^n$, $n > 1$, with C^∞-boundary $\partial\Omega$ and consider the expression

$$\mathcal{L} = -\sum_{j,k=1}^{n} \partial_j\, a_{jk}\, \partial_k + a$$

on Ω with real-valued coefficients $a_{jk} \in C^\infty(\overline{\Omega})$, $a \in L^\infty(\Omega)$ such that $a_{jk} = a_{kj}$ for all $j, k = 1, \ldots, n$. In addition, $\mathcal{L}$ is assumed to be elliptic, that is,

$$\sum_{j,k=1}^{n} a_{jk}(x)\xi_j\xi_k \geq C\sum_{k=1}^{n} \xi_k^2, \qquad \xi = (\xi_1, \ldots, \xi_n)^\top \in \mathbb{R}^n, \ x \in \overline{\Omega},$$

holds for some constant $C > 0$. In the following, the spaces

$$H^s_{\mathcal{L}}(\Omega) := \{f \in H^s(\Omega) \colon \mathcal{L}(f) \in L^2(\Omega)\}, \qquad s \in [\tfrac{3}{2}, 2],$$

are used as domains for the boundary mappings; the cases $s = 2$ and $s = \frac{3}{2}$ were already studied in [Behrndt and Langer, 2007]. The spaces $H^s_{\mathcal{L}}(\Omega)$ are frequently used in the theory of elliptic operators; see, e.g. [Grubb, 1968, 1971; Lions and Magenes, 1972] and are usually defined for all $s \in [0, \infty)$. Then, in particular, $H^0_{\mathcal{L}}(\Omega)$ coincides with the maximal domain $\mathfrak{D}_{\max}$ and $H^s_{\mathcal{L}}(\Omega) = H^s(\Omega)$ for $s \geq 2$. In the following we deal with the family of differential operators T_s, $s \in [\frac{3}{2}, 2]$, defined by

$$T_s f = \mathcal{L}(f), \qquad \operatorname{dom} T_s = H^s_{\mathcal{L}}(\Omega).$$

Recall that the minimal operator associated with $\mathcal{L}$ in $L^2(\Omega)$ is the densely defined closed symmetric operator $Sf = \mathcal{L}(f)$, $\operatorname{dom} S = H^2_0(\Omega)$, that S has equal and infinite deficiency indices, and that the adjoint S^* of S coincides with the maximal realization of $\mathcal{L}$ in $L^2(\Omega)$ defined on $\mathfrak{D}_{\max}$; see (6.11). The self-adjoint realizations of $\mathcal{L}$ in $L^2(\Omega)$ with Dirichlet or Neumann boundary conditions are denoted by A_{D} and A_{N}, respectively, i.e.

$$\begin{aligned} A_{\mathrm{D}} f &= \mathcal{L}(f), &\quad \operatorname{dom} A_{\mathrm{D}} &= \{f \in H^2(\Omega) \colon f|_{\partial\Omega} = 0\}, \\ A_{\mathrm{N}} f &= \mathcal{L}(f), &\quad \operatorname{dom} A_{\mathrm{N}} &= \left\{f \in H^2(\Omega) \colon \frac{\partial f}{\partial \nu_{\mathcal{L}}}\Big|_{\partial\Omega} = 0\right\}, \end{aligned}$$

where $\frac{\partial f}{\partial \nu_{\mathcal{L}}}\big|_{\partial\Omega}$ is defined as in (6.12).

The proof of the next proposition consists in principle of applying Theorem 6.11. However, we provide a short proof here for the convenience of the reader; cf. [Behrndt and Langer, 2007, Proposition 4.6] for a similar consideration.

Proposition 6.22 *For each $s \in [\frac{3}{2}, 2]$ the triple $\{L^2(\partial\Omega), \Gamma_0, \Gamma_1\}$, where*

$$\Gamma_0 f := f|_{\partial\Omega} \qquad \text{and} \qquad \Gamma_1 f := -\frac{\partial f}{\partial \nu_{\mathcal{L}}}\Big|_{\partial\Omega}, \quad f \in H^s_{\mathcal{L}}(\Omega), \tag{6.34}$$

is a quasi boundary triple for the maximal operator $\overline{T}_s = S^*$ such that

$$A_{\mathrm{D}} = T_s \upharpoonright \ker \Gamma_0 \qquad \text{and} \qquad A_{\mathrm{N}} = T_s \upharpoonright \ker \Gamma_1.$$

Proof We apply Theorem 6.11. Since $H^2(\Omega) \subset H^s_{\mathcal{L}}(\Omega)$ for all $s \in [\frac{3}{2}, 2]$, the restriction of T_s to

$$\ker \Gamma_0 = \big\{ f \in H^s_{\mathcal{L}}(\Omega) \colon f|_{\partial\Omega} = 0 \big\}$$

is an extension of the self-adjoint Dirichlet operator A_{D}, i.e. condition (i) of Theorem 6.11 is satisfied. In order to verify condition (ii), note first that for $s \in [\frac{3}{2}, 2]$ and $f \in H^s_{\mathcal{L}}(\Omega)$ we have $f|_{\partial\Omega} \in L^2(\partial\Omega)$, and for $s \in (\frac{3}{2}, 2]$ and $f \in H^s_{\mathcal{L}}(\Omega)$ we have $\frac{\partial f}{\partial \nu_{\mathcal{L}}}|_{\partial\Omega} \in L^2(\partial\Omega)$. According to [Grubb, 1968, Theorem I.3.3] and [Lions and Magenes, 1972], $\frac{\partial f}{\partial \nu_{\mathcal{L}}}|_{\partial\Omega} \in L^2(\partial\Omega)$ holds also for $s = \frac{3}{2}$ and $f \in H^s_{\mathcal{L}}(\Omega)$. Hence Γ_0, Γ_1 are well defined. Since the map

$$H^2(\Omega) \ni f \mapsto \left\{ f|_{\partial\Omega}, \frac{\partial f}{\partial \nu_{\mathcal{L}}}\Big|_{\partial\Omega} \right\} \in H^{3/2}(\partial\Omega) \times H^{1/2}(\partial\Omega)$$

is surjective onto the dense subset $H^{3/2}(\partial\Omega) \times H^{1/2}(\partial\Omega)$ of the space $L^2(\partial\Omega) \times L^2(\partial\Omega)$, see e.g. [Lions and Magenes, 1972, Theorem 1.8.3], and $H^2(\Omega) \subset H^s_{\mathcal{L}}(\Omega)$, $s \in [\frac{3}{2}, 2]$, it follows that $\operatorname{ran}(\Gamma_0, \Gamma_1)^\top$ is dense in $L^2(\partial\Omega) \times L^2(\partial\Omega)$, i.e. the first condition in (ii) of Theorem 6.11 is satisfied; the second condition follows from $C_0^\infty(\Omega) \subset \ker \Gamma_0 \cap \ker \Gamma_1$. With Γ_0 and Γ_1 from (6.34), Green's identity reads as

$$\begin{aligned}(T_s f, g) - (f, T_s g) &= \left(f|_{\partial\Omega}, \frac{\partial g}{\partial \nu_{\mathcal{L}}}\Big|_{\partial\Omega} \right) - \left(\frac{\partial f}{\partial \nu_{\mathcal{L}}}\Big|_{\partial\Omega}, g|_{\partial\Omega} \right) \\ &= (\Gamma_1 f, \Gamma_0 g) - (\Gamma_0 f, \Gamma_1 g)\end{aligned}$$

with the inner product $(\cdot, -)$ in $L^2(\partial\Omega)$ on the right-hand side. Hence also condition (iii) in Theorem 6.11 is fulfilled. Therefore the operator

$$T \upharpoonright \ker \Gamma_0 \cap \ker \Gamma_1 \tag{6.35}$$

is a densely defined closed symmetric operator in $L^2(\Omega)$. Furthermore, $\{L^2(\partial\Omega), \Gamma_0, \Gamma_1\}$ is a quasi boundary triple for its adjoint and $T \upharpoonright \ker \Gamma_0$ is equal to the Dirichlet operator A_{D}. Since $\ker \Gamma_0 = \operatorname{dom} A_{\mathrm{D}} \subset H^2(\Omega)$, we have

$$\begin{aligned}\ker \Gamma_0 \cap \ker \Gamma_1 &= \{f \in H^s_{\mathcal{L}}(\Omega) : \Gamma_0 f = 0,\, \Gamma_1 f = 0\} \\ &= \{f \in H^2(\Omega) : \Gamma_0 f = 0,\, \Gamma_1 f = 0\} = H^2_0(\Omega),\end{aligned}$$

which shows that the operator in (6.35) is the minimal operator S associated with $\mathcal{L}$. □

Remark We point out that for the statements in the above proposition the scale $[\frac{3}{2}, 2]$ cannot be enlarged. The upper bound 2 is necessary in order to ensure that the self-adjoint Dirichlet operator is contained in (and hence equal to) $T \upharpoonright \ker \Gamma_0$, whereas the lower bound $\frac{3}{2}$ is necessary to ensure Green's identity with boundary terms in $L^2(\partial\Omega)$. However, Green's identity could also be considered, e.g. for functions $f, g \in H^1_{\mathcal{L}}(\Omega)$ so that on the right-hand side the extension of the $L^2(\partial\Omega)$ inner product to $H^{1/2}(\partial\Omega) \times H^{-1/2}(\partial\Omega)$ appears, which makes it necessary to modify the boundary mappings by isomorphisms $\iota_\pm$ as in Proposition 6.9. In this case the corresponding Weyl function is not the Dirichlet-to-Neumann map in $L^2(\partial\Omega)$. For completeness we also mention that for $s = \frac{3}{2}$ the quasi boundary triple $\{L^2(\partial\Omega), -\Gamma_1, \Gamma_0\}$ is a generalized boundary triple in the sense of [Derkach and Malamud, 1995]; see also [Behrndt and Langer, 2007, Section 4.2].

In the next proposition the γ-field and Weyl function corresponding to the quasi boundary triples $\{L^2(\partial\Omega), \Gamma_0, \Gamma_1\}$ from Proposition 6.22 are specified.

Proposition 6.23 *Let $s \in [\frac{3}{2}, 2]$ and let $\{L^2(\partial\Omega), \Gamma_0, \Gamma_1\}$ be the quasi boundary triple for the maximal operator $\overline{T}_s = S^*$ from Proposition 6.22 with $A_{\mathrm{D}} = T_s \upharpoonright \ker \Gamma_0$. Then the following statements are true for all $\lambda \in \rho(A_{\mathrm{D}})$:*

(i) *For $y \in H^{s-1/2}(\partial\Omega)$ there exists a unique function $f_\lambda(y)$ in $H^s_{\mathcal{L}}(\Omega)$ that solves the boundary value problem*

$$\mathcal{L}(u) = \lambda u, \qquad u|_{\partial\Omega} = y. \tag{6.36}$$

(ii) *The γ-field of* $\{L^2(\partial\Omega), \Gamma_0, \Gamma_1\}$ *is given by*

$$\gamma_s(\lambda)\colon L^2(\partial\Omega) \to L^2(\Omega), \qquad y \mapsto f_\lambda(y),$$

with $\operatorname{dom}\gamma_s(\lambda) = H^{s-1/2}(\partial\Omega)$.

(iii) *The Weyl function of* $\{L^2(\partial\Omega), \Gamma_0, \Gamma_1\}$ *is given by*

$$M_s(\lambda)\colon L^2(\partial\Omega) \to L^2(\partial\Omega), \qquad y \mapsto -\frac{\partial f_\lambda(y)}{\partial \nu_{\mathcal{L}}}\Big|_{\partial\Omega},$$

with $\operatorname{dom} M_s(\lambda) = H^{s-1/2}(\partial\Omega)$ *and* $\operatorname{ran} M_s(\lambda) \subset H^{s-3/2}(\partial\Omega)$. *If, in addition,* $\lambda \in \rho(A_{\mathrm{N}})$, *then* $\operatorname{ran} M_s(\lambda) = H^{s-3/2}(\partial\Omega)$ *and* $\overline{M_s(\lambda)^{-1}}$ *is a compact operator in* $L^2(\partial\Omega)$ *with values in* $H^1(\partial\Omega)$.

Proof (i) Problem (6.36) is equivalent to

$$u \in \ker(T_s - \lambda), \qquad \Gamma_0 u = y.$$

Proposition 6.13 (i) shows that this boundary value problem has a unique solution, namely $u = \gamma_s(\lambda)y$.

(ii) is a consequence of (i). Observe that the domain of $\gamma_s(\lambda)$ is equal to $\operatorname{ran}\Gamma_0 = H^{s-1/2}(\partial\Omega)$.

(iii) The asserted form of the Weyl function M_s follows immediately by the definition. That $\operatorname{ran} M_s(\lambda) = H^{s-3/2}(\partial\Omega)$ for $\lambda \in \rho(A_{\mathrm{N}})$ is clear from Proposition 6.14 (i). Using duality and interpolation arguments one can show that $M_s(\lambda)^{-1}$ can be extended to a bounded operator from $H^r(\partial\Omega)$ to $H^{r+1}(\partial\Omega)$ for $r \in [-\frac{3}{2}, \frac{1}{2}]$ and $\lambda \in \rho(A_{\mathrm{D}})$; for details see, e.g. [Behrndt, Langer and Lotoreichik, 2011; Lions and Magenes, 1972; Seeley, 1969]. In particular, $\overline{M_s(\lambda)^{-1}}$ is a bounded mapping from $L^2(\partial\Omega)$ to $H^1(\partial\Omega)$. Since $H^1(\partial\Omega)$ is compactly embedded in $L^2(\partial\Omega)$ (see, e.g. [Wloka, 1987, Theorem 7.10]), this shows the compactness of $\overline{M_s(\lambda)^{-1}}$. □

The operator $\gamma_s(\lambda)$ in the previous proposition is often called Poisson operator. The Weyl function M_s is (up to a minus sign) the Dirichlet-to-Neumann operator connected with $\mathcal{L}$. It has been used, e.g. to solve inverse problems, see [Astala and Päivärinta, 2006; Nachman, 1988, 1996; Nachman, Sylvester and Uhlmann, 1988; Sylvester and Uhlmann, 1987], to detect spurious eigenvalues in numerical calculations, see [Brown and Marletta, 2004; Marletta, 2004, 2010] and to prove inequalities between Dirichlet and Neumann eigenvalues, see [Filonov, 2004; Friedlander, 1991; Safarov, 2008]. See also [Amrein and Pearson, 2004] where a Weyl function for elliptic operators was constructed. Moreover, Dirichlet-to-Neumann maps on rough domains were defined and studied in [Arendt and ter Elst, 2011].

As a consequence of Theorem 6.21, the remark below that theorem and Proposition 6.23 (iii), we obtain the following sufficient condition for a self-adjoint, maximal dissipative or maximal accumulative parameter Θ in $L^2(\partial\Omega)$ to determine a self-adjoint, maximal dissipative or maximal accumulative realization A_Θ of $\mathcal{L}$ in $L^2(\Omega)$.

Theorem 6.24 *Let $s \in [\frac{3}{2}, 2]$ and let $\{L^2(\partial\Omega), \Gamma_0, \Gamma_1\}$ be the quasi boundary triple for $\overline{T}_s = S^*$ from Proposition 6.22. For every bounded self-adjoint (dissipative, accumulative) operator Θ in $L^2(\partial\Omega)$ such that*

$$\Theta\big(H^1(\partial\Omega)\big) \subset H^{s-3/2}(\partial\Omega) \tag{6.37}$$

is satisfied, the differential operator

$$A_\Theta f = \mathcal{L}(f), \ \mathrm{dom}\, A_\Theta = \left\{ f \in H^s_{\mathcal{L}}(\Omega) \colon \Theta\, f|_{\partial\Omega} = -\frac{\partial f}{\partial \nu_{\mathcal{L}}}\Big|_{\partial\Omega} \right\} \tag{6.38}$$

is a self-adjoint (maximal dissipative, maximal accumulative, respectively) realization of $\mathcal{L}$ in $L^2(\Omega)$.

Since for $s = \frac{3}{2}$ the condition (6.37) in the above theorem reduces to $\Theta(H^1(\partial\Omega)) \subset L^2(\partial\Omega)$, which is trivially satisfied, we obtain the following corollary; cf. [Behrndt and Langer, 2007, Theorem 4.8].

Corollary 6.25 *Let $s = \frac{3}{2}$ and let $\{L^2(\partial\Omega), \Gamma_0, \Gamma_1\}$ be the quasi boundary triple for $\overline{T}_{3/2} = S^*$ from Proposition 6.22. For every bounded self-adjoint (dissipative, accumulative) operator Θ in $L^2(\partial\Omega)$, the differential operator A_Θ in (6.38) is a self-adjoint (maximal dissipative, maximal accumulative, respectively) realization of $\mathcal{L}$ in $L^2(\Omega)$.*

As a consequence of Theorem 6.16 and its corollaries, we obtain a variant of Krein's formula for the self-adjoint (maximal dissipative, maximal accumulative) realizations of $\mathcal{L}$ in Theorem 6.24 and the above corollary.

Theorem 6.26 *Let $s \in [\frac{3}{2}, 2]$ and let $\{L^2(\partial\Omega), \Gamma_0, \Gamma_1\}$ be the quasi boundary triple for $\overline{T}_s = S^*$ from Proposition 6.22 with γ-field γ_s and Weyl function M_s from Proposition 6.23. Let Θ be a self-adjoint (dissipative, accumulative) operator in $L^2(\partial\Omega)$ such that (6.37) is satisfied. Then for all $\lambda \in \rho(A_\Theta) \cap \rho(A_{\mathrm{D}})$*

$$(A_\Theta - \lambda)^{-1} = (A_{\mathrm{D}} - \lambda)^{-1} + \gamma_s(\lambda)\big(\Theta - M_s(\lambda)\big)^{-1}\gamma_s(\bar{\lambda})^*. \tag{6.39}$$

Krein's formula (6.39) in the above theorem allows the following interpretation: since A_Θ and A_{D} act formally in the same way (as both operators are realizations of the same differential expression $\mathcal{L}$), only their domains are different, and since $\operatorname{dom} A_\Theta$ and $\operatorname{dom} A_{\mathrm{D}}$ are specified by boundary conditions on functions from $H^s_{\mathcal{L}}(\Omega)$, the resolvent difference $(A_\Theta - \lambda)^{-1} - (A_{\mathrm{D}} - \lambda)^{-1}$ can be "localized" on the boundary $\partial\Omega$, that is, as the perturbation term $(\Theta - M_s(\lambda))^{-1}$ on the right-hand side of (6.39).

If the Dirichlet and Neumann boundary mappings are swapped and the quasi boundary triple $\{L^2(\partial\Omega), -\Gamma_1, \Gamma_0\}$ is considered, one obtains a variant of Krein's formula where the operators A_Θ as above and A_{N}, the self-adjoint Neumann realization, are compared. The next result on spectral estimates for singular values of resolvent differences from [Behrndt et al., 2010, Theorem 3.5 and Remark 3.7] is essentially based on this idea.

Theorem 6.27 *Let $s = \frac{3}{2}$ and let $\{L^2(\partial\Omega), \Gamma_0, \Gamma_1\}$ be the quasi boundary triple for $\overline{T}_{3/2} = S^*$ from Proposition 6.22. Let Θ be a bounded self-adjoint (dissipative, accumulative) operator in $L^2(\partial\Omega)$ and let A_Θ be the corresponding self-adjoint (maximal dissipative, maximal accumulative, respectively) realization of $\mathcal{L}$ in $L^2(\Omega)$. Then for all $\lambda \in \rho(A_\Theta) \cap \rho(A_{\mathrm{N}})$ the singular values s_k of the resolvent difference*

$$(A_\Theta - \lambda)^{-1} - (A_{\mathrm{N}} - \lambda)^{-1} \tag{6.40}$$

satisfy $s_k = O(k^{-\frac{3}{n-1}})$, $k \to \infty$, and hence the expression in (6.40) belongs to the Schatten–von Neumann ideal $\mathfrak{S}_p(L^2(\Omega))$ for all $p > \frac{n-1}{3}$.

Proof First we express the resolvent difference (6.40) in a similar form as in Theorem 6.26. Observe that the γ-field and Weyl function corresponding to the quasi boundary triple $\{L^2(\partial\Omega), -\Gamma_1, \Gamma_0\}$ are given by

$$-\gamma_{3/2}(\lambda) M_{3/2}(\lambda)^{-1} \quad \text{and} \quad -M_{3/2}(\lambda)^{-1}, \quad \lambda \in \rho(A_{\mathrm{D}}) \cap \rho(A_{\mathrm{N}}), \tag{6.41}$$

and that the operator A_Θ is self-adjoint (maximal dissipative, maximal accumulative, respectively) by the corollary below Theorem 6.26. Observe that the boundary parameter Θ with respect to the triple $\{L^2(\partial\Omega), \Gamma_0, \Gamma_1\}$ has to be replaced by the self-adjoint (maximal dissipative, maximal accumulative) relation $\widetilde{\Theta} := -\Theta^{-1}$ when one expresses the boundary condition in terms of the quasi boundary triple $\{L^2(\partial\Omega), -\Gamma_1, \Gamma_0\}$. Now it follows from Theorem 6.16 and its corollaries applied to the quasi boundary triple $\{L^2(\partial\Omega), -\Gamma_1, \Gamma_0\}$ and (6.41) that for $\lambda \in \rho(A_\Theta) \cap \rho(A_{\mathrm{N}}) \cap \rho(A_{\mathrm{D}})$ the resolvent difference in (6.40) coincides with

$$\overline{\gamma_{3/2}(\lambda)} M_{3/2}(\lambda)^{-1} \big(\widetilde{\Theta}_{\mathrm{op}} + P_{\mathrm{op}} M_{3/2}(\lambda)^{-1}|_{\mathcal{G}_{\mathrm{op}}}\big)^{-1} P_{\mathrm{op}} M_{3/2}(\lambda)^{-1} \gamma_{3/2}(\bar{\lambda})^*, \tag{6.42}$$

where $\widetilde{\Theta}$ is decomposed into $\widetilde{\Theta}_{\rm op} \oplus \widetilde{\Theta}_\infty$ with $P_{\rm op}$ being the projection onto $\mathcal{G}_{\rm op}$. Note that $M_{3/2}(\lambda)^{-1}$ is defined on the whole space $\mathcal{G}$ and that in fact in (6.42) $\overline{\gamma_{3/2}(\lambda)}$ (as an operator from $L^2(\partial\Omega)$ to $L^2(\Omega)$) can be replaced by $\gamma_{3/2}(\lambda)$. Since $M_{3/2}(\lambda)^{-1}$ is a compact operator in $L^2(\partial\Omega)$ (see Proposition 6.23 (iii)), a Fredholm argument for $\widetilde{\Theta}_{\rm op}$ shows that

$$\left(\widetilde{\Theta}_{\rm op} + P_{\rm op} M_{3/2}(\lambda)^{-1}|_{\mathcal{G}_{\rm op}}\right)^{-1}$$

is a bounded and everywhere defined operator in $\mathcal{G}_{\rm op}$; cf. the proof of [Behrndt and Langer, 2007, Theorem 4.8]. Moreover, by Proposition 6.13 (iii) the operator $\gamma_{3/2}(\bar{\lambda})^*$ is bounded from $L^2(\Omega)$ into $L^2(\partial\Omega)$ with range in $H^{1/2}(\partial\Omega)$. Hence it is closed from $L^2(\Omega)$ to $H^{1/2}(\partial\Omega)$ and therefore bounded by the closed graph theorem. It follows from [Behrndt et al., 2010, Lemma 3.4] (and its proof) that the singular values of $\gamma_{3/2}(\lambda)^*$ satisfy $O(k^{-\frac{1}{2(n-1)}})$, $k \to \infty$. A similar argument using [Behrndt et al., 2010, Lemma 3.4] shows that the singular values of $M_{3/2}(\lambda)^{-1}$ satisfy $O(k^{-\frac{1}{n-1}})$, $k \to \infty$, and hence the singular values of the operators

$$M_{3/2}(\lambda)^{-1}\gamma_{3/2}(\lambda)^* \quad \text{and} \quad \overline{\gamma_{3/2}(\lambda)}M_{3/2}(\lambda)^{-1}$$

in (6.42) both satisfy $O(k^{-\frac{3}{2(n-1)}})$, $k \to \infty$. This implies the spectral estimates in Theorem 6.27. □

The asymptotics of singular values of the resolvent differences of the Dirichlet, Neumann and Robin realizations of $\mathcal{L}$ have been studied already in [Birman, 1962] and later among others in [Grubb, 1974; Birman and Solomjak, 1980; Grubb, 1984a,b; Malamud, 2010; Grubb, 2011] and in Section 8.9 of Chapter 8. For usual Robin boundary conditions, Theorem 6.27 reads as follows.

Corollary 6.28 *Assume that the values of $\beta \in L^\infty(\partial\Omega)$ are real (or have positive, negative imaginary parts) and let A_β be the self-adjoint (maximal dissipative, maximal accumulative, respectively) realization of $\mathcal{L}$ defined on*

$$\operatorname{dom} A_\beta = \left\{ f \in H^{3/2}_{\mathcal{L}}(\Omega) \colon \beta f|_{\partial\Omega} = -\frac{\partial f}{\partial \nu_{\mathcal{L}}}\Big|_{\partial\Omega} \right\}.$$

Then for all $\lambda \in \rho(A_\beta) \cap \rho(A_{\rm N})$ the singular values s_k of the resolvent difference

$$(A_\beta - \lambda)^{-1} - (A_{\rm N} - \lambda)^{-1} \tag{6.43}$$

satisfy $s_k = O(k^{-\frac{3}{n-1}})$, $k \to \infty$, *and hence the expression in* (6.43) *belongs to the Schatten–von Neumann ideal* $\mathfrak{S}_p(L^2(\Omega))$ *for all* $p > \frac{n-1}{3}$.

Finally, we want to relate the quasi boundary triple from this section to the boundary triple from Section 6.2. Denote the boundary mappings and the Weyl function for the boundary triple for the maximal operator S^* from Section 6.2 by $\{L^2(\partial\Omega), \widetilde{\Gamma}_0, \widetilde{\Gamma}_1\}$, where

$$\widetilde{\Gamma}_0 f = \iota_- f|_{\partial\Omega} \qquad \text{and} \qquad \widetilde{\Gamma}_1 f = -\iota_+ \frac{\partial f_{\mathrm{D}}}{\partial \nu_{\mathcal{L}}}\Big|_{\partial\Omega}, \tag{6.44}$$

with $f = f_{\mathrm{D}} + f_\eta$, $f_{\mathrm{D}} \in \operatorname{dom} A_{\mathrm{D}}$, $f_\eta \in \ker(S^* - \eta)$, and η is some fixed point in $\mathbb{R} \cap \rho(A_{\mathrm{D}})$. The corresponding Weyl function is denoted by $\widetilde{M}$.

Proposition 6.29 *Let* $s \in [\frac{3}{2}, 2]$ *and let* $\{L^2(\partial\Omega), \Gamma_0, \Gamma_1\}$ *be the quasi boundary triple for* $\overline{T}_s = S^*$ *from Proposition 6.22 with Weyl function* M_s*. The following relations hold for all* $f = f_{\mathrm{D}} + f_\eta \in \operatorname{dom} T_s$ *with* $f_{\mathrm{D}} \in \operatorname{dom} A_{\mathrm{D}}$ *and* $f_\eta \in \ker(T_s - \eta)$*:*

(i) $\widetilde{\Gamma}_0 f = \iota_- \Gamma_0 f$ *and* $\widetilde{\Gamma}_1 f = \iota_+ \big(\Gamma_1 f - M_s(\eta)\Gamma_0 f\big)$*;*

(ii) $\widetilde{\Gamma}_1 f = \Theta \widetilde{\Gamma}_0 f$ *if and only if* $\Gamma_1 f = \big(\iota_+^{-1}\Theta\iota_- + M_s(\eta)\big)\Gamma_0 f$*;*

(iii) $\widetilde{M}(\lambda)\iota_-\Gamma_0 f = \iota_+\big(M_s(\lambda) - M_s(\eta)\big)\Gamma_0 f$.

In particular, the Weyl functions $\widetilde{M}$ *and* M_s *are connected via*

$$\iota_+^{-1}\widetilde{M}(\lambda)\iota_- = M_s(\lambda) - M_s(\eta), \qquad \lambda \in \rho(A_{\mathrm{D}}). \tag{6.45}$$

Proof (i) The first relation is immediate from (6.44) and (6.34) in Proposition 6.22. The second statement follows with $f = f_{\mathrm{D}} + f_\eta$ from

$$\widetilde{\Gamma}_1 f = -\iota_+ \frac{\partial f_{\mathrm{D}}}{\partial \nu_{\mathcal{L}}}\Big|_{\partial\Omega} = \iota_+\Gamma_1 f_{\mathrm{D}} = \iota_+(\Gamma_1 f - \Gamma_1 f_\eta) = \iota_+\big(\Gamma_1 f - M_s(\eta)\Gamma_0 f\big).$$

(ii) is a simple consequence of (i).
(iii) Let $f \in \ker(T_s - \lambda)$ and $\lambda \in \rho(A_{\mathrm{D}})$. Then

$$\begin{aligned}\widetilde{M}(\lambda)\iota_-\Gamma_0 f = \widetilde{M}(\lambda)\widetilde{\Gamma}_0 f = \widetilde{\Gamma}_1 f &= \iota_+\big(\Gamma_1 f - M_s(\eta)\Gamma_0 f\big) \\ &= \iota_+\big(M_s(\lambda)\Gamma_0 f - M_s(\eta)\Gamma_0 f\big).\end{aligned}$$

implies the third assertion.

Relation (6.45) is an immediate consequence of (iii). □

Note that the operators on both sides of (6.45) are bounded operators in $L^2(\partial\Omega)$. For the right-hand side this follows from Proposition 6.14 (iv). We remark that relation (6.45) also shows that the operator on the right-hand side can be extended to a bounded operator from $H^{-1/2}(\partial\Omega)$ to

$H^{1/2}(\partial\Omega)$. Let us also mention that relation (6.45) implies that the Weyl function of the boundary triple from Section 6.2 is a regularization of the Dirichlet-to-Neumann map. Such regularizations were also considered in, e.g. [Grubb, 1968; Malamud, 2010; Ryzhov, 2007; Višik, 1952], and Chapters 7 and 8.

References

Allakhverdiev, B. P. 1991. On the theory of dilatation and on the spectral analysis of dissipative Schrödinger operators in the case of the Weyl limit circle. *Math. USSR-Izv.*, **36**, 247–262.

Alpay, D., and Behrndt, J. 2009. Generalized Q-functions and Dirichlet-to-Neumann maps for elliptic differential operators. *J. Funct. Anal.*, **257**, 1666–1694.

Amrein, W. O., and Pearson, D. B. 2004. M-operators: a generalisation of Weyl–Titchmarsh theory. *J. Comput. Appl. Math.*, **171** , 1–26.

Arendt, W., and ter Elst, A. F. M. 2011. The Dirichlet-to-Neumann operator on rough domains. *J. Differential Equations*, **251**, 2100–2124.

Arlinskiĭ, Yu. 2000. Abstract boundary conditions for maximal sectorial extensions of sectorial operators. *Math. Nachr.*, **209**, 5–36.

Ashbaugh, M. S., Gesztesy, F., Mitrea, M., and Teschl, G. 2010. Spectral theory for perturbed Krein Laplacians in nonsmooth domains. *Adv. Math.* **223**, 1372–1467.

Astala, K., and Päivärinta, L. 2006. Calderón's inverse conductivity problem in the plane. *Ann. of Math. (2)*, **163**, 265–299.

Azizov, T. Ya., and Iokhvidov, I. S. 1989. *Linear Operators in Spaces with an Indefinite Metric.* Pure and Applied Mathematics (New York), John Wiley and Sons, Chichester.

Behrndt, J. 2010. Elliptic boundary value problems with λ-dependent boundary conditions. *J. Differential Equations*, **249**, 2663–2687.

Behrndt, J., Hassi, S., de Snoo, H.S.V., and Wietsma, H.L. 2011. Square-integrable solutions and Weyl functions for singular canonical systems. *Math. Nachr.*, **284**, 1334–1384.

Behrndt, J., and Langer, M. 2007. Boundary value problems for elliptic partial differential operators on bounded domains. *J. Funct. Anal.*, **243**, 536–565.

Behrndt, J., and Langer, M. 2010. On the adjoint of a symmetric operator. *J. London Math. Soc. (2)*, **82**, 563–580.

Behrndt, J., Langer, M., Lobanov, I., Lotoreichik, V., and Popov, I. Yu. 2010. A remark on Schatten–von Neumann properties of resolvent differences of generalized Robin Laplacians on bounded domains. *J. Math. Anal. Appl.*, **371**, 750–758.

Behrndt, J., Langer, M., and Lotoreichik, V. 2011. Spectral estimates for resolvent differences of self-adjoint elliptic operators. Submitted; preprint: arXiv:1012.4596v1 [math.SP]

Behrndt, J., Malamud, M. M., Neidhardt, H. 2008. Scattering matrices and Weyl functions. *Proc. London Math. Soc. (3)*, **97**, 568–598.

Behrndt, J., and Trunk, C. 2007. On the negative squares of indefinite Sturm–Liouville operators. *J. Differential Equations*, **238**, 491–519.

Birman, M. Sh. 1962. Perturbations of the continuous spectrum of a singular elliptic operator by varying the boundary and the boundary conditions (Russian). *Vestnik Le. Univ.*, **17**, 22–55, (translation in *Amer. Math. Soc. Transl.*, **225** (2008), 19–53).

Birman, M. Š., and Solomjak, M. Z. 1980. Asymptotic behavior of the spectrum of variational problems on solutions of elliptic equations in unbounded domains (Russian). *Funktsional. Anal. i Prilozhen.*, **14**, 27–35 (translation in *Funct. Anal. Appl.*, **14** (1981), 267–274).

Brasche, J., Malamud, M. M., and Neidhardt, H. 2002. Weyl function and spectral properties of selfadjoint extensions. *Integral Equations Operator Theory*, **43**, 264–289.

Brown, B. M., Grubb, G., and Wood, I. G. 2009. M-functions for closed extensions of adjoint pairs of operators with applications to elliptic boundary problems. *Math. Nachr.*, **282**, 314–347.

Brown, M., Hinchcliffe, J., Marletta, M., Naboko, S., and Wood, I. 2009. The abstract Titchmarsh–Weyl M-function for adjoint operator pairs and its relation to the spectrum. *Integral Equations Operator Theory*, **63**, 297–320.

Brown, B. M., and Marletta, M. 2004. Spectral inclusion and spectral exactness for PDEs on exterior domains. *IMA J. Numer. Anal.*, **24**, 21–43.

Brown, M., Marletta, M., Naboko, S., and Wood, I. 2008. Boundary triplets and M-functions for non-selfadjoint operators, with applications to elliptic PDEs and block operator matrices. *J. Lond. Math. Soc. (2)*, **77**, 700–718.

Bruk, V. M. 1976. A certain class of boundary value problems with a spectral parameter in the boundary condition (Russian). *Mat. Sb.*, **100 (142)**, 210–216.

Brüning, J., Geyler, V., and Pankrashkin, K. 2008. Spectra of self-adjoint extensions and applications to solvable Schrödinger operators. *Rev. Math. Phys.*, **20**, 1–70.

Derkach, V. A., Hassi, S., Malamud, M. M., and de Snoo, H. S. V. 2006. Boundary relations and their Weyl families. *Trans. Amer. Math. Soc.*, **358**, 5351–5400.

Derkach, V. A., Hassi, S., Malamud, M. M., and de Snoo, H. S. V. 2009. Boundary relations and generalized resolvents of symmetric operators. *Russ. J. Math. Phys.*, **16**, 17–60.

Derkach, V. A., Hassi, S., and de Snoo, H. S. V. 2003. Singular perturbations of self-adjoint operators. *Math. Phys. Anal. Geom.*, **6**, 349–384.

Derkach, V. A., and Malamud, M. M. 1991. Generalized resolvents and the boundary value problems for Hermitian operators with gaps. *J. Funct. Anal.*, **95**, 1–95.

Derkach, V. A., and Malamud, M. M. 1995. The extension theory of Hermitian operators and the moment problem. *J. Math. Sciences*, **73**, 141–242.

Edmunds, D. E., and Evans, W. D. 1987. *Spectral Theory and Differential Operators.* The Clarendon Press, Oxford University Press, New York.

Everitt, W. N., and Markus, L. 2003. Elliptic partial differential operators and symplectic algebra. *Mem. Amer. Math. Soc.*, **162**, no. 770, 111 pp.

Everitt, W. N., Markus, L., and Plum, M. 2005. An unusual self-adjoint linear partial differential operator. *Trans. Amer. Math. Soc.*, **357**, 1303–1324.

Filonov, N. 2004. On an inequality for the eigenvalues of the Dirichlet and Neumann problems for the Laplace operator (Russian). *Algebra i Analiz*, **16**, 172–176 (translation in *St. Petersburg Math. J.*, **16** (2005), 413–416).

Friedlander, L. 1991. Some inequalities between Dirichlet and Neumann eigenvalues. *Arch. Rational Mech. Anal.*, **116**, 153–160.

Gesztesy, F., Makarov, K. A., and Tsekanovskii, E. 1998. An addendum to Krein's formula. *J. Math. Anal. Appl.*, **222**, 594–606.

Gesztesy, F., and Mitrea, M. 2008. Generalized Robin boundary conditions, Robin-to-Dirichlet maps, and Krein-type resolvent formulas for Schrödinger operators on bounded Lipschitz domains. Pages 105–173 of: *Perspectives in Partial Differential Equations, Harmonic Analysis and Applications. Proc. Sympos. Pure Math.*, Vol. 79, Amer. Math. Soc., Providence, RI.

Gesztesy, F., and Mitrea, M. 2009. Robin-to-Robin maps and Krein-type resolvent formulas for Schrödinger operators on bounded Lipschitz domains. Pages 81–113 of: *Modern Analysis and Applications. The Mark Krein Centenary Conference. Vol. 2: Differential Operators and Mechanics. Oper. Theory Adv. Appl.*, Vol. 191, Birkhäuser Verlag, Basel.

Gesztesy, F., and Mitrea, M. 2011. A description of all self-adjoint extensions of the Laplacian and Kreĭn-type resolvent formulas on non-smooth domains. *J. Anal. Math.*, **113**, 53–172.

Gesztesy, F., and Tsekanovskii, E. 2000. On matrix-valued Herglotz functions. *Math. Nachr.*, **218**, 61–138.

Gorbachuk, V. I., and Gorbachuk, M. L. 1991. *Boundary Value Problems for Operator Differential Equations.* Mathematics and its Applications (Soviet Series), Vol. 48, Kluwer Academic Publishers, Dordrecht.

Grubb, G. 1968. A characterization of the non-local boundary value problems associated with an elliptic operator. *Ann. Scuola Norm. Sup. Pisa (3)*, **22**, 425–513.

Grubb, G. 1971. On coerciveness and semiboundedness of general boundary problems. *Israel J. Math.*, **10**, 32–95.

Grubb, G. 1974. Properties of normal boundary problems for elliptic even-order systems. *Ann. Scuola Norm. Sup. Pisa Cl. Sci. (4)*, **1**, 1–61.

Grubb, G. 1983. Spectral asymptotics for the "soft" selfadjoint extension of a symmetric elliptic differential operator. *J. Operator Theory*, **10**, 9–20.

Grubb, G. 1984a. Singular Green operators and their spectral asymptotics. *Duke Math. J.*, **51**, 477–528.

Grubb, G. 1984b. Remarks on trace estimates for exterior boundary problems. *Comm. Partial Differential Equations*, **9**, 231–270.

Grubb, G. 2006. Known and unknown results on elliptic boundary problems. *Bull. Amer. Math. Soc.*, **43**, 227–230.

Grubb, G. 2008. Krein resolvent formulas for elliptic boundary problems in nonsmooth domains. *Rend. Semin. Mat. Univ. Politec. Torino* **66**, 13–39.

Grubb, G. 2009. *Distributions and Operators.* Graduate Texts in Mathematics 252, Springer, New York.

Grubb, G. 2011. Spectral asymptotics for Robin problems with a discontinuous coefficient. *J. Spectral Theory*, **1**, 155–177.

Kac, I. S., and Kreĭn, M. G. 1974. *R*-functions—analytic functions mapping the upper halfplane into itself. *Amer. Math. Soc. Transl. (2)*, **103**, 1–18.

Karabash, I. M., Kostenko, A. S., and Malamud, M. M. 2009. The similarity problem for *J*-nonnegative Sturm–Liouville operators. *J. Differential Equations*, **246**, 964–997.

Kochubei, A. N. 1975. On extensions of symmetric operators and symmetric binary relations (Russian). *Mat. Zametki*, **17**, 41–48.

Kostenko, A. S., and Malamud, M. M. 2010. 1-D Schrödinger operators with local point interactions on a discrete set. *J. Differential Equations*, **249**, 253–304.

Kopachevskiĭ, N. D., and Kreĭn, S. G. 2004. Abstract Green formula for a triple of Hilbert spaces, abstract boundary-value and spectral problems (Russian). *Ukr. Mat. Visn.*, **1**, 69–97 (translation in *Ukr. Math. Bull.*, **1** (2004), 77–105).

Kreĭn, M. G. 1946. Concerning the resolvents of an Hermitian operator with the deficiency-index (m, m). *C. R. (Doklady) Acad. Sci. URSS (N.S.)*, **52**, 651–654.

Langer, H., and Textorius, B. 1977. On generalized resolvents and Q-functions of symmetric linear relations (subspaces) in Hilbert space. *Pacific J. Math.*, **72**, 135–165.

Lions, J., and Magenes, E. 1972. *Non-Homogeneous Boundary Value Problems and Applications I.* Springer Verlag, New York – Heidelberg.

Malamud, M. M. 1992. On a formula for the generalized resolvents of a non-densely defined Hermitian operator (Russian). *Ukraïn. Mat. Zh.*, **44**, 1658–1688 (translation in *Ukrainian Math. J.*, **44** (1992), 1522–1547 (1993)).

Malamud, M. M. 2010. Spectral theory of elliptic operators in exterior domains. *Russ. J. Math. Phys.*, **17**, 96–125.

Malinen, J., and Staffans, O. J. 2007. Impedance passive and conservative boundary control systems. *Complex Anal. Oper. Theory*, **1**, 279–300.

Marletta, M. 2004. Eigenvalue problems on exterior domains and Dirichlet to Neumann maps. *J. Comput. Appl. Math.*, **171**, 367–391.

Marletta, M. 2010. Neumann–Dirichlet maps and analysis of spectral pollution for non-self-adjoint elliptic PDEs with real essential spectrum. *IMA J. Numer. Anal.* **30**, 917–939.

Mogilevskiĭ, V. 2006. Boundary triplets and Krein type resolvent formula for symmetric operators with unequal defect numbers. *Methods Funct. Anal. Topology*, **12**, 258–280.

Mogilevskiĭ, V. 2009. Boundary triplets and Titchmarsh–Weyl functions of differential operators with arbitrary deficiency indices. *Methods Funct. Anal. Topology*, **15**, 280–300.

Nachman, A. I. 1988. Reconstructions from boundary measurements. *Ann. of Math. (2)*, **128**, 531–576.

Nachman, A. I. 1996. Global uniqueness for a two-dimensional inverse boundary value problem. *Ann. of Math. (2)* **143**, 71–96.

Nachman, A. I., Sylvester, J., and Uhlmann, G. 1988. An n-dimensional Borg–Levinson theorem. *Comm. Math. Phys.*, **115**, 595–605.

Pankrashkin, K. 2006. Resolvents of self-adjoint extensions with mixed boundary conditions. *Rep. Math. Phys.*, **58**, 207–221.

Posilicano, A. 2004. Boundary triples and Weyl functions for singular perturbations of self-adjoint operators. *Methods Funct. Anal. Topology*, **10**, 57–63.

Posilicano, A. 2008. Self-adjoint extensions of restrictions. *Oper. Matrices*, **2**, 483–506.

Posilicano, A., and Raimondi, L. 2009. Krein's resolvent formula for self-adjoint extensions of symmetric second-order elliptic differential operators. *J. Phys. A*, **42**, 015204, 11 pp.

Post, O. 2007. First-order operators and boundary triples. *Russ. J. Math. Phys.*, **14**, 482–492.

Ryzhov, V. 2007. A general boundary value problem and its Weyl function. *Opuscula Math.*, **27**, 305–331.

Ryzhov, V. 2009. Weyl–Titchmarsh function of an abstract boundary value problem, operator colligations, and linear systems with boundary control. *Complex Anal. Oper. Theory*, **3**, 289–322.

Saakjan, Š. N. 1965. Theory of resolvents of a symmetric operator with infinite defect numbers (Russian). *Acad. Nauk Armjan. SSR Dokl.*, **41**, 193–198.

Safarov, Y. 2008. On the comparison of the Dirichlet and Neumann counting functions. *Amer. Math. Soc. Transl. Ser. 2*, 225, pp. 191–204, Amer. Math. Soc., Providence, RI.

Seeley, R. 1969. The resolvent of an elliptic boundary problem. *Amer. J. Math.*, **91**, 889–920.

Šmuljan, Ju. L. 1976. Theory of linear relations, and spaces with indefinite metric (Russian). *Funkcional. Anal. i Priložen.*, **10**, 67–72.

Sylvester, J., and Uhlmann, G. 1987. A global uniqueness theorem for an inverse boundary value problem. *Ann. of Math. (2)* **125**, 153–169.

Višik, M. I. 1952. On general boundary problems for elliptic differential equations (Russian). *Trudy Moskov. Mat. Obšč.*, **1**, 187–246 (translation in *Amer. Math. Soc. Transl.*, **24** (1963), 107–172).

Wloka, J. 1987. *Partial Differential Equations.* Cambridge University Press, Cambridge.

7

Boundary triplets and Weyl functions. Recent developments

Vladimir Derkach, Seppo Hassi, Mark Malamud and Henk de Snoo

Abstract Selfadjoint extensions of a closed symmetric operator in a Hilbert space with equal deficiency indices are described by means of ordinary boundary triplets. In certain problems the more general notion of a boundary triplet of bounded type is needed. It will be shown that such triplets correspond in a certain way with the, in general infinite dimensional, graph perturbations of selfadjoint operators or relations. However, when considering selfadjoint exit spaces extensions for a symmetric operator by means of the so-called Kreĭn's formula one meets the notion of boundary relation (even when the deficiency indices are finite). Whereas ordinary boundary triplets and boundary triplets of bounded type correspond to bounded unitary or unitary operators in a Kreĭn space, respectively, boundary relations correspond to general unitary relations in a Kreĭn space, which are not necessarily single-valued. It is shown that the study of isometric relations in a Kreĭn space has useful applications. This present overview of recent developments includes illustrations, for instance, by means of elliptic differential operators and Schrödinger operators with local point interactions.

7.1 Introduction

The extension theory of densely defined symmetric operators was developed in the 1930s by J. von Neumann. A complete description was given for all selfadjoint extensions in terms of the defect subspaces; see [von Neumann, 1932; Stone, 1932]. Then M.H. Stone suggested to J.W. Calkin to develop another approach based on the notion of "abstract boundary conditions", which reduces the extension problem to a description of the hyper-maximal symmetric subspaces of some auxiliary Hilbert space. This resulted in the seminal paper [Calkin, 1939]; see

The research was supported by the South Ostrobothnia Regional fund of the Finnish Cultural Foundation.

Chapter 2 for a discussion. However Calkin did not pursue his work in this area and his paper did not receive the attention which it deserved.

A revival of interest to applications of this approach to boundary value problems can be found in [Kreĭn, 1947; Višik, 1952; Birman, 1956] and later in [Phillips, 1959, 1961; Grubb, 1968; Rofe-Beketov, 1969; Gorbachuk, 1971]. Influenced by these works [Kochubei, 1975; Bruk, 1976] introduced the concept of a boundary triplet; cf. [Gorbachuk and Gorbachuk, 1984]. The main ingredient of this concept is an abstract version of Green's identity for a densely defined symmetric operator S in a Hilbert space $\mathfrak{H}$ with equal deficiency indices:

$$(S^*f, g)_{\mathfrak{H}} - (f, S^*g)_{\mathfrak{H}} = (\Gamma_1 f, \Gamma_0 g)_{\mathcal{H}} - (\Gamma_0 f, \Gamma_1 g)_{\mathcal{H}},$$

for all $f, g \in \operatorname{dom} S^*$, where $\mathcal{H}$ is an auxiliary Hilbert space and the mapping $\Gamma : f \to (\Gamma_0 f \ \Gamma_1 f)^\top$ from $\operatorname{dom} S^*$ to $\mathcal{H}^2 = \mathcal{H} \times \mathcal{H}$ is surjective. This mapping coincides with the operator which Calkin called the reduction operator. It establishes a one-to-one correspondence between the factor-space $\operatorname{dom} S^*/\operatorname{dom} S$ and $\mathcal{H}^2$. It gives rise to a parametrization of the set of selfadjoint extensions of S in terms of abstract boundary conditions. For instance, the selfadjoint extensions $\widetilde{A}$ which are disjoint from the selfadjoint restriction A_0 of S^*, defined by $\operatorname{dom} A_0 = \ker \Gamma_0$, are parametrized by

$$\widetilde{A} \subset A^*, \quad \operatorname{dom} \widetilde{A} = \{f \in \operatorname{dom} A^* : \Gamma_1 f = B\Gamma_0 f\}; \tag{7.1}$$

here B runs over the set of all selfadjoint operators in $\mathcal{H}$. Note that the description in (7.1) does not require knowledge of the defect subspaces and parameterizes the extensions directly in terms of abstract boundary conditions.

The Weyl function corresponding to the boundary triplet Π for the adjoint S^*, was introduced and investigated in [Derkach and Malamud, 1985, 1987]. It is an abstract version of the well-known Titchmarsh-Weyl function associated with Sturm-Liouville operators. The abstract Weyl function M determines (the simple symmetric part of) the operator S, as well as the boundary triplet Π itself, uniquely, up to the unitary equivalence. A connection between boundary triplets and the Kreĭn's formula for resolvents has also been discovered in [Derkach and Malamud, 1985, 1987]. The latter result makes it possible to apply Kreĭn type formulas to boundary value problems. Such applications can be found in numerous papers.

For every ordinary boundary triplet of a densely defined symmetric operator S the selfadjoint extensions A_0 and A_1 of S with the domains

$\operatorname{dom} A_j = \ker \Gamma_j$, $j = 0, 1$, are *transversal*, that is A_0 and A_1 are disjoint and $\operatorname{dom} A_0 + \operatorname{dom} A_1 = \operatorname{dom} S^*$. This fact is due to the surjectivity of the mapping Γ. However, in some problems of analysis one needs to consider boundary triplets with selfadjoint extensions A_0 and A_1 which are not transversal. For instance, let S be a nonnegative densely defined symmetric operator. In [Kreĭn and Ovcarenko, 1977, 1978] the concepts of Q_F- and Q_K-functions corresponding to the Friedrichs and Kreĭn extensions S_F and S_K of S were introduced. It turned out that Q_F and Q_K are the Q-functions of S in the sense of [Kreĭn and Langer, 1971], as well as the Weyl functions of the respective boundary triplets for S^*, if and only if S_F and S_K are transversal. In order to extend this approach to the case when S_F and S_K are not transversal and to realize Q_F- and Q_K-functions as the Weyl functions of S a new notion of a generalized boundary triplet (of bounded type) was introduced in [Derkach and Malamud, 1995]. The corresponding reduction operator Γ need not be surjective, but instead Γ_0 should be surjective and $A_0 = A_0^*$. As was shown in [Derkach and Malamud, 1995] the Weyl function corresponding to the generalized boundary triplet takes values in the set of bounded operators and is a strict Nevanlinna function (i.e. $\ker \operatorname{Im} M(\mathrm{i}) = \{0\}$). Conversely, every strict Nevanlinna function was shown to be realizable as the Weyl function of a generalized boundary triplet for some S^*.

The next step in the investigation of the connection between boundary triplets and their analytical counterparts (the Weyl functions) has been motivated partially by the the inverse problem for operator-valued Nevanlinna functions. In [Derkach et al., 2006] the concept of boundary relation (generalized boundary pair) was introduced and it was shown that any operator-valued Nevanlinna function and even any Nevanlinna family can be realized as the Weyl function of some boundary relation. Another motivation to introduce this concept was provided by the Kreĭn-Naĭmark formula for generalized resolvents of a symmetric operator. The latter plays an important role in modern analysis. A new proof of this formula was given in [Derkach et al., 2000] based on a coupling construction and the notion of induced boundary triplet in the exit space. The treatment of the spectral parameter in the Kreĭn-Naĭmark formula as the Weyl function of the induced boundary triplet plays a crucial role in (and is the main ingredient of) the coupling construction.

The concept of boundary relation has been introduced in [Derkach et al., 2006]; at that time the authors were not familiar with [Calkin, 1939]. It turned out that, in the case that the symmetric operator is densely defined and the boundary mapping is a (single-valued) operator, the concept of boundary relation is contained in Calkin's notion of reduction operator. However, it must be emphasized that even starting with the construction of an exit space extension for an operator S with deficiency indices $n_\pm(S) = 2$, one necessarily arrives at a multi-valued reduction mapping, which is not covered by Calkin's definition.

The present chapter gives a short review of the different steps in the development of boundary triplets, from ordinary boundary triplets to boundary relations. In Section 7.2 an introduction to linear relations in Hilbert and in Kreĭn spaces and Calkin's approach by means of reduction operators can be found. The necessary facts about ordinary boundary triplets are given in Section 7.3. In Section 7.4 the main facts concerning generalized boundary triplets (of bounded type) are presented and the spectrum of proper extensions is characterized in terms of the boundary operator and the Weyl function. An explicit formula for the resolvent of the extension $\widetilde{A}$ of the form (7.1) is presented. Infinite dimensional restrictions of selfadjoint relations are introduced and studied in Section 7.5. They are shown to be general models for generalized boundary triplets of bounded type. In Section 7.6 results concerning generalized boundary pairs and generalized boundary triplets are reviewed and a new criterion for the extension A_0 to be selfadjoint is proved. In particular, special cases of the so-called unitary boundary triplets are considered. Schrödinger operators with point interactions are shown to be examples of this situation. In Section 7.7 induced boundary triplets are treated; a sketch of the proof of the Kreĭn's formula for generalized resolvents is included. An example of the induced boundary triplet is calculated for the Laplace operator in the interior domain. Isometric boundary triplets are introduced and studied in Section 7.8. Special attention is paid to a class of isometric boundary triplets, which was studied in [Behrndt and Langer, 2007] under the name of quasi-boundary triplets; see also Chapter 6.

During the last decade many papers have appeared, in which different kinds of ordinary and generalized boundary triplets were used in the theory of elliptic boundary value problems and in the theory of local point interactions; see for instance [Albeverio and Kurasov, 2000; Amrein and Pearson, 2004; Behrndt and Langer, 2007; Brown, Grubb, and Wood, 2009; Bruning et al., 2007; Derkach and Malamud, 1991; Gorbachuk et al., 1989; Grubb, 2009; Kostenko and Malamud, 2010; Malamud, 2010; Malamud and Neidhardt, 2011; Posilicano, 2008; Ryzhov, 2009], and the references therein.

The main results of the theory of boundary triplets including Kreĭn type formulas for resolvents have been extended for dual pairs of operators [Malamud and Mogilevskiĭ, 2002; Hassi et al., 2005; Brown et al., 2008, 2009; Brown, Grubb, and Wood, 2009]. For related work concerning an extension of boundary triplets to the case of symmetric operators with not necessarily equal deficiency indices, see [Mogilevskiĭ, 2006, 2009, 2010].

Another interesting point of contact is with system theory, see Chapters 4, 5 and 9.

7.2 Preliminaries

Linear relations in Hilbert spaces Here is a short introduction to linear relations (multi-valued operators) in a Hilbert space $\mathfrak{H}$; cf. [Arens, 1961; Bennewitz, 1972; Coddington, 1973; Dijksma and de Snoo, 1974]. Let H be a linear relation in $\mathfrak{H}$, i.e., H is a linear subspace of the product space $\mathfrak{H} \times \mathfrak{H}$. The symbols $\operatorname{dom} H$, $\ker H$, $\operatorname{ran} H$, and $\operatorname{mul} H$ denote the domain, kernel, range, and the multi-valued part, respectively. The inverse

$$H^{-1} = \{ \{f', f\} : \{f, f'\} \in H \}$$

is a relation in $\mathfrak{H}$ and the adjoint H^* defined by

$$H^* = \{ \{h, k\} \in \mathfrak{H} \oplus \mathfrak{H} : (k, f) = (h, g), \{f, g\} \in H \}$$

is a closed linear relation in $\mathfrak{H}$. The closure of a linear relation H (in $\mathfrak{H} \times \mathfrak{H}$) will be denoted by $\operatorname{clos} H$, so that $\operatorname{clos} H = H^{**}$. The sum $H_1 + H_2$ of two linear relations H_1 and H_2 is defined by

$$H_1 + H_2 = \{ \{f, g + h\} : \{f, g\} \in H_1, \{f, h\} \in H_2 \},$$

whereas the componentwise sum $H_1 \widehat{+} H_2$ is defined by

$$H_1 \widehat{+} H_2 = \{ \{f + h, g + k\} : \{f, g\} \in H_1, \{h, k\} \in H_2 \}.$$

Recall that a linear relation H in $\mathfrak{H}$ is called *symmetric* (*dissipative*) if $\operatorname{Im}(h', h) = 0$ (≥ 0, respectively) for all $\{h, h'\} \in H$. These properties remain invariant under closures. A linear relation H in $\mathfrak{H}$ is called *selfadjoint* if $H = H^*$ and it is called *essentially selfadjoint* if $\operatorname{clos} H = H^*$. A symmetric (dissipative) linear relation H in $\mathfrak{H}$ is called maximal symmetric (maximal dissipative) if it has no proper symmetric (dissipative) extensions.

Assume that H is closed. If H is dissipative, then $\operatorname{mul} H \subset \operatorname{mul} H^*$. In this case the orthogonal decomposition $\mathfrak{H} = (\operatorname{mul} H)^{\perp} \oplus \operatorname{mul} H$ induces an orthogonal decomposition of H as

$$H = H_{\rm s} \oplus H_{\rm mul}, \tag{7.2}$$

where $H_{\rm mul} = \{0\} \times \operatorname{mul} H$ is a selfadjoint relation in $\operatorname{mul} H$ and

$$H_{\rm s} = \{\, \{f, g\} \in H : g \perp \operatorname{mul} H \,\},$$

is a closed dissipative operator in $\mathfrak{H} \ominus \operatorname{mul} H$ with

$$\overline{\operatorname{dom}} \, H_{\rm s} = \overline{\operatorname{dom}} \, H = (\operatorname{mul} H^*)^{\perp}.$$

Moreover, if the relation H is maximal dissipative, then $\operatorname{mul} H = \operatorname{mul} H^*$ and H_s is a densely defined maximal dissipative operator in $(\operatorname{mul} H)^{\perp}$. In particular, if H is a self-adjoint relation, then H_s is a self-adjoint operator in $(\operatorname{mul} H)^{\perp}$. In the following $\sigma_p(H)$, $\sigma_c(H)$ and $\rho(H)$ stand for the point spectrum, continuous spectrum and the resolvent set of H, respectively.

Let S be a closed symmetric relation in a Hilbert space $\mathfrak{H}$. Then $\ker\,(S - \lambda) = \{0\}$ and $\operatorname{ran}\,(S - \bar{\lambda})$ is closed for $\lambda \in \mathbb{C} \setminus \mathbb{R}$. Define the *defect subspace* of S by $\mathfrak{N}_{\lambda}(S^*) := \mathfrak{H} \ominus \operatorname{ran}\,(S - \bar{\lambda}) = \ker\,(S^* - \lambda)$ and define, correspondingly,

$$\widehat{\mathfrak{N}}_{\lambda}(S^*) = \{\, \{f_{\lambda}, \lambda f_{\lambda}\} : f_{\lambda} \in \mathfrak{N}_{\lambda}(S^*) \,\}.$$

The numbers $n_{\pm}(S) = \dim \mathfrak{N}_{\mu}(S^*)$, $\mu \in \mathbb{C}^{\mp}$, are the *deficiency indices* of S. In terms of relations the first formula of von Neumann reads as

$$S^* = S \,\widehat{+}\, \widehat{\mathfrak{N}}_{\lambda}(S^*) \,\widehat{+}\, \widehat{\mathfrak{N}}_{\bar{\lambda}}(S^*), \quad \lambda \in \mathbb{C} \setminus \mathbb{R},$$

where the sums are orthogonal for $\lambda = \pm \mathrm{i}$. The relation S admits self-adjoint extensions in the Hilbert space $\mathfrak{H}$ if and only if $n_+(S) = n_-(S)$. In this case the second formula of von Neumann

$$H = S \,\widehat{+}\, (I - U)\widehat{\mathfrak{N}}_{\mu}(S^*), \quad \mu \in \mathbb{C} \setminus \mathbb{R},$$

establishes a one-to-one correspondence between the self-adjoint extensions H of S and the unitary operators U from $\widehat{\mathfrak{N}}_{\mu}(S^*)$ onto $\widehat{\mathfrak{N}}_{\bar{\mu}}(S^*)$. A similar one-to-one correspondence holds also for symmetric and dissipative extensions of S.

A linear extension H of S is called *proper* if $S \subset H \subset S^*$. Clearly symmetric and selfadjoint extensions of S are proper. Two proper extensions H_0 and H_1 of S are called *disjoint* if $H_0 \cap H_1 = S$, and *transversal*, if $H_0 \widehat{+} H_1 = S^*$. A closed symmetric relation is called *simple*, if there is no non-trivial subspace in $\mathfrak{H}$ which reduces it to a self-adjoint relation.

By definition a simple symmetric relation is an operator. It is known that a symmetric relation S is simple if and only if

$$\mathfrak{H} = \overline{\text{span}}\ \{\mathfrak{N}_\lambda(S^*) : \lambda \in \mathbb{C} \setminus \mathbb{R}\}.$$

Linear relations between Kreĭn spaces A Kreĭn space $(\mathfrak{H}, J)$ is a Hilbert space $\mathfrak{H}$ with a canonical symmetry J, which induces the indefinite inner product; see [Bognar, 1974; Azizov and Iokhvidov, 1989]. The following terminology plays a natural role in extension theory. Let $\mathfrak{L}$ be a linear subspace of a Kreĭn space $(\mathfrak{H}, J)$. Then the orthogonal complement $\mathfrak{L}^{[\perp]}$ in $\mathfrak{H}$ is given by $\mathfrak{L}^{[\perp]} = J\mathfrak{L}^\perp = (J\mathfrak{L})^\perp$. The subspace $\mathfrak{L}$ is called

(a) *neutral* if $\mathfrak{L} \subset \mathfrak{L}^{[\perp]}$;
(b) *maximal neutral* if $\mathfrak{L}$ has no non-trivial neutral extensions;
(c) *hyper-maximal neutral* if $\mathfrak{L} = \mathfrak{L}^{[\perp]}$.

The *isotropic part* of a closed linear subspace $\mathfrak{L}$ is defined as $\mathfrak{L} \cap \mathfrak{L}^{[\perp]}$; the isotropic part is a neutral subspace.

This terminology is useful in the following way. Let $\mathfrak{H}$ be a Hilbert space and let T be a linear relation in $\mathfrak{H}$ or, equivalently, let T be a linear subspace of $\mathfrak{H} \times \mathfrak{H}$. Provide the product space $\mathfrak{H}^2 = \mathfrak{H} \times \mathfrak{H}$ with the signature operator

$$J_{\mathfrak{H}} = \begin{pmatrix} 0 & -\mathrm{i}\, I_{\mathfrak{H}} \\ \mathrm{i}\, I_{\mathfrak{H}} & 0 \end{pmatrix}, \tag{7.3}$$

which leads to a Kreĭn space $(\mathfrak{H}^2, J_{\mathfrak{H}})$. Note that the adjoint T^* of a linear relation T in the Hilbert space $\mathfrak{H}$ can be written as

$$T^* = J_{\mathfrak{H}} T^\perp = (J_{\mathfrak{H}} T)^\perp = T^{[\perp]}, \tag{7.4}$$

where the orthogonal complement $[\perp]$ is in $(\mathfrak{H}^2, J_{\mathfrak{H}})$. Then

(a') T is symmetric in $\mathfrak{H} \iff T$ is neutral in $(\mathfrak{H}^2, J_{\mathfrak{H}})$;
(b') T is maximal symmetric in $\mathfrak{H} \iff T$ is maximal neutral in $(\mathfrak{H}^2, J_{\mathfrak{H}})$;
(c') T is selfadjoint in $\mathfrak{H} \iff T$ is hyper-maximal neutral in $(\mathfrak{H}^2, J_{\mathfrak{H}})$.

Let T be a linear relation from a Kreĭn space $(\mathfrak{H}_1, J_1)$ to a Kreĭn space $(\mathfrak{H}_2, J_2)$. Its adjoint $T^{[*]}$ is a relation from $(\mathfrak{H}_2, J_2)$ to $(\mathfrak{H}_1, J_1)$ given by

$$T^{[*]} = \{\, \{h, h'\} \in \mathfrak{H}_2 \times \mathfrak{H}_1 : (J_2 h', f)_{\mathfrak{H}_1} = (J_1 h, f')_{\mathfrak{H}_2},\ \{f, f'\} \in T \,\}.$$

Hence, $\{h, h'\} \in T^{[*]}$ if and only if $\{J_1 h, J_2 h'\} \in T^*$, which implies that $T^{[*]} := J_1 T^* J_2$. It follows directly from the definition that

$$(\operatorname{dom} T)^{[\perp]} = \operatorname{mul} T^{[*]}, \quad (\operatorname{ran} T)^{[\perp]} = \ker T^{[*]}. \tag{7.5}$$

A linear relation $T : (\mathfrak{H}_1, J_1) \to (\mathfrak{H}_2, J_2)$ is said to be *unitary* if

$$T^{-1} = T^{[*]}. \tag{7.6}$$

A unitary relation is automatically closed. The definition (7.6) and the following proposition go back to [Shmuljan, 1976]; cf. [Derkach et al., 2006, Proposition 2.3]. Its proof follows directly from the definition (7.6) and the identities (7.5). The last statement expresses the fact that for a closed relation its range and the range of its adjoint are simultaneously closed.

Proposition 7.1 *Let T be a unitary relation from $(\mathfrak{H}_1, J_1)$ to $(\mathfrak{H}_2, J_2)$. Then*

$$\operatorname{dom} T = \operatorname{ran} T^{[*]}, \quad \operatorname{ran} T = \operatorname{dom} T^{[*]},$$

and

$$\ker T = (\operatorname{dom} T)^{[\perp]}, \quad \operatorname{mul} T = (\operatorname{ran} T)^{[\perp]}.$$

Moreover, $\operatorname{dom} T$ *is closed if and only if* $\operatorname{ran} T$ *is closed.*

Note that if T is unitary, then $\ker T$ equals the isotropic part of $\operatorname{dom} T$ and $\operatorname{mul} T$ equals the isotropic part of $\operatorname{ran} T$. A unitary relation T is an operator if and only if $\overline{\operatorname{ran}}\, T = \mathfrak{H}$. For examples of unitary relations and unbounded unitary operators, see [Derkach et al., 2006].

Corollary 7.2 *Let T be a unitary relation from $(\mathfrak{H}_1, J_1)$ to $(\mathfrak{H}_2, J_2)$. Then the following statements are equivalent:*

(i) $\operatorname{ran} T = \mathfrak{H}$;
(ii) *T is a bounded linear operator (with* $\operatorname{dom} T = (\ker T)^{[\perp]}$*).*

A linear relation T from $(\mathfrak{H}_1, J_1)$ to $(\mathfrak{H}_2, J_2)$ is said to be *isometric* if

$$T^{-1} \subset T^{[*]}, \tag{7.7}$$

or equivalently,

$$(J_2 f', g')_{\mathfrak{H}_2} = (J_1 f, g)_{\mathfrak{H}_1} \text{ for all } \{f, f'\}, \{g, g'\} \in T. \tag{7.8}$$

The closure of an isometric relation is automatically isometric. The following observations follow directly from the definition (7.7) and the identities (7.5).

Lemma 7.3 *Let T be an isometric relation from $(\mathfrak{H}_1, J_1)$ to $(\mathfrak{H}_2, J_2)$. Then*

$$\operatorname{dom} T \subset \operatorname{ran} T^{[*]}, \quad \operatorname{ran} T \subset \operatorname{dom} T^{[*]},$$

and

$$\ker T \subset \ker(\operatorname{clos} T) \subset (\operatorname{dom} T)^{[\perp]}, \quad \operatorname{mul} T \subset \operatorname{mul}(\operatorname{clos} T) \subset (\operatorname{ran} T)^{[\perp]}.$$

Hence, if T is isometric then $\ker T$ is contained in the isotropic part of $\operatorname{dom} T$ and $\operatorname{mul} T$ is contained in the isotropic part of $\operatorname{ran} T$. An isometric relation with dense range is automatically single-valued and so is its closure; i.e. it is closable. The following result is a combination of Proposition 7.1 and Lemma 7.3; it is also proved in [Sorjonen, 1980, Proposition 2.3.1].

Proposition 7.4 *Let T be a linear relation from $(\mathfrak{H}_1, J_1)$ to $(\mathfrak{H}_2, J_2)$. Then the following statements are equivalent:*

(i) $T^{-1} = T^{[*]}$;
(ii) $T^{-1} \subset T^{[*]}$, $\ker T = (\operatorname{dom} T)^{[\perp]}$, *and* $\operatorname{ran} T = \operatorname{dom} T^{[*]}$;
(iii) $T^{-1} \subset T^{[*]}$, $\operatorname{mul} T = (\operatorname{ran} T)^{[\perp]}$, *and* $\operatorname{dom} T = \operatorname{ran} T^{[*]}$.

Proof By symmetry it suffices to show that *(i)* $\Leftrightarrow$ *(ii)*.
(i) $\Longrightarrow$ (ii) This follows from Proposition 7.1.
(ii) $\Longrightarrow$ (i) See [Derkach et al., 2006, Proposition 2.5]. □

Corollary 7.5 *Let T be a linear relation from $(\mathfrak{H}_1, J_1)$ to $(\mathfrak{H}_2, J_2)$. Then the following statements are equivalent:*

(i) $T^{-1} = T^{[*]}$ *and* $\operatorname{ran} T = \mathfrak{H}_2$;
(ii) $T^{-1} \subset T^{[*]}$, $(\operatorname{dom} T)^{[\perp]} = \ker T$, *and* $\operatorname{ran} T = \mathfrak{H}_2$;
(iii) $T^{-1} \subset T^{[*]}$, $(\ker T)^{[\perp]} = \operatorname{dom} T$, *and* $\overline{\operatorname{ran}}\, T = \mathfrak{H}_2$.

Proof The implications (i) $\Rightarrow$ (ii), (iii) are clear.
(ii) $\Rightarrow$ (i) Observe that $\operatorname{ran} T \subset \operatorname{dom} T^{[*]}$ together with $\operatorname{ran} T = \mathfrak{H}_2$ implies that $\operatorname{ran} T = \operatorname{dom} T^{[*]}$. Then apply Proposition 7.4.
(iii) $\Rightarrow$ (i) Observe that $\overline{\operatorname{ran}}\, T = \mathfrak{H}_2$ implies that $\operatorname{mul} T = (\operatorname{ran} T)^{[\perp]}$. Furthermore, observe

$$\operatorname{ran} T^{[*]} \subset \overline{\operatorname{ran}}\, T^{[*]} = (\ker T)^{\perp} = \operatorname{dom} T.$$

Again apply Proposition 7.4. □

Lemma 7.6 *Let T be an isometric linear relation from $(\mathfrak{H}_1, J_1)$ to $(\mathfrak{H}_2, J_2)$. If $(\operatorname{ran} T)^{[\perp]} = \operatorname{mul} T$, then*

$$\ker T = (\operatorname{dom} T)^{[\perp]} \quad \Longleftrightarrow \quad (\operatorname{dom} T)^{[\perp]} \subset \operatorname{dom} T.$$

Morover, if $(\operatorname{dom} T)^{[\perp]} = \ker T$, then

$$\operatorname{mul} T = (\operatorname{ran} T)^{[\perp]} \quad \Longleftrightarrow \quad (\operatorname{ran} T)^{[\perp]} \subset \operatorname{ran} T.$$

Proof It suffices to consider the case $(\operatorname{ran} T)^{[\perp]} = \operatorname{mul} T$. The implication $(\Rightarrow)$ is clear. For the implication $(\Leftarrow)$ see [Derkach et al., 2006, Proposition 2.5]. □

Remark 7.7 Let T be a relation from $(\mathfrak{H}_1, J_1)$ to $(\mathfrak{H}_2, J_2)$ which is isometric. If T satisfies $(\operatorname{ran} T)^{[\perp]} = \operatorname{mul} T$ and if there exists a neutral closed linear subspace $\mathfrak{L}$ in $\mathfrak{H}$, such that $\mathfrak{L}^{[\perp]} \subset \operatorname{dom} T$, then

$$\ker T = (\operatorname{dom} T)^{[\perp]} \subset \mathfrak{L} \subset \mathfrak{L}^{[\perp]} \subset \operatorname{dom} T. \tag{7.9}$$

Likewise, if T satisfies $(\operatorname{dom} T)^{[\perp]} = \ker T$ and if there exists a neutral closed linear subspace $\mathfrak{L}$ in $\mathfrak{H}$, such that $\mathfrak{L}^{[\perp]} \subset \operatorname{ran} T$, then

$$\operatorname{mul} T = (\operatorname{ran} T)^{[\perp]} \subset \mathfrak{L} \subset \mathfrak{L}^{[\perp]} \subset \operatorname{ran} T.$$

To see for instance (7.9), observe that $\mathfrak{L}$ being neutral, one has $\mathfrak{L} \subset \mathfrak{L}^{[\perp]} \subset \operatorname{dom} T$. Hence $(\operatorname{dom} T)^{[\perp]} \subset \mathfrak{L}^{[\perp]} \subset \operatorname{dom} T$, which by Lemma 7.6 leads to (7.9).

Calkin's approach Calkin's approach [Calkin, 1939] to extension theory was formulated for densely defined symmetric operators. In the context of symmetric relations the definition takes the following form.

Definition 7.8 [Calkin, 1939] Let S be a closed symmetric relation in a Hilbert space $\mathfrak{H}$ and let $\mathfrak{M}$ be a Hilbert space. A linear operator $\Gamma : S^* \to \mathfrak{M}$ is called a *reduction operator* for S^* if

(R1) $\operatorname{dom} \Gamma$ is dense in S^*;
(R2) Γ is closed;
(R3) there is a signature operator J in $\mathfrak{M}$ ($J = J^* = J^{-1}$), such that

$$(\mathfrak{H} \oplus \mathfrak{H} \oplus \mathfrak{M}) \ominus \Gamma = \{\{f' \oplus -f \oplus \mathrm{i}\, J\Gamma \widehat{f}\} \;:\; \widehat{f} := \{f, f'\} \in \operatorname{dom} \Gamma\}.$$

The lefthand side in the defining identity in (R3) denotes the orthogonal complement of the graph of Γ in $\mathfrak{H} \times \mathfrak{H} \times \mathfrak{M}$. This identity gives rise to the abstract form of Green's (or Lagrange's) identity:

$$(f', g)_{\mathfrak{H}} - (f, g')_{\mathfrak{H}} = \mathrm{i}(J\Gamma \widehat{f}, \Gamma \widehat{g})_{\mathfrak{M}}, \quad \widehat{f}, \widehat{g} \in \operatorname{dom} \Gamma, \tag{7.10}$$

since any element in Γ has the form $\{f, f', \Gamma\widehat{f}\}$, $\widehat{f} \in \operatorname{dom}\Gamma \subset S^*$.

The Hilbert space $\mathfrak{M}$ together with its signature operator J gives rise to a Kreĭn space $(\mathfrak{M}, J)$ with canonical symmetry J, whereas the product space $\mathfrak{H}^2 = \mathfrak{H} \times \mathfrak{H}$ with $J_{\mathfrak{H}}$ as in (7.3) gives rise to a Kreĭn space $(\mathfrak{H}^2, J_{\mathfrak{H}})$ with canonical symmetry $J_{\mathfrak{H}}$. With these canonical symmetries Green's identity (7.10) can be rewritten as

$$(J_{\mathfrak{H}}\widehat{f}, \widehat{g}) = (J\Gamma\widehat{f}, \Gamma\widehat{g}), \quad \widehat{f}, \widehat{g} \in \operatorname{dom}\Gamma,$$

in other words, the operator Γ from $(\mathfrak{H}^2, J_{\mathfrak{H}})$ to $(\mathfrak{M}, J)$ is isometric; cf. (7.8). In fact, it follows from (R2) and (R3) that the operator Γ is unitary; cf. (7.4), (7.6). In particular, $\operatorname{ran}\Gamma$ is dense in $\mathfrak{M}$. By (7.4) and Proposition 7.1 one obtains $\ker\Gamma = (\operatorname{dom}\Gamma)^{[\perp]} = (S^*)^{[\perp]} = (S^{[\perp]})^{[\perp]}$. Hence, $\ker\Gamma = S$. For the next result, use Corollary 7.2.

Corollary 7.9 *Let Γ be a reduction operator for S^*. Then the following statements are equivalent:*

(i) Γ *is bounded unitary operator;*
(ii) $\operatorname{dom}\Gamma = S^*$;
(iii) $\operatorname{ran}\Gamma = \mathfrak{M}$.

Let S be a closed symmetric relation in a Hilbert space $\mathfrak{H}$ and let Γ be a reduction operator for S^*. If A is a proper extension of S, such that $A \subset \operatorname{dom}\Gamma$, then $\Theta := \Gamma(A)$ is a subspace of $\mathfrak{M}$ such that $\Theta \subset \operatorname{ran}\Gamma$. Conversely, if Θ is a subspace of $\mathfrak{M}$ such that $\Theta \subset \operatorname{ran}\Gamma$, then the formula

$$A_\Theta := \Gamma^{-1}\Theta \tag{7.11}$$

determines a proper extension of A, such that $A_\Theta \subset \operatorname{dom}\Gamma$. The correspondence in (7.11) simplifies under the conditions of Corollary 7.9.

Theorem 7.10 [Calkin, 1939] *Assume that S^* has a bounded reduction operator Γ. Then the formula* (7.11) *establishes a one-to-one correspondence between all linear subspaces Θ of $\mathfrak{M}$ and all proper extensions A_Θ of A. Moreover,*

(i) A_Θ *is closed* $\Longleftrightarrow$ Θ *is closed;*
(ii) A_Θ *is symmetric* $\Longleftrightarrow$ Θ *is neutral in* $(\mathfrak{M}, J)$*;*
(iii) A_Θ *is maximal symmetric* $\Longleftrightarrow$ Θ *is maximal neutral in* $(\mathfrak{M}, J)$*;*
(iv) A_Θ *is self-adjoint* $\Longleftrightarrow$ Θ *is hyper-maximal neutral in* $(\mathfrak{M}, J)$.

In general reduction operators need not be bounded. If the reduction operator Γ is unbounded then there are still maximal symmetric (not necessarily selfadjoint) extensions A of S, such that $A \subset \operatorname{dom}\Gamma$ and, hence, $A = A_\Theta$ for some $\Theta \subset \mathfrak{M}$; see [Calkin, 1939, Theorem 4.3]. However, there are also maximal symmetric extensions $\widetilde{A}$ of S, such that $\widetilde{A} \cap \operatorname{dom}\Gamma = S$; see [Calkin, 1939, Theorem 4.6]. For a detailed discussion on Calkin's main results on unbounded reduction operators, see Chapter 2.

For the connection of the present paper with Calkin's work one assumes that in Definition 7.8 $\mathfrak{M} = \mathcal{H} \times \mathcal{H}$ with a Hilbert space $\mathcal{H}$ so that $(\mathfrak{M}, J_\mathcal{H})$ is a Kreĭn space with the canonical symmetry $J_\mathcal{H}$ given by

$$J_\mathcal{H} = \begin{pmatrix} 0 & -\mathrm{i}\, I_\mathcal{H} \\ \mathrm{i}\, I_\mathcal{H} & 0 \end{pmatrix}.$$

7.3 Ordinary boundary triplets

The following definition (under the title "boundary value spaces") was introduced in [Kochubei, 1975; Bruk, 1976] for the case of a densely defined symmetric operator with equal deficiency indices. Due to its symplectic structure the definition is a special case of Calkin's definition; cf. Corollary 7.9. It has been extended to the case of a non-densely defined operator in [Malamud, 1992].

Definition 7.11 Let S be a closed symmetric relation in a Hilbert space $\mathfrak{H}$ with equal deficiency indices and let S^* be the adjoint linear relation. Then the triplet $\{\mathcal{H}, \Gamma_0, \Gamma_1\}$, where $\mathcal{H}$ is a Hilbert space and $\Gamma = \{\Gamma_0, \Gamma_1\}$ is a single-valued linear mapping from S^* to $\mathcal{H}^2$, is said to be an *ordinary boundary triplet* for S^* if:

(O1) $\operatorname{dom}\Gamma = S^*$;

(O2) the abstract Green's identity

$$(f', g)_\mathfrak{H} - (f, g')_\mathfrak{H} = (\Gamma_1 \widehat{f}, \Gamma_0 \widehat{g})_\mathcal{H} - (\Gamma_0 \widehat{f}, \Gamma_1 \widehat{g})_\mathcal{H}, \tag{7.12}$$

holds for all $\widehat{f} = \{f, f'\}$, $\widehat{g} = \{g, g'\} \in S^*$;

(O3) the mapping $\Gamma : \widehat{f} \to (\Gamma_0 \widehat{f}\ \Gamma_1 \widehat{f})^\top$ from S^* to $\mathcal{H}^2$ is surjective.

Remark The above definition grew out of long line of investigations. The first steps toward a description of self-adjoint extensions of a minimal symmetric operator generated by ordinary differential operators on a finite interval in the regular case have been initiated in [Graff, 1946;

Kreĭn, 1947] (see also [Naĭmark, 1969; Achieser and Glasmann, 1981]). All self-adjoint extensions of a singular even order ordinary differential operator on a semiaxis were described in [Glazman, 1950]. Boundary value problems for elliptic operators of the second order and of even order, respectively, were investigated in [Višik, 1952; Grubb, 1968]. In [Rofe-Beketov, 1969; Gorbachuk, 1971] the boundary value approach was used to get descriptions of self-adjoint extensions of minimal symmetric operators generated by ordinary differential operators with operator potentials, both bounded and unbounded.

By (7.12) the mapping Γ: $(\mathfrak{H}^2, J_{\mathfrak{H}})$ to $(\mathcal{H}^2, J_{\mathcal{H}})$ is isometric and it follows from Corollary 7.2 and Corollary 7.5 that $\Gamma_i \in \boldsymbol{B}(S^*, \mathcal{H})$, $i = 0, 1$. Associated with this ordinary boundary triplet Π are the following selfadjoint extensions of the symmetric operator S:

$$A_0 := \ker \Gamma_0, \quad A_1 := \ker \Gamma_1, \tag{7.13}$$

which are transversal. Conversely, given two transversal self-adjoint extensions A_0 and A_1 of a closed symmetric relation S, there is an ordinary boundary triplet $\Pi = \{\mathcal{H}, \Gamma_0, \Gamma_1\}$ for S^*, such that (7.13) holds; see [Derkach and Malamud, 1985, 1987]. In the case of ordinary boundary triplets Theorem 7.10 can be reformulated as follows

Proposition 7.12 *Let $\Pi = \{\mathcal{H}, \Gamma_0, \Gamma_1\}$ be an ordinary boundary triplet for S^*. Then the mapping Γ establishes a one-to-one correspondence*

$$\Gamma : A_\Theta \leftrightarrow \Theta$$

between the proper extensions A_Θ of S and the linear relations Θ in $\mathcal{H}$. Moreover,

(i) *A_Θ is closed $\Longleftrightarrow$ Θ is closed;*
(ii) *A_Θ is symmetric $\Longleftrightarrow$ Θ is symmetric;*
(iii) *A_Θ is self-adjoint $\Longleftrightarrow$ Θ is self-adjoint;*
(iv) *A_Θ is maximal dissipative $\Longleftrightarrow$ Θ is maximal dissipative;*
(v) *A_Θ is disjoint with A_0 $\Longleftrightarrow$ Θ is a closed linear operator;*
(vi) *A_Θ is transversal with A_0 $\Longleftrightarrow$ Θ is an everywhere defined bounded linear operator.*

The main analytical tool in the description of the spectral properties of self-adjoint extensions of a symmetric operator or relation is the abstract Weyl function, introduced and investigated in [Derkach and Malamud, 1985, 1987]. It plays a role similar to that of the classical

Titchmarsh-Weyl coefficient in the spectral theory of singular Sturm-Liouville operator on a semiaxis.

Definition 7.13 The operator-valued function defined by the equality

$$\Gamma_1\widehat{f}_\lambda = M(\lambda)\Gamma_0\widehat{f}_\lambda, \quad \widehat{f}_\lambda \in \widehat{\mathfrak{N}}_\lambda,\ \lambda \in \rho(A_0), \tag{7.14}$$

is called the abstract *Weyl function* of S, corresponding to the ordinary boundary triplet $\Pi = \{\mathcal{H}, \Gamma_0, \Gamma_1\}$.

It is easily seen that the operator $\Gamma_0{\upharpoonright}\widehat{\mathfrak{N}}_\lambda : \widehat{\mathfrak{N}}_\lambda \to \mathcal{H}$ is invertible for all $\lambda \in \rho(A_0)$ and that

$$\widehat{\gamma}(\lambda) := (\Gamma_0{\upharpoonright}\widehat{\mathfrak{N}}_\lambda)^{-1} \in \boldsymbol{B}(\mathcal{H}, \widehat{\mathfrak{N}}_\lambda).$$

Therefore, the Weyl function M is correctly defined for all $\lambda \in \rho(A_0)$ by the equality

$$M(\lambda)u = \Gamma_1\widehat{\gamma}(\lambda)u, \quad u \in \mathcal{H}. \tag{7.15}$$

Let π_1 be the orthogonal projection onto the first component of $\mathfrak{H}^2$. The operator-valued function

$$\lambda \in \rho(A_0) \mapsto \gamma(\lambda) := \pi_1\widehat{\gamma}(\lambda) \tag{7.16}$$

is called the γ-field, corresponding to $\Pi = \{\mathcal{H}, \Gamma_0, \Gamma_1\}$. The following statement connects the γ-field to the Weyl function and the selfadjoint extension A_0.

Proposition 7.14 [Derkach and Malamud, 1985] *Let* $\Pi = \{\mathcal{H}, \Gamma_0, \Gamma_1\}$ *be an ordinary boundary triplet for* S^*. *Then the* γ*-field* γ *and the Weyl function* M *corresponding to the boundary triplet* Π *satisfy the identities*

$$\gamma(\lambda) - \gamma(\mu) = (\lambda - \mu)(A_0 - \lambda)^{-1}\gamma(\mu), \quad \lambda, \mu \in \rho(A_0), \tag{7.17}$$

$$M(\lambda) - M(\mu)^* = (\lambda - \bar{\mu})\gamma(\mu)^*\gamma(\lambda), \quad \lambda, \mu \in \rho(A_0). \tag{7.18}$$

In particular, this proposition implies that the Weyl function M belongs to the class $R[\mathcal{H}]$ of Nevanlinna functions, characterized by the conditions

(i) $M(\lambda) : \mathbb{C} \setminus \mathbb{R} \to \boldsymbol{B}(\mathcal{H})$ is holomorphic;
(ii) $M(\lambda)^* = M(\bar{\lambda})$ for all $\lambda \in \rho(A_0)$;
(iii) $\operatorname{Im} M(\lambda)\operatorname{Im}(\lambda) \geq 0$ for all $\lambda \in \rho(A_0)$.

Since the γ-field γ is boundedly invertible, it follows from (7.18) that, in fact, M belongs to the class $R^u[\mathcal{H}]$, i.e., the class of all uniformly strict functions in $R[\mathcal{H}]$, which satisfy

$$0 \in \rho(\mathrm{Im}\, M(\lambda)) \quad \text{for all } \lambda \in \rho(A_0).$$

Note that the equality (7.18) means that the Weyl function M is the Q-function of the pair (A, A_0) in the sense of [Kreĭn, 1944, 1946] (see also [Saakyan, 1965; Kreĭn and Langer, 1971]).

Theorem 7.15 [Langer and Textorius, 1977; Derkach and Malamud, 1995] *Every uniformly strict Nevanlinna function is the Weyl function of a closed simple symmetric operator S, corresponding to an ordinary boundary triplet Π.*

In the following theorem the spectrum and the resolvent of any proper extension A_Θ as in Proposition 7.12 are characterized.

Theorem 7.16 [Derkach and Malamud, 1985, 1995] *Let $\Pi = \{\mathcal{H}, \Gamma_0, \Gamma_1\}$ be an ordinary boundary triplet for S^*, $\lambda \in \rho(A_0)$, and Θ be a closed linear relation in $\mathcal{H}$. Then:*

(i) $\lambda \in \sigma_p(A_\Theta) \Longleftrightarrow 0 \in \sigma_p(\Theta - M(\lambda))$;
(ii) $\lambda \in \sigma_c(A_\Theta) \Longleftrightarrow 0 \in \sigma_c(\Theta - M(\lambda))$;
(iii) $\lambda \in \rho(A_\Theta) \Longleftrightarrow 0 \in \rho(\Theta - M(\lambda))$.

In the last case one has for all $\lambda \in \rho(A_\Theta) \cap \rho(A_0)$

$$(A_\Theta - \lambda)^{-1} = (A_0 - \lambda)^{-1} + \gamma(\lambda)(\Theta - M(\lambda))^{-1}\gamma(\bar{\lambda})^*. \tag{7.19}$$

Formula (7.19) establishes a one-to-one correspondence between the class of proper extensions of S with non-empty resolvent set and a subclass of linear relations in $\mathcal{H}$. Note that all objects in (7.19) are expressed in terms of the operators Γ_0 and Γ_1 (see (7.13), (7.14), (7.16)).

7.4 Boundary triplets of bounded type

Definition and block representation The notion of the ordinary boundary triplet was extended in [Derkach and Malamud, 1995] by weakening the surjectivity assumption in Definition 7.11

Definition 7.17 [Derkach and Malamud, 1995] Let S be a closed symmetric operator in a Hilbert space $\mathfrak{H}$ with equal deficiency indices and let T be a linear relation in $\mathfrak{H}$ such that $S \subset T \subset \operatorname{clos} T = S^*$. Then the triplet $\{\mathcal{H}, \Gamma_0, \Gamma_1\}$, where $\mathcal{H}$ is a Hilbert space and $\Gamma = \{\Gamma_0, \Gamma_1\}$ is a single-valued linear mapping from T to $\mathcal{H}^2$, is said to be a *boundary triplet of bounded type* (or *generalized boundary triplet) for* S^*, *if:*

(B1) *Green's identity* (7.12) *holds for all* $\widehat{f} = \{f, f'\}$, $\widehat{g} = \{g, g'\} \in T$;
(B2) $\operatorname{ran} \Gamma_0 = \mathcal{H}$;
(B3) $A_0 := \ker \Gamma_0$ *is a selfadjoint relation in* $\mathfrak{H}$.

By definition $A_0 \subset \operatorname{dom} \Gamma = T$, which implies that A_0 is a selfadjoint extension of S. The term "boundary triplet of bounded type" is used here to indicate that the Weyl function M corresponding to the boundary triplet in Definition 7.17 is bounded, as is stated in the next proposition.

Proposition 7.18 [Derkach and Malamud, 1995] *Let* $\{\mathcal{H}, \Gamma_0, \Gamma_1\}$ *be a boundary triplet of bounded type for* S^*. *Then:*

(i) $T = A_0 \widehat{+} \widehat{\mathfrak{N}}_\lambda(T)$, *where* $\widehat{\mathfrak{N}}_\lambda(T) = \widehat{\mathfrak{N}}_\lambda(S^*) \cap T$ *is dense in* $\widehat{\mathfrak{N}}_\lambda(S^*)$ *for every* $\lambda \in \mathbb{C} \setminus \mathbb{R}$;
(ii) $\operatorname{clos} \Gamma_1(A_0) = \mathcal{H}$ *and* $\overline{\operatorname{ran}}\, \Gamma = \mathcal{H}^2$;
(iii) *the restriction* $\Gamma_0 : \widehat{\mathfrak{N}}_\lambda(T) \to \mathcal{H}$ *is a closed mapping for every* $\lambda \in \mathbb{C} \setminus \mathbb{R}$;
(iv) *the equalities* (7.16) *and* (7.15) *define a* $\boldsymbol{B}(\mathcal{H}, \widehat{\mathfrak{N}}_\lambda)$*-valued function* $\widehat{\gamma}$, *a* $\boldsymbol{B}(\mathcal{H}, \mathfrak{N}_\lambda)$*-valued function* γ, *and a* $\boldsymbol{B}(\mathcal{H})$*-valued function* M, *which are holomorphic on* $\mathbb{C} \setminus \mathbb{R}$ *and satisfy the identities* (7.17), (7.18).

The operator-valued functions M and γ defined in item (iv) are called the abstract *Weyl function* of S and the γ*-field* corresponding to the boundary triplet of bounded type $\Pi = \{\mathcal{H}, \Gamma_0, \Gamma_1\}$. It follows from the identity (7.18) that the Weyl function M belongs to the class $R[\mathcal{H}]$.

Denote by $R^s[\mathcal{H}]$ the class of strict Nevanlinna operator-valued functions with values in $\boldsymbol{B}(\mathcal{H})$, that is

$$M \in R^s[\mathcal{H}] \Longleftrightarrow M \in R[\mathcal{H}] \text{ and } 0 \notin \sigma_p(\operatorname{Im} M(\lambda)) \text{ for all } \lambda \in \mathbb{C} \setminus \mathbb{R}.$$

The next proposition shows that the class $R^s[\mathcal{H}]$ of bounded strict Nevanlinna functions in fact characterizes boundary triplets of bounded type; see [Derkach et al., 2006] for further details.

Proposition 7.19 *The Weyl function* M *corresponding to a boundary triplet of bounded type* $\{\mathcal{H}, \Gamma_0, \Gamma_1\}$ *belongs to the class* $R^s[\mathcal{H}]$. *Conversely, every* $R^s[\mathcal{H}]$*-function is the Weyl function of some boundary triplet of bounded type* $\{\mathcal{H}, \Gamma_0, \Gamma_1\}$.

The next result collects some further properties of boundary triplets of bounded type. Note that the statement (ii) was obtained in [Derkach and Malamud, 1995] under the additional assumption that $A_1 = A_1^*$.

Proposition 7.20 [Derkach and Malamud, 1995] *Let $\{\mathcal{H}, \Gamma_0, \Gamma_1\}$ be a boundary triplet of bounded type for S^* and let M be the corresponding Weyl function. Then:*

(i) $0 \in \rho(\operatorname{Im}(M(\mathrm{i})))$ *if and only if* $\operatorname{ran}\Gamma = \mathcal{H} \oplus \mathcal{H}$;
(ii) $0 \in \rho(M(\mathrm{i}))$ *if and only if* $\operatorname{ran}\Gamma_1 = \mathcal{H}$.

In case (ii) $A_1 = A_1^*$, the transposed boundary triplet $\{\mathcal{H}, -\Gamma_1, \Gamma_0\}$ is also a boundary triplet of bounded type and the corresponding Weyl function is $-M^{-1}$.

Proof (i) This statement was proved in [Derkach and Malamud, 1995].

(ii) The transposed boundary triplet $\Pi^\top = \{\mathcal{H}, -\Gamma_1, \Gamma_0\}$ is a unitary boundary triplet for S^* in the sense of Definition 7.34; see [Derkach et al., 2006, p. 5370]. Since $0 \in \rho(M(\mathrm{i}))$, then $-M^{-1} \in R^s[\mathcal{H}]$ and, by Proposition 7.19, $\Pi^\top$ is a boundary triplet of bounded type for S^*. This implies that $A_1 = A_1^*$. □

The next remark indicates how one can obtain boundary triplets of bounded type by transforming ordinary boundary triplets.

Remark 7.21 Let $\{\mathcal{H}, \Gamma_0^0, \Gamma_1^0\}$ be an ordinary boundary triplet for S^* with $A_0 = \ker\,\Gamma_0^0$ and $A_1 = \ker\,\Gamma_1^0$, and let K be a bounded linear operator in $\mathcal{H}$ with $\ker\,K = \ker\,K^* = \{0\}$. Consider the transform

$$\begin{pmatrix} \widetilde{\Gamma}_0 \\ \widetilde{\Gamma}_1 \end{pmatrix} = \begin{pmatrix} K^{-1} & 0 \\ 0 & K^* \end{pmatrix} \begin{pmatrix} \Gamma_0^0 \\ \Gamma_1^0 \end{pmatrix}. \tag{7.20}$$

Then $T := \operatorname{dom}\widetilde{\Gamma} = \{\widehat{f} \in S^* : \Gamma_0^0 \widehat{f} \in \operatorname{ran} K\}$ and

$$\begin{aligned} \operatorname{ran}\widetilde{\Gamma}_0 &= \mathcal{H}, & \widetilde{A}_0 &= \ker\,\widetilde{\Gamma}_0 = A_0 \subset T, \\ \operatorname{ran}\widetilde{\Gamma}_1 &= \operatorname{ran} K^*, & \widetilde{A}_1 &= \ker\,\widetilde{\Gamma}_1 = A_1 \cap T. \end{aligned}$$

It follows that the triplet $\{\mathcal{H}, \widetilde{\Gamma}_0, \widetilde{\Gamma}_1\}$ is a boundary triplet of bounded type for S^*. It is an ordinary boundary triplet for S^* if and only if K is surjective. The Weyl functions M_0 and $\widetilde{M}$, corresponding to boundary triplets $\{\mathcal{H}, \Gamma_0^0, \Gamma_1^0\}$ and $\{\mathcal{H}, \widetilde{\Gamma}_0, \widetilde{\Gamma}_1\}$, are connected by the equality

$$\widetilde{M}(\lambda) = K^* M_0(\lambda) K, \quad \lambda \in \rho(A_0). \tag{7.21}$$

Although the mapping Γ from S^* to $\mathcal{H}^2$ of a boundary triplet of bounded type need not be bounded, there is some replacement of this property, which can be formulated in terms of rigged Hilbert spaces. With $M \in R^s[\mathcal{H}]$ one can associate the following rigging $\mathcal{H}_+ \subset \mathcal{H} \subset \mathcal{H}_-$ of the space $\mathcal{H}$, where $\mathcal{H}_+ = \mathrm{ran}\,(\mathrm{Im}\,(M(\mathrm{i}))^{1/2}$ is a Hilbert space with the norm

$$\|k\|_{\mathcal{H}_+} = \|K^{-1}k\|, \quad K := (\mathrm{Im}\,(M(\mathrm{i}))^{1/2}, \quad k \in \mathrm{ran}\,K. \tag{7.22}$$

Then the dual space $\mathcal{H}_-$ can be identified with the completion of $\mathcal{H}$ with respect to the norm

$$\|h\|_{\mathcal{H}_-} = \|Kh\|, \quad h \in \mathcal{H}. \tag{7.23}$$

The closure of an operator $T : \mathcal{H} \to \mathcal{H}$ will be called the $(0,0)$−closure of T and denoted by $\overline{T}_{00}$. Similarly, the $(-,+)$−closure of T is defined as the closure of the operator T acting from $\mathcal{H}_-$ to $\mathcal{H}_+$ and it is denoted by $\overline{T}_{-,+}$. The following lemma provides some continuation results; a proof of the lemma is included for the convenience of the reader.

Lemma 7.22 [Derkach and Malamud, 1995] *Let $\{\mathcal{H}, \Gamma_0, \Gamma_1\}$ be a boundary triplet of bounded type for S^*, let M be the corresponding Weyl function, and let $\lambda \in \mathbb{C} \setminus \mathbb{R}$. Then:*

(i) *$\gamma(\lambda)$ admits a continuation to a bounded operator $\overline{\gamma(\lambda)}$ from $\mathcal{H}_-$ onto $\mathfrak{N}_\lambda(S^*)$;*

(ii) *Γ_0 admits a continuation to a bounded operator $\overline{\Gamma}_0$ from S^* onto $\mathcal{H}_-$;*

(iii) *$\mathrm{Im}\,M(\lambda)$ admits a continuation to a bounded operator $\overline{\mathrm{Im}\,M(\lambda)}_{-,+}$ from $\mathcal{H}_-$ onto $\mathcal{H}_+$;*

(iv) *if, in addition, $M(\mathrm{i})$ is a bounded operator from $\mathcal{H}_-$ to $\mathcal{H}_+$, then Γ_1 admits a continuation to a bounded operator $\overline{\Gamma}_1$ from S^* onto $\mathcal{H}_+$.*

Proof (i) It follows from (7.22), (7.23) and (7.18) that

$$\|h\|_-^2 = (\mathrm{Im}\,M(\mathrm{i})h, h) = \|\gamma(\mathrm{i})h\|^2, \quad h \in \mathcal{H},$$

and, hence, the operator $\gamma(\mathrm{i})$ admits a continuation to an isometry $\overline{\gamma(\mathrm{i})}$ from $\mathcal{H}_-$ onto $\mathfrak{N}_\mathrm{i}(S^*)$. Now the statement (i) is implied by the identity (7.17).

(ii) Recall that $T := \mathrm{dom}\,\Gamma = A_0 \,\widehat{+}\, \mathfrak{N}_\lambda(T)$. By item (i) the closure of the operator $\Gamma_0 \upharpoonright \mathfrak{N}_\lambda(T) = (\gamma(\lambda))^{-1}$ is bounded from $\mathfrak{N}_\lambda(S^*)$ onto $\mathcal{H}_-$. Since $\Gamma_0 \upharpoonright A_0 = 0$ and the angle between $\mathfrak{N}_\lambda(S^*)$ and A_0 is positive, it follows that the closure of Γ_0 is also bounded from S^* onto $\mathcal{H}_-$.

(iii) This statement is implied by (i) and the identity (7.18).

(iv) Since the angle between $\mathfrak{N}_i(S^*)$ and A_0 is positive and

$$\Gamma_1\widehat{f} = \gamma(\mathrm{i})^*(f' + \mathrm{i}\, f), \quad \Gamma_1\widehat{f}_{\mathrm{i}} = M(\mathrm{i})\gamma(\mathrm{i})^{-1},$$

when $\{f, f'\} \in A_0$, $\{f_{\mathrm{i}}, \mathrm{i}\, f_{\mathrm{i}}\} \in \widehat{\mathfrak{N}}_i(T)$, the conditions $\gamma(\mathrm{i})^* \in \boldsymbol{B}(\mathfrak{N}_{\mathrm{i}}, \mathcal{H}_+)$, $M(\mathrm{i}) \in \boldsymbol{B}(\mathcal{H}_-, \mathcal{H}_+)$ guarantee that the operator Γ_1 is bounded from S^* into $\mathcal{H}_+$. Moreover, it follows from

$$\overline{\Gamma}_1\widehat{f} = (\overline{\gamma(\mathrm{i})})^*(f' + \mathrm{i}\, f), \quad \widehat{f} = \{f, f'\} \in A_0,$$

and item (i) that $\operatorname{ran}\overline{\Gamma}_1 = \mathcal{H}_+$. □

The Weyl function $\widetilde{M}$ appearing in Remark 7.21 is a bounded operator from $\mathcal{H}_-$ to $\mathcal{H}_+$ for all $\lambda \in \mathbb{C} \setminus \mathbb{R}$. This property turns out to be characteristic for boundary triplets of bounded type of the form (7.20).

Lemma 7.23 *Let $\{\mathcal{H}, \Gamma_0, \Gamma_1\}$ be a boundary triplet of bounded type for S^* with the Weyl function $\widetilde{M}$ and assume that $\widetilde{M}(\mathrm{i})$ is a bounded operator from $\mathcal{H}_-$ to $\mathcal{H}_+$. Then there exists an ordinary boundary triplet $\{\mathcal{H}, \Gamma_0^0, \Gamma_1^0\}$ such that (7.20) and (7.21) hold.*

Proof By Lemma 7.22 the mapping Γ_0 admits the continuation $\overline{\Gamma}_0$ from S^* onto $\mathcal{H}_-$, and the mapping Γ_1 admits the continuation $\overline{\Gamma}_1$ from S^* onto $\mathcal{H}_+$. Since $K : \mathcal{H}_- \to \mathcal{H}$ in (7.22) is isometric, see (7.23), the equality

$$\begin{pmatrix} \Gamma_0^0 \\ \Gamma_1^0 \end{pmatrix} = \begin{pmatrix} \overline{K} & 0 \\ 0 & K^{-1} \end{pmatrix} \begin{pmatrix} \overline{\Gamma}_0 \\ \overline{\Gamma}_1 \end{pmatrix} \tag{7.24}$$

correctly defines the operators Γ_0^0, Γ_1^0 and the triplet $\{\mathcal{H}, \Gamma_0^0, \Gamma_1^0\}$ is also a boundary triplet of bounded type for S^*. If M_0 is the corresponding Weyl function then it follows from (7.24) that the Weyl functions $\widetilde{M}$ and M_0 are connected by the equality

$$\widetilde{M}(\lambda) = KM_0(\lambda)K$$

and $\operatorname{Im} M_0(\mathrm{i}) = I$. Hence, $\{\mathcal{H}, \Gamma_0^0, \Gamma_1^0\}$ is an ordinary boundary triplet; see Proposition 7.20. □

Theorem 7.24 *Let $\{\mathcal{H}, \Gamma_0, \Gamma_1\}$ be a boundary triplet of bounded type for S^* and let M be the corresponding Weyl function. Then there exists an ordinary boundary triplet $\{\mathcal{H}, \Gamma_0^0, \Gamma_1^0\}$ and operators $K = K^*, R = R^* \in \boldsymbol{B}(\mathcal{H})$ such that $0 \notin \sigma_p(K)$, and*

$$\begin{pmatrix} \Gamma_0 \\ \Gamma_1 \end{pmatrix} = \begin{pmatrix} K^{-1} & 0 \\ RK^{-1} & K \end{pmatrix} \begin{pmatrix} \Gamma_0^0 \\ \Gamma_1^0 \end{pmatrix}. \tag{7.25}$$

If M_0 is the Weyl function corresponding $\{\mathcal{H}, \Gamma_0^0, \Gamma_1^0\}$, then

$$M(\lambda) = K^* M_0(\lambda) K + R, \quad \lambda \in \rho(A_0).$$

Proof Define $R = \operatorname{Re} M(\mathrm{i})$, $K = (\operatorname{Im} M(\mathrm{i}))^{1/2}$, and introduce a new boundary triplet of bounded type $\{\mathcal{H}, \widetilde{\Gamma}_0, \widetilde{\Gamma}_1\}$ by the transform

$$\begin{pmatrix} \widetilde{\Gamma}_0 \\ \widetilde{\Gamma}_1 \end{pmatrix} = \begin{pmatrix} I & 0 \\ -R & I \end{pmatrix} \begin{pmatrix} \Gamma_0 \\ \Gamma_1 \end{pmatrix}. \tag{7.26}$$

The corresponding Weyl function $\widetilde{M}$ is given by

$$\widetilde{M}(\lambda) = M(\lambda) - R.$$

Then $\widetilde{M}(\mathrm{i}) = \mathrm{i} \operatorname{Im} M(\mathrm{i}) = \mathrm{i} K^2$ and hence in view of (7.22) and (7.23) $\widetilde{M}(\mathrm{i}) : \mathcal{H}_- \to \mathcal{H}_+$ is bounded. Thus by Lemma 7.23 there exists an ordinary boundary triplet $\{\mathcal{H}, \Gamma_0^0, \Gamma_1^0\}$, which is connected to the boundary triplet of bounded type $\{\mathcal{H}, \widetilde{\Gamma}_0, \widetilde{\Gamma}_1\}$ by the equality (7.20). Now (7.25) is obtained by combining (7.26) and (7.20). □

Corollary 7.25 *Under the assumptions of Theorem 7.24 the mapping $\widetilde{\Gamma}_1 := \Gamma_1 - R\Gamma_0$ with $R = \operatorname{Re} M(\mathrm{i})$ admits a continuation*

$$G_1 = \overline{\Gamma_1 - R\Gamma_0} \in \boldsymbol{B}(S^*, \mathcal{H}_+)$$

with $\operatorname{ran} G_1 = \mathcal{H}_+$.

A Kreĭn type resolvent formula for proper extensions The resolvent formula (7.19) appearing in Theorem 7.16 was stated in terms of an ordinary boundary triplet for S^*. There is an analog of this resolvent formula, which can be expressed by means of a boundary triplet of bounded type for S^*.

Theorem 7.26 *Let the assumptions in Theorem 7.24 be satisfied, let*

$$G_0 = \overline{\Gamma}_0 \in \boldsymbol{B}(S^*, \mathcal{H}_-), \quad G_1 = \overline{\Gamma_1 - R\Gamma_0} \in \boldsymbol{B}(S^*, \mathcal{H}_+)$$

be the continuations as in Lemma 7.22 (ii) and Corollary 7.25, and let $\lambda \in \rho(A_0)$. Then:

(i) *there is a one-to-one correspondence between closed proper extensions $\widetilde{A}$ of S disjoint with A_0 (i.e. $\widetilde{A} \cap A_0 = S$) and $(-,+)$-closed operators $\mathcal{B}$, from $\mathcal{H}_-$ to $\mathcal{H}_+$, given by*

$$\widetilde{A} = \left\{ \widehat{f} \in S^* : G_1 \widehat{f} - \mathcal{B} G_0 \widehat{f} = 0,\, G_0 \widehat{f} \in \operatorname{dom} \mathcal{B} \right\}; \tag{7.27}$$

(ii) $\lambda \in \sigma_p(\widetilde{A}) \Leftrightarrow \ker\,(\mathcal{B} + (\overline{R - M(\lambda)})_{-+}) \neq \{0\}$ *and, moreover,*

$$\ker\,(\widetilde{A} - \lambda) = \overline{\gamma}(\lambda)\ker\,(\mathcal{B} + (\overline{R - M(\lambda)})_{-+});$$

(iii) $\lambda \in \rho(\widetilde{A}) \Leftrightarrow (\mathcal{B} + (\overline{R - M(\lambda)})_{-+})^{-1} \in \boldsymbol{B}(\mathcal{H}_+, \mathcal{H}_-)$;

Furthermore, if $\lambda \in \rho(A_0) \cap \rho(\widetilde{A})$, *then the resolvent of* $\widetilde{A}$ *has the form*

$$(\widetilde{A} - \lambda)^{-1} = (A_0 - \lambda)^{-1} - \overline{\gamma}(\lambda)(\mathcal{B} + (\overline{R - M(\lambda)})_{-+})^{-1}\gamma(\bar{\lambda})^*, \quad (7.28)$$

where $\overline{\gamma}(\lambda)$ *is the continuation as in Lemma 7.22.*

Proof Consider the ordinary boundary triplet $\{\mathcal{H}, \Gamma_0^0, \Gamma_1^0\}$ from Theorem 7.24. Due to Proposition 7.12 every closed proper extension $\widetilde{A}$ of S disjoint with A_0 can be determined by the equality

$$\widetilde{A} = \left\{\widehat{f} \in A^* : \Gamma_1^0\widehat{f} - B_0\Gamma_0^0\widehat{f} = 0,\ \Gamma_0^0\widehat{f} \in \mathrm{dom}\,B_0\right\}, \quad (7.29)$$

where B_0 is a closed linear operator in $\mathcal{H}$. Since the operators

$$K : \mathcal{H} \to \mathcal{H}_+, \quad \overline{K} : \mathcal{H}_- \to \mathcal{H}$$

are isometries (see (7.22), (7.23)), the operator $\mathcal{B} = KB_0\overline{K}$ is a closed operator from $\mathcal{H}_-$ to $\mathcal{H}_+$ precisely when the operator B_0 is closed in $\mathcal{H}$. Using the equalities (7.24) and (7.26) one can rewrite (7.29) in the form (7.27).

By Theorem 7.16 $\lambda \in \sigma_p(\widetilde{A})$ if and only if

$$\ker\,(B_0 - M_0(\lambda)) \neq \{0\}. \quad (7.30)$$

It follows from $\mathcal{B} = KB_0\overline{K}$ and $(\overline{M(\lambda) - R})_{-+} = KM_0(\lambda)\overline{K}$ that

$$\mathcal{B} + (\overline{R - M(\lambda)})_{-+} = K(B_0 - M_0(\lambda))\overline{K}. \quad (7.31)$$

Hence (7.30) can be rewritten as

$$\ker\,(\mathcal{B} + (\overline{R - M(\lambda)})_{-+}) \neq \{0\}.$$

Moreover, $\ker\,(\mathcal{B} + (\overline{R - M(\lambda)})_{-+}) = \overline{K}^{-1}\ker\,(B_0 - M_0(\lambda))$ and since $\gamma_0(\lambda)\overline{K} = \overline{\gamma}(\lambda)$, one obtains

$$\ker\,(\widetilde{A} - \lambda) = \gamma_0(\lambda)\ker\,(B_0 - M_0(\lambda)) = \overline{\gamma}(\lambda)\ker\,(\mathcal{B} + (\overline{R - M(\lambda)})_{-+}).$$

Next, due to Theorem 7.16, observe that $\lambda \in \rho(\widetilde{A}_B)$ if and only if $0 \in \rho(B_0 - M_0(\lambda))$. In view of (7.31) the latter condition is equivalent to

$$(\mathcal{B} + (\overline{R - M(\lambda)})_{-+})^{-1} \in \boldsymbol{B}(\mathcal{H}_+, \mathcal{H}_-),$$

and for every $\lambda \in \rho(A_0) \cap \rho(\widetilde{A})$ the resolvent of $\widetilde{A}$ takes the form

$$\begin{aligned}(\widetilde{A} - \lambda)^{-1} &= (A_0 - \lambda)^{-1} - \gamma_0(\lambda)(\mathcal{B}_0 - M_0(\lambda))^{-1}\gamma_0(\bar{\lambda})^* \\ &= (A_0 - \lambda)^{-1} - \gamma_0(\lambda)\overline{K}(\mathcal{B} + \overline{(R - M(\lambda))}_{-+})^{-1}K\gamma_0(\bar{\lambda})^*.\end{aligned}$$

Since $\gamma(\bar{\lambda}) = \gamma_0(\bar{\lambda})K$, one sees that $\gamma(\bar{\lambda})^* = K\gamma_0(\bar{\lambda})^*$ maps into $\mathcal{H}_+$. This proves (7.28). □

Under additional assumptions on the parameter $\mathcal{B}$ corresponding to the extension $\widetilde{A}$ in Theorem 7.26, the resolvent formula (7.28) can be rewritten in a more familiar form: cf. [Behrndt and Langer, 2007, Theorem 2.8] and Theorem 6.16.

Proposition 7.27 *Let $\{\mathcal{H}, \Gamma_0, \Gamma_1\}$ be a boundary triplet of bounded type for S^* with domain $T = \operatorname{dom}\Gamma$ and let M be the corresponding Weyl function. Let B be a closed operator in $\mathcal{H}$, such that $0 \in \rho(B - M(\lambda))$ for some $\lambda \in \rho(A_0)$. Then the linear relation*

$$\widetilde{A}_B = \left\{ \widehat{f} \in T : \Gamma_1\widehat{f} - B\Gamma_0\widehat{f} = 0 \right\} \tag{7.32}$$

is a closed proper extension of S whose domain belongs to T, $\lambda \in \rho(\widetilde{A}_B)$, and the resolvent of $\widetilde{A}_B$ with $\lambda \in \rho(A_0) \cap \rho(\widetilde{A}_B)$ has the form

$$(\widetilde{A}_B - \lambda)^{-1} = (A_0 - \lambda)^{-1} - \gamma(\lambda)(B - M(\lambda))^{-1}\gamma(\bar{\lambda})^*. \tag{7.33}$$

Proof With the notations as in Theorem 7.26 consider a linear manifold

$$\mathcal{D} = \{h \in \mathcal{H} : (B - M(\lambda))h \in \mathcal{H}_+\} = (B - M(\lambda))^{-1}\mathcal{H}_+,$$

where $\lambda \in \rho(A_0)$. Then it follows from

$$B - M(\lambda) = (B - R) + (R - M(\lambda)), \quad \operatorname{ran}(R - M(\lambda)) \subset \mathcal{H}_+,$$

that $\mathcal{D} = \{u \in \mathcal{H} : (B - R)u \in \mathcal{H}_+\}$, and therefore the definition of $\mathcal{D}$ does not depend on the choice of $\lambda \in \rho(A_0)$. Since $0 \in \rho(B - M(\lambda))$, the restriction $L := ((B - M(\lambda)){\restriction}\mathcal{D})^{-1}$ is bounded as an operator in $\mathcal{H}$ with domain $\mathcal{H}_+$. This implies that L is closed, and thus also bounded, as an operator from $\mathcal{H}_+$ to $\mathcal{H}_-$, and hence the inverse $(B - M(\lambda)){\restriction}\mathcal{D}$ is closed as an operator from $\mathcal{H}_-$ to $\mathcal{H}_+$. Since $M(\lambda) - R$ is a $(-,+)$-bounded operator, also $(B - R){\restriction}\mathcal{D}$ is closed as an operator from $\mathcal{H}_-$ to $\mathcal{H}_+$.

Let $\widetilde{A}$ be the extension of S given by (7.27) with $\mathcal{B} := (B - R){\restriction}\mathcal{D}$. Since $\operatorname{dom}\mathcal{B} \subset \mathcal{H}$, one has for every $\widehat{f} \in \widetilde{A}$ that $G_0\widehat{f} = \Gamma_0\widehat{f} \in \mathcal{H}$ and hence $\widehat{f} \in T = \operatorname{dom}\Gamma$. Therefore, the conditions in (7.27) can be rewritten in the form

$$(G_1 - \mathcal{B}G_0)\widehat{f} = (\Gamma_1 - R\Gamma_0)\widehat{f} - (B - R)\Gamma_0\widehat{f} = 0, \ \Gamma_0\widehat{f} \in \operatorname{dom}\mathcal{B} = \mathcal{D}.$$

The latter condition is satisfied automatically, since $(\Gamma_1 - R\Gamma_0)\widehat{f} \in \mathcal{H}_+$; see Theorem 7.26. This proves (7.32). The formula (7.33) is implied by (7.28) and the identities

$$\mathcal{B} + \overline{(R - M(\lambda))}_{-+} = (B - R)\restriction\mathcal{D} + (R - M(\lambda)) = (B - M(\lambda))\restriction\mathcal{D},$$

and $\overline{\gamma}(\lambda)\restriction\mathcal{H} = \gamma(\lambda)$. Recall that $\gamma(\bar{\lambda})^*$ maps into $\mathcal{H}_+$. □

Example 7.28 Let Ω be a bounded domain in $\mathbb{R}^3$ with a smooth boundary $\partial\Omega$. Consider the differential expression $\ell := -\Delta$, where Δ is a Laplacian operator in Ω and denote by $A_{\min}$ and $A_{\max}$ the minimal and the maximal differential operators generated in $L^2(\Omega)$ by the differential expression ℓ. Let γ_{D} and γ_{N} be the trace operators defined by

$$\gamma_{\mathrm{D}} : f \mapsto f\restriction\partial\Omega \in W_2^{3/2}(\partial\Omega), \quad \gamma_{\mathrm{N}} : f \mapsto \frac{\partial f}{\partial n}\restriction\partial\Omega \in W_2^{1/2}(\partial\Omega), \quad (7.34)$$

for any $f \in W_2^2(\mathbb{R}^n)$; see [Triebel, 1978; Lions and Magenes, 1972]). It is known (see, for instance, [Berezanskii, 1965]) that $A_{\min} = A_{\max}^*$ and

$$\operatorname{dom} A_{\min} = \{y \in W_2^2(\Omega) : \gamma_{\mathrm{D}} y = \gamma_{\mathrm{N}} y = 0\}.$$

The differential expression ℓ admits two classical selfadjoint realizations: the Dirichlet Laplacian $-\Delta_{\mathrm{D}}$ and the Neumann Laplacian $-\Delta_{\mathrm{N}}$ whose domains are given by

$$\operatorname{dom}\Delta_{\mathrm{D}} = \{y \in W_2^2(\Omega) : \gamma_{\mathrm{D}} y = 0\}, \quad \operatorname{dom}\Delta_{\mathrm{N}} = \{y \in W_2^2(\Omega) : \gamma_{\mathrm{N}} y = 0\}.$$

General, not necessarily local, boundary value problems for elliptic operators have been studied in [Višik, 1952; Grubb, 1968] (see also [Grubb, 2009]) where the regularized trace operators are defined as follows

$$\widetilde{\Gamma}_{0,\Omega} y = \gamma_{\mathrm{D}} y, \quad \widetilde{\Gamma}_{1,\Omega} y = (\gamma_{\mathrm{N}} - \Lambda(0)\gamma_{\mathrm{D}})y, \quad y \in \operatorname{dom} A_{\max}. \quad (7.35)$$

Here $\Lambda(z)$ is the Poincaré-Steklov operator defined for any $z \in \rho(-\Delta_D)$ by

$$\Lambda(z) : W_2^{-1/2}(\partial\Omega) \to W_2^{-3/2}(\partial\Omega), \quad u \to \gamma_N f,$$

where $f \in \operatorname{dom} A_{\max}$ is the unique solution of the Dirichlet boundary value problem

$$-\Delta f - zf = 0, \quad \gamma_D f = u.$$

It is shown in [Grubb, 1968] that the mappings $\widetilde{\Gamma}_{0,\Omega}$ and $\widetilde{\Gamma}_{1,\Omega}$ are well defined and

$$\widetilde{\Gamma}_{0,\Omega} : \operatorname{dom} A_{\max} \to W_2^{-1/2}(\partial\Omega), \quad \widetilde{\Gamma}_{1,\Omega} : \operatorname{dom} A_{\max} \to W_2^{1/2}(\partial\Omega).$$

This property makes it possible to extend the (regularized) Green formula for $\operatorname{dom} A_{\max}$ by using the pairing between the Sobolev spaces $W_2^{-1/2}(\partial\Omega)$ and $W_2^{1/2}(\partial\Omega)$ (see [Grubb, 1968]). To adapt this construction in the context of boundary triplets introduce the Laplace-Beltrami operator $\Delta_{\partial\Omega}$ on $\partial\Omega$ and put $K = (-\Delta_{\partial\Omega} + I)^{-1/4}$. Then with

$$\mathcal{H} = L^2(\partial\Omega), \quad \Gamma_0 = K\gamma_{\rm D}, \quad \Gamma_1 = K^{-1}(\gamma_{\rm N} - \Lambda(0)\gamma_{\rm D}), \tag{7.36}$$

one obtains an ordinary boundary triplet $\Pi = \{\mathcal{H}, \Gamma_0, \Gamma_1\}$ for $A_{\max}$, see [Malamud, 2010]. Now introduce the boundary mappings $\widetilde{\Gamma}_0$ and $\widetilde{\Gamma}_1$ via

$$\begin{pmatrix} \widetilde{\Gamma}_0 \\ \widetilde{\Gamma}_1 \end{pmatrix} := \begin{pmatrix} K^{-1} & 0 \\ 0 & K \end{pmatrix} \begin{pmatrix} \Gamma_0 \\ \Gamma_1 \end{pmatrix} \subset \begin{pmatrix} \widetilde{\Gamma}_{0,\Omega} \\ \widetilde{\Gamma}_{1,\Omega} \end{pmatrix},$$

see (7.35) and (7.36). Since K by definition is a bounded operator in $\mathcal{H} = L^2(\partial\Omega)$ and Π is an ordinary boundary triplet for $A_{\max}$, Remark 7.21 shows that $\widetilde{\Pi} = \{L^2(\partial\Omega), \widetilde{\Gamma}_0, \widetilde{\Gamma}_1\}$ is a boundary triplet of bounded type for $A_{\max}$; in fact, the domain $\operatorname{dom}\widetilde{\Gamma}$ is given by

$$\operatorname{dom}\widetilde{\Gamma} = \{y \in L^2(\Omega) : \Delta y \in L^2(\Omega), \widetilde{\Gamma}_{0,\Omega} y \in L^2(\partial\Omega), \widetilde{\Gamma}_{1,\Omega} y \in L^2(\partial\Omega)\}.$$

The Weyl functions corresponding to Π and $\widetilde{\Pi}$ are given by

$$M(z) = K^{-1}(\Lambda(z) - \Lambda(0))K^{-1}, \quad \widetilde{M}(z) = \Lambda(z) - \Lambda(0).$$

It was shown in [Bade and Freeman, 1962; Beals, 1965] that the natural domain for the trace operators is the Beals space

$$\mathcal{D}_1(\Omega) = \{\, y \in L^2(\Omega) : \Delta y \in L^2(\Omega), \frac{\partial y}{\partial x_j} \in L^2(\partial\Omega) \,\},$$

and that the operators $\Gamma_0^{(1)} = \gamma_{\rm N} \restriction \mathcal{D}_1(\Omega)$ and $\Gamma_1^{(1)} = -\gamma_{\rm D} \restriction \mathcal{D}_1(\Omega)$ satisfy

$$\Gamma_0^{(1)} : \mathcal{D}_1(\Omega) \to L^2(\partial\Omega), \quad \Gamma_1^{(1)} : \mathcal{D}_1(\Omega) \to W_2^1(\partial\Omega),$$

and are surjective. The triplet $\{L^2(\partial\Omega), \Gamma_0^{(1)}, \Gamma_1^{(1)}\}$ is a boundary triplet of bounded type for a realization of ℓ with the domain $\mathcal{D}_1(\Omega)$ and was considered in [Amrein and Pearson, 2004] (see also [Behrndt and Langer, 2007]). The corresponding Weyl function $M^{(1)}$ (M-function in [Amrein and Pearson, 2004]) coincides with the so-called Neumann to Dirichlet map, that is $M^{(1)}(z)u = v := -\gamma_D f$ where $f(\in \mathcal{D}_1(\Omega))$ is the unique solution of the Neumann problem

$$-\Delta f - zf = 0, \quad \gamma_N f = u.$$

It follows from Proposition 7.27 that for each closed operator B in

$L^2(\partial\Omega)$, such that $0 \in \rho(B - M^{(1)}(z))$ for some $z \in \mathbb{C}_\pm$, the non-local boundary condition

$$\gamma_D y = B\gamma_N y, \quad y \in \mathcal{D}_1(\Omega),$$

determines a closed realization $\widetilde{A}_B$ of ℓ. Results of this type for non-local boundary value problems were obtained in [Grubb, 1968; Amrein and Pearson, 2004; Behrndt and Langer, 2007]. If, in particular, B is selfadjoint and $0 \notin \sigma_{ess}(B)$, then the extension $\widetilde{A}_B$ is also selfadjoint, see [Behrndt and Langer, 2007]. See also Chapters 6 and 8.

7.5 Boundary triplets of bounded type and infinite dimensional graph perturbations

The graph perturbation of a selfadjoint operator or relation gives rise to a boundary triplet of bounded type. Moreover every boundary triplet of bounded type can be seen to originate from such a construction. The present section contains a number of results about graph perturbations. For further results, applications, and proofs the reader is referred to [Derkach et al., 2012].

Let A_0 be a selfadjoint relation in a Hilbert space and let Z be a closed linear subspace of $\mathfrak{H}^2$ such that $A_0 \cap Z = \{0, 0\}$. Then $S = A_0 \cap Z^*$ is a closed symmetric restriction of A_0. Note that S need not be densely defined even if A_0 is a selfadjoint operator and that S can be an operator, while A_0 is multi-valued.

Lemma 7.29 *Let A_0 be a selfadjoint relation in the Hilbert space $\mathfrak{H}$ and let Z be a linear manifold in $\mathfrak{H}^2$ such that $A_0 \cap Z = \{0, 0\}$. Then $S = A_0 \cap Z^*$ is a closed symmetric relation with deficiency indices $n_\pm(S) =$ $\dim(\operatorname{clos} Z)$ and the adjoint given by*

$$S^* = \operatorname{clos}(A_0 \widehat{+} Z).$$

Proof Since $A_0 = A_0^*$ and Z^* are closed subspaces, S is also a closed subspace of $\mathfrak{H}^2$ and $S = A_0 \cap Z^* = (A_0 \widehat{+} Z)^*$. Now taking adjoints on both sides leads to

$$S^* = (A_0 \cap Z^*)^* = \operatorname{clos}\left(A_0 \widehat{+} Z\right).$$

By construction A_0 is a selfadjoint extension of S and hence S has equal deficiency indices $n_\pm(S) = \dim(S^*/\operatorname{clos} Z)$ (by the first von Neumann formula). □

Note that Z is symmetric if $Z \subset Z^*$. Two manifolds Z_1 and Z_2 satisfying $Z_j \cap A_0 = \{0,0\}$ are said to be *equivalent* if $A_0 \widehat{+} Z_1 = A_0 \widehat{+} Z_2$; this direct sum will be denoted by

$$T = A_0 \widehat{+} Z, \quad A_0 \cap Z = \{0,0\}. \tag{7.37}$$

Note that T is a dense subspace of S^*. Each element $\{f, f'\} \in T$ can be decomposed as

$$\{f, f'\} = \{f_0, f_0'\} + \{h, h'\}, \quad \{f_0, f_0'\} \in A_0, \quad \{h, h'\} \in Z. \tag{7.38}$$

By the assumption $A_0 \cap Z = \{0,0\}$ the decomposition in (7.38) is unique.

Next the aim is to define a pair of boundary mappings on S^* whose domain will coincide with T. For this purpose, it is assumed in addition that Z is an *operator range*, i.e. that there is a Hilbert space $\mathcal{H}$ and an everywhere defined bounded operator $G : \mathcal{H} \to \mathfrak{H}^2$ such that

$$Z = \operatorname{ran} G, \quad \ker G = \{0\}. \tag{7.39}$$

Then G maps $\mathcal{H}$ injectively onto $Z \subset S^*$, considered as a subspace of $\mathfrak{H}^2$. Decompose G according to $\mathfrak{H}^2 = \mathfrak{H} \times \mathfrak{H}$ into a block operator

$$G = \begin{pmatrix} \Phi \\ \Psi \end{pmatrix} \in \boldsymbol{B}(\mathcal{H}, \mathfrak{H}^2), \tag{7.40}$$

so that $G^* = (\Phi^* \ \Psi^*) \in \boldsymbol{B}(\mathfrak{H}^2, \mathcal{H})$.

Decompose $\{f, f'\} \in T$ as in (7.38) with $\{f_0, f_0'\} \in A_0$ and $\{h, h'\} \in Z$, and define $u = G^{-1}\{h, h'\} \in \mathcal{H}$, so that $\{f, f'\}$ has the unique decomposition

$$\widehat{f} = \{f, f'\} = \{f_0, f_0'\} + Gu, \quad \{f_0, f_0'\} \in A_0, \quad u \in \mathcal{H},$$

due to the assumption $A_0 \cap Z = \{0,0\}$. Now define the mappings Γ_0^G and Γ_0^G from T to $\mathcal{H}$ by

$$\begin{aligned} &\Gamma_0^G\{f, f'\} := u, \\ &\Gamma_1^G\{f, f'\} := \mathrm{i}\, G^* J_{\mathfrak{H}}\{f_0, f_0'\} + \Phi^*\Psi u = -\Psi^* f_0 + \Phi^* f_0' + \Phi^*\Psi u. \end{aligned} \tag{7.41}$$

Then the following proposition is easily established.

Proposition 7.30 *Let A_0 be a selfadjoint relation in a Hilbert space $\mathfrak{H}$ and let $Z = \operatorname{ran} G$ be as in (7.39) with T and G as in (7.37) and (7.40). Moreover, let $S = A_0 \cap Z^*$ and let the mappings Γ_0^G, Γ_1^G be defined as in (7.41). Then:*

(i) *$\Pi = \{\mathcal{H}, \Gamma_0^G, \Gamma_1^G\}$ is a boundary triplet of bounded type for S^* with*

$$\operatorname{dom} \Gamma^G = T, \quad \ker \Gamma_0^G = A_0;$$

(ii) *the corresponding γ-field is given by*

$$\gamma_G(\lambda) = \Phi + (A_0 - \lambda)^{-1}(\lambda\Phi - \Psi), \quad \lambda \in \rho(A_0);$$

(iii) *the corresponding Weyl function is given by*

$$M_G(\lambda) = \lambda\Phi^*\Phi + (\Phi^*\lambda - \Psi^*)(A_0 - \lambda)^{-1}(\lambda\Phi - \Psi), \quad \lambda \in \rho(A_0).$$

Remark The formulas for the γ-field and the Weyl function (as a Q-function) in Proposition 7.30 for an arbitrary symmetric relation S with deficiency indices $(1, 1)$ were established in [Hassi and de Snoo, 1997]. The case of symmetric relations with finite defect numbers (n, n), $n \in \mathbb{N}$, was treated in [Hassi et al., 2007], where also boundary mappings of the form (7.41) were constructed under the assumption that Z is an n-dimensional symmetric subspace (the difference being a selfadjoint constant in the mapping Γ_1^G and in the corresponding Weyl function).

If E is a bounded selfadjoint operator in $\mathcal{H}$ then replacing Γ_1^G with $\Gamma_1^G + E\Gamma_0^G$ gives a new boundary triplet $\{\mathcal{H}, \Gamma_0^G, \Gamma_1^G + E\Gamma_0^G\}$ of bounded type with the same domain $T = A_0 \widehat{+} Z$ and with the γ-field γ_E and the Weyl function M_E given by

$$\gamma_E(\lambda) = \gamma_G(\lambda), \quad M_E(\lambda) = M_G(\lambda) + E, \quad \lambda \in \rho(A_0).$$

The next theorem shows that, in fact, the class of boundary triplets as constructed above via operator ranges Z of the form (7.39) can be considered as a model for the class of all boundary triplets of bounded type; in particular, every $R^s[\mathcal{H}]$-function can be obtained as a Weyl function in this manner.

Theorem 7.31 *Let $\{\mathcal{H}, \Gamma_0, \Gamma_1\}$ be a boundary triplet of bounded type for S^* with the domain $T \subset S^*$ and let $A_0 = \ker \Gamma_0$. Then there is a subspace Z satisfying the conditions* (7.37) *and* (7.39)*, and a bounded selfadjoint operator E in $\mathcal{H}$, such that the triplet $\{\mathcal{H}, \Gamma_0^G, \Gamma_1^G + E\Gamma_0^G\}$ with Γ_0^G and Γ_1^G defined as in* (7.41)*, coincides with $\{\mathcal{H}, \Gamma_0, \Gamma_1\}$.*

Next two special cases of Proposition 7.30 are given; the first one corresponds to the case of infinite dimensional range perturbations and the second one to the case of infinite dimensional domain perturbations.

Corollary 7.32 *Let A_0, Z, and G be as in Proposition 7.30 and let the boundary triplet of bounded type $\Pi = \{\mathcal{H}, \Gamma_0^G, \Gamma_1^G\}$ for S^* be defined by* (7.41)*.*

(i) *If* A_0 *is a selfadjoint operator and* $\operatorname{ran}\Phi \subset \operatorname{dom} A_0$, *or equivalently* $\operatorname{ran}\gamma(\lambda) \subset \operatorname{dom} A_0$ *for all* $\lambda \in \rho(A_0)$, *then the formulas for the* γ*-field and Weyl function in Proposition 7.30 take the form*

$$\gamma_G(\lambda) = (A_0 - \lambda)^{-1}\widetilde{\Psi}, \quad M_G(\lambda) = E + \widetilde{\Psi}^*(A_0 - \lambda)^{-1}\widetilde{\Psi},$$

where $\lambda \in \rho(A_0)$, $\widetilde{\Psi} = A_0\Phi - \Psi \in \boldsymbol{B}(\mathcal{H}, \mathfrak{H})$, $\ker \widetilde{\Psi} = \{0\}$, *and* $E = \Phi^*\Psi + \Psi^*\Phi - \Phi^*A_0\Phi$ *is a bounded selfadjoint operator in* $\mathcal{H}$.

(ii) *If the inverse* A_0^{-1} *is a selfadjoint operator and* $\operatorname{ran}\Psi \subset \operatorname{ran} A_0$, *or equivalently* $\operatorname{ran}\gamma(\lambda) \subset \operatorname{ran} A_0$ *for all* $\lambda \in \rho(A_0)$, *then the* γ*-field and Weyl function can be rewritten in the form*

$$\gamma_G(\lambda) = (I - \lambda A_0^{-1})^{-1}\widetilde{\Phi}, \quad M_G(\lambda) = F + \lambda\widetilde{\Phi}^*(I - \lambda A_0^{-1})^{-1}\widetilde{\Phi},$$

where $\lambda \in \rho(A_0)$, $\widetilde{\Phi} = \Phi - A_0^{-1}\Psi \in \boldsymbol{B}(\mathcal{H}, \mathfrak{H})$, $\ker \widetilde{\Phi} = \{0\}$, *and* $F = \Psi^*A_0^{-1}\Psi$ *is a bounded selfadjoint operator in* $\mathcal{H}$.

Remark If $\operatorname{ran}\Phi \subset \operatorname{dom} A_0$ then the pairs $\{\Phi, \Psi\}$ and $\{0, -\widetilde{\Psi}\}$ are equivalent, since their difference is $\{\Phi, A_0\Phi\}$. In this case the formulas for the boundary mappings in (7.41) simplify accordingly. In fact, there is no essential difference, if one actually uses in this case a pair $\{0, \Psi\}$ (i.e. one can take $\Phi = 0$): this means that $S = A_0 \cap Z^* = A_0 \upharpoonright \ker \Psi^*$ is nondensely defined. If, in particular, $\operatorname{ran} G$ is closed, then all the selfadjoint extensions of S can be seen as range perturbations of A_0 of the form $A_0 + \Psi\Theta\Psi^*$, $\Theta = \Theta^*$. Similar observations can be made in the case that $\ker A_0 = \{0\}$ and $\operatorname{ran}\Psi \subset \operatorname{ran} A_0$. Then one can replace the pair $\{\Phi, \Psi\}$ with the equivalent pair $\{\widetilde{\Phi}, 0\}$, since their difference is $\{A_0^{-1}\Psi, \Psi\}$. Then all the above formulas, including (7.41), simplify accordingly by using a pair $\{\Phi, 0\}$ (i.e. one can take $\Psi = 0$). In this case S is a range restriction of A_0 and if $\operatorname{ran}\Psi$ is closed then all selfadjoint extensions of S can be interpreted as domain perturbations of A_0. For further details and results see [Hassi and de Snoo, 1997; Hassi et al., 1998] for the scalar case, [Derkach et al., 2003] for the matrix case, and [Hassi et al., 2012] for the operator case.

If $\operatorname{ran}\Phi \subset \operatorname{dom} |A_0|^{\frac{1}{2}}$ then there are similar simplifications in the formulas as in the case $\operatorname{ran}\Phi \subset \operatorname{dom} A_0$. In this case one needs to replace $\mathfrak{H}$ by the rigged space $\mathfrak{H}_+ \subset \mathfrak{H} \subset \mathfrak{H}_-$, where $\mathfrak{H}_+ = \operatorname{dom} |A_0|^{\frac{1}{2}}$ is equipped with the graph norm and $\mathfrak{H}_-$ is the corresponding dual space. Then A_0 becomes bounded as an operator from $\mathfrak{H}_+$ to $\mathfrak{H}_-$ and one replaces A_0 by its unique continuation $\widetilde{A}_0 \in \boldsymbol{B}(\mathfrak{H}_+, \mathfrak{H}_-)$ (which is in fact selfdual). Similarly, if $\ker A_0 = \{0\}$ and $\operatorname{ran}\Psi \subset \operatorname{ran} |A_0|^{\frac{1}{2}}$ then one can use the rigging corresponding to the inverse A_0^{-1} and replace A_0^{-1} by its continuation $\widetilde{A}_0^{-1}$ in the rigging $\mathfrak{H}^+ \subset \mathfrak{H} \subset \mathfrak{H}^-$ associated with A_0^{-1}, so that $\mathfrak{H}^+ = \operatorname{ran} |A_0|^{\frac{1}{2}}$. In this way one arrives at the following formulas for the γ-field and Weyl function in Proposition 7.30.

Corollary 7.33 *Let A_0, Z, and G be as in Proposition 7.30 and let the boundary triplet of bounded type $\Pi = \{\mathcal{H}, \Gamma_0^G, \Gamma_1^G\}$ for S^* be defined by* (7.41).

(i) *If A_0 is a selfadjoint operator and* $\operatorname{ran}\Phi \subset \operatorname{dom}|A_0|^{\frac{1}{2}}$, *or equivalently* $\operatorname{ran}\gamma(\lambda) \subset \operatorname{dom}|A_0|^{\frac{1}{2}}$ *for all $\lambda \in \rho(A_0)$, then the formulas for the γ-field and Weyl function in Proposition 7.30 take the form*

$$\gamma_G(\lambda) = (\widetilde{A}_0 - \lambda)^{-1}\widetilde{\Psi}, \quad M_G(\lambda) = E + \widetilde{\Psi}^*(\widetilde{A}_0 - \lambda)^{-1}\widetilde{\Psi},$$

where $\lambda \in \rho(A_0)$, $\widetilde{A}_0 \in \boldsymbol{B}(\mathfrak{H}_+, \mathfrak{H}_-)$, $\widetilde{\Psi} = \widetilde{A}_0\Phi - \Psi \in \boldsymbol{B}(\mathcal{H}, \mathfrak{H}_-)$, $\ker \widetilde{\Psi} = \{0\}$, *and E is a bounded selfadjoint operator in $\mathcal{H}$.*

(ii) *If the inverse A_0^{-1} is a selfadjoint operator and* $\operatorname{ran}\Psi \subset \operatorname{ran}|A_0|^{\frac{1}{2}}$, *or equivalently* $\operatorname{ran}\gamma(\lambda) \subset \operatorname{ran}|A_0|^{\frac{1}{2}}$ *for all $\lambda \in \rho(A_0)$, then the γ-field and Weyl function can be rewritten in the form*

$$\gamma_G(\lambda) = (I - \lambda\widetilde{A}_0^{-1})^{-1}\widetilde{\Phi}, \quad M_G(\lambda) = F + \lambda\widetilde{\Phi}^*(I - \lambda\widetilde{A}_0^{-1})^{-1}\widetilde{\Phi},$$

where $\lambda \in \rho(A_0)$, $\widetilde{A}_0 \in \boldsymbol{B}(\mathfrak{H}^+, \mathfrak{H}^-)$, $\widetilde{\Phi} = \Phi - \widetilde{A}_0^{-1}\Psi \in \boldsymbol{B}(\mathcal{H}, \mathfrak{H}^-)$, $\ker \widetilde{\Phi} = \{0\}$, *and F is a bounded selfadjoint operator in $\mathcal{H}$.*

Corollary 7.33 is connected with form-bounded range and domain perturbations; see [Hassi et al., 1997, 1998; Derkach et al., 2003; Behrndt et al., 2010]. For further details and results, see [Hassi et al., 1995; Hassi and de Snoo, 1997; Hassi et al., 1997, 1998; Arlinskii et al., 2011; Hassi et al., 2012].

7.6 Unitary boundary relations and Weyl families

Unitary boundary relations The notion of unitary boundary relation will be introduced with a number of properties of the corresponding Weyl family.

Definition 7.34 [Derkach et al., 2006] Let S be a closed symmetric relation in a Hilbert space $\mathfrak{H}$. A relation Γ from $\mathfrak{H}^2$ to $\mathcal{H}^2$, where $\mathcal{H}$ is a Hilbert space is said to be a *unitary boundary relation* for S^* if

(R1) Γ is a unitary relation from $(\mathfrak{H}^2, J_{\mathfrak{H}})$ into $(\mathcal{H}^2, J_{\mathcal{H}})$;
(R2) $\operatorname{dom}\Gamma$ is dense in S^*.

The pair $(\mathcal{H}^2, \Gamma)$ will be called a *unitary boundary pair* for S^*.

Since $\overline{\mathrm{dom}\,\Gamma} = S^*$ and Γ is unitary it follows from Proposition 7.1 that $\ker \Gamma = (\mathrm{dom}\,\Gamma)^{[\perp]} = (S^*)^{[\perp]}$. Hence by (7.4) one gets

$$\ker \Gamma = S.$$

Define the relation $T := \mathrm{dom}\,\Gamma$ so that T is dense S^* and, in general T is not closed. Furthermore, define

$$\mathfrak{N}_\lambda(T) = \ker\,(T - \lambda), \quad \widehat{\mathfrak{N}}_\lambda(T) = \{\,\{f, \lambda f\} \in T : f \in \mathfrak{N}_\lambda(T)\,\},$$

which are dense in $\mathfrak{N}_\lambda(S^*)$ and $\widehat{\mathfrak{N}}_\lambda(S^*)$, respectively.

Definition 7.35 The Weyl family M of S corresponding to a unitary boundary pair $(\mathcal{H}^2, \Gamma)$ is defined by

$$M(\lambda) := \Gamma(\widehat{\mathfrak{N}}_\lambda(T)), \quad \lambda \in \mathbb{C} \setminus \mathbb{R}. \tag{7.42}$$

The γ-field γ corresponding to $(\mathcal{H}^2, \Gamma)$ is defined by

$$\gamma(\lambda) := \left\{ \{h, f_\lambda\} \in \mathcal{H} \times \mathfrak{H} : \{\widehat{f}_\lambda, \widehat{h}\} \in \Gamma \text{ for some } \widehat{f}_\lambda = \{f, \lambda f\} \in \mathfrak{H}^2 \right\},$$

and moreover, $\widehat{\gamma}(\lambda)$ stands for $\widehat{\gamma}(\lambda) = \{\gamma(\lambda), \lambda\gamma(\lambda)\}$, $\lambda \in \mathbb{C} \setminus \mathbb{R}$.

It was shown in [Derkach et al., 2006] that the Weyl family M belongs to the class $\widetilde{R}(\mathcal{H})$ of Nevanlinna families, i.e.,

(N1) for every $\lambda \in \mathbb{C}_+(\mathbb{C}_-)$ $M(\lambda)$ is maximal dissipative (accumulative, respectively);
(N2) $M(\lambda)^* = M(\bar{\lambda})$, $\lambda \in \mathbb{C} \setminus \mathbb{R}$;
(N3) for some, and hence for all, $\mu \in \mathbb{C}_+(\mathbb{C}_-)$ the operator family $(M(\lambda) + \mu)^{-1} (\in \boldsymbol{B}(\mathcal{H}))$ is holomorphic for all $\lambda \in \mathbb{C}_+(\mathbb{C}_-)$.

The Weyl family M satisfies $\mathrm{mul}\,\Gamma = M(\lambda) \cap M(\lambda)^*$, $\lambda \in \mathbb{C} \setminus \mathbb{R}$. The subclass $R(\mathcal{H})$ of $\widetilde{R}(\mathcal{H})$ consists of single-valued Nevanlinna families M ($\mathrm{mul}\,M(\lambda) = \{0\}$ for all $\lambda \in \mathbb{C}_+ \cup \mathbb{C}_-$) and $R^s(\mathcal{H})$ is the set of all $M \in \widetilde{R}(\mathcal{H})$ for which $M(\lambda) \cap M(\lambda)^* = \{0\}$ for all $\lambda \in \mathbb{C} \setminus \mathbb{R}$. These classes satisfy the following inclusions

$$R^u[\mathcal{H}] \subset R^s[\mathcal{H}] \subset R^s(\mathcal{H}) \subset R(\mathcal{H}) \subset \widetilde{R}(\mathcal{H}).$$

Theorem 7.15 and Proposition 7.19 describe the boundary triplets, whose Weyl functions belong to the classes $R^u[\mathcal{H}]$ and $R^s[\mathcal{H}]$. In the next theorem a similar result is formulated for the classes $R^s(\mathcal{H})$, $R(\mathcal{H})$, $\widetilde{R}(\mathcal{H})$. A unitary boundary pair $\{\mathcal{H}^2, \Gamma\}$ is called *minimal*, if

$$\mathfrak{H} = \mathfrak{H}_{\min} := \overline{\mathrm{span}}\,\{\,\mathfrak{N}_\lambda(T) : \lambda \in \mathbb{C}_+ \cup \mathbb{C}_-\,\}.$$

Theorem 7.36 [Derkach et al., 2006, Theorem 3.9] *For every* $M \in \widetilde{R}(\mathcal{H})$ *there is a simple symmetric operator* S *and a unique (up to a unitary equivalence) minimal unitary boundary pair* $\{\mathcal{H}^2, \Gamma\}$ *for* S^**, such that the corresponding Weyl family coincides with* M*. Moreover,*

(i) $M \in R(\mathcal{H})$ *if and only if* $\operatorname{mul}\Gamma \cap (\{0\} \times \mathcal{H}) = \{0\}$*;*
(ii) $M \in R^s(\mathcal{H})$ *if and only if* $\operatorname{mul}\Gamma = \{0\}$.

Unitary boundary triplets The unitary boundary relations in Definition 7.34 which are actually single-valued are characterized analytically in Theorem 7.36. A single-valued unitary boundary relation Γ can be written in the block form $\Gamma = (\Gamma_0\ \Gamma_1)^{\mathsf{T}}$; and the triplet $\{\mathcal{H}, \Gamma_0, \Gamma_1\}$ is called a *unitary boundary triplet* for S^*. These unitary boundary triplets coincide with Calkin's reduction operators in Definition 7.8 (in the context of symmetric relations), assuming that Calkin's boundary value space has a symplectic structure. Recall that [Lyantse and Storozh, 1983] consider the more special situation of a pair $(\mathfrak{M}, \Gamma)$ of a Hilbert space $\mathfrak{M}$ and a mapping $\Gamma : S^* \to \mathfrak{M}$ such that $\ker \Gamma = S$ and $\operatorname{ran}\Gamma = \mathfrak{M}$.

In this subsection it will be supposed that $\{\mathcal{H}, \Gamma_0, \Gamma_1\}$ is a unitary boundary triplet and thus, by Theorem 7.36, the corresponding Weyl function belongs to $R^s(\mathcal{H})$. The γ-field corresponding to the unitary boundary triplet $\{\mathcal{H}, \Gamma_0, \Gamma_1\}$ is defined by the equalities

$$\widehat{\gamma}(\lambda) = (\Gamma_0 \restriction \widehat{\mathfrak{N}}_\lambda(T))^{-1}, \quad \gamma(\lambda) = \pi_1 \widehat{\gamma}(\lambda),$$

and $M(\lambda) = \Gamma_1 \widehat{\gamma}(\lambda)$. If $M \in R^s(\mathcal{H})$, then

$$\operatorname{dom} M(\lambda) = \operatorname{dom} \gamma(\lambda) = \Gamma_0 \widehat{\mathfrak{N}}_\lambda(T) \quad \text{for every } \lambda \in \mathbb{C}_+ \cup \mathbb{C}_-.$$

Moreover, in this case $\ker\,(M(\lambda) - M(\lambda)^*) = \{0\}$; see [Derkach et al., 2006, Lemma 4.1]. For unitary boundary triplets the identity (7.18) can be extended as follows; see [Derkach et al., 2006, Proposition 4.8].

Proposition 7.37 *If* $\{\mathcal{H}, \Gamma_0, \Gamma_1\}$ *is a unitary boundary triplet, then for all* $\lambda, \mu \in \mathbb{C} \setminus \mathbb{R}$

$$\frac{(M(\lambda)h, k)_{\mathcal{H}} - (h, M(\mu)k)_{\mathcal{H}}}{\lambda - \bar{\mu}} = (\gamma(\lambda)h, \gamma(\mu)k)_{\mathfrak{H}}, \tag{7.43}$$

where $h \in \operatorname{dom} M(\lambda)$, $k \in \operatorname{dom} M(\mu)$.

Let $\{\mathcal{H}, \Gamma_0, \Gamma_1\}$ be a unitary boundary triplet for S^* and let $A_0 =$

$\ker \Gamma_0$. Since A_0 is a, not necessarily closed, symmetric relation, one can write for every $\lambda \in \mathbb{C} \setminus \mathbb{R}$

$$A_0 = \left\{ \{(A_0 - \lambda)^{-1}h, h + \lambda(A_0 - \lambda)^{-1}h\} : h \in \operatorname{ran}(A_0 - \lambda) \right\}.$$

The linear mapping $H(\lambda) : h \to \{(A_0 - \lambda)^{-1}h, h + \lambda(A_0 - \lambda)^{-1}h\}$ from $\operatorname{ran}(A_0 - \lambda)$ onto A_0 is clearly bounded with bounded inverse.

Lemma 7.38 *Let $\{\mathcal{H}, \Gamma_0, \Gamma_1\}$ be a unitary boundary triplet. Then the following statements hold for every $\lambda \in \mathbb{C} \setminus \mathbb{R}$ fixed:*

(i) $\Gamma_1 H(\lambda) \subset \gamma(\bar{\lambda})^*$ *and, in particular,* $\Gamma_1 H(\lambda)$ *is closable; if* $\gamma(\bar{\lambda})$ *is bounded then* $\Gamma_1 H(\lambda)$ *is bounded;*
(ii) $\Gamma_1 H(\lambda)$ *is bounded* $\iff$ $A_0 = \ker \Gamma_0$ *is closed.*

Proof (i) Let $h \in \operatorname{dom} \gamma(\bar{\lambda}) = \operatorname{dom} M(\bar{\lambda})$ and let $k_\lambda \in \operatorname{ran}(A_0 - \lambda)$. Then $\widehat{\gamma}(\bar{\lambda})h \in \widehat{\mathfrak{N}}_{\bar{\lambda}}(T)$, $T = \operatorname{dom} \Gamma$, and since $\Gamma_0 \widehat{\gamma}(\bar{\lambda})h = h$ and $H(\lambda)k_\lambda \in A_0 = \ker \Gamma_0$, it follows from the isometry of Γ (see (7.12)) that

$$\begin{aligned}(\gamma(\bar{\lambda})h, k_\lambda)_{\mathfrak{H}} &= (\gamma(\bar{\lambda})h, (I + \lambda(A_0 - \lambda)^{-1})k_\lambda - \lambda(A_0 - \lambda)^{-1}k_\lambda)_{\mathfrak{H}} \\ &= (h, \Gamma_1 H(\lambda)k_\lambda)_{\mathcal{H}}.\end{aligned}$$

This proves the inclusion in (i). Since $\overline{\operatorname{dom}}\, \gamma(\bar{\lambda}) = \overline{\operatorname{ran}}\, \Gamma_0 = \mathcal{H}$, the adjoint $\gamma(\bar{\lambda})^*$ is a closed operator and thus the operator $\Gamma_1 H(\lambda)$ is closable. If $\gamma(\bar{\lambda})$ is bounded then so is its adjoint $\gamma(\bar{\lambda})^*$ and hence $\Gamma_1 H(\lambda)$ is bounded, too.

(ii) The operator $H(\lambda)$ is bounded with bounded inverse. Therefore $\operatorname{dom} H(\lambda) = \operatorname{dom}(\Gamma_1 H(\lambda))$ is closed precisely when A_0 is closed. Hence, if A_0 is closed then $\Gamma_1 H(\lambda)$, being closable with closed domain, is closed and bounded. To prove the converse, assume that $k_n \in \operatorname{dom}(\Gamma_1 H(\lambda))$ with $k_n \to k$. Then $\Gamma_1 H(\lambda)k_n \to g$ by boundedness of $\Gamma_1 H(\lambda)$. Moreover, since $H(\lambda)$ is bounded, $H(\lambda)k_n \to H(\lambda)k \in \operatorname{clos} A_0$, and

$$\Gamma H(\lambda)k_n = \{0, \Gamma_1 H(\lambda)k_n\} \to \{0, g\}.$$

Since Γ is closed (as a unitary operator), it follows that $\{H(\lambda)k, \{0, g\}\} \in \Gamma$, which shows that $\{H(\lambda)k, g\} \in \Gamma_1$, in particular, $k \in \operatorname{dom} \Gamma_1 H(\lambda)$. Therefore, $\operatorname{dom} \Gamma_1 H(\lambda)$ is closed and then, equivalently, $A_0 = \operatorname{ran} H(\lambda)$ is closed. □

A Weyl function $M \in R^s(\mathcal{H})$ is said to belong to the class $R^s_{\mathrm{inv}}(\mathcal{H})$, if there exists a linear (not necessarily closed) subspace $\mathcal{H}_0 \subset \mathcal{H}$ such that

$$\operatorname{dom} M(\lambda) = \Gamma_0(\widehat{\mathfrak{N}}_\lambda(T)) = \mathcal{H}_0, \quad \lambda \in \mathbb{C} \setminus \mathbb{R}.$$

It was shown in [Derkach et al., 2006] that if the Weyl function M

corresponding to a unitary boundary triplet belongs to $R^s_{\rm inv}(\mathcal{H})$, then the identity (7.18) holds and $A_0 = \ker \Gamma_0$ is essentially self-adjoint. In this case the γ-field satisfies

$$\gamma(\lambda) = (I + (\lambda - \mu)(A_0 - \lambda)^{-1})\gamma(\mu), \quad \lambda, \mu \in \mathbb{C} \setminus \mathbb{R}. \tag{7.44}$$

By means of Lemma 7.38 it is possible to characterize selfadjointness of A_0 by the boundedness of $\operatorname{Im} M(\lambda)$. In fact, the following result improves some related results established in [Derkach et al., 2006, Section 4].

Theorem 7.39 *Let $\{\mathcal{H}, \Gamma_0, \Gamma_1\}$ be a unitary boundary triplet for S^* with Weyl function M. Then the following statements are equivalent:*

(i) $\operatorname{Im} M(\lambda)$ *is bounded and densely defined for some* $\lambda \in \mathbb{C} \setminus \mathbb{R}$;
(ii) $\gamma(\lambda)$ *is bounded and* $\operatorname{dom} \gamma(\lambda) \cap \operatorname{dom} \gamma(\bar\lambda)$ *is dense in* $\mathcal{H}$ *for some* $\lambda \in \mathbb{C} \setminus \mathbb{R}$;
(iii) $\operatorname{ran} \Gamma_0 = \operatorname{dom} M(\lambda) = \operatorname{dom} M(\bar\lambda)$ *for some* $\lambda \in \mathbb{C} \setminus \mathbb{R}$;
(iv) $M(\lambda) = E + M_0(\lambda)$, *where* $E = E^*$ *and* $M_0 \in R^s[\mathcal{H}]$;
(v) A_0 *is selfadjoint.*

In this case the corresponding Weyl function M belongs to the class $R^s_{\rm inv}(\mathcal{H})$ and the properties in (i)–(iii) *hold for all* $\lambda \in \mathbb{C} \setminus \mathbb{R}$.

Proof The assumption $\operatorname{mul} \Gamma = \{0\}$ implies that $M \in R^s(\mathcal{H})$. In particular, $\operatorname{dom} \gamma(\lambda) = \operatorname{dom} M(\lambda)$ is dense in $\mathcal{H}$; see [Derkach et al., 2006, Corollary 4.3].

Recall that $M(\lambda)^* = M(\bar\lambda)$ and hence

$$\operatorname{dom} (\operatorname{Im} M(\lambda)) = \operatorname{dom} M(\lambda) \cap \operatorname{dom} M(\bar\lambda).$$

Thus, the equivalence of (i) and (ii) is immediate from Proposition 7.37 (for invariance on $\lambda \in \mathbb{C} \setminus \mathbb{R}$ see also (7.44)).

(ii) $\Rightarrow$ (v) If $\gamma(\lambda)$ is bounded, then by Lemma 7.38 A_0 is closed. By assumption the set $\mathcal{M}_\lambda := \operatorname{dom} \gamma(\lambda) \cap \operatorname{dom} \gamma(\bar\lambda)$ is dense in $\mathcal{H}$ and it is easy to check that $\gamma(\lambda)h \in \operatorname{ran} (A_0 - \bar\lambda)$ for all $h \in \mathcal{M}_\lambda$ (see the proof of [Derkach et al., 2006, Proposition 4.11]). Since $\gamma(\lambda)$ is bounded the image $\gamma(\lambda)(\mathcal{M}_\lambda)$ is dense in $\mathfrak{N}_\lambda(S^*)$ and, consequently,

$$\mathfrak{H} = \mathfrak{N}_\lambda(S^*) \oplus \operatorname{ran} (S - \bar\lambda) \subset \operatorname{clos} (\operatorname{ran} (A_0 - \bar\lambda)) = \operatorname{ran} (A_0 - \bar\lambda).$$

By symmetry in (7.43) $\gamma(\bar\lambda)$ is also bounded on the dense set $\mathcal{M}_\lambda$, and therefore the above reasoning applied to $\bar\lambda$ gives $\operatorname{ran} (A_0 - \lambda) = \mathfrak{H}$. Thus, A_0 is selfadjoint.

(v) $\Rightarrow$ (ii), (iv) If A_0 is selfadjoint, then $\operatorname{dom} \Gamma_1 H(\lambda) = \operatorname{ran} (A_0 - \lambda) = \mathfrak{H}$ and $\Gamma_1 H(\lambda)$ is bounded by Lemma 7.38. Now, it follows from

Lemma 7.38 (i) that $\Gamma_1 H(\lambda) = \gamma(\bar{\lambda})^*$, which implies that the closure $\gamma(\bar{\lambda})^{**}$ is also bounded. Moreover, by selfadjointness of A_0, $\operatorname{ran}\Gamma_0 = \Gamma_0(\widehat{\mathfrak{N}}_\lambda(T)) = \operatorname{dom}\gamma(\lambda)$ holds for all $\lambda \in \mathbb{C}\setminus\mathbb{R}$; see [Derkach et al., 2006, Theorem 4.13]. Hence, (ii) holds in fact for all $\lambda \in \mathbb{C}\setminus\mathbb{R}$. This also proves the last statement $M \in R^s_{\text{inv}}(\mathcal{H})$ of the theorem. Furthermore, according to [Derkach et al., 2006, Proposition 4.25] (cf. (7.43), (7.44)) the Weyl function has the following representation

$$M(\lambda) = M(\mu)^* + (\lambda-\bar{\mu})\gamma(\mu)^*[I+(\lambda-\mu)(A_0-\lambda)^{-1}]\gamma(\mu), \quad \lambda,\mu \in \mathbb{C}\setminus\mathbb{R}.$$

By taking $E := \operatorname{Re} M(\mu)$ and replacing $\gamma(\mu)$ by its bounded closure (and thus also $\operatorname{Im} M(\mu)$ by its bounded closure $\operatorname{Im}\mu\gamma(\mu)^*\gamma(\mu)^{**}$) yields a desired representation for M.

(iv) $\Rightarrow$ (i) This is clear.

(iii) $\Leftrightarrow$ (v) See [Derkach et al., 2006, Corollary 4.14]. □

Thus, if $\operatorname{Im} M(\lambda)$ is unbounded, then $A_0 = \ker\Gamma_0$ cannot be selfadjoint. Moreover, if $\operatorname{Im} M(\lambda)$ is bounded, then $M(\lambda)$ itself need not be bounded.

Example 7.40 The simplest example of a unitary boundary triplet whose γ-field is unbounded can be found in [Derkach et al., 2006, Example 6.5]). The corresponding γ-field and Weyl function are of the form $\gamma(\lambda) = \frac{1}{\lambda} B$ and $M(\lambda) = -\frac{1}{\lambda} B^* B$, where B is a closed densely defined operator in a Hilbert space $\mathcal{H}$. Observe, that $M \in R^s(\mathcal{H})$ precisely when $\ker B = \{0\}$. Thus, if B is an unbounded operator with $\ker B = \{0\}$, then the associated boundary triplet (see [Derkach et al., 2006, Example 6.5] for the definition of $\{\mathcal{H}, \Gamma_0, \Gamma_1\}$) is unitary (by Theorem 7.36). Here $\operatorname{Im} M(\lambda)$ is bounded if and only if B is bounded or, equivalently, $M \in R^s[\mathcal{H}]$, which means that $\{\mathcal{H}, \Gamma_0, \Gamma_1\}$ becomes a boundary triplet of bounded type under the equivalent conditions in Theorem 7.39.

The next result is an extension of Corollary 7.32 (ii), which was connected with infinite dimensional domain perturbations. In fact, in the last section of this paper a more general situation will be treated. Although this result can be derived from an analogous range perturbation result appearing in [Derkach et al., 2006, Example 6.6] (by suitably applying inverses), a direct self-contained proof is given.

Proposition 7.41 *Let A_0^{-1} and E be selfadjoint operators in $\mathfrak{H}$ and $\mathcal{H}$, respectively, and let the operator $G : \mathcal{H} \to \mathfrak{H}$ be bounded and everywhere defined with* $\ker G = \{0\}$. *Moreover, let*

$$T = \{\{A_0^{-1} f' + G\varphi, f'\} : f' \in \operatorname{ran} A_0,\ \varphi \in \operatorname{dom} E\}$$

and define the operators $\Gamma_0, \Gamma_1 : T \to \mathcal{H}$ *by*

$$\Gamma_0 \widehat{f} = \varphi, \quad \Gamma_1 \widehat{f} = G^* f' + E\varphi; \quad \widehat{f} = \{A_0^{-1} f' + G\varphi, f'\} \in T.$$

Then $\{\mathcal{H}, \Gamma_0, \Gamma_1\}$ *is a unitary boundary triplet for* $\operatorname{clos} T$. *If* $\lambda \in \rho(A_0)$ *and* $\varphi \in \operatorname{dom} E$ *then the corresponding* γ*-field and Weyl function are given by*

$$\gamma(\lambda)\varphi = (I - \lambda A_0^{-1})^{-1} G\varphi, \quad M(\lambda)\varphi = E\varphi + \lambda G^*(I - \lambda A_0^{-1})^{-1} G\varphi.$$

Proof The abstract Green's identity is checked with a direct calculation. This means that Γ is isometric, i.e., $\Gamma^{-1} \subset \Gamma^{[*]}$; cf. (7.6). To check the reverse inclusion, let $g, g' \in \mathfrak{H}$, $k, k' \in \mathcal{H}$, and assume that for all $\{A_0^{-1} f' + G\varphi, f'\} \in T$ the identity

$$(f', g)_{\mathfrak{H}} - (A_0^{-1} f' + G\varphi, g')_{\mathfrak{H}} = (G^* f' + E\varphi, k)_{\mathcal{H}} - (\varphi, k')_{\mathcal{H}} \qquad (7.45)$$

holds. By taking $f' = 0$ in (7.45) gives for all $\varphi \in \operatorname{dom} E$,

$$(E\varphi, k)_{\mathcal{H}} = (\varphi, k')_{\mathcal{H}} - (G\varphi, g')_{\mathfrak{H}} = (\varphi, k' - G^* g')_{\mathcal{H}},$$

and thus, $k \in \operatorname{dom} E$, $Ek = k' - G^* g'$, and $k' = G^* g' + Ek$. By taking $\varphi = 0$ in (7.45) gives

$$(A_0^{-1} f', g')_{\mathfrak{H}} = (f', g)_{\mathfrak{H}} - (G^* f', k)_{\mathcal{H}} = (f', g - Gk)_{\mathfrak{H}}, \quad f' \in \operatorname{ran} A_0.$$

Thus $g' \in \operatorname{ran} A_0$, $A_0^{-1} g' = g - Gk$, and $g = A_0^{-1} g' + Gk \in \operatorname{dom} T$. Hence, $\widehat{g} = \{A_0^{-1} g' + Gk, g'\} \in T$ and $\Gamma_0 \widehat{g} = k$, $\Gamma_1 \widehat{g} = k'$, which proves the inclusion $\Gamma^{[*]} \subset \Gamma^{-1}$. Thus, Γ is unitary. Since $\ker G = \{0\}$, the mappings $\Gamma_j : T \to \mathcal{H}$, $j = 0, 1$, are (single-valued) operators.

It remains to note that $\{f, \lambda f\} \in T$ if and only if $f = A_0^{-1} f' + G\varphi$ with $f' = \lambda(I - \lambda A_0^{-1})^{-1} G\varphi$ and $\varphi \in \operatorname{dom} E$, $\lambda \in \rho(A_0)$. This yields the stated formulas for the Weyl function M and the γ-field γ. □

The linear relations T and A_0 in Proposition 7.41 need not be (single-valued) operators: in fact, T is an operator if and only if

$$\operatorname{dom} A_0 \cap \operatorname{ran} G = \{0\}.$$

The Weyl function M in Proposition 7.41 is unbounded precisely when E is unbounded, while the γ-field is bounded. Thus Proposition 7.41 defines a class of boundary triplets where the equivalent conditions in Theorem 7.39 are satisfied, in particular, $\ker \Gamma_0 = A_0$ is a selfadjoint relation. Note that if E is bounded, then Proposition 7.41 reduces to Corollary 7.32 (ii).

Remark A special case of boundary mappings appearing in Proposition 7.41 has been constructed in another way in some recent papers of V. Ryzhov under the additional assumption that A_0 is a selfadjoint operator with a bounded inverse $A_0^{-1} \in \boldsymbol{B}(\mathfrak{H})$; see [Ryzhov, 2009] and the further references therein. In fact, it was only proved that Γ is isometric, i.e., that Green's identity is satisfied.

Now assume that B is a linear operator in $\mathcal{H}$ such that $\operatorname{dom} B \supset \operatorname{dom} E$ and $B - M(\lambda)$ has a bounded inverse, which is densely defined in $\mathcal{H}$. As was shown in [Ryzhov, 2009, Theorem 5.5] in this case the formula

$$R_\lambda = (A_0 - \lambda)^{-1} - \gamma(\lambda)(\overline{B - M(\lambda)})^{-1}\gamma(\bar{\lambda})^*$$

with $\lambda \in \rho(A_0)$ determines a resolvent of some closed extension $\widetilde{A}$ of $S = T^*$. However, the domain of $\widetilde{A}$ was not specified.

In order to describe the extension $\widetilde{A}$, let $(\mathcal{H}, \widetilde{\Gamma}_0, \widetilde{\Gamma}_1)$ be a boundary triplet of bounded type with the domain

$$\widetilde{T} := \{\{A_0^{-1}f' + G\varphi, f'\} : f' \in \operatorname{ran} A_0,\ \varphi \in \mathcal{H}\}$$

and the mappings defined by

$$\widetilde{\Gamma}_0 \widehat{f} = \varphi, \quad \widetilde{\Gamma}_1 \widehat{f} = G^* f'; \quad \widehat{f} = \{A_0^{-1}f' + G\varphi, f'\} \in \widetilde{T}.$$

Then the operators $\widetilde{\Gamma}_0$, $\widetilde{\Gamma}_1$ can be treated as extensions of the operators $\Gamma_0' = \Gamma_0$, $\Gamma_1' = \Gamma_1 - E\Gamma_0$; see Proposition 7.41. In view of Proposition 7.27 R_λ is the resolvent of the closed extension

$$\widetilde{A} = \left\{ \widehat{f} \in \widetilde{T} : \widetilde{\Gamma}_1 \widehat{f} - \mathcal{B}\widetilde{\Gamma}_0 \widehat{f} = 0 \right\}$$

of the operator S, where $\mathcal{B}$ is the closure of the operator $B - E$. Notice that the formula (5.3) in [Ryzhov, 2009] is not correct, since $\Gamma_1 u$ has no sense for $u \in \operatorname{dom} \mathcal{B}$.

Orthogonal sums of ordinary boundary triplets This subsection is based on [Kostenko and Malamud, 2010]. For each $n \in \mathbb{N}$ let S_n be a closed symmetric operator in a Hilbert space $\mathfrak{H}_n$ with equal, not necessarily finite, deficiency indices and let $\Pi_n = \{\mathcal{H}_n, \Gamma_0^n, \Gamma_1^n\}$ be an ordinary boundary triplet for S_n^* with the Weyl function $M_n \in R_u[\mathcal{H}]$. The Hilbert space $\mathfrak{H} = \bigoplus_{n=1}^\infty \mathfrak{H}_n$ and the symmetric operator $S = \bigoplus_{n=1}^\infty S_n$ are defined in the usual way, so that the adjoint of S is given by

$$S^* = \bigoplus_{n=1}^\infty S_n^* = \left\{ \sum_{n=1}^\infty \{f_n, f_n'\} \in \mathfrak{H}^2 : \{f_n, f_n'\} \in S_n^* \right\}. \tag{7.46}$$

Let $\mathcal{H} = \bigoplus_{n=1}^{\infty} \mathcal{H}_n$ and define $\Gamma_j : \mathfrak{H}^2 \to \mathcal{H}$ by $\Gamma_i = \bigoplus_{n=1}^{\infty} \Gamma_i^n$ with

$$\operatorname{dom} \Gamma_i = \left\{ \sum_{n=1}^{\infty} \{f_n, f_n'\} \in S^* : \sum_{n=1}^{\infty} |\Gamma_i^n \{f_n, f_n'\}|^2 < \infty \right\}. \qquad (7.47)$$

This construction gives rise to a triplet $\Pi = \{\mathcal{H}, \Gamma_0, \Gamma_1\}$. Define

$$M = \bigoplus_{n=1}^{\infty} M_n, \qquad (7.48)$$

and it is clear from (7.46) and (7.47) that M satisfies (7.42).

The next lemma about Π and M can be seen as a special case of [Kostenko and Malamud, 2010, Theorem 3.2]. For the convenience of the reader a short proof is provided.

According to [Derkach et al., 2006, Proposition 3.6] for Γ to be a unitary boundary triplet it suffices to show that: (i) $\operatorname{dom} \Gamma$ is dense in S^*; (ii) Γ satisfies the Green's identity; and (iii) M in (7.48) satisfies $\operatorname{ran}(M(\lambda)+\lambda) = \mathcal{H}$ for some $\lambda \in \mathbb{C}_+$ and some $\lambda \in \mathbb{C}_-$; cf. Theorem 7.51.

Lemma 7.42 *The direct sum $\Pi = \bigoplus_{n=1}^{\infty} \Pi_n$ of the ordinary boundary triplets Π_n is a unitary boundary triplet for S^* with the Weyl function $M \in R^s(\mathcal{H})$.*

Proof It follows from (7.46) and (7.47) that $\operatorname{dom} \Gamma = \operatorname{dom} \Gamma_0 \cap \operatorname{dom} \Gamma_1$ is dense in S^*, since the set of vectors $\bigoplus_{n=1}^{\infty} \{f_n, f_n'\} \subset S^*$ with finitely many nonzero entries is dense in S^* and, in addition, belongs to $\operatorname{dom} \Gamma$.

For each $n \in \mathbb{N}$ the pair $\{\Gamma_0^n, \Gamma_1^n\}$ satisfies Green's identity; hence the same is true for $\Gamma = \{\Gamma_0, \Gamma_1\}$.

The condition $\operatorname{ran}(M(\lambda)+\lambda) = \mathcal{H}$ for all $\lambda \in \mathbb{C} \setminus \mathbb{R}$ is immediate from the corresponding property of the Weyl functions M_n. □

The first example of a direct sum $\bigoplus_{n=1}^{\infty} \Pi_n$ of ordinary boundary triplets which is not ordinary, was constructed in [Kochubei, 1979]. The question when the boundary triplet is ordinary or of bounded type is answered in [Kostenko and Malamud, 2010].

Schrödinger operators with local point interactions Differential operators with point interactions appear in various physical applications as exactly solvable models that describe complicated physical phenomena, see [Albeverio et al., 2005; Albeverio and Kurasov, 2000; Exner, 2004]. The best known examples are the operators $H_{X,\alpha,q}$ and $H_{X,\beta,q}$

associated with the formal differential expressions

$$\begin{aligned} \ell_{X,\alpha,q} &:= -\frac{\mathrm{d}^2}{\mathrm{d}x^2} + q(x) + \sum_{x_n \in X} \alpha_n \delta(x - x_n), \\ \ell_{X,\beta,q} &:= -\frac{\mathrm{d}^2}{\mathrm{d}x^2} + q(x) + \sum_{x_n \in X} \beta_n \delta'(x - x_n). \end{aligned} \tag{7.49}$$

These operators describe δ- and δ'-interactions, respectively, on a discrete set $X = \{x_n\}_{n \in I} \subset \mathbb{R}$, and the coefficients $\alpha_n, \beta_n \in \mathbb{R}$ are called the strengths of the interaction at the point $x = x_n$; cf. [Albeverio et al., 2005]. In particular, the "Kronig–Penney model" ($\ell_{X,\alpha,q}$ with $X = \mathbb{Z}$, $\alpha_n \equiv \alpha$, and $q \equiv 0$) provides a simple model for a nonrelativistic electron moving in a fixed crystal lattice. The present exposition follows [Kostenko and Malamud, 2010].

Assume that $X = \{x_n\}_{n=0}^{\infty} \subset \mathbb{R}_+$ is a strictly increasing sequence, such that

$$0 = x_0 < x_1 < x_2 < \cdots < x_n < \cdots < +\infty, \qquad \lim_{n \to \infty} x_n = \infty,$$

and denote $d_n := x_n - x_{n-1}$. The symmetric operator $H_{X,\alpha,q}$ is naturally associated with (7.49) in $L^2(\mathbb{R}_+)$. Namely, define the operator $H^0_{X,\alpha,q}$ by the differential expression

$$\tau_q := -\frac{\mathrm{d}^2}{\mathrm{d}x^2} + q(x)$$

on $\operatorname{dom} H^0_{X,\alpha,q}$ given by all $f \in W^{2,2}_{\mathrm{comp}}(\mathbb{R}_+ \setminus X)$ which satisfy for all $n \in \mathbb{N}$;

$$f'(0) = 0, \ f(x_n+) = f(x_n-), \ f'(x_n+) - f'(x_n-) = \alpha_n f(x_n).$$

Let $H_{X,\alpha,q}$ be the closure of $H^0_{X,\alpha,q}$. In general, the operator $H_{X,\alpha,q}$ is symmetric but not automatically self-adjoint, even in the case $q \equiv 0$. To include the operator $H_{X,\alpha,q}$ in the extension theory consider it as an extension of the following closed symmetric operator in $L^2(\mathbb{R}_+)$

$$H_{\min} = -\frac{\mathrm{d}^2}{\mathrm{d}x^2}, \quad \operatorname{dom} H_{\min} = W^{2,2}_0(\mathbb{R}_+ \setminus X),$$

and it is clear that

$$H_{\min} = \bigoplus_{n=1}^{\infty} H_n, \quad H_n = -\frac{\mathrm{d}^2}{\mathrm{d}x^2}, \quad \operatorname{dom} H_n = W^{2,2}_0[x_{n-1}, x_n].$$

Note that $H_{\min} \geq 0$ and the Friedrichs extension H_n^{F} of H_n is defined by the Dirichlet boundary conditions, i.e.,

$$\operatorname{dom} H_n^{\mathrm{F}} = \{f \in W^{2,2}[x_{n-1}, x_n] : \; f(x_{n-1}+) = f(x_n-) = 0\}.$$

Thus the Friedrichs extension of $H_{\min}$ is given by $H^{\mathrm{F}} = \bigoplus_{n=1}^{\infty} H_n^{\mathrm{F}}$, that is $H_{\mathrm{F}} = -\frac{\mathrm{d}^2}{\mathrm{d}x^2}$ with $\operatorname{dom} H_F$ given by

$$\{f \in W_2^2(\mathbb{R}_+ \setminus X) : \; f(0) = f(x_n+) = f(x_n-) = 0, \, n \in \mathbb{N}\}.$$

Define the linear mappings $\widetilde{\Gamma}_i^{(n)} : W_2^2[x_{n-1}, x_n] \to \mathbb{C}^2$, $i = 0, 1$, by

$$\widetilde{\Gamma}_0^{(n)} f := \begin{pmatrix} f(x_{n-1}+) \\ -f(x_n-) \end{pmatrix}, \quad \widetilde{\Gamma}_1^{(n)} f := \begin{pmatrix} f'(x_{n-1}+) \\ f'(x_n-) \end{pmatrix}.$$

Then $\widetilde{\Pi}_n = \{\mathbb{C}^2, \widetilde{\Gamma}_0^{(n)}, \widetilde{\Gamma}_1^{(n)}\}$ forms a boundary triplet for H_n^* satisfying

$$\ker \widetilde{\Gamma}_0^{(n)} = \operatorname{dom} H_n^F,$$

and, by means of Definition 7.13,

$$\widetilde{M}_n(\lambda) = -\frac{\sqrt{\lambda}}{\sin(\sqrt{\lambda} d_n)} \begin{pmatrix} \cos(\sqrt{\lambda} d_n) & 1 \\ 1 & \cos(\sqrt{\lambda} d_n) \end{pmatrix}, \quad \lambda \in \mathbb{C}_+.$$

Moreover, $H_n = d_n^{-2} U_n^{-1} S_1 U_n$, where $S_1 := -\frac{\mathrm{d}^2}{\mathrm{d}x^2}$, $\operatorname{dom} S_1 = W_0^{2,2}[0,1]$, and $(U_n f)(x) := \sqrt{d_n} f(d_n x + x_{n-1})$. Clearly, U_n maps $L^2[x_{n-1}, x_n]$ isometrically onto $L^2[0,1]$.

According to Lemma 7.42 the direct sum $\widetilde{\Pi} = \bigoplus_{n \in \mathbb{N}} \widetilde{\Pi}_n$ forms a unitary boundary triplet for $H_{\min}^* := (H_{\min})^* = H_{\max}$. Following [Malamud and Neidhardt, 2011; Kostenko and Malamud, 2010] the triplet $\widetilde{\Pi}$ can be regularized to obtain an ordinary triplet for $H_{\min}^*$, when

$$d^* = \sup_{n \in \mathbb{N}} d_n \; < \; +\infty. \tag{7.50}$$

Proposition 7.43 [Kostenko and Malamud, 2010] *Assume* (7.50) *and define the mappings* $\Gamma_j^{(n)} : W_2^2[x_{n-1}, x_n] \to \mathbb{C}^2$, $n \in \mathbb{N}$, $j \in \{0, 1\}$, *by*

$$\Gamma_0^{(n)} f = \begin{pmatrix} d_n^{1/2} f(x_{n-1}+) \\ -d_n^{1/2} f(x_n-) \end{pmatrix}, \; \Gamma_1^{(n)} f = \begin{pmatrix} \frac{d_n f'(x_{n-1}+) + (f(x_{n-1}+) - f(x_n-))}{d_n^{3/2}} \\ \frac{d_n f'(x_n-) + (f(x_{n-1}+) - f(x_n-))}{d_n^{3/2}} \end{pmatrix}.$$

Then

(i) $\Pi_n = \{\mathbb{C}^2, \Gamma_0^{(n)}, \Gamma_1^{(n)}\}$ *is an ordinary boundary triplet for* H_n^*;
(ii) $\Pi = \bigoplus_{n=1}^{\infty} \Pi_n$ *is an ordinary boundary triplet for* $H_{\min}^*$.

Sketch of the proof. According to Lemma 7.42 it suffices to check that the Weyl function M corresponding to Π belongs to the class $R^u[\mathcal{H}]$. For this observe that the mappings $\Gamma^{(n)}$ and $\widetilde{\Gamma}^{(n)}$, $n \in \mathbb{N}$, are connected by a triangular J-unitary transform of the form

$$\Gamma^{(n)} = \begin{pmatrix} R_n & 0 \\ R_n^{-1}Q_n & R_n^{-1} \end{pmatrix} \widetilde{\Gamma}^{(n)}; \ R_n = R_n^* = \sqrt{d_n} I_2, \ Q_n = \frac{1}{d_n}\begin{pmatrix} 1 & 1 \\ 1 & 1 \end{pmatrix}.$$

Thus, $\Gamma^{(n)}$ is an ordinary boundary triplet for H_n^*, whose Weyl function M_n is given by

$$M_n(\lambda) = R_n^{-1}(\widetilde{M}_n(\lambda) + Q_n)R_n^{-1}.$$

The Weyl function of $\Pi = \bigoplus_{n=1}^{\infty} \Pi_n$ equals $M = \bigoplus_{n=1}^{\infty} M_n$. Note that $\widetilde{M}_n(\lambda)$ admits a holomorphic continuation to $\lambda = 0$ with

$$\widetilde{M}_n(0) = -Q_n, \quad \widetilde{M}_n'(0) = d_n \begin{pmatrix} 1/3 & -1/6 \\ -1/6 & 1/3 \end{pmatrix} =: d_n B_0.$$

Due to (7.50) the functions $\widetilde{M} = \bigoplus_{n=1}^{\infty} \widetilde{M}_n$ and $M = \bigoplus_{n=1}^{\infty} M_n$ are also holomorphic at $\lambda = 0$ (in the sense of Definition 7.35). Moreover, since $M_n(0) = 0$ and $M_n'(0) = B_0 > 0$ it follows that $M \in R^u[\mathcal{H}]$; cf. [Kostenko and Malamud, 2010, Theorem 3.3]. □

The formulas in the proof of Proposition 7.43 show that the Weyl functions $\widetilde{M} = \bigoplus_{n=1}^{\infty} \widetilde{M}_n$ for $\widetilde{\Pi}$ and $M = \bigoplus_{n=1}^{\infty} M_n$ for Π are connected by

$$\widetilde{M}(\lambda) = -Q + RM(\lambda)R; \quad Q = \bigoplus_{n=1}^{\infty} Q_n, \quad R = \bigoplus_{n=1}^{\infty} R_n. \qquad (7.51)$$

Here Q and R are selfadjoint operators in $\mathcal{H}$. In addition, R is bounded due to condition $d^* < \infty$ in (7.50), while Q is unbounded if $d^* = 0$. Now Theorem 7.39 yields the following improvement for $\widetilde{\Pi} = \bigoplus_{n=1}^{\infty} \widetilde{\Pi}_n$.

Corollary 7.44 *Let the condition* (7.50) *be satisfied. Then*

(i) $\widetilde{M}$ *belongs to* $R^s_{\text{inv}}(\mathcal{H})$ *with* $\operatorname{Im} M(\lambda)$, $\lambda \in \mathbb{C} \setminus \mathbb{R}$, *bounded;*
(ii) $\widetilde{A}_0 := \ker \widetilde{\Gamma}_0$ *is selfadjoint and coincides with the Friedrichs extension* $H^F = \bigoplus_{n=1}^{\infty} H_n^F$ *of* $H_{\min}$*;*
(iii) $\widetilde{\Pi} = \bigoplus_{n=1}^{\infty} \widetilde{\Pi}_n$ *is a boundary triplet of bounded type for* $H_{\min}^*$ *if and only if* $d_* = \inf_{n \in \mathbb{N}} d_n > 0$*; in this case* $\widetilde{\Pi}$ *is necessarily an ordinary boundary triplet for* $H_{\min}^*$.

Proof Due to $\ker R = \{0\}$ one has $M_0(\cdot) := RM(\cdot)R \in R^s[\mathcal{H}]$ and therefore (7.51) shows that condition (iv) of Theorem 7.39 is satisfied; this proves (i) and (ii). As to (iii) note that $\widetilde{M}(\lambda)$ is bounded if and only if the operator Q in (7.51) is bounded. Clearly, this holds if and only if $d_* > 0$. Moreover, if $d_* > 0$ then R has a bounded inverse, which implies that $M_0 \in R^u[\mathcal{H}]$, or equivalently, $\widetilde{M} \in R^u[\mathcal{H}]$. □

In [Kochubei, 1989] it was shown that $\widetilde{\Pi} = \bigoplus_{n\in\mathbb{N}} \widetilde{\Pi}_n$ forms an ordinary boundary triplet for $H^*_{\min}$ whenever

$$0 < d_* = \inf_{n\in\mathbb{N}} d_n \le d^* = \sup_{n\in\mathbb{N}} d_n < +\infty.$$

Under this condition spectral properties of the corresponding Hamiltonians have been investigated in [Kochubei, 1989; Mikhailets, 1994]. Proposition 7.43 provides the tool to treat Hamiltonians $H_{X,\alpha}$ within the boundary triplets approach in the case $d_* = 0$.

The following result gives a connection with Jacobi matrices. According to Proposition 7.12 the symmetric extension $H_{X,\alpha}$ of $H_{\min}$ admits a representation

$$\begin{aligned} H_{X,\alpha} &= H_\Theta := H^*_{\min}\restriction \operatorname{dom} H_\Theta, \\ \operatorname{dom} H_\Theta &= \{f \in \operatorname{dom} H^*_{\min} : \{\Gamma_0 f, \Gamma_1 f\} \in \Theta\}, \end{aligned} \tag{7.52}$$

for some closed symmetric relation Θ.

Proposition 7.45 [Kostenko and Malamud, 2010] *Let $\Pi = \{\mathcal{H}, \Gamma_0, \Gamma_1\}$ be the boundary triplet constructed in Proposition 7.43 and let the closed linear relation Θ be defined by (7.52). Then Θ admits a representation of the form (7.2), where the operator part Θ_s is unitarily equivalent to the minimal Jacobi operator $B_{X,\alpha}$ generated in $l^2(\mathbb{N})$ by the following Jacobi matrix*

$$\begin{pmatrix} r_1^{-2}(\alpha_1 + \frac{1}{d_1} + \frac{1}{d_2}) & (r_1r_2d_2)^{-1} & 0 & \dots \\ (r_1r_2d_2)^{-1} & r_2^{-2}(\alpha_2 + \frac{1}{d_2} + \frac{1}{d_3}) & (r_2r_3d_3)^{-1} & \\ 0 & (r_2r_3d_3)^{-1} & r_3^{-2}(\alpha_3 + \frac{1}{d_3} + \frac{1}{d_4}) & \ddots \\ \vdots & & \ddots & \ddots \end{pmatrix}$$

with $r_n = \sqrt{d_n + d_{n+1}},\ n \in \mathbb{N}$.

The following result shows that spectral properties of the Hamiltonian $H_{X,\alpha}$ are closely related with those of the boundary operator $B_{X,\alpha}$ (the Jacobi matrix) in Proposition 7.45.

Theorem 7.46 [Kostenko and Malamud, 2010] *Let* $\sup_{n\in\mathbb{N}} d_n < \infty$. *Then:*

(i) $n_\pm(H_{X,\alpha}) = n_\pm(B_{X,\alpha}) \leq 1$;

(ii) $H_{X,\alpha}$ *is lower semibounded (nonnegative) if and only if* $B_{X,\alpha}$ *is lower semibounded (nonnegative);*

(iii) *if* $H_{X,\alpha} = H^*_{X,\alpha}$, *then* $\sigma(H_{X,\alpha})$ *is discrete if and only if* $d_n \to 0$ *and* $\sigma(B_{X,\alpha})$ *is discrete;*

(iv) *if* $H_{X,\alpha} = H^*_{X,\alpha}$, *then the negative part of* $\sigma(H_{X,\alpha})$ *is discrete (finite) precisely when the negative part of* $\sigma(B_{X,\alpha})$ *is discrete (finite). In particular,* $\sigma_c(H_{X,\alpha}) \subseteq \mathbb{R}_+$ *if and only if* $\sigma_c(B_{X,\alpha}) \subseteq \mathbb{R}_+$.

Applications of Theorem 7.46 as well as further connections between Hamiltonians $H_{X,\alpha}$ and Jacobi matrices $B_{X,\alpha}$ are given in [Kostenko and Malamud, 2010; Albeverio et al., 2010], [Ismagilov and Kostjuchenko, 2010], [Mirzoev and Safonova, 2011].

7.7 Generalized resolvents and unitary boundary triplets

Coupling of boundary triplets Couplings of boundary triplets were introduced and studied in [Derkach et al., 2000, 2004, 2009] in connection with the theory of generalized resolvents of symmetric operators.

Theorem 7.47 *Let* S_+ *be a closed symmetric operator in a Hilbert space* $\mathfrak{H}_+$ *and let* $\Pi^+ = \{\mathcal{H}, \Gamma_0^+, \Gamma_1^+\}$ *be an ordinary boundary triplet for* S_+^* *with the Weyl function* M. *Let* S_- *be a closed symmetric operator in a Hilbert space* $\mathfrak{H}_-$ *and let* $\Pi^- = \{\mathcal{H}, \Gamma_0^-, \Gamma_1^-\}$ *be a unitary boundary triplet for* S_-^* *with* $T_- := \operatorname{dom}\Gamma^-$ *and with the Weyl function* M_-. *Then:*

(i) *With* $\Gamma_i^{(c)} : S_+^* \oplus T_- \to \mathcal{H}$, $i = 0, 1$, *defined by*

$$\Gamma_0^{(c)}(\widehat{f}_+ \oplus \widehat{f}_-) = \begin{pmatrix} \Gamma_1^{(-)}\widehat{f}_- + \Gamma_1^{(+)}\widehat{f}_+ \\ \Gamma_0^{(-)}\widehat{f}_- - \Gamma_0^{(+)}\widehat{f}_+ \end{pmatrix}, \quad \Gamma_1^{(c)}(\widehat{f}_+ \oplus \widehat{f}_-) = \begin{pmatrix} -\Gamma_0^{(+)}\widehat{f}_+ \\ \Gamma_1^{(-)}\widehat{f}_- \end{pmatrix}$$

where $\widehat{f}_+ \in S_+^*$, $\widehat{f}_- \in T_-$, *the triplet* $\Pi^{(c)} = \{\mathcal{H}, \Gamma_0^{(c)}, \Gamma_1^{(c)}\}$ *is a boundary triplet of bounded type for the relation* $S_+^* \oplus T_-$;

(ii) *the corresponding Weyl function* $M^{(c)}$ *is given by*

$$M^{(c)} = \begin{pmatrix} -(M_+ + M_-)^{-1} & I - (M_+ + M_-)^{-1}M_+ \\ I - M_+(M_+ + M_-)^{-1} & (M_+^{-1} + M_-^{-1})^{-1} \end{pmatrix}.$$

The boundary triplet $\Pi^{(c)}$ is usually called the *coupling* of the boundary triplets $\Pi_{\pm}$. Note that the selfadjoint linear relation $\widetilde{A} := \ker \Gamma_0^{(c)}$ is given by

$$\widetilde{A} = \left\{ \widehat{f}_+ \oplus \widehat{f}_- \in S_+^* \oplus T_- : \Gamma_0^+ \widehat{f}_+ - \Gamma_0^- \widehat{f}_- = \Gamma_1^+ \widehat{f}_+ + \Gamma_1^- \widehat{f}_- = 0 \right\}. \tag{7.53}$$

It can be called the coupling of the symmetric operators $S_{\pm}$, corresponding to the triplet $\Pi^{(c)}$. The Weyl function $M^{(c)}$ of $\Pi^{(c)}$ is frequently encountered in boundary-eigenvalue problems with boundary conditions depending on the eigenvalue parameter (see e.g. [Kac, 1963; Dijksma et al., 1987, 1988]). The statements of Theorem 7.47 remain valid for a unitary boundary pair $\{\mathcal{H}, \Gamma^-\}$ with a multi-valued mapping Γ^-, see [Derkach et al., 2009, Theorem 5.12].

Two other couplings of the boundary triplets $\Pi_{\pm}$ were considered in [Derkach et al., 2000], given by

$$\begin{aligned} T^{(1)} &= \left\{ \widehat{f} = \widehat{f}_+ \oplus \widehat{f}_- \in S_+^* \oplus T_- : \Gamma_0^+ \widehat{f}_+ - \Gamma_0^- \widehat{f}_- = 0 \right\}, \\ \Gamma_0^{(1)} \widehat{f} &= \Gamma_1^+ \widehat{f}_+ + \Gamma_1^- \widehat{f}_-, \quad \Gamma_1^{(1)} \widehat{f} = -\Gamma_0^+ \widehat{f}_+, \quad \widehat{f} \in T^{(1)}; \end{aligned} \tag{7.54}$$

$$\begin{aligned} T^{(2)} &= \left\{ \widehat{f} = \widehat{f}_+ \oplus \widehat{f}_- \in S_+^* \oplus T_- : \Gamma_1^+ \widehat{f}_+ + \Gamma_1^- \widehat{f}_- = 0 \right\}, \\ \Gamma_0^{(2)} \widehat{f} &= \Gamma_0^+ \widehat{f}_+ - \Gamma_0^- \widehat{f}_-, \quad \Gamma_1^{(1)} \widehat{f} = \Gamma_1^+ \widehat{f}_+, \quad \widehat{f} \in T^{(2)}. \end{aligned} \tag{7.55}$$

These boundary triplets $\Pi^{(1)}$ and $\Pi^{(2)}$ can be treated as projections of $T^{(c)}$ to the first and the second component of $\mathcal{H}^2$. The corresponding Weyl functions have the forms

$$M^{(1)} = -(M_+ + M_-)^{-1}, \; M^{(2)} = (M_+^{-1} + M_-^{-1})^{-1}, \tag{7.56}$$

associated with the intermediate extensions $(T^{(1)})^*$ and $(T^{(2)})^*$.

Generalized resolvents of symmetric operators Let S be a symmetric operator in a Hilbert space $\mathfrak{H}$ with equal deficiency indices. Let $\widetilde{A}$ be a selfadjoint extension of S in a Hilbert space $\widetilde{\mathfrak{H}}$ containing $\mathfrak{H}$ as a closed subspace. The compression $\boldsymbol{R}_\lambda = P_{\mathfrak{H}}(\widetilde{A} - \lambda)^{-1} \upharpoonright \mathfrak{H}$ of the resolvent of $\widetilde{A}$ to $\mathfrak{H}$ is said to be a *generalized resolvent* of S. Here $P_{\mathfrak{H}}$ is the orthogonal projection from $\widetilde{\mathfrak{H}}$ onto $\mathfrak{H}$.

The description of all generalized resolvents was originally given in different forms in [Kreĭn, 1944; Naĭmark, 1943]. It has been extended to the case of infinite indices in [Saakyan, 1965]; see [Kreĭn and Langer, 1971; Derkach and Malamud, 1991] and references therein. Another description of generalized resolvents was given in [Štrauss, 1954]. Kreĭn type formulas for elliptic boundary value problems and singular perturbations of selfadjoint operators have been studied in [Albeverio and Kurasov, 2000; Derkach et al., 2003; Gesztesy and Mitrea, 2008, 2011; Ashbaugh et al., 2010; Malamud, 2010; Posilicano, 2001, 2008].

Theorem 7.48 [Kreĭn and Langer, 1971] *Let S be a symmetric operator in $\mathfrak{H}$ with $n_+(S)=n_-(S)$, let $\Pi_+=\{\mathcal{H}_1,\Gamma_0,\Gamma_1\}$ be a boundary triplet for S^*, and let M_+ and γ_+ be the corresponding Weyl function and the γ-field. Then, with $\lambda\in\rho(A_0)\cap\rho(\widetilde{A})$, the formula*

$$\boldsymbol{R}_\lambda=(A_0-\lambda)^{-1}-\gamma_+(\lambda)(M_+(\lambda)+\tau(\lambda))^{-1}\gamma_+(\bar\lambda)^*, \tag{7.57}$$

where $A_0=\ker\Gamma_0$, establishes a bijective correspondence between the generalized resolvents $\boldsymbol{R}_\lambda$ of S and the Nevanlinna families $\tau\in\widetilde{R}(\mathcal{H})$.

Using the coupling method from [Derkach et al., 2000] one part of the proof will be sketched now. In the particular case when $\tau\in R^u[\mathcal{H}]$ there exist a symmetric operator S_- in a Hilbert space $\mathfrak{H}_-$ and an ordinary boundary triplet $\Pi_-=\{\mathcal{H},\Gamma_0^-,\Gamma_1^-\}$ for $T_-=S_-^*$, such that the corresponding Weyl function M_- coincides with τ. Let $S^{(1)}(\supset S\oplus S_-)$ be a symmetric operator in $\mathfrak{H}\oplus\mathfrak{H}_-$ given by

$$S^{(1)}=\left\{\widehat{f}_+\oplus\widehat{f}_-\in S_+^*\oplus T_-:\Gamma_0^+\widehat{f}_+=\Gamma_0^-\widehat{f}_-=0,\,\Gamma_1^+\widehat{f}_++\Gamma_1^-\widehat{f}_-=0\right\}.$$

Then $S^{(1)}=(T^{(1)})^*$, where $T^{(1)}$ is as described in (7.54). Let $\Pi^{(1)}$ be the boundary triplet for $T^{(1)}$ determined by (7.54). Then the Weyl function of $S^{(1)}$ corresponding to the boundary triplet $\Pi^{(1)}$ takes the form (7.56), and the corresponding γ-field takes the form

$$\gamma^{(1)}(\lambda)=\begin{pmatrix}\gamma_+(\lambda)\\ \gamma_-(\lambda)\end{pmatrix}(\tau(\lambda)+M_+(\lambda))^{-1}.$$

Let $\widetilde{A}$ be a selfadjoint extension of H given by (7.53) and let $\widetilde{A}_1=\operatorname{diag}(A_0,A_0^-)$, $A_0^-=\ker\Gamma_0^-$. Since $\widetilde{A}=\ker\Gamma_0^{(1)}$ and $\widetilde{A}_1=\ker\Gamma_1^{(1)}$ it follows from (7.19) that the resolvent $(\widetilde{A}-\lambda)^{-1}$ is given by

$$(\widetilde{A}_1-\lambda)^{-1}-\begin{pmatrix}\gamma_+(\lambda)\\ \gamma_--(\lambda)\end{pmatrix}(\tau(\lambda)+M(\lambda))^{-1}\begin{pmatrix}\gamma_+(\bar\lambda)^* & \gamma_-(\bar\lambda)^*\end{pmatrix}. \tag{7.58}$$

The compression of (7.58) to $\mathfrak{H}$ gives (7.57).

In the general case when τ is a Nevanlinna family one can use another coupling $\{\mathcal{H}^2,\Gamma_0^{(c)},\Gamma_1^{(c)}\}$ from Theorem 7.47 to prove the formula (7.57) in a similar way. Note that the extension $\widetilde{A}^{(c)}:=\ker\Gamma_0^{(c)}$ coincides with $\widetilde{A}$ from (7.53) and its resolvent can be calculated explicitly via the formula (7.19); see [Derkach et al., 2000, 2009] for details.

The proof of the converse statement can be also given via a coupling argument. This will be considered in the next subsection.

Observe that the above construction relies on the direct realization of the parameter τ in the formula (7.57); see for the realization in terms of reproducing kernel Hilbert spaces [Behrndt et al., 2009] and [Derkach, 2009]. In [Kreĭn and Langer, 1971] another approach was followed: the exit space was constructed via a reproducing kernel Hilbert space associated with the generalized resolvent. The present approach remains valid in the context of Nevanlinna families with a finite number of negative squares; see [Behrndt et al., 2011] for the construction of the corresponding Pontryagin space. The coupling method in its general form for constructing generalized resolvents was developed in [Derkach et al., 2009], where also applications to solve, for instance, some problems concerning admissibility of exit space extensions and characterization of Naĭmark extensions are given; see also Chapter 10 for an application of Naĭmark extensions for analyzing properties of representing measures that arise in the connection of moment problems.

Induced boundary triplets Let $\widetilde{A}$ be a selfadjoint exit space extension A of S in a direct sum $\mathfrak{H} = \mathfrak{H}_+ \oplus \mathfrak{H}_-$, where $\mathfrak{H}_+ = \mathfrak{H}$. With $P_\pm$ the orthogonal projections onto $\mathfrak{H}_\pm$, define

$$S_\pm = \widetilde{A} \cap \mathfrak{H}_\perp^2, \quad T_\pm = \left\{ \{P_\pm \varphi, P_\pm \varphi'\} : \{\varphi, \varphi'\} \in \widetilde{A} \right\}.$$

Then $S_\pm$ are closed symmetric relations in $\mathfrak{H}_\pm$ and $S_\pm^* = \overline{T}_\pm$. The relation $\widetilde{A}$ is called a *minimal* selfadjoint extension of S_+ if

$$\widetilde{\mathfrak{H}} = \overline{\operatorname{span}} \left\{ \mathfrak{H}_+ + (\widetilde{A} - \lambda)^{-1} \mathfrak{H}_+ : \lambda \in \rho(\widetilde{A}) \right\}.$$

Let $\Pi = \{\mathcal{H}, \Gamma_0, \Gamma_1\}$ be an ordinary boundary triplet for S_+^* and define the induced triplet by

$$\Gamma_0^- \widehat{f}_- = \Gamma_0 \widehat{f}_+, \quad \Gamma_1^- \widehat{f}_- = -\Gamma_1 \widehat{f}_+, \quad \widehat{f}_+ \oplus \widehat{f}_- \in \widetilde{A}. \tag{7.59}$$

Theorem 7.49 [Derkach et al., 2009] *The triplet $\Pi^- = \{\mathcal{H}, \Gamma_0^-, \Gamma_1^-\}$ is a unitary boundary triplet for S_-^* if $S_+ = S$. Moreover,*

(i) *Π^- is an ordinary boundary triplet if and only if T_- is closed;*
(ii) *Π^- is minimal if and only if the extension $\widetilde{A}$ is minimal.*

In the case $S_+ = S$ the extension $\widetilde{A}$ can be recovered via the formula (7.53), which leads to a proof of the converse statement of Theorem 7.48. In the case $S_+ \neq S$ the formula (7.59) defines a unitary boundary pair with a multi-valued mapping Γ^-, see [Derkach et al., 2009, Theorem 5.3] for the treatment in this case.

Example 7.50 Let $\widetilde{A}$ be a selfadjoint operator generated in $L^2(\mathbb{R}^3)$ by the differential expression $\ell = -\Delta$; cf. Example 7.28. As is known, $\operatorname{dom} \widetilde{A} = W_2^2(\mathbb{R}^3)$. Let again Ω be a bounded domain in $\mathbb{R}^3$ with a smooth boundary $\partial\Omega$ and let $\Omega_+ := \Omega$, $\Omega_- := \mathbb{R}^3 \setminus \overline{\Omega}_+$, and let $A^{\pm}_{\rm min}$ and $A^{\pm}_{\rm max}$ be the minimal and the maximal differential operators generated in $L^2(\Omega_\pm)$ by the differential expression ℓ. Then

$$\operatorname{dom} A^{\pm}_{\rm min} = \overset{\circ}{W}{}_2^2(\Omega_\pm),$$

see [Berezanskii, 1965; Grubb, 2009]. Decompose

$$\widetilde{\mathfrak{H}} = L_2(\mathbb{R}^n) = L_2(\Omega_+) \oplus L_2(\Omega_-) =: \mathfrak{H}_+ \oplus \mathfrak{H}_-,$$

and define the operators $S_\pm$ and $T_\pm$ as above, so that $S_\pm = A^{\pm}_{min}$. Furthermore, $P_\pm W_2^2(\mathbb{R}^3) = W_2^2(\Omega_\pm)$. Therefore, $T_\pm$ is the elliptic operator generated in $L_2(\Omega_\pm)$ by the differential expression ℓ with the domain $\operatorname{dom} T_\pm = W_2^2(\Omega_\pm)$, which is a proper subset of $\operatorname{dom} A^{\pm}_{\rm max}$ and $\overline{T}_\pm = A^{\pm}_{\rm max}$.

Let $\Pi_+ = \{\mathcal{H}, \Gamma_0, \Gamma_1\}$ be the ordinary boundary triplet for A^*_+ defined in (7.36). Since for every $y \in \operatorname{dom} \widetilde{A} = W_2^2(\mathbb{R}^3)$

$$\gamma^+_{\rm D} y = \gamma^-_{\rm D} y, \quad \gamma^+_{\rm N} y = -\gamma^-_{\rm N} y,$$

the induced boundary triplet $\Pi_- = \{\mathcal{H}, \Gamma^-_0, \Gamma^-_1\}$ takes the form

$$\mathcal{H} = L^2(\partial\Omega), \quad \Gamma^-_0 = K\gamma^-_{\rm D}, \quad \Gamma^-_1 = K^{-1}(\gamma^-_{\rm N} + \Lambda^+(0)\gamma^-_{\rm D}),$$

where $\Lambda^+(0)$ is the Steklov-Poincaré operator for the domain Ω_+. It follows from (7.34), cf. (7.36), that

$$\Gamma^-_0 : W_2^2(\Omega_-) \to W_2^2(\partial\Omega), \quad \Gamma^-_1 : W_2^2(\Omega_-) \to L^2(\partial\Omega).$$

Furthermore, the mapping Γ^-_1 is surjective and it follows from Definition 7.17 that the transposed boundary triplet $\Pi^\top_-$ for $\operatorname{clos} T_-$ is of bounded type. In contrast to Π_+ here the boundary triplets Π_- and $\Pi^\top_-$ are not ordinary, since T_- is not the maximal operator and Γ^-_0 is not onto $L^2(\partial\Omega)$.

Let $S^{(2)}$ be the symmetric extension of $S_+ \oplus S_-$, which is the Laplacian restricted to the domain $\operatorname{dom} S^{(2)}$ given by

$$\{\, y \in W_2^2(\Omega_+) \oplus W_2^2(\Omega_-) : \gamma^+_{\rm D} y_+ = \gamma^-_{\rm D} y_- = \gamma^+_{\rm N} y_+ + \gamma^-_{\rm N} y_- = 0 \,\},$$

and let the triplet $\Pi^{(2)}$ be defined by (7.55). Then the Weyl function of $S^{(2)}$ corresponding to the boundary triplet $\Pi^{(2)}$ has the form

$$M^{(2)}(\lambda) = (M_+(\lambda)^{-1} + M_-(\lambda)^{-1})^{-1},$$

and $M^{(2)}$ is the Weyl function of the pair $(S^{(2)}, \widetilde{A})$. Such a transformation of Weyl functions (M-functions) appears in the literature; see, for instance [Amrein and Pearson, 2004], where M_+ and M_- serve as M-functions of the Schrödinger operator in interior and exterior domains, respectively, and $M^{(2)}$ is interpreted as an analytic family of Poincaré-Steklov operators.

7.8 Isometric boundary mappings

Some definitions and properties Let Γ be an isometric relation from the Kreĭn space $(\mathfrak{H}^2, J_{\mathfrak{H}})$ to the Kreĭn space $(\mathcal{H}^2, J_{\mathcal{H}})$. This implies that the abstract Green's identity

$$(f', g)_{\mathfrak{H}} - (f, g')_{\mathfrak{H}} = (h', k)_{\mathcal{H}} - (h, k')_{\mathcal{H}}, \quad \{\widehat{f}, \widehat{h}\}, \{\widehat{g}, \widehat{k}\} \in \Gamma, \tag{7.60}$$

holds, where

$$\widehat{f} = \{f, f'\}, \widehat{g} = \{g, g'\} \in \mathfrak{H}^2, \ \widehat{h} = \{h, h'\}, \widehat{k} = \{k, k'\} \in \mathcal{H}^2.$$

Since Γ is isometric one has $\ker \Gamma \subset (\operatorname{dom} \Gamma)^{[\perp]}$ and $\operatorname{mul} \Gamma \subset (\operatorname{ran} \Gamma)^{[\perp]}$; cf. Lemma 7.3. Note that these inclusions need not hold as equalities, if Γ is not unitary. In this general context an isometric relation $\Gamma : \mathfrak{H}^2 \to \mathcal{H}^2$ can be also viewed as an *isometric boundary relation for the closure of* $T = \operatorname{dom} \Gamma$.

Associate with Γ the component mappings according to the product decomposition $\mathcal{H} \times \mathcal{H}$ of $\mathcal{H}^2$ via $\Gamma_i = \pi_i \Gamma$, $i = 0, 1$, where π_0 and π_1 stand for the orthogonal projection onto the first and the second component of $\mathcal{H} \times \mathcal{H}$, respectively. Then the kernels $A_0 := \ker \Gamma_0$ and $A_1 := \ker \Gamma_1$ are contained in $\operatorname{dom} \Gamma$ and, in general, they are non-closed symmetric extensions of $\ker \Gamma$; see [Derkach et al., 2006, Proposition 2.13].

To every isometric relation $\Gamma : \mathfrak{H}^2 \to \mathcal{H}^2$ one can associate the *Weyl family* in a similar way as in the unitary case:

$$M(\lambda) := \Gamma \widehat{\mathfrak{N}}_\lambda(T), \quad \lambda \in \mathbb{C},$$

where $\widehat{\mathfrak{N}}_\lambda(T) = \{\widehat{f}_\lambda : \widehat{f}_\lambda = \{f_\lambda, \lambda f_\lambda\} \in \operatorname{dom} \Gamma\}$. Let $\widehat{h} \in M(\lambda)$

and $\widehat{k} \in M(\mu)$ with $\lambda, \mu \in \mathbb{C}$, then there exist $\widehat{f}_\lambda, \widehat{g}_\mu \in T$, such that $\{\widetilde{h}, \widehat{f}_\lambda\}$, $\{\widehat{k}, \widehat{g}_\mu\} \in T$. Green's identity (7.60) then gives

$$(h', k)_{\mathcal{H}} - (h, k')_{\mathcal{H}} = (\lambda - \bar{\mu})(f_\lambda, g_\mu)_{\mathfrak{H}}. \tag{7.61}$$

In particular, with $\mu = \bar{\lambda}$ (7.61) implies that

$$M(\lambda) \subset M(\bar{\lambda})^*, \quad \lambda \in \mathbb{C}.$$

With $\mu = \lambda \in \mathbb{C} \setminus \mathbb{R}$ (7.61) implies that, for instance, $\ker (\Gamma_0 \upharpoonright \widehat{\mathfrak{N}}_\lambda(T)) = \{0\}$. Therefore,

$$\widehat{\gamma}(\lambda) = (\Gamma_0 \upharpoonright \widehat{\mathfrak{N}}_\lambda(T))^{-1}$$

is a single-valued mapping from $\operatorname{dom} M(\lambda)$ onto $\widehat{\mathfrak{N}}_\lambda(T)$ and, thus, as in the unitary case one can define the *γ-field* as the first component of the mapping $\widehat{\gamma}(\lambda)$, $\lambda \in \mathbb{C} \setminus \mathbb{R}$. Furthermore, (7.61) shows that $M(\lambda)$ is dissipative (accumulative) for $\lambda \in \mathbb{C}_+$ (for $\lambda \in \mathbb{C}_-$, respectively). However, observe that by definition $M(\lambda) \subset \operatorname{ran} \Gamma$, while in general $M(\lambda)^* \not\subset \operatorname{ran} \Gamma$.

The next result gives a criterion for an isometric relation $\Gamma : \mathfrak{H}^2 \to \mathcal{H}^2$ to be a boundary relation for S^* which is based on the properties of the Weyl function M. The result stated here is a slight (but useful) extension of [Derkach et al., 2006, Proposition 3.6].

Theorem 7.51 *The linear relation $\Gamma : \mathfrak{H}^2 \mapsto \mathcal{H}^2$ is a unitary boundary relation for S^* if and only if the following conditions hold:*

(i) $\operatorname{dom} \Gamma$ *is dense in* S^**;*
(ii) Γ *is closed and isometric from* $(\mathfrak{H}^2, J_{\mathfrak{H}})$ *to* $(\mathcal{H}^2, J_{\mathcal{H}})$*;*
(iii) $\operatorname{ran}(M(\lambda) + \lambda)$ *is dense in* $\mathcal{H}$ *for some (and, hence, for all)* $\lambda \in \mathbb{C}_+$ *and for some (and, hence, for all)* $\lambda \in \mathbb{C}_-$*.*

Proof ($\Rightarrow$) This part is clear from the definition of a boundary relation and the fact that its Weyl function belongs to the class $\widetilde{R}(\mathcal{H})$ of Nevanlinna families (see [Derkach et al., 2006, Theorem 3.9]).

($\Leftarrow$) The proof given in [Derkach et al., 2006, Proposition 3.6] will be modified. Assume that Γ satisfies (i), (ii), and (iii); and let $\widetilde{A} = \mathcal{J}(\Gamma)$ be the main transform of Γ. Then $\widetilde{A}$ is closed and symmetric; see [Derkach et al., 2006, Proposition 2.10]. Hence, to prove that $\widetilde{A}$ is selfadjoint it suffices to show that $\operatorname{ran}(\widetilde{A} - \lambda)$ is dense for some $\lambda \in \mathbb{C}_+$ and for some $\lambda \in \mathbb{C}_-$. It follows from the formula

$$\widetilde{A} - \lambda = \left\{ \left\{ \begin{pmatrix} f \\ h \end{pmatrix}, \begin{pmatrix} f' - \lambda f \\ -h' - \lambda h \end{pmatrix} \right\} : \left\{ \begin{pmatrix} f \\ f' \end{pmatrix}, \begin{pmatrix} h \\ h' \end{pmatrix} \right\} \in \Gamma \right\}$$

that $\{0\} \times \operatorname{ran}(M(\lambda)+\lambda) \subset \operatorname{ran}(\widetilde{A}-\lambda)$ for some $\lambda \in \mathbb{C}_+$ and for some $\lambda \in \mathbb{C}_-$, which are dense in $\{0\} \times \mathcal{H}$ by assumption (iii). Thus, if $\operatorname{ran}(\widetilde{A}-\lambda)$ is not dense in $\mathfrak{H} \times \mathcal{H}$, then there is an element $\varphi \in \mathfrak{H}$ which is orthogonal to $\operatorname{ran}(\widetilde{A}-\lambda)$. This implies that $\{\varphi, \bar{\lambda}\varphi\} \in T^*$, where $T = \operatorname{dom}\Gamma$. By (i) T is dense in S^* and hence $S = T^*$ and $\{\varphi, \bar{\lambda}\varphi\} \in S$. Since S is symmetric this yields $\varphi = 0$ and completes the proof. □

Remark A triplet $\{\mathcal{H}, \Gamma_0, \Gamma_1\}$ is an ordinary boundary triplet for S^* if $\Gamma = \{\Gamma_0, \Gamma_1\}$ is an isometric relation with $\ker \Gamma = S$ and dense domain in S^* such that

$$\operatorname{ran}\Gamma = \mathcal{H}^2.$$

This is an immediate consequence of Corollary 7.5. It is a slightly weaker characterization for ordinary boundary triplets than the one in [Derkach et al., 2006, Proposition 5.3].

The following lemma is proved in exactly the same way as in [Derkach et al., 2006, Lemma 4.1].

Lemma 7.52 *Let $\Gamma : \mathfrak{H}^2 \to \mathcal{H}^2$ be an isometric relation with the Weyl family M. Then the following equalities hold for all $\lambda \in \mathbb{C} \setminus \mathbb{R}$:*

(i) $M(\lambda) \cap M(\lambda)^* = \operatorname{mul}\Gamma$;
(ii) $\ker M(\lambda) \times \{0\} = \operatorname{mul}\Gamma \cap (\mathcal{H} \times \{0\})$;
(iii) $\{0\} \times \operatorname{mul} M(\lambda) = \operatorname{mul}\Gamma \cap (\{0\} \times \mathcal{H})$;
(iv) $\ker\,(M(\lambda) - M(\lambda)^*) = \operatorname{mul}\Gamma_0$;
(v) $\ker\,(M(\lambda)^{-1} - M(\lambda)^{-*}) = \operatorname{mul}\Gamma_1$.

The proof of the next lemma is based on the identities $\operatorname{dom}\gamma(\lambda) = \operatorname{dom} M(\lambda) = \Gamma_0(\widehat{\mathfrak{N}}_\lambda(T))$; cf. [Derkach et al., 2006, Proposition 4.11].

Lemma 7.53 *Let $\Gamma : \mathfrak{H}^2 \to \mathcal{H}^2$ be an isometric relation with the γ-field γ and let $A_0 = \ker \Gamma_0$. Then for all $h \in \operatorname{dom}\gamma(\lambda) \cap \operatorname{dom}\gamma(\mu)$ one has $\gamma(\mu)h \in \operatorname{dom}(A_0-\lambda)^{-1}$ and*

$$\gamma(\lambda)h = [I + (\lambda-\mu)(A_0-\lambda)^{-1}]\gamma(\mu)h, \quad \lambda, \mu \in \mathbb{C} \setminus \mathbb{R}.$$

In what follows it is assumed that the range of Γ is dense in $\mathcal{H}^2$, so that Γ is single-valued. In this case Lemma 7.52 shows that the Weyl family M is an operator-valued function which is strict in the sense that $\ker\,(M(\lambda) - M(\lambda)^*) = \{0\}$. The following useful relations are also satisfied for all $\lambda \in \mathbb{C} \setminus \mathbb{R}$:

$$\Gamma_0\widehat{\gamma}(\lambda)h = h, \quad \Gamma_1\widehat{\gamma}(\lambda)h = M(\lambda)h, \quad h \in \operatorname{dom} M(\lambda).$$

With $\widehat{f}_\lambda = \widehat{\gamma}(\lambda)h$ and $\widehat{g}_\mu = \widehat{\gamma}(\mu)k$ one can rewrite (7.61) as follows:

$$(\lambda - \bar{\mu})(\gamma(\lambda)h, \gamma(\mu)k)_{\mathfrak{H}} = (M(\lambda)h, k)_{\mathcal{H}} - (h, M(\mu)k)_{\mathcal{H}}, \tag{7.62}$$

where $h \in \operatorname{dom} M(\lambda)$ and $k \in \operatorname{dom} M(\mu)$, $\lambda, \mu \in \mathbb{C} \setminus \mathbb{R}$. Also the formula in Lemma 7.38 (i) remains true in this setting; the proof is the same.

Proposition 7.54 *Let $\Gamma = \{\Gamma_0, \Gamma_1\} : \mathfrak{H}^2 \to \mathcal{H}^2$ be an isometric relation with dense range, let γ be its γ-field, and let $H(\lambda)$ be as in Lemma 7.38. Then*

$$\Gamma_1 H(\lambda) \subset \gamma(\bar{\lambda})^*, \quad \lambda \in \mathbb{C} \setminus \mathbb{R}. \tag{7.63}$$

If, in particular, $\operatorname{dom} \gamma(\lambda)$ $(= \operatorname{dom} M(\bar{\lambda}))$ is dense in $\mathcal{H}$, then (7.63) shows that $\gamma(\bar{\lambda})^*$ is an operator and, consequently, $\Gamma_1 H(\lambda)$ is a closable operator. If, in addition, $A_0 = \ker \Gamma_0$ is closed or equivalently $\operatorname{dom}(\Gamma_1 H(\lambda))$ is closed, then $\Gamma_1 H(\lambda)$ is bounded. Note also that, if $\Gamma_1 H(\lambda)$ is bounded and Γ is closed, then A_0 is also closed (see the proof of Lemma 7.38).

On the other hand, if A_0 is essentially selfadjoint, then by definition $\Gamma_1 H(\lambda)$ is densely defined and, thus, $\gamma(\bar{\lambda})$ is a closable operator for all $\lambda \in \mathbb{C} \setminus \mathbb{R}$. If, in addition, A_0 is selfadjoint, then $\operatorname{dom} \Gamma = A_0 \widehat{+} \widehat{\mathfrak{N}}_\lambda(T)$, which implies that $\operatorname{dom} \gamma(\lambda) = \operatorname{dom} M(\lambda) = \operatorname{ran} \Gamma_0$ is dense in $\mathfrak{H}$. Thus, in this case the closure of the γ-field is a bounded everywhere defined operator for all $\lambda \in \mathbb{C} \setminus \mathbb{R}$.

Coupling of symmetric operators A particular class of isometric relations is provided when the orthogonal sum of two, not necessarily closed, symmetric operators, A_0 in a Hilbert space $\mathfrak{H}$ and E in a Hilbert space $\mathcal{H}$, are perturbed by an operator $B \in \boldsymbol{B}(\mathcal{H}, \mathfrak{H})$. Then

$$\widetilde{A} = \left\{ \left\{ \begin{pmatrix} f \\ h \end{pmatrix}, \begin{pmatrix} A_0 f + Bh \\ B^* f - Eh \end{pmatrix} \right\} : f \in \operatorname{dom} A_0,\ h \in \operatorname{dom} E \right\}.$$

defines a symmetric operator $\widetilde{A}$ in $\mathfrak{H} \times \mathcal{H}$. With $\Gamma = \mathcal{J}\widetilde{A}$, where $\mathcal{J}$ is the so-called main transform, see [Derkach et al., 2006, Proposition 2.10], it follows that

$$\Gamma = \left\{ \left\{ \begin{pmatrix} f \\ A_0 f + Bh \end{pmatrix}, \begin{pmatrix} h \\ Eh - B^* f \end{pmatrix} \right\} : f \in \operatorname{dom} A_0,\ h \in \operatorname{dom} E \right\} \tag{7.64}$$

is an isometric relation from $(\mathfrak{H}^2, J_{\mathfrak{H}})$ to $(\mathcal{H}^2, J_{\mathcal{H}})$. Moreover, it is clear that

$$\begin{aligned} \operatorname{ran}\Gamma &= E \,\widehat{+}\, (\{0\} \times B^*(\operatorname{dom} A_0)), \\ \operatorname{mul}\Gamma &= \{\{h, Eh\} : h \in \ker B\}, \end{aligned} \tag{7.65}$$

so that in particular $\operatorname{ran}\Gamma = \operatorname{dom} E \times B^*(\operatorname{dom} A_0)$. Furthermore,

$$\operatorname{ran}\Gamma_0 = \operatorname{dom} E, \quad \operatorname{ran}\Gamma_1 = B^*(\operatorname{dom} A_0) + \operatorname{ran} E.$$

Proposition 7.55 *Let A_0 and E be, not necessarily closed, symmetric operators in $\mathfrak{H}$ and $\mathcal{H}$, respectively, and let $B \in \mathbf{B}(\mathcal{H}, \mathfrak{H})$. Then Γ in (7.64) is an isometric relation from $(\mathfrak{H}^2, J_{\mathfrak{H}})$ to $(\mathcal{H}^2, J_{\mathcal{H}})$. With $T = \operatorname{dom}\Gamma$ and $S = \ker\Gamma$ one has*

$$T = A_0 \,\widehat{+}\, (\{0\} \times B(\operatorname{dom} E)), \quad S = A_0 \restriction \ker B^*. \tag{7.66}$$

The symmetric extensions $\ker\Gamma_0$ *and* $\ker\Gamma_1$ *of S are given by*

$$\begin{aligned} &\ker\Gamma_0 = A_0, \\ &\ker\Gamma_1 = \{\{f, A_0 f + Bh\} : f \in \operatorname{dom} A_0,\ h \in \operatorname{dom} E,\ B^* f = Eh\}. \end{aligned}$$

The corresponding γ-field and the Weyl family are given by

$$\begin{aligned} \gamma(\lambda) &= -(A_0 - \lambda)^{-1} B \restriction \operatorname{dom} E, \\ M(\lambda) &= B^*(A_0 - \lambda)^{-1} B \restriction \operatorname{dom} E + E. \end{aligned} \tag{7.67}$$

Proof The formulas (7.66) for $T = \operatorname{dom}\Gamma$ and $S = \ker\Gamma$ follow directly from (7.64). Note $\{f, \lambda f\} \in T$ if and only if $Bh = -(A_0 - \lambda) f$ for $h \in \operatorname{dom} E$, so that

$$\mathfrak{N}_\lambda(T) = \operatorname{span}\{-(A_0 - \lambda)^{-1} Bh : h \in \operatorname{dom} E\}.$$

It follows from this result that the corresponding γ-field and the Weyl family are given by (7.67). □

Note that Γ is an operator if and only if $\operatorname{dom} E \cap \ker B = \{0\}$, see (7.65). In particular, if $\ker B = \{0\}$, and $\operatorname{dom} A_0$ and $\operatorname{dom} E$ are dense, then it is clear from (7.65) that $\operatorname{mul}\Gamma = \{0\}$ and $\operatorname{ran}\Gamma$ is dense.

It follows from (7.65) that

$$\operatorname{mul}\Gamma \cap (\{0\} \times \mathcal{H}) = \{0, 0\},$$

and that

$$\operatorname{mul}\Gamma \cap (\mathcal{H} \times \{0\}) = (\ker B \cap \ker E) \times \{0\}.$$

This implies that $M(\lambda)$ is an operator (which already was stated in the proposition) and that $\ker M(\lambda) = \ker B \cap \ker E$; see Lemma 7.52.

Note that $\ker \Gamma_0 = A_0$ is a, not necessarily closed, symmetric extension of S. Moreover, $\ker \Gamma_1 = A_1$ is a symmetric relation with

$$\operatorname{mul} A_1 = \{Bh : h \in \operatorname{dom} E\}.$$

It is clear that the following identities hold:

$$\Gamma_0(\widehat{\mathfrak{N}}_\lambda(T)) = \{h \in \operatorname{dom} E : Bh \in \operatorname{ran}(A_0 - \lambda)\},$$
$$\Gamma_1(\widehat{\mathfrak{N}}_\lambda(T)) = \operatorname{ran}\left(B^*(A_0 - \lambda)^{-1}B{\restriction}\operatorname{dom} E + E\right).$$

Hence, in general,

$$\operatorname{dom}\gamma(\lambda) = \operatorname{dom} M(\lambda) = \Gamma_0(\widehat{\mathfrak{N}}_\lambda(T))$$

is depending on λ. However, if A_0 is maximal symmetric, then $\Gamma_0(\widehat{\mathfrak{N}}_\lambda(T))$ is independent of $\lambda \in \mathbb{C}_+$ or of $\lambda \in \mathbb{C}_-$.

Remark Observe the general fact that $S \subset T^*$, since $\ker \Gamma \subset (\operatorname{dom}\Gamma)^{[\perp]}$, cf. Lemma 7.3. However, the extension T^* of S may be proper. To show this, assume that the symmetric operator A_0 is not densely defined, so that $\operatorname{mul} A_0^*$ is not trivial, and assume that $\operatorname{dom} E$ is dense. Then it follows from (7.66) that

$$S = A_0 \cap (\ker B^* \times \mathfrak{H}), \quad T^* = A_0^* \cap (\ker B^* \times \mathfrak{H}).$$

Thus, if $\{0, \varphi\} \in A_0^*$, then $\{0, \varphi\} \in T^*$, and if $\varphi \neq 0$, then $\{0, \varphi\} \notin S$. Hence T^* is a proper extension of S.

Remark The transform

$$\begin{pmatrix} I & 0 \\ -E & I \end{pmatrix} \begin{pmatrix} h \\ Eh - B^*f \end{pmatrix} = \begin{pmatrix} h \\ -B^*f \end{pmatrix}$$

gives an isometric relation whose γ-field and Weyl function are given by

$$-(A_0 - \lambda)^{-1}B{\restriction}\operatorname{dom} E, \quad B^*(A_0 - \lambda)^{-1}B{\restriction}\operatorname{dom} E.$$

which are bounded but whose domains still depend on λ.

Quasi-boundary triplets The following definition can be seen as a modification of the notion of a boundary triplet of bounded type, where the second condition in Definition 7.17 is weakened; see also Chapter 6.

Definition 7.56 [Behrndt and Langer, 2007] Let S be a closed symmetric relation in a Hilbert space $\mathfrak{H}$ with equal deficiency indices. Then $\{\mathcal{H}, \Gamma_0, \Gamma_1\}$ is said to be a *quasi-boundary triplet* for S^* if Γ_0 and Γ_1 are linear mappings defined on a dense subspace $T = \operatorname{dom}\Gamma$ of S^* with values in $\mathcal{H}$ such that

(Q1) Green's identity (7.60) holds for all $\widehat{f} = \{f, f'\}$, $\widehat{g} = \{g, g'\} \in T$;
(Q2) the range of $\Gamma := \{\Gamma_0, \Gamma_1\}$ is dense in $\mathcal{H}^2$;
(Q3) $A_0 := \ker \Gamma_0$ is a selfadjoint linear relation in $\mathfrak{H}$.

Thus Γ is isometric with dense range by (Q1) and (Q2). The condition (Q3) implies that $A_0 \subset T = \operatorname{dom}\Gamma$. Hence, by Remark 7.7 it follows that $\ker \Gamma = (\operatorname{dom}\Gamma)^{[\perp]} \subset A_0 \subset \operatorname{dom}\Gamma$, which implies that

$$S = \ker \Gamma = T^*.$$

Clearly, the closure of a quasi-boundary triplet is again a quasi-boundary triplet with the same kernel. An application of Corollary 7.5 shows that for a quasi-boundary triplet the following statements are equivalent:

(i) $\operatorname{dom}\Gamma = S^*$;
(ii) $\operatorname{ran}\Gamma = \mathcal{H}^2$;
(iii) Γ is a bounded unitary operator,

cf. [Behrndt and Langer, 2007, Theorem 2.3]. It is possible to consider the above also in the context of isometric relations, but this will not be pursued here.

Remark Assume that in Proposition 7.55 $\ker B = \{0\}$, A_0 is selfadjoint, and that $\operatorname{dom} E$ is dense. Then Γ is a quasi-boundary triplet. The corresponding γ-field and the Weyl family are defined on $\operatorname{dom} M(\lambda) = \operatorname{dom} E$. In addition, $M(\lambda) - E = B^*(A_0-\lambda)^{-1}B$ is bounded on the dense domain $\operatorname{dom} E$. In fact, all quasi-boundary triplets have a similar behavior.

The following result gives a complete description for the class of quasi-boundary triplets, describes their closures, and expresses their connection to boundary triplets of bounded type via simple (isometric) triangular transformations on the boundary space $\mathcal{H} \times \mathcal{H}$.

Theorem 7.57 *Let $\{\mathcal{H}, \Gamma_0, \Gamma_1\}$ be a boundary triplet of bounded type for S^* and let E be a symmetric densely defined operator in $\mathcal{H}$. Then the transform*

$$\begin{pmatrix} \widetilde{\Gamma}_0 \\ \widetilde{\Gamma}_1 \end{pmatrix} = \begin{pmatrix} I & 0 \\ E & I \end{pmatrix} \begin{pmatrix} \Gamma_0 \\ \Gamma_1 \end{pmatrix} \tag{7.68}$$

is a quasi-boundary triplet for S^*. Furthermore, $\widetilde{\Gamma} := \{\widetilde{\Gamma}_0, \widetilde{\Gamma}_1\}$ in (7.68) *is closed if and only if E is a closed symmetric operator in $\mathcal{H}$, in particular, the closure of $\widetilde{\Gamma}$ is given by* (7.68) *with E replaced by its closure E^{**}.*

Conversely, if $\widetilde{\Gamma}$ is a quasi-boundary triplet for S^* then there exists *a boundary triplet of bounded type* $\Gamma = \{\Gamma_0, \Gamma_1\}$ *and a densely defined symmetric operator* E *in* $\mathcal{H}$ *such that* $\widetilde{\Gamma}$ *is given by* (7.68).

Proof (⇒) The mapping $\widetilde{\Gamma}$ is isometric, since the block triangular transformation acting on $\mathcal{H}\times\mathcal{H}$ is an isometric operator precisely when E is a symmetric operator. It is clear from (7.68) that $A_0 := \ker\Gamma_0 \subset \ker\widetilde{\Gamma}_0$, which by symmetry of $\ker\widetilde{\Gamma}_0$ implies that $\ker\widetilde{\Gamma}_0 = A_0$. Moreover, $\operatorname{ran}\widetilde{\Gamma}$ is dense in $\mathcal{H}\times\mathcal{H}$, since $\operatorname{ran}\Gamma_0 = \mathcal{H}$ and $\operatorname{ran}\Gamma$ is dense in $\mathcal{H}\times\mathcal{H}$ as a single-valued unitary relation. Now, it follows from Remark 7.7 that $S = \ker\widetilde{\Gamma} = (\operatorname{dom}\widetilde{\Gamma})^*$, i.e., $\operatorname{dom}\widetilde{\Gamma}$ is dense in S^* and thus $\widetilde{\Gamma}$ is a quasi-boundary triplet for S^*.

By Theorem 7.24 there exist operators K and $R = R^*$ in $\boldsymbol{B}(\mathcal{H})$ with $\ker K = \ker K^* = \{0\}$, such that Γ can be obtained as the transform (7.25) of an ordinary boundary triplet $\Gamma^0 = \{\Gamma^0_0, \Gamma^0_1\}$. Thus,

$$\begin{pmatrix}\widetilde{\Gamma}_0\\ \widetilde{\Gamma}_1\end{pmatrix} = \begin{pmatrix} K^{-1} & 0\\ (E+R)K^{-1} & K^*\end{pmatrix}\begin{pmatrix}\Gamma^0_0\\ \Gamma^0_1\end{pmatrix}. \tag{7.69}$$

Thus $\widetilde{\Gamma}$ is closed precisely when the block operator in (7.69) is closed. Hence, since K^* is bounded, $\widetilde{\Gamma}$ is closed precisely when the column operator

$$\begin{pmatrix} K^{-1}\\ (E+R)K^{-1}\end{pmatrix} \tag{7.70}$$

is closed. The domain of this operator is $\mathcal{L}_0 := K(\operatorname{dom} E)$, which is dense in $\mathcal{H}$ since K is bounded; note that $\operatorname{dom}\widetilde{\Gamma} = (\Gamma^0)^{-1}(\mathcal{L}_0\times\mathcal{H})$, see (7.69). Next observe that

$$\operatorname{clos}\begin{pmatrix} K^{-1}\\ (E+R)K^{-1}\end{pmatrix} = \begin{pmatrix} K^{-1}\\ (\operatorname{clos}E+R)K^{-1}\end{pmatrix}. \tag{7.71}$$

Indeed, let $f_n \in \mathcal{L}_0$ be such that

$$f_n \to f, \quad K^{-1}f_n \to g_0, \quad (E+R)K^{-1}f_n \to g_1,$$

then clearly $\{f, g_0\} \in K^{-1}$ and $\{g_0, g_1\} \in \operatorname{clos}(E+R) = \operatorname{clos}(E) + R$. This proves the inclusion $\subset$ in (7.71). Conversely, if $K^{-1}f = g_0$ and $\{g_0, g_1\} \in \operatorname{clos}(E) + R$, then there is a sequence $g_n \in \operatorname{dom} E$ such that

$$g_n \to g_0, \quad (E+R)g_n \to g_1.$$

Then $f_n := Kg_n \in \mathcal{L}_0$ and $f_n \to Kg_0 = f$, since K is bounded. Hence, $K^{-1}f_n = g_n \to g_0$ and $(E+R)K^{-1}f_n = (E+R)g_n \to g_1$, which proves the reverse inclusion $\supset$ in (7.71). In particular (7.71) shows that the column operator in (7.70) is closed if and only if E is closed. In addition, note that E is closable as a densely defined symmetric operator.

($\Leftarrow$) Let $\widetilde{\Gamma}$ be a quasi-boundary triplet. Since $A_0 := \ker \widetilde{\Gamma}_0$ is self-adjoint, it follows from Proposition 7.54 and the remarks following it, that the γ-field is bounded with $\operatorname{dom}\gamma(\lambda) = \operatorname{ran}\Gamma_0$ for all $\lambda \in \mathbb{C}\setminus\mathbb{R}$. Now it follows from (7.62) and Lemma 7.53 that the corresponding Weyl function $\widetilde{M}$ is of the form $\widetilde{M} = E + M$, where M is a bounded Nevanlinna function with the imaginary part $\operatorname{Im}\widetilde{M}(\lambda) = (\operatorname{Im}\lambda)\,\gamma(\lambda)^*\gamma(\lambda)$ and $E\,(=\operatorname{Re}M(\mu))$ is a symmetric densely defined operator in $\mathcal{H}$ (see also [Behrndt and Langer, 2007, Proposition 2.6]). In particular, the closure of M belongs to the class $R^s[\mathcal{H}]$. Now consider the mapping

$$\begin{pmatrix} \widehat{\Gamma}_0 \\ \widehat{\Gamma}_1 \end{pmatrix} := \begin{pmatrix} I & 0 \\ -E & I \end{pmatrix} \begin{pmatrix} \widetilde{\Gamma}_0 \\ \widetilde{\Gamma}_1 \end{pmatrix}. \tag{7.72}$$

Since $\widetilde{M}(\lambda) = \widetilde{\Gamma}(\widehat{\mathfrak{N}}_\lambda(\widetilde{\mathcal{D}})) \subset \operatorname{ran}\widetilde{\Gamma}$, where $\widetilde{\mathcal{D}} := \operatorname{dom}\widetilde{\Gamma}$, and $\operatorname{dom}\widetilde{M}(\lambda) = \operatorname{dom}E$, it follows that the graph of $\widetilde{M}(\lambda)$ belongs to the domain of the block operator

$$\begin{pmatrix} I & 0 \\ E & I \end{pmatrix},$$

i.e., $\widehat{\mathfrak{N}}_\lambda(\widetilde{\mathcal{D}}) \subset \operatorname{dom}\widehat{\Gamma}$ for all $\lambda \in \mathbb{C}\setminus\mathbb{R}$. Moreover,

$$\widehat{\Gamma}(\widehat{\mathfrak{N}}_\lambda(\widetilde{\mathcal{D}})) = -E + \widetilde{M}(\lambda) = M(\lambda) \upharpoonright \operatorname{dom}E \subset \operatorname{ran}\widehat{\Gamma}$$

for all $\lambda \in \mathbb{C}\setminus\mathbb{R}$. Since $\operatorname{clos}M \in R^s[\mathcal{H}]$ this implies that $\operatorname{ran}\widehat{\Gamma}$ is dense in $\mathcal{H}^2$. Clearly, $\ker\widehat{\Gamma}_0 = \ker\widetilde{\Gamma}_0 = A_0$ and since $\widetilde{\mathcal{D}} = \operatorname{dom}\widetilde{\Gamma} = A_0 \widehat{+} \widehat{\mathfrak{N}}_\lambda(\widetilde{\mathcal{D}})$ one concludes that $\operatorname{dom}\widehat{\Gamma} = \operatorname{dom}\widetilde{\Gamma}$ is dense in S^*. Thus, $\widehat{\Gamma}$ is also a quasi-boundary triplet for S^*, in particular, it is closable. Let $\Gamma^{(0)}$ be the closure of $\widehat{\Gamma}$. Then the corresponding Weyl function $M^{(0)}$ is an extension of M and its closure is equal to $\operatorname{clos}M$. Since $\Gamma^{(0)}$ is closed, it must be unitary by Theorem 7.51. In particular, $M^{(0)}$ is also closed, i.e., $M^{(0)} = \operatorname{clos}M \in R^s[\mathcal{H}]$. Thus, $\Gamma^{(0)}$ is a boundary triplet of bounded type for S^*. Finally, in view of (7.72) one has

$$\begin{pmatrix} \widetilde{\Gamma}_0 \\ \widetilde{\Gamma}_1 \end{pmatrix} = \begin{pmatrix} I & 0 \\ E & I \end{pmatrix} \begin{pmatrix} \widehat{\Gamma}_0 \\ \widehat{\Gamma}_1 \end{pmatrix} \subset \begin{pmatrix} I & 0 \\ E & I \end{pmatrix} \Gamma^{(0)} =: \widetilde{\Gamma}^{(0)}; \tag{7.73}$$

note that $\operatorname{ran}\widetilde{\Gamma}_0 = \operatorname{ran}\widehat{\Gamma}_0 = \operatorname{dom}E$. Since $\ker\widetilde{\Gamma}_0^{(0)} = A_0$ and $\operatorname{dom}\widetilde{\Gamma}_0^{(0)} = \operatorname{dom}\widetilde{\Gamma}^{(0)}$, the equality $\operatorname{ran}\widehat{\Gamma}_0 = \operatorname{dom}E$ implies that the inclusion in (7.73) is in fact also an equality. The equality $\widetilde{\Gamma} = \widetilde{\Gamma}^{(0)}$ can be also seen from $\ker \widetilde{\Gamma}_0 = \ker \widetilde{\Gamma}_0^{(0)}$ and the equality of the corresponding Weyl functions: $\widetilde{M}^{(0)} = T + \operatorname{clos} M = E + M = \widetilde{M}$. □

Corollary 7.58 *The class of quasi-boundary triplets coincides with the class of isometric boundary triplets whose Weyl function is of the form*

$$\widetilde{M}(\lambda) = E + M(\lambda), \tag{7.74}$$

with E a symmetric densely defined operator in $\mathcal{H}$ and $M \in R^s[\mathcal{H}]$.

Proof The Weyl function $\widetilde{M}$ of the quasi-boundary triplet $\{\mathcal{H}, \widetilde{\Gamma}_0, \widetilde{\Gamma}_1\}$ and the Weyl function M of the boundary triplet of bounded type $\{\mathcal{H}, \Gamma_0, \Gamma_1\}$ in (7.68) are connected by (7.74) where M is bounded Nevanlinna function belonging to the class $R^s[\mathcal{H}]$. The class $R^s[\mathcal{H}]$ characterizes the class boundary triplets of bounded type. □

In particular $\operatorname{Im} \widetilde{M}(\lambda)$ (or the γ-field $\gamma(\lambda)$) is bounded, see [Behrndt and Langer, 2007, Proposition 2.6]. Since M is bounded, the function $\widetilde{M}$ is closed if and only if E is closed, which according to Theorem 7.57 is equivalent for $\widetilde{\Gamma}$ itself to be closed.

Corollary 7.59 *Let $\{\mathcal{H}, \widetilde{\Gamma}_0, \widetilde{\Gamma}_1\}$ be a quasi-boundary triplet for S^* and let E be a symmetric densely defined operator in $\mathcal{H}$ as in* (7.68). *Then:*

- (i) *$\{\mathcal{H}, \widetilde{\Gamma}_0, \widetilde{\Gamma}_1\}$ is a unitary boundary triplet for S^* if and only if the operator E is selfadjoint;*
- (ii) *$\{\mathcal{H}, \widetilde{\Gamma}_0, \widetilde{\Gamma}_1\}$ has an extension to a unitary boundary triplet for S^* if and only if the operator E has equal deficiency indices and in this case the formula* (7.68) *defines a unitary extension of $\widetilde{\Gamma}$ when E is replaced by some selfadjoint extension E_0 of E;*
- (iii) *$\{\mathcal{H}, \widetilde{\Gamma}_0, \widetilde{\Gamma}_1\}$ is a boundary triplet of bounded type for S^* if and only if the operator E is bounded and selfadjoint.*

Proof (i) First note that since $E = E^*$ is closed, $\widetilde{\Gamma}$ is closed by Theorem 7.57. Now the statement follows from Theorem 7.51 (or also from [Derkach et al., 2006, Proposition 3.6]).

(ii) This is clear from part (i) and Theorem 7.57.

(iii) This is obtained by combining part (i) with the equalities $\operatorname{ran} \widetilde{\Gamma}_0 = \operatorname{dom} \widetilde{M} = \mathcal{H}$. □

References

Achieser, N.I., and Glasmann, I.M. 1981. *Theorie der linearen Operatoren im Hilbertraum*, 8th edition, Akademie Verlag, Berlin.

Albeverio, S., Gesztesy, F., Hoegh-Krohn, R., and Holden, H. 2005. *Solvable Models in Quantum Mechanics*, Second Edition, AMS Chelsea Publ.

Albeverio, S., Kostenko, A.S., and Malamud, M.M. 2010. Spectral theory of semibounded Sturm-Liouville operators with local interactions on a discrete set. *J. Math. Physics*, **51**, 1–24.

Albeverio, S., and Kurasov, P. 2000. *Singular Perturbations of Differential Operators and Schrödinger Type Operators*, Cambridge University Press.

Amrein, W.O., and Pearson, D.B. 2004. M-operator: a generalisation of Weyl-Titchmarsh theory. *J. Comp. Appl. Math.*, **171**, 1–26.

Arens, R. 1961. Operational calculus of linear relations. *Pacific J. Math.*, **11**, 9–23.

Arlinskii, Y., Belyi, S., and Tsekanovskii, E. 2011. *Conservative realizations of Herglotz-Nevanlinna functions*, Birkhäuser Verlag.

Ashbaugh, M.S., Gesztesy, F., Mitrea, M., Shterenberg, R., and Teschl, G. 2010. The Krein-von Neumann extension and its connection to an abstract buckling problem. *Math. Nachr.*, **210**, 165–179.

Azizov, T. Ya., and Iokhvidov, I.S. 1989. *Linear operators in spaces with indefinite metric*, John Wiley and Sons, New York.

Bade, W.G., and Freeman, R.S. 1962. Closed extensions of the Laplace operator by a general class of boundary conditions. *Pacific J. Math.*, **12**, 395–410.

Beals, R. 1965. Non-local boundary value problems for elliptic operators. *Amer. J. Math.*, **87**, 315–362.

Behrndt, J., Derkach, V.A., Hassi, S., and de Snoo, H.S.V. 2011. A realization theorem for generalized Nevanlinna pairs. *Operators and Matrices*, **5**, 679–706

Behrndt, J., Hassi, S., and de Snoo, H.S.V. 2009. Boundary relations, unitary colligations, and functional models. *Complex Analysis Operator Theory*, **3**, 57–98.

Behrndt, J., Hassi, S., de Snoo, H.S.V., and Wietsma, H.L. 2010. Monotone convergence theorems for semibounded operators and forms with applications. *Proc. Royal Soc. Edinburgh*, **140A**, 927–951.

Behrndt, J. and Langer, M. 2007 Boundary value problems for elliptic partial differential operators on bounded domains. *J. Functional Analysis*, **243**, 536–565.

Bennewitz, C. 1972. Symmetric relations on a Hilbert space. *Lect. Notes Math.*, **280**, 212–218.

Berezanskii, Yu. M. 1965. *Expansions in eigenfunctions of selfadjoint operators*, Naukova Dumka, Kiev (Russian). English translation in Amer. Math. Soc. Providence, RI, 1968.

Birman, M.Sh. 1956. On the self-adjoint extensions of positive definite operators. *Mat. Sb.*, **38**, 431-450 (Russian).

Bognar, J. 1974. *Indefinite inner product spaces*, Ergebnisse der Mathematik und ihrer Grenzgebiete, **78**, Springer-Verlag, New York-Heidelberg.

Brown, B.M., Grubb, G., and Wood, I. 2009. M-functions for closed extensions of adjoint pairs of operators with applications to elliptic boundary problems. *Math. Nachr.*, **282**, 314–347.

Brown, M., Hinchcliffe, J., Marletta, M., Naboko, S., and Wood, I. 2009. The abstract Titchmarsh-Weyl M-function for adjoint operator pairs and its relation to the spectrum. *Integral Equations Operator Theory*, **63**, 297–320.

Brown, M., Marletta, M., Naboko, S., and Wood, I. 2008. Boundary triplets and M-functions for non-selfadjoint operators, with applications to elliptic PDEs and block operator matrices. *J. London Math. Soc.*, **77**, 700–718.

Bruk, V.M. 1976. On a class of problems with the spectral parameter in the boundary conditions. *Mat. Sb.*, **100**, 210–216.

Bruning, J., Geyler, V., and Pankrashkin, K. 2007. Cantor and band spectra for periodic quantum graphs with magnetic fields. *Commun. Math. Phys.*, **269**, 87–105.

Calkin, J.W. 1939. Abstract symmetric boundary conditions. *Trans. Amer. Math. Soc.*, **45**, 369–442

Coddington, E.A. 1973. Extension theory of formally normal and symmetric subspaces. *Mem. Amer. Math. Soc.*, **134**, 1–80.

Derkach, V.A. 2009. Abstract interpolation problem in Nevanlinna classes. *Oper. Theory Adv. Appl.*, **190**, 197–236.

Derkach, V.A., Hassi, S., Malamud, M.M., and de Snoo, H.S.V. 2000. Generalized resolvents of symmetric operators and admissibility. *Methods Funct. Anal. Topology*, **6**, 24–55.

Derkach, V.A., Hassi, S., Malamud, M.M., and de Snoo, H.S.V. 2004. Boundary relations and their Weyl families. *Doklady Russian Akad. Nauk*, **39**, 151–156.

Derkach, V.A., Hassi, S., Malamud, M.M., and de Snoo, H.S.V. 2006. Boundary relations and Weyl families. *Trans. Amer. Math. Soc.*, **358**, 5351–5400.

Derkach, V.A., Hassi, S., Malamud, M.M., and de Snoo, H.S.V. 2009. Boundary relations and generalized resolvents of symmetric operators. *Russ. J. Math. Phys.*, **16**, 17–60.

Derkach, V.A., Hassi, S., Malamud, M.M., and de Snoo, H.S.V. 2012. Graph perturbations of selfadjoint operators and relations, and boundary triplets.

Derkach, V.A., Hassi, S., and de Snoo, H.S.V. 2003. Singular perturbations of selfadjoint operators. *Mathematical Physics, Analysis and Geometry*, **6**, 349–384.

Derkach, V.A., and Malamud, M.M. 1985. Weyl function of Hermitian operator and its connection with characteristic function. *Preprint 85-9 (104) Donetsk Fiz-Techn. Institute AN Ukrain. SSR*, (Russian).

Derkach, V.A., and Malamud, M.M. 1987. On Weyl function and Hermitian operators with gaps. *Doklady Akad. Nauk SSSR*, **293**, 1041–1046.

Derkach, V.A., and Malamud, M.M. 1991. Generalized resolvents and the boundary value problems for hermitian operators with gaps. *J. Funct. Anal.*, **95**, 1–95.

Derkach, V.A., and Malamud, M.M. 1995. The extension theory of hermitian operators and the moment problem. *J. Math. Sciences*, **73**, 141–242.

Dijksma, A., Langer, H., and de Snoo, H.S.V. 1987. Symmetric Sturm-Liouville operator with eigenvalue depending boundary conditions. *Canadian Math. Soc. Conference Proceedings*, **8**, 87–116.

Dijksma, A., Langer, H., and de Snoo, H.S.V. 1988. Hamiltonian systems with eigenvalue depending boundary conditions. *Oper. Theory Adv. Appl.*, **35**, 37–83.

Dijksma, A., and de Snoo, H.S.V. 1974. Self-adjoint extensions of symmetric subspaces. *Pacific J. Math.*, **54**, 71–100.

Exner, P. 2004. Seize ans après. *Appendix K to* "Solvable Models in Quantum Mechanics" by Albeverio S., Gesztesy F., Hoegh-Krohn R., and Holden H., Second Edition, AMS Chelsea Publ.

Gesztesy, F., and Mitrea, M. 2008. Robin-to-Robin maps and Kreĭn type resolvent formulas for Schrödinger operators on bounded Lipschitz domains, arXiv.org:0803.3072, 2008, 1–26.

Gesztesy, F., and Mitrea, M. 2011. A description of selfadjoint extensions of the Laplacian and Krein-type resolvent formulas on non-smooth domains. *J. Math. Anal, Appl.*, **113**, 53–172.

Glazman, I.M. 1950. On the theory of singular differential operators. *Uspekhi Matematicheskikh Nauk*, **5**, 102–135.

Gorbachuk, M.L. 1971. Self-adjoint boundary value problems for differential equation of the second order with unbounded operator coefficient. *Functional Anal. Appl.*, **5**, (1971), 10–21.

Gorbachuk, V.I., and Gorbachuk, M.L. 1984. *Boundary problems for differential operator equations*, Naukova Dumka, Kiev (Russian).

Gorbachuk, M.L., Gorbachuk, V.I., and Kochubei, A.N. 1989. The theory of extensions of symmetric operators, and boundary value problems for differential equations. *Ukrain. Math. Zh.*, **41**, 1299–1313 (Russian).

Graff, A.A. 1946. To the theory of linear differential systems in one-dimensional domain. *Mat. Sb.*, **18**, 305–327

Grubb, G. 1968. A characteriz264 f the non local boundary value problems associated with an elliptic operator. *Ann. Scuola Normale Superiore de Pisa*, **22**, 425–513.

Grubb, G. 2009. *Distributions and Operators*, Vol 552, Graduate Texts in Mathematics, **552**, Springer, New York.

Hassi, S., Malamud, M.M., and Mogilevskiĭ, V. I. 2005. Generalized resolvents and boundary triplets for dual pairs of linear relations. *Meth. Funct. Anal. Topology*, **11**, 170–187.

Hassi, S., Kaltenbäck, M., and de Snoo, H.S.V. 1997. Triplets of Hilbert spaces and Friedrichs extensions associated with the subclass $\mathbf{N}_1$ of Nevanlinna functions. *J. Operator Theory*, **37**, 155–181.

Hassi, S., Kaltenbäck, M., and de Snoo, H.S.V. 1998. Generalized Kreĭn-von Neumann extensions and associated operator models. *Acta Sci. Math. (Szeged)*, **64**, 627–655.

Hassi, S., Langer, H., and de Snoo, H.S.V. 1995. Selfadjoint extensions for a class of symmetric operators with defect numbers (1,1). *15th Oper. Theory Conf. Proc.*, 115–145.

Hassi, S., and de Snoo, H.S.V. 1997. One-dimensional graph perturbations of selfadjoint relations. *Ann. Acad. Sci. Fenn. Ser. A I Math.*, **22**, 123–164.

Hassi, S., de Snoo, H.S.V., Sterk, A.E., and Winkler, H. 2007. Finite-dimensional graph perturbations of selfadjoint Sturm-Liouville operators, in: *Operator Theory, Structured Matrices, and Dilations* (Tiberiu Constantinescu Memorial Volume), Theta Series in Advanced Mathematics, 205–228.

Hassi, S., de Snoo, H.S.V., and Szafraniec, F. H. 2012. Infinite-dimensional perturbations, maximally nondensely defined symmetric operators, and some matrix representations.

Ismagilov, R.S., and Kostjuchenko, A.G. 2010. Asymptotics of the spectrum of Sturm-Liouville operator with point interaction. *Funct. Analysis Appl.*, **44**, 14–20.

Kaç, I.S. 1963. Spectral multiplicity of a second-order differential operator and expansion in eigenfunctions. *Izv. Akad Nauk. SSSR Ser. Mat*, **27**,1081–1112.

Kochubei, A.N. 1975. On extentions of symmetric operators and symmetric binary relations. *Matem. Zametki*, **17**, 41–48.

Kochubei, A.N. 1979. Symmetric operators and nonclassical spectral problems. *Math. Notes*, **25**, 425–434.

Kochubei, A.N. 1989. One-dimensional point interactions, *Ukrain. Math.J.*, **41**, 1391–1395.

Kostenko, A.S., and Malamud, M.M. 2010. 1–D Schrödinger operators with local point interactions on a discrete set. *J. Differential Equations*, **249**, 253–304.

Kreĭn, M.G. 1944. On hermitian operators with defect indices (1, 1). *Dokl. Akad. Nauk SSSR*, **43**, 339–342.

Kreĭn, M.G. 1946. On resolvents of Hermitian operator with deficiency index (m, m). *Dokl. Akad. Nauk SSSR*, **52**, 657–660.

Kreĭn, M.G. 1947. Theory of self-adjoint extensions of semibounded hermitian operators and applications, II. *Mat. Sb.*, **21**, 365–404.

Kreĭn, M.G., and Langer, H. 1971. On defect subspaces and generalized resolvents of Hermitian operator in Pontryagin space. *Funkts. Anal. i Prilozhen*, **5**, 59–71; ibid. **5**, 54–69 (Russian). English translation in *Funct. Anal. Appl.*, **5**, (1971), 136–146; ibid. **5**, (1971), 217–228.

Kreĭn, M. G., and Ovcarenko, I. E. 1977. Q-functions and sc-resolvents of non-densely defined Hermitian contractions (Russian). *Sibirsk. Mat. Zh.*, **18**, 1032–1056, 1206.

Kreĭn, M. G., and Ovcarenko, I. E. 1978. Inverse problems for Q-functions and resolvent matrices of positive Hermitian operators (Russian). *Dokl. Akad. Nauk SSSR*, **242**, 521–524.

Langer, H., and Textorius, B. 1977. On generalized resolvents and Q-functions of symmetric linear relations (subspaces) in Hilbert space. *Pacific J. Math.*, **72**, 135–165.

Lions, J.L., and E. Magenes, E. 1972. *Non-homogeneous boundary value problems and applications, Vol. 1*, Springer, Berlin, 1972

Lyantse, V.E., and Storozh, O.G. 1983. *Methods of the Theory of Unbounded Operators*, Nauk Dumka, Kiev (Russian).

Malamud, M.M. 1992. On the formula of generalized resolvents of a nondensely defined Hermitian operator. *Ukr. Mat. Zh.*, **44**, 1658–1688.

Malamud, M.M. 2010. Spectral theory of elliptic operators in exterior domains. *Russ. J. Math. Phys.*, **17**, 96–125

Malamud, M.M., and Mogilevskiĭ, V.I. 2002. Krein type formula for canonical resolvents of dual pairs of linear relations. *Meth. Funct. Anal. Topology*, **8**, 72–100.

Malamud, M.M., and Neidhardt, H. 2011. On the unitary equivalence of absolutely continuous parts of self-adjoint extensions. *J. Funct. Anal.*, **260**, 613–638.

Mikhailets, V.A. 1994. One-dimensional Schrödinger operator with point interactions. *Dokl. Math.*, **335**, 421–423.

Mirzoev, K. A., and Safonova, T. A. 2011 Singular Sturm-Liouville operators with potential distribution in space of vector functions, *Dokl Russian Academy.*, **441**, 165–168.

Mogilevskiĭ, V.I. 2006. Boundary triplets and Kreĭn type resolvent formula for symmetric operators with unequal defect numbers. *Meth. Funct. Anal. Topology*, **12**, 258–280.

Mogilevskiĭ, V.I. 2009. Boundary triplets and Titchmarsh-Weyl functions of differential operators with arbitrary deficiency indices. *Meth. Funct. Anal. Topology*, **15**, 280–300.

Mogilevskiĭ, V.I. 2010. Description of generalized resolvents and characteristic matrices of differential operators by means of a boundary parameter. *Mat. Zametki*, **90**, 558–583.

Naĭmark, M.A. 1943. On spectral functions of a symmetric operator. *Izv. Akad. Nauk SSSR, Ser. Matem.*, **7**, 285–296.

Naĭmark, M.A. 1969. *Linear Differential Operators*, Nauka, Moscow (Russian). English translation by F. Ungar Pub. Co., New York, 1967.

von Neumann, J. 1932. Über adjungierte Operatoren. *Ann. Math.*, **33**, 294–310.

Phillips, R.S. 1959. Dissipative operators and hyperbolic systems of partial differential equations. *Trans. Amer. Math. Soc.*, **90**, 193–254.

Phillips, R.S. 1961. The extension of dual subspaces invariant under an algebra. *Proc. Internat. Symp. Linear Spaces, Jerusalem 1960*, Academic Press, 363–398.

Posilicano, A. 2001. A Krein-like formula for singular perturbations of selfadjoint operators and applications. *J. Funct. Anal.*, **183**, 109–147.

Posilicano, A. 2008. Self-adjoint extensions of restrictions. *Oper. Matrices*, **2**, 483–506.

Rofe-Beketov, F.S. 1969. On selfadjoint extensions of differential operators in a space of vector-functions. *Teor. Funkts., Funkts. Anal. i Prilozhen*, **8**, 3–24.

Ryzhov, V. 2009. Spectral boundary value problems and their linear operators. arXiv:0904.0276v1, (2009), 1–38.

Saakyan, Sh. N. 1965. Theory of resolvents of a symmetric operator with infinite defect numbers. *Akad. Nauk Armjan. SSR Dokl.*, **41**, 193–198 (Russian).

Shmuljan, Yu.L. 1976. Theory of linear relations, and spaces with indefinite metric. *Funkcional. Anal. i Priložen*, **10**, 67–72 (Russian).

Sorjonen, P. 1980. Extensions of isometric and symmetric linear relations in a Krein space. *Ann. Acad. Sci. Fenn. Ser. A I Math.*, **5**, 355–375.

Stone, M.H. 1932. Linear transformations in Hilbert space and their applications to analysis. *Amer. Math. Soc. Colloquium Publ.*, **15**, New York.

Štrauss, A.V. 1954. Generalized resolvents of symmetric operators. *Izv. Akad. Nauk. SSSR, Ser. Mat.*, **18**, 51-86 (Russian). English translation in *Math. USSR-Izvestija*, **4**, (1970), 179–208.

Triebel, H. 1978. *Interpolation Theory, Function Spaces, Differential Operators*, Berlin.

Višik, M. I. 1952. On general boundary problems for elliptic differential equations. *Trudy Moskov. Mat. Obšc.*, **1**, 187–246 (Russian).

8

Extension theory for elliptic partial differential operators with pseudodifferential methods

Gerd Grubb

Abstract This is a short survey on the connection between general extension theories and the study of realizations of elliptic operators A on smooth domains in $\mathbb{R}^n$, $n \geq 2$. The theory of pseudodifferential boundary problems has turned out to be very useful here, not only as a formulational framework, but also for the solution of specific questions. We recall some elements of that theory, and show its application in several cases (including new results), namely to the lower boundedness question, and the question of spectral asymptotics for differences between resolvents.

8.1 Introduction

The general theory of extensions of a symmetric operator (or a dual pair of operators) in a Hilbert space, originating in the mid-1900's, has been applied in numerous works to ordinary differential equations (ODE), and also in a (smaller) number of works to partial differential equations (PDE).

There is a marked difference between the two cases: In ODE, the playground for boundary conditions is usually finite-dimensional vector spaces, where linear conditions can be expressed by the help of matrices. Moreover, the domains of differential operators defined by closure in L_2-based Hilbert spaces can usually all be expressed in terms of functions with the relevant number of absolutely continuous derivatives.

In contrast, boundary conditions for PDE (in space dimensions $n \geq 2$) are prescribed on infinite-dimensional vector spaces. Moreover, the domains of differential operators in L_2-based spaces will contain functions with distribution derivatives, not continuous and possibly highly irregular.

Whereas extensions of ODEs can usually be described in terms of matrices, the tools to interpret extensions in PDE cases are therefore much more complicated. We shall give a survey of some tools developed through the years, and their applications, emphasizing the use of pseudodifferential operators.

Outline In Section 8.2 we recall the basic issues of elliptic boundary value problems. Pseudodifferential operators (ψdo's) are introduced in Section 8.3, and Section 8.4 introduces pseudodifferential boundary operators (ψdbo's). In Section 8.5 we recall the elements of a general abstract extension theory, and in Section 8.6 we show how this is implemented for realizations $\widetilde{A}$ of an elliptic operator A on a domain $\Omega \subset \mathbb{R}^n$. Section 8.7 focuses on the resolvent formulas that can be obtained via the general theory. In the last sections we go through several cases where pseudodifferential techniques have proved extremely useful (some of the results here are quite recent): In Section 8.8 it is the question of whether lower boundedness holds simultaneously for a realization $\widetilde{A}$ and the operator L over the boundary that enters in the corresponding boundary condition; the new results deal with unbounded domains. In Section 8.9 it is the question of showing Weyl-type spectral asymptotics formulas for differences between resolvents. The results there go back to the early theory, and Section 8.10 presents some additional new results.

8.2 Elliptic boundary value problems

In the following we use the customary multi-index notation for differential operators: $\partial = \partial_x = (\partial_1, \dots, \partial_n)$, $\partial_j = \partial_{x_j} = \partial/\partial x_j$, and $D = D_x = (D_1, \dots, D_n)$, $D_j = D_{x_j} = -i\partial/\partial x_j$; then $\partial^\alpha = \partial_1^{\alpha_1} \cdots \partial_n^{\alpha_n}$, $D^\alpha = D_1^{\alpha_1} \cdots D_n^{\alpha_n}$, for $\alpha \in \mathbb{N}_0^n$; here $|\alpha| = \alpha_1 + \cdots + \alpha_n$.

A differential operator of order $m > 0$,

$$A = \sum_{|\alpha| \le m} a_\alpha(x) D^\alpha$$

is said to be *elliptic*, resp. *strongly elliptic*, on a set $U \subset \mathbb{R}^n$, when the principal symbol

$$a_m(x, \xi) = \sum_{|\alpha| = m} a_\alpha(x) \xi^\alpha$$

satisfies

$$a_m(x, \xi) \ne 0, \text{ resp. } \operatorname{Re} a_m(x, \xi) > 0,$$

for $x \in U$, $\xi \in \mathbb{R}^n \setminus \{0\}$. As a basic example, the Laplacian $\Delta = \partial_1^2 + \cdots + \partial_n^2$ has principal symbol (and symbol) equal to $-|\xi|^2$, so it is elliptic, and $-\Delta$ is strongly elliptic. The Laplacian has been studied for several hundred years, and the problems around it solved by explicit solution formulas. It is the cases with variable (x-dependent) coefficients, and domains more general than simple geometric figures, that have been a challenge in more modern times.

The problem

$$Au = f \tag{8.1}$$

for a given function f on a subset Ω of $\mathbb{R}^n$ usually has infinitely many solutions. To get a problem with unique solvability, we must adjoin extra conditions such as suitable boundary conditions. We can consider A with a domain consisting of the functions satisfying the boundary condition, as an operator $\widetilde{A}$ acting between suitable spaces.

Then the question of *existence* of a solution corresponds to the question of whether $\widetilde{A}$ is *surjective*, and the question of *uniqueness* of a solution corresponds to the question of whether $\widetilde{A}$ is *injective*. In this way, the question of solvability of differential equations is turned into a question of properties of specific operators. The operator point of view became particularly fruitful when it was combined with appropriate scales of function spaces, such as the *Sobolev spaces*, [Sobolev, 1950], and with *Distribution theory*, [Schwartz, 1950].

When Ω is a smooth open subset of $\mathbb{R}^n$ with boundary $\partial\Omega = \Sigma$, we refer to the standard L_2-Sobolev spaces, with the following notation: $H^s(\mathbb{R}^n)$ ($s \in \mathbb{R}$) has the norm $\|v\|_s = \|\mathcal{F}^{-1}(\langle\xi\rangle^s \mathcal{F}v)\|_{L_2(\mathbb{R}^n)}$; here $\mathcal{F}$ is the Fourier transform

$$\mathcal{F}\colon u(x) \mapsto (\mathcal{F}u)(\xi) = \hat{u}(\xi) = \int_{\mathbb{R}^n} e^{-ix\cdot\xi} u(x)\,dx,$$

and $\langle\xi\rangle = (1 + |\xi|^2)^{\frac{1}{2}}$. Next, with r_Ω denoting restriction to Ω,

$$H^s(\Omega) = r_\Omega H^s(\mathbb{R}^n),$$

provided with the norm $\|u\|_s = \inf\{\|v\|_s \mid v \in H^s(\mathbb{R}^n),\, u = r_\Omega v\}$. Moreover,

$$H_0^s(\overline{\Omega}) = \{u \in H^s(\mathbb{R}^n) \mid \operatorname{supp} u \subset \overline{\Omega}\},$$

closed subspace of $H^s(\mathbb{R}^n)$; it identifies with the antidual space of $H^{-s}(\Omega)$ (the space of antilinear, i.e., conjugate linear, functionals), with a duality consistent with the L_2 scalar product. For s equal to a nonnegative integer k, $H_0^k(\overline{\Omega})$ is usually written $H_0^k(\Omega)$. Spaces over the boundary, $H^s(\Sigma)$, are defined by local coordinates from $H^s(\mathbb{R}^{n-1})$, $s \in \mathbb{R}$. (There are many equally justified equivalent choices of norms there; one can choose a particular norm when convenient.) When $s > 0$, there are dense continuous embeddings

$$H^s(\Sigma) \subset L_2(\Sigma) \subset H^{-s}(\Sigma),$$

and there is an identification of $H^{-s}(\Sigma)$ with the antidual space of $H^s(\Sigma)$, such that the duality $(\varphi, \psi)_{-s,s}$ coincides with the $L_2(\Sigma)$-scalar product when the elements lie there. Detailed explanations are found in many books, e.g. [Lions and Magenes, 1968; Hörmander, 1963; Grubb, 2009]. (There is a difference of notation: For $k \in \mathbb{N}$, the spaces denoted $H_0^{k+\frac{1}{2}}(\Omega)$ in [Lions and Magenes, 1968] are not the same as our $H_0^{k+\frac{1}{2}}(\overline{\Omega})$ that are consistent with [Hörmander, 1963]; the latter have the best duality and interpolation properties.)

Consider the case where A is defined on a smooth open subset Ω of $\mathbb{R}^n$ and has coefficients in $C^\infty(\overline{\Omega})$, and assume that A is elliptic on $\overline{\Omega}$. The results in this case are a model for results under weaker smoothness hypotheses. One defines the maximal realization $A_{\max}$ as the operator acting like A in the distribution sense with domain

$$D(A_{\max}) = \{u \in L_2(\Omega) \mid Au \in L_2(\Omega)\};$$

it is a closed, unbounded operator in $L_2(\Omega)$. The minimal realization $A_{\min}$ is defined as the closure of A acting on $C_0^\infty(\Omega)$ (the compactly supported C^∞-functions on Ω). When Ω is bounded, or is unbounded and there are suitable bounds on the coefficients of A,

$$D(A_{\min}) = H_0^m(\Omega).$$

The formal adjoint A' of A is the differential operator acting as follows:

$$A'u = \sum_{|\alpha| \le m} D^\alpha(\overline{a}_\alpha(x)u).$$

By definition, $A'_{\min}$ and $A_{\max}$ are adjoints of one another (as unbounded operators in $L_2(\Omega)$).

The linear operators $\widetilde{A}$ satisfying

$$A_{\min} \subset \widetilde{A} \subset A_{\max}$$

are called *realizations* of A.

Generally, $A_{\max}$ is far from being injective, whereas $A_{\min}$ is far from being surjective, but it may be possible to find realizations $\widetilde{A}$ that are bijective from $D(\widetilde{A})$ to $L_2(\Omega)$.

We see that the theory of distributions (which in this context was preceded by the definition of differential operators acting in the *weak sense*) allows defining operators representing the action of A in a generalized sense. Here invertibility can sometimes be achieved by methods of functional analysis. A fundamental example is the Dirichlet problem (where $\gamma_0 u = u|_\Sigma$)

$$Au = f \text{ in } \Omega, \quad \gamma_0 u = \varphi \text{ on } \Sigma, \tag{8.2}$$

for a strongly elliptic second-order operator having $\operatorname{Re}(Av, v) \geq c\|v\|^2_{L_2(\Omega)}$ with $c > 0$ for $v \in C_0^\infty(\Omega)$. By use of the so-called Lax-Milgram lemma one could define a realization A_γ of A with $D(A_\gamma) \subset H_0^1(\Omega)$, such that $A_\gamma \colon D(A_\gamma) \to L_2(\Omega)$ bijectively. (Details are found in many books, e.g. [Grubb, 2009], Ch. 12.)

But then the question was: How close is A_γ^{-1} to solving the problem in a more classical sense? Second-order derivatives have a meaning on $H^2(\Omega)$, by closure of the definition on $C^2(\overline{\Omega})$, so one can ask:

- If $f \in L_2(\Omega)$, is $u \in H^2(\Omega)$?
- More generally, if $f \in H^k(\Omega)$ for some $k \in \mathbb{N}_0$, is $u \in H^{k+2}(\Omega)$?

The answer was first found for the behavior of u in the interior of Ω: Indeed, when $f \in H^k(\Omega)$, u is in H^{k+2} over subsets of Ω with positive distance from the boundary. This is the so-called *interior regularity*.

There remained the question of *regularity at the boundary*. It was answered positively in [Nirenberg, 1955] and by Ladyzhenskaya (see the account in [Ladyzhenskaya, 1985]). This was followed up by research on higher-order operators A and more general boundary conditions $Tu = \varphi$ (possibly vector valued), where results on interior regularity and regularity at the boundary were established under suitable conditions. Besides ellipticity of the operator A one needs a condition on how the boundary condition fits together with A. Some authors called it the "covering condition" or the "complementing condition", but the name "the Shapiro-Lopatinskiĭ condition" (after [Shapiro, 1953; Lopatinskii, 1953]) has been more generally used. It is also customary to call the system $\{A, T\}$ *elliptic* when it holds (this was suggested in [Hörmander, 1963]; we return to a motivation in Section 8.4). A fundamental reference in this connection is in [Agmon, Douglis and Nirenberg, 1959] that collects and expands the knowledge on elliptic boundary value problems. An important point of view was to obtain so-called "à priori estimates"

(estimates of a Sobolev norm on u by norms on Au and Tu plus a lower order norm on u), shown for smooth functions at first, and extended to the considered solution.

Important monographs exposing the theories and the various authors' own contributions were written by [Agmon, 1965; Lions and Magenes, 1968]; the latter moreover contains valuable information on the surrounding literature. The early theory is exposed in [Courant and Hilbert, 1953, 1962].

In the works at that time, although the striving to show existence of a solution operator was always in the picture, the emphasis was more on showing qualitative properties of the unknown function u in terms of properties of the given data f and φ, regardless of whether u could be described by an operator acting on $\{f, \varphi\}$ or not.

Direct machinery to construct approximate solution operators in general came into the picture with the advent of pseudodifferential methods.

8.3 Pseudodifferential operators

One of the few cases where an elliptic differential operator has an explicit solution operator is the case of $I - \Delta$ on $\mathbb{R}^n$, whose action can be described by $(1-\Delta)u = \mathcal{F}^{-1}\big((1+|\xi|^2)\mathcal{F}u\big)$, where $\mathcal{F}$ is the Fourier transform, and whose solution operator is

$$\operatorname{Op}\Big(\frac{1}{1+|\xi|^2}\Big)u = \mathcal{F}^{-1}\Big(\frac{1}{1+|\xi|^2}\mathcal{F}u\Big).$$

A variable-coefficient elliptic differential operator on $\mathbb{R}^n$ can also be described by the help of the Fourier transform,

$$\begin{aligned} Au &= \sum_{|\alpha|\le m} a_\alpha(x) D^\alpha u = \sum_{|\alpha|\le m} a_\alpha(x)\mathcal{F}^{-1}(\xi^\alpha \mathcal{F}u) \\ &= \mathcal{F}^{-1} a(x,\xi)\mathcal{F}u = \operatorname{Op}(a(x,\xi))u, \end{aligned}$$

where $a(x,\xi) = \sum_{|\alpha|\le m} a_\alpha(x)\xi^\alpha$ is the symbol. But even when the symbol satisfies $a(x,\xi) \neq 0$ for all x, ξ, the operator

$$\operatorname{Op}(a(x,\xi)^{-1}) = \mathcal{F}^{-1} a(x,\xi)^{-1}\mathcal{F}$$

is not an exact inverse. Nevertheless, it is useful in the discussion of solutions, since one can show that

$$\operatorname{Op}(a(x,\xi))\operatorname{Op}(a(x,\xi)^{-1}) = I + \mathcal{R},$$

where the remainder $\mathcal{R}$ is of order -1 (lifts the exponent of a Sobolev space by 1).

A thorough treatment of $\mathrm{Op}(a(x,\xi)^{-1})$ and suitable generalizations that come closer to being an inverse of A (such "almost-inverses" are called parametrices) was obtained with the invention of *pseudo-differential operators*, ψdo's. Some of the initiators were [Mihlin, 1948; Calderón and Zygmund, 1957; Seeley, 1965; Kohn and Nirenberg, 1965; Hörmander, 1965, 1971].

General ψdo's are defined from general symbols $p(x,\xi)$ as

$$\mathrm{Op}(p(x,\xi))u = \mathcal{F}^{-1}p(x,\xi)\mathcal{F}u = (2\pi)^{-n}\int_{\mathbb{R}^n} e^{ix\cdot\xi}p(x,\xi)\hat{u}(\xi)\,d\xi;$$

here $p(x,\xi)$ is required to belong to a suitable class of functions.

Not only do the operators make sense on $\mathbb{R}^n$ where the Fourier transform acts, they are also given a meaning on manifolds, by use of coordinate change formulas and cutoff functions. The theory is not altogether easy; it uses concepts from distribution theory in a refined way. Moreover, it is not exact but qualitative in many statements, so it can be something of a challenge to derive good results from its use. A fine achievement is that it leads to Fredholm operators, when applied to elliptic operators on compact manifolds. Here one is just a small step away from having invertible operators; this can sometimes be achieved by relying on additional knowledge of the situation.

A so-called "classical" ψdo is an operator defined from a symbol that has an asymptotic series expansion in homogeneous terms (a polyhomogeneous symbol):

$$p(x,\xi) \sim \sum_{j\in\mathbb{N}_0} p_{m-j}(x,\xi), \text{ where}$$
$$p_{m-j}(x,t\xi) = t^{m-j}p_{m-j}(x,\xi) \text{ for } |\xi|\geq 1,\ t\geq 1.$$

It is said to be of order m, and $\mathrm{Op}(p)$ maps H^s to H^{s-m} for all $s\in\mathbb{R}$. p_m is called the principal symbol, and p is said to be elliptic when $p_m(x,\xi)\neq 0$ for $|\xi|\geq 1$. Here one has that

$$\mathrm{Op}(p)\,\mathrm{Op}(p') = \mathrm{Op}(pp') + \mathcal{R}_1 = \mathrm{Op}(p_m p'_{m'}) + \mathcal{R}_2,$$

where $\mathcal{R}_1$ and $\mathcal{R}_2$ are of order $m+m'-1$. In this way, the principal part *dominates the behavior*. When p is elliptic, the principal part of a parametrix is found as p_m^{-1} (for $|\xi|\geq 1$, extended smoothly to $\xi\in\mathbb{R}^n$). Also the notation p^0 is used for the principal symbol.

There exist many different symbol classes with generalizations of the above properties, designed for particular purposes.

The very attractive feature of classical ψdo's is that they form a scale of operators of *all integer orders*, including differential operators among those of positive order, and including parametrices and inverses of elliptic differential operators among those of negative order. Moreover, it is an "algebra", in the sense that the elements by composition (and by addition) lead to other classical ψdo's.

The calculus is explained in the original papers and in several subsequent books such as [Treves, 1980; Hörmander, 1985]; a detailed introduction can be also found in [Grubb, 2009, Chapters 7–8].

The ψdo theory gives (after one has done the work to set it up) an easy proof of *interior regularity* of solutions to elliptic problems.

8.4 Pseudodifferential boundary operators

When an elliptic differential operator A is considered on a subset of $\mathbb{R}^n$ or on a manifold with boundary — let us here for simplicity just consider the case of a smooth bounded open subset Ω of $\mathbb{R}^n$ — we must impose boundary conditions to get uniquely solvable problems. Let us assume that we are in a case where the boundary condition $Tu = \varphi$ together with (8.1) gives a uniquely solvable problem; here T is a *trace operator* mapping functions on Ω into M-tuples of functions on $\Sigma = \partial\Omega$. We can formulate this in terms of matrices:

$$\begin{gathered}\begin{pmatrix} A \\ T \end{pmatrix} : C^\infty(\overline{\Omega}) \to \begin{matrix} C^\infty(\overline{\Omega}) \\ \times \\ C^\infty(\Sigma)^M \end{matrix} \\ \text{has an inverse} \\ \begin{pmatrix} R & K \end{pmatrix} : \begin{matrix} C^\infty(\overline{\Omega}) \\ \times \\ C^\infty(\Sigma)^M \end{matrix} \to C^\infty(\overline{\Omega}). \end{gathered} \tag{8.3}$$

Here K is called a *Poisson operator*; it solves the semi-homogeneous problem

$$Av = 0 \text{ in } \Omega, \quad Tv = \varphi \text{ on } \Sigma.$$

The operator R solves the other semi-homogeneous problem

$$Aw = f \text{ in } \Omega, \quad Tw = 0 \text{ on } \Sigma.$$

In a closer analysis of R, we can write it as a sum of two terms:

$$R = Q_+ + G, \tag{8.4}$$

where Q is the ψdo A^{-1} on $\mathbb{R}^n$, Q_+ is its *truncation* r^+Qe^+ to Ω, and G is a supplementing operator adapted to the specific boundary condition, called a *singular Green operator* (s.g.o.). The operator e^+ stands for extension by 0 (to functions on $\mathbb{R}^n$), and the operator r^+ stands for restriction to Ω.

The calculus of pseudodifferential boundary operators (ψdbo's) was initated by Boutet de Monvel [Boutet de Monvel, 1971], who introduced operator systems encompassing both the systems $\binom{A}{T}$ and their solution operators $\begin{pmatrix} R & K \end{pmatrix}$. The original presentation is somewhat brief, and was followed up by extended expositions, in the detailed book [Rempel and Schulze, 1982], which elaborated the index theory, and in the paper [Grubb, 1984a] which completed some proofs of composition rules (with new points of view), and showed spectral asymptotic estimates for singuar Green operators. The book [Grubb, 1996, also 1986 edition] developed a calculus of parameter-dependent ψdbo's, leading to resolvent and heat operator constructions. The recent book [Grubb, 2009] gives a full introduction to the theory.

In the systematic calculus of Boutet de Monvel one considers systems (called Green operators):

$$\mathcal{A} = \begin{pmatrix} P_+ + G & K \\ T & S \end{pmatrix} : \begin{matrix} C^\infty(\overline{\Omega})^N \\ \times \\ C^\infty(\Sigma)^M \end{matrix} \to \begin{matrix} C^\infty(\overline{\Omega})^{N'} \\ \times \\ C^\infty(\Sigma)^{M'} \end{matrix},$$

where

- P is a ψdo on $\mathbb{R}^n$, satisfying the so-called *transmission condition* at Σ (always true for operators stemming from elliptic differential operators);
- $P_+ = r^+Pe^+$ is the truncation to Ω (the transmission condition assures that P_+ maps $C^\infty(\overline{\Omega})$ into $C^\infty(\overline{\Omega})$);
- T is a *trace operator* from Ω to Σ, K is a *Poisson operator* from Σ to Ω, S is a ψdo on Σ;
- G is a *singular Green operator*, e.g. of the type KT.

The composition of two such systems is again a system belonging to the calculus.

The operators extend to act on Sobolev spaces. For T and G there is a condition expressing which differential trace operators $\gamma_j u = (\partial/\partial n)^j u|_\Sigma$

that enter: T or G is said to be of class r when only γ_j's with $j < r$ enter; and then they act on $H^s(\Omega)$ for $s > r - \frac{1}{2}$. The class 0 case is the case where they are purely integral operators, well-defined on $L_2(\Omega)$.

All entries can be matrix-formed. They are defined in local coordinates by formulas involving Fourier transformation and polyhomogeneous symbols. The idea is as follows: In local coordinates at the boundary, where Ω and Σ are replaced by $\mathbb{R}^n_+ = \{x \in \mathbb{R}^n \mid x_n > 0\}$ and $\mathbb{R}^{n-1}$ (with points $x' = (x_1, \dots, x_{n-1})$), the system has for each (x', ξ') a *boundary symbol operator* acting in the x_n-variable:

$$a(x', \xi', D_n) = \begin{pmatrix} p(x', 0, \xi', D_n) + g(x', \xi', D_n) & k(x', \xi', D_n) \\ t(x', \xi', D_n) & s(x', \xi') \end{pmatrix} : \begin{matrix} H^m(\mathbb{R}_+)^N \\ \times \\ \mathbb{C}^M \end{matrix} \to \begin{matrix} L_2(\mathbb{R}_+)^{N'} \\ \times \\ \mathbb{C}^{M'} \end{matrix} \tag{8.5}$$

Here m is the order of the operator. Each entry in a acts in a specific way. E.g., when the matrix is $\binom{A}{T}$ in (8.3), the boundary symbol operator is the *model operator* obtained by freezing the coefficients at x' and replacing derivatives $D^\alpha_{x'}$ by their Fourier transforms $(\xi')^\alpha$ (with respect to $x' \in \mathbb{R}^{n-1}$). The principal boundary symbol operator $a^0(x', \xi', D_n)$ is formed of the top order terms. The principal boundary symbol operator for $\begin{pmatrix} R & K \end{pmatrix}$ is the inverse of the principal boundary symbol operator for $\binom{A}{T}$. (For (8.5), g and t must be of class $\le m$.)

From the boundary symbol operator one defines a full operator by applying the ψdo definition in the x'-variable,

$$\mathrm{Op}'(a(x', \xi', D_n))u = (2\pi)^{1-n} \int e^{ix' \cdot \xi'} a(x', \xi', D_n)(\mathcal{F}_{y' \to \xi'} u(y', x_n))\, d\xi'.$$

The symbols have asymptotic series of terms that are homogeneous in (ξ', ξ_n) (different rules apply to the different ingredients, and we must refer to the mentioned references for further details). One then defines $\mathcal{A}$ to be *elliptic*, when

(a) P is elliptic, i.e. its principal symbol $p^0(x, \xi)$ is invertible at each (x, ξ) with $|\xi| \ge 1$;
(b) the principal boundary symbol operator $a^0(x', \xi', D_n)$ is invertible at each (x', ξ') with $|\xi'| \ge 1$.

For a system $\binom{A}{T}$ formed of an elliptic differential operator A and a differential trace operator T, (b) is precisely the old covering/-complementing/Shapiro-Lopatinskiĭ condition for $\{A, T\}$.

In the elliptic case, one can construct a parametrix $\mathcal{B}^0$ from the inverses of the symbols in (a)–(b), such that $\mathcal{A}\mathcal{B}^0 - I$ and $\mathcal{B}^0\mathcal{A} - I$ have order ≤ -1, and the construction can be refined to give errors of arbitrarily low order. With supplementing information it can be possible to obtain an inverse.

For example, there holds a a solvability theorem for an elliptic differential operator problem as in (8.3), formulated in this framework as follows:

Theorem 8.1 *Let $\Omega \subset \mathbb{R}^n$ be a smooth, bounded open set, denote $\partial\Omega = \Sigma$, and let $A = \sum_{|\alpha| \leq 2m} a_\alpha(x) D^\alpha$ with $a_\alpha \in C^\infty(\overline{\Omega})$ be elliptic on $\overline{\Omega}$, i.e., $\sum_{|\alpha|=2m} a_\alpha(x)\xi^\alpha \neq 0$ for $x \in \overline{\Omega}$, $\xi \in \mathbb{R}^n \setminus \{0\}$. Let $T = (T_j)_{j=1}^m$ be a column vector of trace operators $T_j = \gamma_0 B_j$, where the B_j are differential operators of order m_j with C^∞-coefficients, $0 \leq m_1 < \cdots < m_m \leq 2m - 1$. (Then T is of class $r = m_m + 1 \leq 2m$.) Assume that $\{A, T\}$ is elliptic.*

The operator $\mathcal{A} = \binom{A}{T}$ defines a continuous mapping

$$\mathcal{A} = \begin{pmatrix} A \\ T \end{pmatrix} : H^{2m+s}(\Omega) \to \begin{matrix} H^s(\Omega) \\ \times \\ \prod_{j=1}^m H^{2m+s-m_j-\frac{1}{2}}(\Sigma) \end{matrix}, \quad s > r - 2m - \tfrac{1}{2}, \tag{8.6}$$

and there is a system $\mathcal{B} = (R \quad K)$ (a parametrix) belonging to the calculus and continuous in the opposite direction, such that

$$\mathcal{A}\mathcal{B} = \begin{pmatrix} I & 0 \\ 0 & I \end{pmatrix} + \mathcal{R}_1, \quad \mathcal{B}\mathcal{A} = I + \mathcal{R}_2,$$

$$\mathcal{R}_1 \colon \begin{matrix} H^s(\Omega) \\ \times \\ \prod_{j=1}^m H^{2m+s-m_j-\frac{1}{2}}(\Sigma) \end{matrix} \to \begin{matrix} H^{s'}(\Omega) \\ \times \\ \prod_{j=1}^m H^{2m+s'-m_j-\frac{1}{2}}(\Sigma) \end{matrix}$$

$$\mathcal{R}_2 \colon H^{2m+s}(\Omega) \to H^{2m+s'}(\Omega)$$

for all s as in (8.6), all $s' \geq s$. Here K is a row vector of Poisson operators $(K_j)_{j=1}^m$ of orders $-m_j$, and $R = Q_+ + G$, where Q is a parametrix of A on a neighborhood of $\overline{\Omega}$, and G is a singular Green operator.

The operator $\mathcal{A}$ in (8.6) is Fredholm for each s, with the same finite dimensional kernel and cokernel in C^∞ for all s.

If $\mathcal{A}$ is bijective, the inverse belongs to the calculus (and is of the same form as $\mathcal{B}$).

When $r = 2m$, the lower limit for s is $-\frac{1}{2}$; cases where it is $< -\frac{1}{2}$ occur for example for the Dirichlet problem, where $r = m$, and s can go down to $-m - \frac{1}{2}$. It is useful to know that the Poisson operator K in fact has the mapping property

$$K\colon \prod_{j=1}^{m} H^{2m+s-m_j-\frac{1}{2}}(\Sigma) \to H^{2m+s}(\Omega)$$

for *all* $s \in \mathbb{R}$. The trace operator T is called *normal*, when $\gamma_0 B_j = b_j \gamma_{m_j} + \sum_{k<m_j} B_{jk}\gamma_k$ with an invertible coefficient b_j for each j. (More general normal boundary value problems are described below in Section 8.9.)

For example, for a second-order strongly elliptic operator with a Dirichlet condition, the operator in the theorem maps as follows:

$$\mathcal{A} = \begin{pmatrix} A \\ \gamma_0 \end{pmatrix} : H^{2+s}(\Omega) \to \begin{matrix} H^s(\Omega) \\ \times \\ H^{\frac{3}{2}+s}(\Sigma) \end{matrix} \quad \text{for } s > -\tfrac{3}{2},$$

with parametrices and solution operators continuous in the opposite direction.

Elliptic operators A of odd order occur mainly as square matrix-formed operators, and there is a similar theorem for such cases, where also the B_j can be matrix-formed. Operators of Dirac-type are a prominent first-order example. The matrix case is also interesting for even-order operators. The results can moreover be worked out for operators defined on manifolds, acting in vector bundles. (See e.g. [Grubb, 1974], on the even-order case, for notation and the appropriate definition of normal boundary conditions.)

8.5 Extension theories

We shall now recall some elements of the functional analysis theory of extensions of given operators. This has a long history, with prominent contributions from [von Neumann, 1929; Friedrichs, 1934; Krein, 1947; Vishik, 1952; Birman, 1956], and others. The present author made a number of contributions in [Grubb, 1968, 1970, 1971, 1973, 1974], completing the preceding theories and working out applications to elliptic boundary value problems; further developments are found e.g. in [Grubb, 1983, 1984a], and in recent works.

At the same time there was another, separate development of abstract extension theories, where the operator concept gradually began to be replaced by the concept of *relations*. This development has been aimed primarily towards applications to ODE, however including operator-valued such equations and Schrödinger operators on $\mathbb{R}^n$; keywords in this connection are: boundary triples theory, Weyl-Titchmarsh m-functions and Kreĭn resolvent formulas. Cf. e.g. [Kochubei, 1975; Vainerman, 1980; Lyantze and Storozh, 1983; Gorbachuk and Gorbachuk, 1991; Derkach and Malamud, 1991; Arlinskii, 1999; Malamud and Mogilevski, 2002; Derkach, Hassi, Malamud and de Snoo, 2006; Brüning, Geyler and Pankrashkin, 2008], and their references. In later years there have also been applications to elliptic boundary value problems, cf. e.g. [Amrein and Pearson, 2004; Kopachevski and Krein, 2004; Behrndt and Langer, 2007; Ryzhov, 2007; Brown, Marletta, Naboko and Wood, 2008; Gesztesy and Mitrea, 2008], and references therein. See also Chapters 3, 6 and 7.

The connection between the two lines of extension theories has been clarified in a recent work [Brown, Grubb and Wood, 2009].

At this point we should also mention the recent efforts for problems on nonsmooth domains: [Posilicano and Raimondi, 2009; Grubb, 2008; Gesztesy and Mitrea, 2008, 2011; Abels, Grubb and Wood]; here [Grubb, 2008; Gesztesy and Mitrea, 2011; Abels, Grubb and Wood] use [Grubb, 1968].

In the following, we shall use the notation from [Grubb, 1968, 1970, 1971, 1973, 1974] and [Brown, Grubb and Wood, 2009].

Let there be given a pair $A_{\min}$, $A'_{\min}$ of closed, densely defined operators in a Hilbert space H, such that the following holds:

$$A_{\min} \subset (A'_{\min})^* =: A_{\max}, \quad A'_{\min} \subset (A_{\min})^* =: A'_{\max}.$$

Let $\mathcal{M} = \{\widetilde{A} \mid A_{\min} \subset \widetilde{A} \subset A_{\max}\}$. Write $\widetilde{A}u$ as Au, when $\widetilde{A} \in \mathcal{M}$.

We assume that there is given an operator $A_\gamma \in \mathcal{M}$, the reference operator, with $0 \in \varrho(A_\gamma)$ (the resolvent set); then

$$A_{\min} \subset A_\gamma \subset A_{\max}, \quad A'_{\min} \subset A^*_\gamma \subset A'_{\max}.$$

The case where $A_{\min} = A'_{\min}$ and A_γ is selfadjoint, is called the *symmetric* case.

Let $Z = \ker A_{\max}$, $Z' = \ker A'_{\max}$, and define the basic non-orthogonal decompositions

$$D(A_{\max}) = D(A_\gamma) \dot{+} Z, \quad D(A'_{\max}) = D(A^*_\gamma) \dot{+} Z',$$

denoted $u = u_\gamma + u_\zeta = \operatorname{pr}_\gamma u + \operatorname{pr}_\zeta u$, where $\operatorname{pr}_\gamma = A_\gamma^{-1} A_{\max}$, with a similar notation with primes.

By $\operatorname{pr}_X u = u_X$ we denote the *orthogonal projection* from H to X. The injection $X \hookrightarrow H$ is denoted i_X (it is the adjoint of $\operatorname{pr}_X \colon H \to X$).

There holds an "abstract Green's formula" for $u \in D(A_{\max})$, $v \in D(A'_{\max})$:

$$(Au, v) - (u, A'v) = ((Au)_{Z'}, v_{\zeta'}) - (u_\zeta, (A'v)_Z). \tag{8.7}$$

It can be used to show that when $\widetilde{A} \in \mathcal{M}$, and we define

$$V = \overline{\mathrm{pr}_\zeta D(\widetilde{A})}, \quad W = \overline{\mathrm{pr}_{\zeta'} D(\widetilde{A}^*)},$$

then

$$\{\{u_\zeta, (Au)_W\} \mid u \in D(\widetilde{A})\} \text{ is a graph,}$$

defining an operator T from $D(T) \subset V$ to W.

Theorem 8.2 [Grubb, 1968] *There is a 1–1 correspondence between the closed operators $\widetilde{A} \in \mathcal{M}$ and the closed densely defined operators $T \colon V \to W$, where $V \subset Z$, $W \subset Z'$ (arbitrary closed subspaces), such that $\widetilde{A}$ corresponds to $T \colon V \to W$ if and only if*

$$D(\widetilde{A}) = \{u \in D(A_{\max}) \mid \mathrm{pr}_\zeta u \in D(T),\ (Au)_W = T \,\mathrm{pr}_\zeta u\}. \tag{8.8}$$

In this correspondence, $V = \overline{\mathrm{pr}_\zeta D(\widetilde{A})}$, $W = \overline{\mathrm{pr}_{\zeta'} D(\widetilde{A}^*)}$, and

- *$\widetilde{A}^*$ corresponds analogously to $T^* \colon W \to V$.*
- $\ker \widetilde{A} = \ker T$; $\quad \operatorname{ran} \widetilde{A} = \operatorname{ran} T + (H \ominus W)$.
- *$\widetilde{A}$ is bijective if and only if T is so, and then*

$$\widetilde{A}^{-1} = A_\gamma^{-1} + \mathrm{i}_V T^{-1} \mathrm{pr}_W .$$

One also has

$$D(\widetilde{A}) = \{u = v + A_\gamma^{-1}(Tz + f) + z \mid v \in D(A_{\min}),\ z \in D(T),\ f \in Z \ominus W\},$$

where v, z and f are uniquely determined from u.

The result builds on the works [Krein, 1947; Birman, 1956] (for selfadjoint operators) and [Vishik, 1952], and completes the latter: In Vishik's paper, the $\widetilde{A}$ were set in relation to operators over the nullspaces going in the opposite direction of our T's, and the results were focused on those $\widetilde{A}$'s that have closed range (the so-called normally solvable realizations). Our analysis covered all closed $\widetilde{A}$.

The condition in (8.8)

$$(Au)_W = T \,\mathrm{pr}_\zeta u \tag{8.9}$$

can be viewed as an "abstract boundary condition".

When $\lambda \in \varrho(A_\gamma)$, one can do the same construction for the operators shifted by subtraction of λ. We denote

$$Z_\lambda = \ker(A_{\max} - \lambda), \quad Z'_{\bar\lambda} = \ker(A'_{\max} - \bar\lambda),$$

and have the decompositions (where $\mathrm{pr}^\lambda_\gamma = (A_\gamma - \lambda)^{-1}(A_{\max} - \lambda)$)

$$D(A_{\max}) = D(A_\gamma) \dot{+} Z_\lambda, \quad u = u^\lambda_\gamma + u^\lambda_\zeta = \mathrm{pr}^\lambda_\gamma u + \mathrm{pr}^\lambda_\zeta u,$$

with a similar notation with primes.

Corollary 8.3 *Let $\lambda \in \varrho(A_\gamma)$. For the closed $\widetilde{A} \in \mathcal{M}$, there is a 1–1 correspondence*

$$\widetilde{A} - \lambda \longleftrightarrow \begin{cases} T^\lambda : V_\lambda \to W_{\bar\lambda}, \text{ closed, densely defined} \\ \text{with } V_\lambda \subset Z_\lambda,\ W_{\bar\lambda} \subset Z'_{\bar\lambda}, \text{ closed subspaces.} \end{cases}$$

Here $D(T^\lambda) = \mathrm{pr}^\lambda_\zeta D(\widetilde{A})$, $V_\lambda = \overline{D(T^\lambda)}$, $W_{\bar\lambda} = \overline{\mathrm{pr}^{\bar\lambda}_{\zeta'} D(\widetilde{A}^*)}$, and $D(\widetilde{A})$ consists of the functions $u \in D(A_{\max})$ such that $u^\lambda_\zeta \in D(T^\lambda)$ and

$$T^\lambda u^\lambda_\zeta = ((A - \lambda)u)_{W_{\bar\lambda}}.$$

Moreover,

- $\ker(\widetilde{A} - \lambda) = \ker T^\lambda$; $\quad \mathrm{ran}(\widetilde{A} - \lambda) = \mathrm{ran}\, T^\lambda + (H \ominus W_{\bar\lambda})$.
- *$\widetilde{A} - \lambda$ is bijective if and only if T^λ is so, and when $\lambda \in \varrho(\widetilde{A}) \cap \varrho(A_\gamma)$,*

$$(\widetilde{A} - \lambda)^{-1} = (A_\gamma - \lambda)^{-1} + \mathrm{i}_{V_\lambda}(T^\lambda)^{-1}\, \mathrm{pr}_{W_{\bar\lambda}}.$$

This gives a Kreĭn-type resolvent formula for any closed $\widetilde{A} \in \mathcal{M}$ with $\varrho(\widetilde{A}) \cap \varrho(A_\gamma) \neq \emptyset$.

The relation between T and T^λ was determined in [Grubb, 1974] in the symmetric case, for real λ, and the proof given there extends immediately to the general situation (as shown in [Brown, Grubb and Wood, 2009]): For $\lambda \in \varrho(A_\gamma)$, define

$$E^\lambda = I + \lambda(A_\gamma - \lambda)^{-1}, \text{ it has the inverse } F^\lambda = I - \lambda A_\gamma^{-1},$$

and similarly $E'^{\bar\lambda} = I + \bar\lambda(A^*_\gamma - \bar\lambda)^{-1}$ has the inverse $F'^{\bar\lambda} = I - \bar\lambda(A^*_\gamma)^{-1}$ on H. Then $E^\lambda F^\lambda = F^\lambda E^\lambda = I$, $E'^{\bar\lambda} F'^{\bar\lambda} = F'^{\bar\lambda} E'^{\bar\lambda} = I$ on H.

Moreover, E^λ and $E'^{\bar\lambda}$ restrict to homeomorphisms

$$E^\lambda_V : V \xrightarrow{\sim} V_\lambda, \quad E'^{\bar\lambda}_W : W \xrightarrow{\sim} W_{\bar\lambda},$$

with inverses F^λ_V resp. $F'^{\bar\lambda}_W$. In particular, $D(T^\lambda) = E^\lambda_V D(T)$.

The operator families derived from E^λ are related to what was called a gamma-field in other works from the 1970's and onwards, as a simple special case.

Theorem 8.4 *Let* $G^\lambda_{V,W} = -\operatorname{pr}_W \lambda E^\lambda \mathrm{i}_V$*; then*

$$(E'^{\bar\lambda}_W)^* T^\lambda E^\lambda_V = T + G^\lambda_{V,W}.$$

In other words, T *and* T^λ *are related by the commutative diagram*

$$\begin{array}{ccc} V_\lambda & \xleftarrow[E^\lambda_V]{\sim} & V \\ \Big\downarrow T^\lambda & & \Big\downarrow T+G^\lambda_{V,W} \\ W_{\bar\lambda} & \xrightarrow[(E'^{\bar\lambda}_W)^*]{\sim} & W \end{array} \qquad D(T^\lambda) = E^\lambda_V D(T).$$

In [Brown, Grubb and Wood, 2009] we moreover showed how the study relates to studies of boundary triples and M-functions by other researchers (as referred to in the start of this section; more references are given in [Brown, Grubb and Wood, 2009]):

Let $V = Z$, $W = Z'$, then, with $\mathcal{H} = Z'$, $\mathcal{K} = Z$, the mappings

$$\begin{pmatrix}\Gamma_1 u\\ \Gamma_0 u\end{pmatrix} = \begin{pmatrix}(Au)_{Z'}\\ u_\zeta\end{pmatrix} : D(A_{\max}) \to \mathcal{H}\times\mathcal{K},$$

$$\begin{pmatrix}\Gamma'_1 v\\ \Gamma'_0 v\end{pmatrix} = \begin{pmatrix}(A'v)_{Z}\\ v_{\zeta'}\end{pmatrix} : D(A'_{\max}) \to \mathcal{K}\times\mathcal{H},$$

form a boundary triple: Both mappings $\binom{\Gamma_1}{\Gamma_0}$ and $\binom{\Gamma'_1}{\Gamma'_0}$ are surjective, their kernels are $D(A_{\min})$ resp. $D(A'_{\min})$, and they satisfy the Green's formula

$$(A_{\max}u, v) - (u, A'_{\max}v) = (\Gamma_1 u, \Gamma'_0 v)_{\mathcal{H}} - (\Gamma_0 u, \Gamma'_1 v)_{\mathcal{K}},$$

which is a rewriting of (8.7).

Here one can consider a boundary condition

$$\Gamma_1 u = T\Gamma_0 u, \tag{8.10}$$

where we allow T to be unbounded (closed densely defined) from $\mathcal{K}$ to $\mathcal{H}$; it defines a restriction $\widetilde{A}$ of $A_{\max}$ by $D(\widetilde{A}) = \{u \in D(A_{\max}) \mid \Gamma_0 u \in D(T), \Gamma_1 u = T\Gamma_0 u\}$. Then it is customary to define an M-function as follows:

Definition 8.5 For $\lambda \in \varrho(\widetilde{A})$, $M(\lambda)\colon \operatorname{ran}(\Gamma_1 - T\Gamma_0) \to \mathcal{K}$ is the operator satisfying

$$M(\lambda)(\Gamma_1 x - T\Gamma_0 x) = \Gamma_0 x, \text{ for all } x \in \ker(A_{\max} - \lambda) = Z_\lambda.$$

The analysis in [Brown, Grubb and Wood, 2009] showed that $M(\lambda)$ is a holomorphic family of operators in $\mathcal{L}(\mathcal{H}, \mathcal{K})$. On the other hand, when $\widetilde{A}$ and its boundary condition (8.10) are considered from the point of view of extensions [Grubb, 1968, 1970, 1971, 1973, 1974] recalled further above, $\widetilde{A}$ is the operator corresponding to $T\colon Z \to Z'$ by Theorem 8.2. Then we find moreover, in terms of the λ-dependent families introduced in that context:

$$M(\lambda) = -(T + G^\lambda_{Z,Z'})^{-1} = -F^\lambda_Z (T^\lambda)^{-1}(F'^{\bar\lambda}_{Z'})^*, \text{ when } \lambda \in \varrho(\widetilde{A}) \cap \varrho(A_\gamma).$$

This gives the Kreĭn resolvent formula in the form

$$(\widetilde{A} - \lambda)^{-1} = (A_\gamma - \lambda)^{-1} - \mathrm{i}_{Z_\lambda} E^\lambda_Z M(\lambda)(E'^{\bar\lambda}_{Z'})^* \operatorname{pr}_{Z'_{\bar\lambda}}.$$

For the case of general V and W, we could likewise construct an M-function from W to V for $\lambda \in \varrho(\widetilde{A})$, and establish a Kreĭn resolvent formula around it. The following result is shown in [Brown, Grubb and Wood, 2009]:

Theorem 8.6 *Let $\widetilde{A}$ correspond to $T\colon V \to W$. For $\lambda \in \varrho(\widetilde{A})$ there is a well-defined holomorphic family $M(\lambda) \in \mathcal{L}(W, V)$:*

$$M(\lambda) = \operatorname{pr}_\zeta \big(I - (\widetilde{A} - \lambda)^{-1}(A_{\max} - \lambda)\big) A_\gamma^{-1} \mathrm{i}_W.$$

When $\lambda \in \varrho(\widetilde{A}) \cap \varrho(A_\gamma)$, then

$$M(\lambda) = -(T + G^\lambda_{V,W})^{-1} = -F^\lambda_V (T^\lambda)^{-1}(F'^{\bar\lambda}_W)^*,$$

and

$$(\widetilde{A} - \lambda)^{-1} = (A_\gamma - \lambda)^{-1} - \mathrm{i}_{V_\lambda} E^\lambda_V M(\lambda)(E'^{\bar\lambda}_W)^* \operatorname{pr}_{W_{\bar\lambda}}.$$

To have both T^λ (for $\lambda \in \varrho(A_\gamma)$) and $M(\lambda)$ (for $\lambda \in \varrho(\widetilde{A})$) available is an advantage, since $\ker(\widetilde{A} - \lambda) = \ker T^\lambda$ and $\operatorname{ran}(\widetilde{A} - \lambda) = \operatorname{ran} T^\lambda + (H \ominus W_{\bar\lambda})$ give straightforward eigenvalue information at the poles of $M(\lambda)$ in $\varrho(A_\gamma)$.

Remark The name M-function is consistent with the notation in some papers that [Brown, Grubb and Wood, 2009] refers to, but possibly diverges from others (one could also use the longer name Weyl-Titchmarsh function). There is a recent publication [Malamud, 2010] that exposes related resolvent formulas on the basis of [Malamud and Mogilevski,

2002]. (Let us remark that [Malamud, 2010] gives the impression that only separate surjectiveness of Γ_0 and Γ_1 is assumed in [Brown, Grubb and Wood, 2009]; this is not so.)

8.6 Implementation of the abstract set-up for elliptic operators

We shall now recall how the abstract theory is applied to a concrete choice of elliptic operator A. Here $A_{\max}$ and $A_{\min}$ are defined as in Section 8.2; they are closed operators in $H = L_2(\Omega)$. In [Grubb, 1968] general even-order operators were considered, and the reference operator (called A_γ in Section 8.5) was taken to represent a general normal boundary condition. To simplify our explanation, we shall here just consider a second-order strongly elliptic operator A and let A_γ stand for the Dirichlet realization, mentioned after (8.2). We have by elliptic regularity that $D(A_\gamma) = H^2(\Omega) \cap H_0^1(\Omega)$, and we can assume that (a constant has been added to A such that) the lower bound $m(A_\gamma)$ is positive. The lower bound $m(P)$ of an operator P is defined by

$$m(P) = \inf\{\operatorname{Re}(Pu, u) \mid u \in D(P),\ \|u\| = 1\} \geq -\infty; \tag{8.11}$$

when it is finite, P is said to be lower bounded.

The trace operator γ_0 defines a continuous mapping $H^s(\Omega) \to H^{s-\frac{1}{2}}(\Sigma)$ for $s > -\frac{1}{2}$. We shall also need a more advanced fact, namely that, as shown by Lions and Magenes (see e.g. [Lions and Magenes, 1968]), γ_0 extends to a mapping $D(A_{\max}) \to H^{-\frac{1}{2}}(\Sigma)$, and defines *homeomorphisms*

$$\gamma_Z \colon Z \xrightarrow{\sim} H^{-\frac{1}{2}}(\Sigma), \quad \gamma_{Z'} \colon Z' \xrightarrow{\sim} H^{-\frac{1}{2}}(\Sigma),$$

where Z and Z' are the nullspaces of $A_{\max}$ and $A'_{\max}$ (not contained in $H^s(\Omega)$ for $s > 0$). The inverse of γ_Z is consistent with the Poisson operator K_γ solving the semi-homogeneous Dirichlet problems (8.2) with $f = 0$, in the sense that

$$K_\gamma = \mathrm{i}_Z \gamma_Z^{-1}.$$

Similarly, the inverse $\gamma_{Z'}^{-1}$ is consistent with the Poisson solution operator K'_γ solving the Dirichlet problem for A' with $f = 0$, $K'_\gamma = \mathrm{i}_{Z'} \gamma_{Z'}^{-1}$. Moreover, with λ-dependence,

$$K_\gamma^\lambda = \mathrm{i}_{Z_\lambda} \gamma_{Z_\lambda}^{-1}, \quad K_\gamma'^{\bar\lambda} = \mathrm{i}_{Z'_{\bar\lambda}} \gamma_{Z'_{\bar\lambda}}^{-1},$$

solve the semi-homogeneous Dirichlet problems for $A - \lambda$, $A' - \bar{\lambda}$, when $\lambda \in \varrho(A_\gamma)$.

For a closed subspace $V \subset Z$, let $X = \gamma_0 V \subset H^{-\frac{1}{2}}(\Sigma)$. Here we denote the restriction of γ_0 by γ_V;

$$\gamma_V : V \overset{\sim}{\to} X, \tag{8.12}$$

with a similar notation for $Y = \gamma_0 W$ and λ-dependent cases. The map $\gamma_V : V \overset{\sim}{\to} X$ has the adjoint $\gamma_V^* : X^* \overset{\sim}{\to} V$. Here X^* denotes the antidual space of X, with a duality coinciding with the scalar product in $L_2(\Sigma)$ when applied to elements that come from $L_2(\Sigma)$. The duality is written $(\psi, \varphi)_{X^*,X}$.

We denote

$$K_{\gamma,X} = \mathrm{i}_V \gamma_V^{-1} : X \to V \subset H; \tag{8.13}$$

it is a Poisson operator when $X = H^{-\frac{1}{2}}(\Sigma)$.

Now a given $T: V \to W$ is carried over to a closed, densely defined operator $L: X \to Y^*$ by the definition

$$L = (\gamma_W^{-1})^* T \gamma_V^{-1}, \quad D(L) = \gamma_V D(T);$$

it is expressed in the diagram

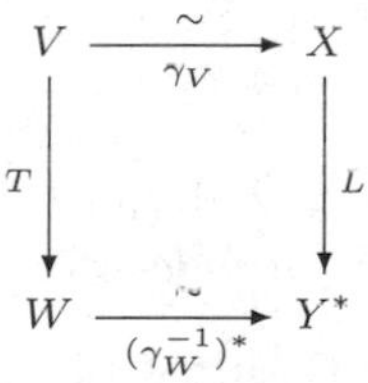

$$\begin{array}{ccc} V & \xrightarrow[\gamma_V]{\sim} & X \\ {\scriptstyle T}\downarrow & & \downarrow{\scriptstyle L} \\ W & \xrightarrow[(\gamma_W^{-1})^*]{\sim} & Y^* \end{array}$$

There is a similar definition in the λ-dependent case.

Before formulating the results in a theorem, we shall interpret the abstract boundary condition (8.9) defining the realization $\widetilde{A}$, as a concrete condition in terms of L.

A has a Green's formula (for sufficiently smooth u, v)

$$(Au, v)_\Omega - (u, A'v)_\Omega = (\nu_1 u, \gamma_0 v)_\Sigma - (\gamma_0 u, \nu_1' v)_\Sigma, \tag{8.14}$$

where

$$\nu_1 = s\gamma_1, \quad \nu_1' = \bar{s}\gamma_1 + \mathcal{A}'\gamma_0,$$

with a nonvanishing smooth function s and a suitable first-order differential operator $\mathcal{A}'$ on Σ.

Let $\lambda \in \varrho(A_\gamma)$. In addition to the Poisson operators K_γ^λ resp. $K_\gamma'^{\bar{\lambda}}$

solving the Dirichlet problems for $A-\lambda$ resp. $A'-\bar\lambda$, we shall need the *Dirichlet-to-Neumann* operators

$$P^{\lambda}_{\gamma_0,\nu_1} = \nu_1 K^{\lambda}_{\gamma}, \quad P'^{\bar\lambda}_{\gamma_0,\nu_1'} = \nu_1' K'^{\bar\lambda}_{\gamma},$$

that map the Dirichlet boundary value into the Neumann boundary value for null-solutions. By the composition rules for ψdbo's, they are pseudodifferential operators of order 1; moreover, it is known that $P^{\lambda}_{\gamma_0,\nu_1}$ is elliptic.

Theorem 8.7 *Define the* **reduced Neumann trace operator** Γ *by*

$$\Gamma = \nu_1 - P^{0}_{\gamma_0,\nu_1}\gamma_0 = \nu_1 A_{\gamma}^{-1} A_{\max} \colon D(A_{\max}) \to H^{\frac12}(\Sigma).$$

It is continuous and surjective, and vanishes on Z. With the analogous definition for A' one has the **reduced Green's formula***:*

$$(Au,v) - (u,A'v) = (\Gamma u, \gamma_0 v)_{\frac12,-\frac12} - (\gamma_0 u, \Gamma' v)_{-\frac12,\frac12},$$

valid for all $u \in D(A_{\max})$, $v \in D(A'_{\max})$. In particular,

$$(Au,w) = (\Gamma u, \gamma_0 w)_{\frac12,-\frac12}, \text{ when } u \in D(A_{\max}), w \in Z'. \tag{8.15}$$

For $\lambda \in \varrho(A_\gamma)$ we similarly define

$$\Gamma^{\lambda} = \nu_1 - P^{\lambda}_{\gamma_0,\nu_1}\gamma_0 = \nu_1 (A_\gamma - \lambda)^{-1}(A_{\max} - \lambda),$$
$$\Gamma'^{\bar\lambda} = \nu_1 - P'^{\bar\lambda}_{\gamma_0,\nu_1'}\gamma_0 = \nu_1' (A_\gamma^* - \bar\lambda)^{-1}(A'_{\max} - \bar\lambda),$$

continuous and surjective from $D(A_{\max})$ resp. $D(A'_{\max})$ to $H^{\frac12}(\Sigma)$; then there holds a reduced Green's formula

$$(Au,v) - (u,A'v) = (\Gamma^{\lambda} u, \gamma_0 v)_{\frac12,-\frac12} - (\gamma_0 u, \Gamma'^{\bar\lambda} v)_{-\frac12,\frac12},$$

for $u \in D(A_{\max})$, $v \in D(A^*_{\max})$.

Here $D(A_{\max})$ is provided with the graph-norm.

Now let $\widetilde A$ be a closed operator lying between $A_{\min}$ and $A_{\max}$, so $\widetilde A \in \mathcal{M}$. The abstract boundary condition (8.9) for $\widetilde A$ may be written:

$$(Au,w) = (Tu_\zeta, w), \text{ all } w \in W. \tag{8.16}$$

The left-hand side equals $(\Gamma u, \gamma_0 w)_{\frac12,-\frac12}$ by (8.15). The right-hand side equals

$$(Tu_\zeta, w) = (T\gamma_V^{-1}\gamma_0 u, \gamma_W^{-1}\gamma_0 w) = (L\gamma_0 u, \gamma_0 w)_{Y^*,Y},$$

by definition of L (it is used that when $u_\zeta \in V$, $u_\zeta = \gamma_V^{-1}\gamma_V u_\zeta = \gamma_V^{-1}\gamma_0 u$).

Hence (8.16) may be rewritten as

$$(\Gamma u, \gamma_0 w)_{\frac{1}{2},-\frac{1}{2}} = (L\gamma_0 u, \gamma_0 w)_{Y^*,Y}, \text{ all } w \in W. \tag{8.17}$$

The injection $\mathrm{i}_Y : Y \to H^{-\frac{1}{2}}(\Sigma)$ has the adjoint $\mathrm{i}_Y^* : H^{\frac{1}{2}}(\Sigma) \to Y^*$ that sends a functional ψ on $H^{-\frac{1}{2}}(\Sigma)$ over into a functional $\mathrm{i}_Y^*\psi$ on Y by:

$$(\mathrm{i}_Y^*\psi, \varphi)_{Y^*,Y} = (\psi, \varphi)_{\frac{1}{2},-\frac{1}{2}} \text{ for all } \varphi \in Y.$$

Then (8.17) may be rewritten as

$$\mathrm{i}_Y^*\Gamma u = L\gamma_0 u,$$

or, when we use that $\Gamma = \nu_1 - P^0_{\gamma_0,\nu_1}\gamma_0$,

$$\mathrm{i}_Y^*\nu_1 u = (L + \mathrm{i}_Y^* P^0_{\gamma_0,\nu_1})\gamma_0 u. \tag{8.18}$$

This is the boundary condition derived from (8.9), when $\widetilde{A}$ corresponds to $T : V \to W$ by Theorem 8.2, carried over to $L : X \to Y^*$ by (8.12).

Then Theorem 8.2 implies:

Theorem 8.8 *There is a 1–1 correspondence between the closed operators $\widetilde{A} \in \mathcal{M}$ and the closed densely defined operators $L : X \to Y^*$, where X and Y are closed subspaces of $H^{-\frac{1}{2}}(\Sigma)$, such that $\widetilde{A}$ corresponds to $L : X \to Y^*$ if and only if $D(\widetilde{A})$ consists of the functions in $D(A_{\max})$ for which*

$$\gamma_0 u \in D(L), \quad \mathrm{i}_Y^*\nu_1 u = (L + \mathrm{i}_Y^* P^0_{\gamma_0,\nu_1})\gamma_0 u. \tag{8.19}$$

In this correspondence, $X = \overline{\gamma_0 D(\widetilde{A})}$, $Y = \overline{\gamma_0 D(\widetilde{A}^*)}$, and

- *$\widetilde{A}^*$ corresponds analogously to $L^* : Y \to X^*$.*
- *$\ker \widetilde{A} = \mathrm{i}_V \gamma_V^{-1} \ker L$; $\quad \operatorname{ran} \widetilde{A} = \gamma_W^* \operatorname{ran} L + (H \ominus W)$, cf. (8.12), (8.13).*
- *$\widetilde{A}$ is bijective if and only if L is so, and then*

$$\widetilde{A}^{-1} = A_\gamma^{-1} + \mathrm{i}_V \gamma_V^{-1} L^{-1} \gamma_W^* \operatorname{pr}_W = A_\gamma^{-1} + K_{\gamma,X} L^{-1} (K'_{\gamma,Y})^*. \tag{8.20}$$

Theorems 8.7 and 8.8 are from [Grubb, 1968], except that we have modified the notation a little.

In [Brown, Grubb and Wood, 2009], the subspace cases are treated with insertion of an *isometry* $\Lambda_{\frac{1}{2}} : L_2(\Sigma) \overset{\sim}{\to} H^{-\frac{1}{2}}(\Sigma)$, that allows replacing X and Y by closed subspaces X_1 and Y_1 of $L_2(\Sigma)$, identified with their dual spaces; then i_Y^* is replaced by an orthogonal projection pr_{Y_1}.

In the case where $Y = H^{-\frac{1}{2}}(\Sigma)$, i.e., $W = Z'$, the map i_Y^* is superfluous, and the second condition in (8.19) takes the form

$$\nu_1 u = (L + P^0_{\gamma_0,\nu_1})\gamma_0 u.$$

When also $X = H^{-\frac{1}{2}}(\Sigma)$, we say that $\widetilde{A}$ represents a *Neumann-type condition*

$$\nu_1 u = C\gamma_0 u; \text{ here } C = L + P^0_{\gamma_0,\nu_1} \text{ on } D(L).$$

In this case, L can act like a pseudodifferential operator, namely when C (in the condition $\nu_1 u = C\gamma_0 u$) is a differential or pseudodifferential operator.

Let us consider a slightly different set-up where C is a *given* first-order differential or pseudodifferential operator on Σ, and we *define* $\widetilde{A}$ by

$$D(\widetilde{A}) = \{u \in D(A_{\max}) \mid \nu_1 u = C\gamma_0 u\}, \tag{8.21}$$

where γ_0 and ν_1 are considered as mappings from $D(A_{\max})$ to $H^{-\frac{1}{2}}(\Sigma)$ resp. $H^{-\frac{3}{2}}(\Sigma)$. We shall discuss the corresponding operator $L\colon X \to Y^*$. Since $\{\gamma_0 u, \nu_1 u\}$ maps $H^2(\Omega)$ onto $H^{\frac{3}{2}}(\Sigma) \times H^{\frac{1}{2}}(\Sigma)$, $D(L) = \gamma_0 D(\widetilde{A}) \supset H^{\frac{3}{2}}(\Sigma)$. Then since $H^{\frac{3}{2}}(\Sigma)$ is dense in $H^{-\frac{1}{2}}(\Sigma)$, $X = H^{-\frac{1}{2}}(\Sigma)$. By use of Green's formula (8.14) it is checked that the adjoint $\widetilde{A}^*$ extends the realization of A' with domain consisting of the functions $v \in H^2(\Omega)$ satisfying

$$\nu_1' v = C^*\gamma_0 v,$$

so also $Y = H^{-\frac{1}{2}}(\Sigma)$. Thus we are in the case of Neumann-type boundary conditions, so by comparison with (8.6), it is seen that L acts like $C - P^0_{\gamma_0,\nu_1}$.

The domain $D(L)$ equals $\{\varphi \in H^{-\frac{1}{2}}(\Sigma) \mid (C - P^0_{\gamma_0,\nu_1})\varphi \in H^{\frac{1}{2}}(\Sigma)\}$; it may not be easy to determine more exactly. Note that L is used as a map from its domain in $H^{-\frac{1}{2}}(\Sigma)$ to $H^{+\frac{1}{2}}(\Sigma)$, although it acts like a ψdo of order 1.

One case is clear, though: If $C - P^0_{\gamma_0,\nu_1}$ is *elliptic* of order 1, then $L\varphi \in H^{\frac{1}{2}}(\Sigma)$ implies $\varphi \in H^{\frac{3}{2}}(\Sigma)$; in this case $D(L) = H^{\frac{3}{2}}(\Sigma)$. Moreover, $D(\widetilde{A}) \subset H^2(\Omega)$. A check of the boundary symbol rules shows that this is precisely the case where the system $\{A, \nu_1 - C\gamma_0\}$ is *elliptic*. Here we have:

Theorem 8.9 *Let C be a first-order differential or pseudodifferential operator on Σ and define the realization $\widetilde{A}$ of A by (8.21). Then if*

$C - P^0_{\gamma_0,\nu_1}$ is elliptic, the operator $L\colon X \to Y^*$ corresponding to $\widetilde{A}$ by Theorem 8.8 *acts like $C - P^0_{\gamma_0,\nu_1}$ and has*

$$X = Y = H^{-\frac{1}{2}}(\Sigma), \quad D(L) = H^{\frac{3}{2}}(\Sigma).$$

Moreover, $D(\widetilde{A}) \subset H^2(\Omega)$. Related statements hold for the adjoint $\widetilde{A}^*$.

A Robin condition $\nu_1 u = b\gamma_0 u$, with a smooth function $b(x)$ on Σ, is elliptic, since L acts like $b - P^0_{\gamma_0,\nu_1}$, where $P^0_{\gamma_0,\nu_1}$ is elliptic of order 1 and b is of order 0.

In the case of Theorem 8.9, when L is bijective, the formula (8.20) has the form

$$\widetilde{A}^{-1} = A_\gamma^{-1} + K_\gamma L^{-1}(K'_\gamma)^*, \tag{8.22}$$

where all ingredients belong to the ψdbo calculus: K_γ is a Poisson operator, L^{-1} is a ψdo on Σ, $(K'_\gamma)^*$ is a trace operator of class 0, and the composition $K_\gamma L^{-1}(K'_\gamma)^*$ is a singular Green operator (of class 0). A_γ^{-1} is the sum $Q_+ + G$ of a truncated ψdo Q on $\mathbb{R}^n$ and a singular Green operator, as in (8.4).

Realizations defined by boundary conditions of the type $\nu_1 u = C\gamma_0 u$ are studied by means of quasi boundary triples techniques for selfadjoint A in Theorem 6.24 ff.; here C is a selfadjoint bounded operator in $L_2(\Sigma)$ and $D(\widetilde{A}) \subset H^{\frac{3}{2}}(\Omega)$.

When A is of order $2m$, there is a Green's formula generalizing (8.14), where γ_0 is replaced by an m vector $\gamma = \{\gamma_0, \dots, \gamma_{m-1}\}$, and ν_1 and ν_1' are replaced by m-vectors of trace operators of orders $m, \dots, 2m-1$, mapping into products of Sobolev spaces of different orders over Σ. One then gets vector versions of the reduced Neumann trace operators Γ and Γ', with matrix-formed versions of the Dirichlet-to-Neumann pseudodifferential operators, but the basic mechanisms in the interpretation are the same. There are interesting cases of subspaces X, Y of the products of Sobolev spaces over Σ, where ellipticity considerations are relevant. Details are given in [Grubb, 1968]–[Grubb, 1974] and [Brown, Grubb and Wood, 2009].

8.7 Resolvent formulas

When $\lambda \in \varrho(A_\gamma)$, there is a similar representation of $\widetilde{A} - \lambda$ in terms of a boundary condition defined from an operator L^λ acting over the boundary. Here it is of particular interest to find the connection between

L and L^λ, just as we found the connection between T and T^λ. It turns out that the relation between L and L^λ is simpler: They both go from X to Y^*, whereas T resp. T^λ map between different spaces due to the shift from Z to Z_λ. This holds, since

$$D(L) = \gamma_0 D(\widetilde{A}) = \gamma_0 D(\widetilde{A} - \lambda) = D(L^\lambda), \quad X = \overline{D(L)} = \overline{D(L^\lambda)},$$

with similar statements for $D(L^*)$, $D((L^\lambda)^*)$ and Y. Then we have:

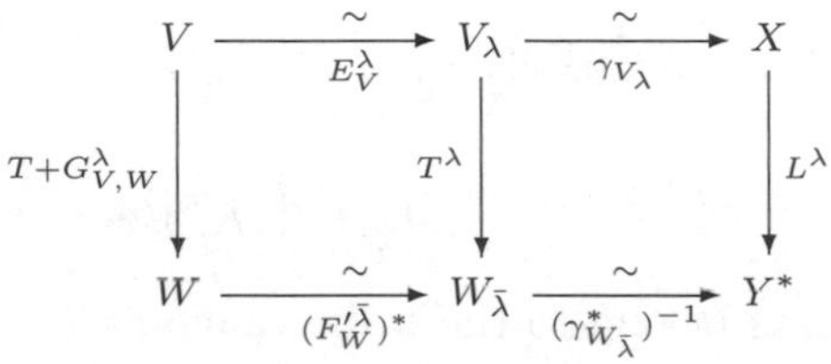

$$\begin{array}{ccccc} V & \xrightarrow[E_V^\lambda]{\sim} & V_\lambda & \xrightarrow[\gamma_{V_\lambda}]{\sim} & X \\ \downarrow {\scriptstyle T+G_{V,W}^\lambda} & & \downarrow {\scriptstyle T^\lambda} & & \downarrow {\scriptstyle L^\lambda} \\ W & \xrightarrow[(F_W^{\prime\bar\lambda})^*]{\sim} & W_{\bar\lambda} & \xrightarrow[(\gamma_{W_{\bar\lambda}}^*)^{-1}]{\sim} & Y^* \end{array}$$

The horizontal maps compose as $\gamma_{V_\lambda} E_V^\lambda = \gamma_V$, $(\gamma_{W_{\bar\lambda}}^*)^{-1}(F_W^{\prime\bar\lambda})^* = (\gamma_W^*)^{-1}$, so

$$L^\lambda = \gamma_V^{-1}(T + G_{V,W}^\lambda)(\gamma_W^*)^{-1}.$$

In terms of L^λ, the boundary condition reads (analogously to (8.18)):

$$\mathrm{i}_Y^* \nu_1 u = (L^\lambda + \mathrm{i}_Y^* P_{\gamma_0,\nu_1}^\lambda)\gamma_0 u, \ \gamma_0 u \in D(L). \tag{8.23}$$

Since $D(\widetilde{A} - \lambda) = D(\widetilde{A})$ is at the same time defined by the boundary condition $\mathrm{i}_Y^* \nu_1 u = (L + \mathrm{i}_Y^* P_{\gamma_0,\nu_1}^0)\gamma_0 u$ for $\gamma_0 u \in D(L)$, we have that $L^\lambda + \mathrm{i}_Y^* P_{\gamma_0,\nu_1}^\lambda = L + \mathrm{i}_Y^* P_{\gamma_0,\nu_1}^0$ on $D(L)$, so

$$L^\lambda = L + \mathrm{i}_Y^*(P_{\gamma_0,\nu_1}^0 - P_{\gamma_0,\nu_1}^\lambda) \text{ on } D(L).$$

The last formula is convenient, since $P_{\gamma_0,\nu_1}^0 - P_{\gamma_0,\nu_1}^\lambda$ can be shown to be *bounded* from $H^{-\frac{1}{2}}(\Sigma)$ to $H^{\frac{1}{2}}(\Sigma)$; hence L^λ is a perturbation of L by a bounded operator.

Also the general M-function defined in Section 8.5 carries over to an M-function on the boundary, a holomorphic family of operators $M_L(\lambda) \in \mathcal{L}(Y^*, X)$ defined for $\lambda \in \varrho(\widetilde{A})$.

The results are collected in the following theorem:

Theorem 8.10 *Let $\widetilde{A}$ correspond to $T\colon V \to W$ as in Theorem 8.2, carried over to $L\colon X \to Y^*$ as in Theorem 8.8. For $\lambda \in \varrho(A_\gamma)$ it is also described by the boundary condition (8.23), and there holds:*

(i) *For $\lambda \in \varrho(A_\gamma)$, $P_{\gamma_0,\nu_1}^0 - P_{\gamma_0,\nu_1}^\lambda \in \mathcal{L}(H^{-\frac{1}{2}}(\Sigma), H^{\frac{1}{2}}(\Sigma))$ and*

$$L^\lambda = L + \mathrm{i}_Y^*(P_{\gamma_0,\nu_1}^0 - P_{\gamma_0,\nu_1}^\lambda) \textit{ on } D(L).$$

(ii) *For* $\lambda \in \varrho(\widetilde{A})$, *there is a related* M*-function* $\in \mathcal{L}(Y^*, X)$,

$$M_L(\lambda) = \gamma_0\big(I - (\widetilde{A} - \lambda)^{-1}(A_{\max} - \lambda)\big)A_\gamma^{-1}\mathrm{i}_W\gamma_W^*.$$

(iii) *For* $\lambda \in \varrho(\widetilde{A}) \cap \varrho(A_\gamma)$,

$$M_L(\lambda) = -(L^\lambda)^{-1} = -\big(L + \mathrm{i}_Y^*(P^0_{\gamma_0,\nu_1} - P^\lambda_{\gamma_0,\nu_1})\mathrm{i}_X\big)^{-1},$$

and we have the Kreĭn-type resolvent formulas:

$$\begin{aligned}(\widetilde{A} - \lambda)^{-1} - (A_\gamma - \lambda)^{-1} &= K^\lambda_{\gamma,X}(L^\lambda)^{-1}(K'^{\bar\lambda}_{\gamma,Y})^* \\ &= -K^\lambda_{\gamma,X}M_L(\lambda)(K'^{\bar\lambda}_{\gamma,Y})^*.\end{aligned}$$

In the case of a Neumann-type boundary condition as in (8.21), $L^\lambda = C - P^\lambda_{\gamma_0,\nu_1}$ on $D(L^\lambda) = D(L)$.

8.8 Applications of pseudodifferential methods I: Conditions for lower boundedness

The formulas we have shown so far use the terminology of ψdbo's mainly as a way to indicate what the ingredients in certain operator compositions are. The next question to consider is how properties of $\widetilde{A}$ are reflected in properties of L. Part of the analysis can be carried out with methods of functional analysis, but there also exist problems that are solved most efficiently by involving deeper pseudodifferential principles.

An example of how functional analytic and pseudodifferential methods are useful together, is the question of lower boundedness inequalities.

We here restrict the attention to the symmetric set-up where A is formally selfadjoint (so $A_{\max} = A^*_{\min}$) and A_γ is selfadjoint; methods for extending the results to nonsymmetric set-ups are found in [Grubb, 1974]. We assume that A_γ has positive lower bound $m(A_\gamma)$ (cf. (8.11)).

In the symmetric case, the general extensions $\widetilde{A}$ can of course be nonsymmetric (since $A_{\max}$ is so). Let us speak of the "selfadjoint case" when only selfadjoint $\widetilde{A}$'s are considered.

In the following, we assume throughout that $\widetilde{A}$ corresponds to $T\colon V \to W$ as in Theorem 8.2, and to $L\colon X \to Y^*$ as in Theorem 8.8. The lower boundedness problem is the problem of how information on lower bounds on $\widetilde{A}$ is related to similar information on T or L. The following was shown in [Grubb, 1970] (together with studies of coerciveness estimates):

Theorem 8.11 *In the symmetric set-up with* A_γ *selfadjoint positive, let* $\widetilde{A}$ *correspond to* $T\colon V \to W$ *as in Theorem* 8.2.

1° *If* $V \subset W$ *and* T *has a lower bound* $m(T)$ *satisfying* $m(T) > -m(A_\gamma)$, *then* $m(\widetilde{A}) \geq m(T)m(A_\gamma)/(m(T)+m(A_\gamma))$.

2° *Assume that* A_γ *is the Friedrichs extension of* $A_{\min}$. *If* $m(\widetilde{A}) > -\infty$, *then* $V \subset W$ *and* $m(T) \geq m(\widetilde{A})$.

In the selfadjoint case these rules go back to [Birman, 1956], preceded by sesquilinear form results of [Krein, 1947]. For studies where $m(A_\gamma)$ is only assumed ≥ 0, see Chapter 3.

The properties of T are easily translated to similar properties of L using the homeomorphism (8.6); here when $X \subset Y$, we set

$$m_{-\frac{1}{2}}(L) = \inf\{\operatorname{Re}(L\varphi,\varphi)_{Y^*,Y} \mid \varphi \in D(L),\ \|\varphi\|_{-\frac{1}{2}} = 1\}, \tag{8.24}$$

for some choice of norm $\|\varphi\|_{-\frac{1}{2}}$ on $H^{-\frac{1}{2}}(\Sigma)$. (One can let γ_Z be an isometry, to carry numerical information over between T and L. Sometimes qualitative objects such as the sign of $m_{-\frac{1}{2}}(L)$ are sufficiently interesting.) To take A_γ as the Friedrichs extension of $A_{\min}$ means that it is taken as the Dirichlet realization.

In 1° we see that the statement "$m(T) > -\infty \implies m(\widetilde{A}) > -\infty$" holds under the additional assumption that $m(T) > -m(A_\gamma)$; there is a nontrivial question of when that assumption can be removed. In [Grubb, 1974] it was shown that when A_γ is the Friedrichs extension and A_γ^{-1} is a *compact* operator, then $m(T) > -\infty$ does imply $m(\widetilde{A}) > -\infty$. This same result was also announced in [Gorbachuk and Mikhailets, 1976] for the selfadjoint case.

In the application to boundary value problems, we therefore have from this early result that lower boundedness of $\widetilde{A}$ and L hold simultaneously when we consider problems on *bounded domains* Ω, for then A_γ^{-1} is indeed compact.

For unbounded domains, the question has, to our knowledge, remained unsolved up until recently. The question is closely connected with the comparison of T with T^λ as in Theorem 8.4. Indeed, as shown in [Grubb, 1974]:

Proposition 8.12 *Let* $G^\lambda_{V,W}$ *be as defined in Theorem* 8.4. *The property*

$$m(G^\mu_{Z,Z}) \to \infty \text{ for } \mu \to -\infty,\ \mu \in \mathbb{R}, \tag{8.25}$$

is necessary and sufficient for the validity of

$$m(T) > -\infty \implies m(\widetilde{A}) > -\infty \tag{8.26}$$

for general closed $\widetilde{A} \in \mathcal{M}$.

The question was also studied later in [Derkach and Malamud, 1991] who worked out an analysis that generalizes Proposition 8.12 and gives further conditions for the validity of the conclusion from $m(T)$ to $m(\widetilde{A})$. However this did not capture elliptic problems on unbounded domains ($n \geq 2$).

Because of the recent interest in the analysis of extensions, we have considered the problem again, and found a solution in [Grubb, 2012] for *exterior domains* (complements in $\mathbb{R}^n$ of bounded domains).

Theorem 8.13 *Let Ω be the complement of a smooth bounded set $\overline{\Omega}_-$ in $\mathbb{R}^n$, and let A be symmetric and uniformly strongly elliptic on Ω with coefficients in $C_b^\infty(\overline{\Omega})$, and with a positive lower bound for A_γ. In the application of the extension theory to this situation,* (8.25) *holds, and hence also* (8.26).

Here $C_b^\infty(\overline{\Omega})$ stands for the C^∞-functions that are bounded with bounded derivatives.

The proof relies on the "translation" of abstract operators $\widetilde{A}$ to concrete operators defined by boundary conditions. Indeed, it turns out that the lower bound of $G'^{\mu}_{Z,Z}$ behaves like the lower bound $m_{-\frac{1}{2}}(Q^\mu)$ (cf. (8.24)) of $Q^\mu - P^0_{\gamma_0,\nu_1} - P^\mu_{\gamma_0,\nu_1}$. Then the deep part of the proof lies in setting the operator Q^μ in relation to the analogous operator for the interior domain Ω_-, Q^μ_-, which does have the desired property, in view of our knowledge of problems on bounded domains. The point is to show that $|((Q^\mu - Q^\mu_-)\varphi, \varphi)|$ is bounded by $c\|\varphi\|^2_{-\frac{1}{2}}$ uniformly for $\mu \to -\infty$, so that addition of $Q^\mu - Q^\mu_-$ to Q^μ_- does not violate the growth of the lower bound. This goes by a delicate application of the ψdbo calculus. Details are in [Grubb, 2012].

Let us mention that there is a considerably easier result that holds regardless of boundedness of $\partial\Omega$ and only requires some uniformity in the estimates of coefficients of the operators, namely preservation of coerciveness inequalities (Gårding inequalities). We here assume that Ω is a subset of $\mathbb{R}^n$ with smooth boundary, admissible as defined in [Grubb, 1996] (besides bounded domains, this allows exterior domains, perturbed halfspaces and other cases that can be covered with a finite system of local coordinates of a relatively simple kind). Moreover, we assume that A is uniformly strongly elliptic on $\overline{\Omega}$ with coefficients in $C_b^\infty(\overline{\Omega})$. The result is that then $\widetilde{A}$ satisfies a Gårding inequality (with $c > 0$, $k \in \mathbb{R}$)

$$\operatorname{Re}(\widetilde{A}u, u) \geq c\|u\|_1^2 - k\|u\|_0^2, \quad u \in D(\widetilde{A}),$$

if and only if $X \subset Y$ and L satisfies an inequality

$$\operatorname{Re}(L\varphi, \varphi) \geq c' \|\varphi\|^2_{H^{\frac{1}{2}}(\Sigma)} - k' \|\varphi\|^2_{H^{-\frac{1}{2}}(\Sigma)}, \quad \varphi \in D(L). \tag{8.27}$$

In the case of differential or pseudodifferential Neumann-type boundary conditions, the inequality (8.27) for L holds precisely when the pseudodifferential operator it acts like, is strongly elliptic. Details for bounded sets are in [Grubb, 1970] for realizations of $2m$-order operators; the extension to unbounded sets is shown in [Grubb, 2012] — the argumentation just involves standard trace theorems and interpolation inequalities.

8.9 Applications of pseudodifferential methods II: Spectral asymptotics

In the symmetric set-up, when Ω is bounded, the eigenvalues of the selfadjoint operator A_γ form a sequence λ_j going to ∞ on $\mathbb{R}$. In [Weyl, 1912] appeared the famous estimate for $A = -\Delta$, for $n = 2, 3$, $m = 2$:

$$\lambda_j(A_\gamma) - c_0 j^{m/n} \text{ is } o(j^{m/n}) \text{ for } j \to \infty,$$

where c_0 is a constant defined from the volume of Ω; the eigenvalues are repeated according to multiplicities. Equivalently, the counting function $N(t; A_\gamma)$ (counting the number of eigenvalues in $[0, t]$), and the eigenvalues μ_j of the inverse A_γ^{-1}, satisfy

$$N(t; A_\gamma) - c_A t^{n/m} \text{ is } o(t^{n/m}) \text{ for } t \to \infty,$$
$$\mu_j(A_\gamma^{-1}) - c_A^{m/n} j^{-m/n} \text{ is } o(j^{-m/n}) \text{ for } j \to \infty,$$

where

$$c_A = (2\pi)^{-n} \int_{x \in \Omega,\, a^0(x,\xi) < 1} dx d\xi$$

(and $c_0 = c_A^{-m/n}$). The estimates have been shown for general n and sharpened since then, with more precision on the remainder, and the validity has been extended to general elliptic operators A and boundary conditions, and to elliptic pseudodifferential operators P on compact manifolds. These improvements have a long history that we shall not try to account for here in detail; they are interesting not only because of the results but even more because of the refined theories that were invented in connection with the proofs (for example: Fourier integral operators).

See e.g. [Hörmander, 1968, 1971; Brüning, 1974; Seeley, 1978; Ivrii, 1982, 1984, 1991; Safarov and Vassiliev, 1997].

We shall here be concerned with a slightly different question, namely the spectral behavior of the difference between the resolvents of two realizations of A.

It is shown in [Birman, 1962] for second-order symmetric uniformly strongly elliptic operators A that the singular numbers $s_j(B) = \mu_j(B^*B)^{\frac{1}{2}}$ of the compact operator $B = \widetilde{A}^{-1} - A_\gamma^{-1}$ satisfy an upper estimate:

$$s_j(\widetilde{A}^{-1} - A_\gamma^{-1}) \le Cj^{-2/(n-1)}, \text{ for all } j, \tag{8.28}$$

when $\widetilde{A}$ is a selfadjoint realization of A defined by a Neumann or Robin condition. In other words, $\widetilde{A}^{-1} - A_\gamma^{-1}$ belongs $\mathfrak{S}_{(n-1)/2,\infty}$, where $\mathfrak{S}_{p,\infty}$ is the space of compact operators B for which $s_j(B)$ is $O(j^{-\frac{1}{p}})$ (often called a weak Schatten class). It is particularly interesting that Birman showed this not just for interior, but also for exterior domains, and under low smoothness assumptions.

We note in passing that the estimate (8.28) implies that

$$\widetilde{A}^{-1} - A_\gamma^{-1} \in \mathcal{C}_p \text{ for } p > (n-1)/2, \tag{8.29}$$

where $\mathcal{C}_p$ is the p-th Schatten class (consisting of the operators B such that$(s_j(B))_{j\in\mathbb{N}} \in \ell_p(\mathbb{N})$); but this is less informative than (8.28).

One of the fundamental ingredients in these studies is embedding properties, more precisely the knowledge that an operator B that is continuous from $L^2(\Omega)$ to $H^s(\Omega)$ for some $s > 0$ (Ω bounded smooth $\subset \mathbb{R}^n$) is compact in $L_2(\Omega)$ and belongs to $\mathfrak{S}_{n/s,\infty}$. However, this alone only gives upper bounds on the behavior of singular numbers. To get Weyl-type limit properties one must know more about the differential or pseudodifferential structure of the operators.

The estimate (8.28) was sharpened to a Weyl-type asymptotic estimate in [Birman and Solomyak, 1977, 1980]; the latter showed a general principle for the spectrum of a ratio of two quadratic forms, implying that

$$s_j(\widetilde{A}^{-1} - A_\gamma^{-1})j^{2/(n-1)} \to c, \text{ for } j \to \infty, \tag{8.30}$$

for interior and exterior smooth domains.

Prior to this, a far-reaching result had been shown in [Grubb, 1974], Section 8: We consider a symmetric, strongly elliptic $2m$-order operator acting in an N-dimensional vector bundle E over a smooth compact Riemanninan manifold $\overline{\Omega}$ with boundary Σ, assuming that the Dirichlet realization A_γ is invertible. Let A_B be a selfadjoint invertible realization defined by a normal boundary condition

$$\sum_{k\le j} B_{jk}\gamma_k u = 0,\ j = 0,1,\dots,2m-1,$$

where the B_{jk} are differential operators of order $j-k$ from $E|_\Sigma$ to given vector bundles F_j over Σ (with $\dim F_j \ge 0$); *normality* of the boundary condition means that the B_{jj} are *surjective morphisms.* (That A acts in the vectorbundle E means that it is locally $(N\times N)$-matrix-formed. In the scalar case, $N=1$ and the F_j are 0- or 1-dimensional, with B_{jj} an invertible function when $\dim F_j = 1$. Ellipticity of the boundary condition requires in particular that $\sum_j \dim F_j = mN$.) Denote $\oplus_{j>m} F_j = F^1$.

Theorem 8.14 *Let $T\colon V\to V$ be the operator corresponding to A_B by Theorem 8.2. There exists an* **isometry** $J\colon L_2(\Sigma,F^1)\xrightarrow{\sim} V$ *with inverse $J^{-1}=J^*$ (in the ψdbo calculus), such that*

$$\mathcal{T}_1 = J^*TJ$$

acts like an elliptic invertible ψdo $\mathcal{T}$ in F^1 of order $2m$, and $D(\mathcal{T}_1) = \{\varphi\in L_2(\Sigma,F^1)\mid \mathcal{T}\varphi\in L_2(\Sigma,F^1)\} = H^{2m}(\Sigma,F^1)$. Here $\mathcal{T}_1$ has the same spectrum as T, and its eigenvectors are mapped to the corresponding eigenvectors of $\mathcal{T}_1$ by the isometry J. Moreover,

$$A_B^{-1} - A_\gamma^{-1} = \mathrm{i}_V J\mathcal{T}_1^{-1}J^*\,\mathrm{pr}_V, \tag{8.31}$$

whereby the positive resp. negative eigenvalues satisfy

$$\mu_j^\pm(A_B^{-1}-A_\gamma^{-1}) = \mu_j^\pm(\mathcal{T}_1^{-1}) \text{ for all } j.$$

It follows that with constants determined from the principal symbols,

$$\begin{aligned} N'^{\pm}(t;A_B^{-1}-A_\gamma^{-1}) &= C^\pm t^{(n-1)/(2m)} + O(t^{(n-1)/(2m)}) \text{ for } t\to\infty,\\ \mu_j^\pm(A_B^{-1}-A_\gamma^{-1}) &= (C^\pm)^{2m/(n-1)} j^{-2m/(n-1)} + O(j^{-(2m+1)/(n-1)}); \end{aligned} \tag{8.32}$$

here $N'^{\pm}(t;S)$ indicates the number of positive, resp. negative eigenvalues of S outside the interval $]-1/t,1/t[$.

The two statements in (8.32) are equivalent, cf. e.g. [Grubb, 1996], Lemma A.5. They follow from [Hörmander, 1968] for elliptic ψdo's, when the principal symbol eigenvalues of $\mathcal{T}$ are simple; this restriction is removed by results of [Ivrii, 1982]. See also Theorem 8.17 below.

We have recently checked that the proof extends to exterior domains, for uniformly strongly elliptic systems.

The fine estimates with remainders depend on the ellipticity of the ψdo. For simple Weyl-type estimates, the ellipticity hypothesis is removed in [Birman and Solomyak, 1977], showing that

$$s_j(P)j^{k/n} \to c(p^0) \text{ for } j \to \infty$$

holds for any classical ψdo P of order $-k < 0$ on a compact manifold of dimension n. They even allowed a certain nonsmoothness of the homogeneous principal symbol, both in x and ξ, needing only a little more than continuity. In the elliptic case, there are recent works of Ivrii dealing with remainder estimates under weak smoothness hypotheses.

In [Grubb, 1984a] an effort was made to increase the accessibility of the ψdbo calculus by publishing an introduction to it with several improvements, and showing as a main result that any singular Green operator of negative order and class 0 has a Weyl-type spectral asymptotics formula:

Theorem 8.15 *Let G be a classical singular Green operator of order $-k < 0$ and class 0 on an n-dimensional compact manifold with boundary. It has a spectral asymptotics behavior*

$$s_j(G)j^{k/(n-1)} \to c(g^0) \textit{ for } j \to \infty, \tag{8.33}$$

where $c(g^0)$ is a constant defined from on the principal symbol of G.

This was moreover used to show asymptotic formulas generalizing (8.30), both for interior and exterior domains, followed up in another study [Grubb, 1984b] including also the dependence on a spectral parameter λ. Indeed, we have as an immediate corollary of Theorem 8.15, also for nonselfadjoint cases:

Corollary 8.16 *Let A be elliptic of order $2m$ with invertible Dirichlet realization and let A_B be an invertible realization defined by a normal elliptic boundary condition. For any positive integer N, $A_B^{-N} - A_\gamma^{-N}$ is a singular Green operator of order $-2mN$ and class 0, and hence satisfies*

$$s_j(A_B^{-N} - A_\gamma^{-N})j^{2mN/(n-1)} \to c_N \textit{ for } j \to \infty, \tag{8.34}$$

for a constant c_N defined from the principal symbols.

Also exterior domains are considered in [Grubb, 1984a,b], where (8.34) is shown for realizations of second-order operators and their iterates. The results apply of course to resolvents by replacement of A by $A-\lambda$; the λ-dependence is studied in [Grubb, 1984b]. It is seen that the ψdbo theory provides a forceful tool for such questions, and we strongly recommend its use.

8.10 New spectral results

Spectral estimates of resolvent differences have been taken up in recent papers [Alpay and Behrndt, 2009; Behrndt, Langer, Lobanov, Lotoreichik and Popov, 2010; Behrndt, Langer and Lotoreichik] for second-order operators and [Malamud, 2010] for $2m$-order operators with normal boundary conditions, based on boundary triples methods. Here Schatten class and weak Schatten class estimates are shown, relying on such estimates for Sobolev space embeddings.

We have returned to the subject in [Grubb, 2011a] where we, besides showing new results on perturbations of essential spectra, have reformulated and extended results in [Grubb, 1984a] on estimates like (8.34), including general differences and exterior domains. The central ingredient is the estimate (8.33) for singular Green operators, plus the fact that s.g.o.s give their essential contribution in a small neighborhood of the boundary, also for exterior domains, allowing cutoffs eliminating infinity.

An inspection of the results of [Grubb, 1974] shows that the spectral estimates in Theorem 8.14 can be further sharpened by use of results of [Ivrii, 1982]:

Theorem 8.17 *In the setting of Theorem* 8.14, *assume in addition that the principal symbol of* $\mathcal{T}$ *satifies Ivrii's conditions* ($\mathrm{H}_\pm$) *from* [Ivrii, 1982] *(the bicharateristics through points of* $T^*(\Sigma)\setminus 0$ *are nonperiodic except for a set of measure zero). Then there are constants* $C_1^\pm$ *such that*

$$N'^{\pm}(t; A_B^{-1} - A_\gamma^{-1}) = C^\pm t^{(n-1)/(2m)} + C_1^\pm t^{(n-2)/(2m)} + o(t^{(n-2)/(2m)}).$$

The proof is a direct application of [Ivrii, 1982] Th. 0.2 to $\mathcal{T}$.

The formula (8.31) is a special type of Kreĭn resolvent formula with isometries, valid for selfadjoint realizations, but the analysis in [Grubb, 1974] also implies Kreĭn formulas in the nonselfadjoint cases. Namely, Th. 6.4 there shows how $T\colon V \to W$ is represented by a realization $\mathcal{L}_1$ of a ψdo $\mathcal{L}$ acting between vector bundles over Σ, and here A_B is elliptic if and only if $\mathcal{L}$ is elliptic (Cor. 6.10). In the invertible elliptic case, formula (8.20) then takes the form

$$A_B^{-1} - A_\gamma^{-1} = \mathrm{i}_V \gamma_V^{-1} \Phi \mathcal{L}_1^{-1} \Psi^* (\gamma_W^*)^{-1} \,\mathrm{pr}_W,$$

with $\mathcal{L}_1$ acting like $-(B^{10} + B^{11} P_{\gamma,\chi})\Phi$ (notation explained in [Grubb, 1974]); the right-hand side is a composition of a Poisson operator, an elliptic ψdo and the adjoint of a Poisson operator, all of mixed order. Its s-numbers can be studied by reduction to an elliptic ψdo over Σ, where Ivrii's sharp results can be used.

Let us just demonstrate this for second-order operators, for the formula (8.22), with L^{-1} elliptic of order -1: Denote $\widetilde{A}^{-1} - A_\gamma^{-1} = S$. Then

$$\begin{aligned} s_j(S)^2 &= s_j(K_\gamma L^{-1}(K'_\gamma)^*)^2 = \mu_j(K_\gamma L^{-1}(K'_\gamma)^* K'_\gamma (L^{-1})^* K_\gamma^*) \\ &= \mu_j(L^{-1}(K'_\gamma)^* K'_\gamma (L^{-1})^* K_\gamma^* K_\gamma), \end{aligned}$$

where we used the general rule $\mu_j(B_1B_2) = \mu_j(B_2B_1)$. Both operators $P_1 = K_\gamma^* K_\gamma$ and $P'_1 = (K'_\gamma)^* K'_\gamma$ are selfadjoint positive elliptic ψdo's of order -1 (cf. e.g. [Grubb, 2011b, proof of Theorem 4.4]). Let $P_2 = P_1^{\frac{1}{2}}$, then we continue the calculations as follows:

$$s_j(S)^2 = \mu_j(L^{-1}P'_1(L^{-1})^*P_1) = \mu_j(P_2 L^{-1} P'_1 (L^{-1})^* P_2) = \mu_j(P_3),$$

where $P_3 = P_2 L^{-1} P'_1 (L^{-1})^* P_2$ is a selfadjoint positive elliptic ψdo on Σ of order -4. Applying Ivrii's theorem to P_3^{-1}, we conclude:

Theorem 8.18 *For the operator considered in Theorem 8.9, the s-numbers satisfy*

$$N'(t; \widetilde{A}^{-1} - A_\gamma^{-1}) = Ct^{(n-1)/2} + O(t^{(n-2)/2}).$$

Moreover, if the principal symbol of P_3^{-1} satifies Ivrii's condition from [Ivrii, 1982] *(the bicharateristics through points of $T^*(\Sigma) \setminus 0$ are non-periodic except for a set of measure zero), there is a constant C_1 such that*

$$N'(t; \widetilde{A}^{-1} - A_\gamma^{-1}) = Ct^{(n-1)/2} + C_1 t^{(n-2)/2} + o(t^{(n-2)/2}).$$

Sharpened asymptotic formulas can also be obtained for differences between resolvents of two realizations that both differ from the Dirichlet realization, by use of the analysis in [Grubb, 1968] with a general invertible realization A_β as reference operator.

To give another example of applications of the ψdbo theory, the following result is found straightforwardly as a consequence of [Grubb, 1984a]:

Theorem 8.19 *Let A_B and $A_{B'}$ be elliptic invertible realizations of A such that B and B' map into the same bundles and have the* **same principal part**. *Then $A_B^{-1} - A_{B'}^{-1}$ is a singular Green operator of order $-2m-1$ (since its principal part is zero), and hence, by* (8.33),

$$s_j(A_B^{-1} - A_{B'}^{-1}) j^{(2m+1)/(n-1)} \to c \text{ for } j \to \infty.$$

The singular Green operator will be of a still lower order $-2m-r$ if the first $r>1$ terms in the symbols of B and B' coincide.

Example 8.20 As a special case, we can compare two Robin conditions for a second-order operator A:

$$\widetilde{A}_1 \text{ defined by } \nu_1 u = b_1\gamma_0 u,$$
$$\widetilde{A}_2 \text{ defined by } \nu_1 u = b_2\gamma_0 u;$$

$b_1, b_2 \in C^\infty(\Sigma)$. When regarded from the point of view of Theorem 8.19, these are normal boundary conditions $\nu_1 u - B_i\gamma_0 u = 0$, where $B_1 = b_1$ and $B_2 = b_2$ considered as first-order operators have principal part 0, so the boundary operators have the same principal part. Then the s.g.o. $\widetilde{A}_1^{-1} - \widetilde{A}_2^{-1}$ is of order -3, and by (8.33),

$$s_j(\widetilde{A}_1^{-1} - \widetilde{A}_2^{-1})j^{3/(n-1)} \to c \text{ for } j\to\infty. \tag{8.35}$$

In [Behrndt, Langer, Lobanov, Lotoreichik and Popov, 2010] upper estimates are shown for this difference in the case $A = -\Delta - \lambda$, namely Schatten class estimates of $s_j(\widetilde{A}_1^{-1} - \widetilde{A}_2^{-1})$ as in (8.29) with $(n-1)/2$ replaced by $(n-1)/3$.

In case the b_i are C^∞, the result is covered by (8.35) as explained above. However, the b_i in [Behrndt, Langer, Lobanov, Lotoreichik and Popov, 2010] are allowed to be nonsmooth, namely to be in $L_\infty(\Sigma)$, which goes outside the range covered by the smooth ψdbo theory.

This led us to investigate how far we could push the proof of asymptotic estimates (8.35) to make them valid for nonsmooth choices of b_i. The outcome is published in [Grubb, 2011b], where it is shown that (8.35) holds for symmetric second-order strongly elliptic operators on smooth domains, when b_1 and b_2 are piecewise C^ε on Σ, having jumps at a smooth hypersurface.

The Schatten class estimates have been followed up by [Behrndt, Langer and Lotoreichik] in a study of selfadjoint realizations; see also Chapter 6 in this volume.

Unsolved questions of asymptotic estimates lie primarily in the range of situations with limited smoothness. Resolvent formulas have been studied in such general cases, [Posilicano and Raimondi, 2009; Grubb, 2008] for $C^{1,1}$-domains, [Gesztesy and Mitrea, 2008, 2011] for Lipschitz and quasi-convex domains, [Abels, Grubb and Wood] for a class of domains containing $C^{3/2+\varepsilon}$, with a nonsmooth generalization of ψdbo's. To our knowledge, spectral asymptotic estimates have not yet been worked out for such resolvent differences. Some upper estimates are in selfadjoint cases known from [Birman, 1962].

A problem with a different flavor is the case of a *mixed boundary condition*, such as prescribing for $-\Delta$ the Dirichlet condition on a part Σ_- of the boundary and a Neumann-type condition on the other part Σ_+. Here there is a jump in the *order* of the boundary condition. The domain of the realization is contained in $H^{\frac{3}{2}-\varepsilon}(\Omega)$ only for $\varepsilon > 0$, so the mixed problem is not covered by those boundary triples methods that require the domain to be in $H^{\frac{3}{2}}(\Omega)$. Spectral upper estimates are known from [Birman, 1962]. A spectral asymptotic estimate was obtained recently in [Grubb, 2011c], based on somewhat intricate applications of results on nonstandard pseudodifferential operators.

There are also other questions that can benefit from pseudodifferential methods, for example the study of spectral asymptotics of the nonelliptic Kreĭn-like extensions, cf. [Grubb, 2012].

References

Abels, H., Grubb, G., and Wood, I. Extension theory and Kreĭn-type resolvent formulas for nonsmooth boundary value problems; arXiv:1008.3281.

Agmon, S. *Lectures on Elliptic Boundary Value Problems*. Van Nostrand Math. Studies, D. Van Nostrand Publ. Co., Princeton 1965.

Agmon, S., Douglis, A., and Nirenberg, L. 1964. Estimates near the boundary for solutions of elliptic partial differential equations satisfying general boundary conditions, I. *Comm. Pure Appl. Math.*, **12**, 623–727.

Alpay, D., and Behrndt, J. 2009. Generalized Q-functions and Dirichlet-to-Neumann maps for elliptic differential operators. *J. Funct. Anal.*, **257**, 1666–1694.

Amrein, W. O., and Pearson, D. B. 2004. *M* operators: a generalisation of Weyl-Titchmarsh theory. *J. Comp. Appl. Math.*, **171**, 1–26.

Arlinskii, Yu. M. 1999. On functions connected with sectorial operators and their extensions. *Integral Equations Operator Theory*, **33**, 125–152.

Behrndt, J., and Langer, M. 2007. Boundary value problems for elliptic partial differential operators on bounded domains. *J. Funct. Anal.*, **243**, 536–565.

Behrndt, J., Langer, M., Lobanov, I., Lotoreichik, V., and Popov, I. 2010. A remark on Schatten-von Neumann properties of resolvent differences of generalized Robin Laplacians on bounded domains. *J. Math. Anal. Appl.*, **371**, 750–758.

Behrndt, J., Langer, M., and Lotoreichik, V. Spectral estimates for differences of resolvents of selfadjoint elliptic operators; arXiv:1012.4596.

Birman, M. S. 1956. On the theory of self-adjoint extensions of positive definite operators. *Mat. Sb. N.S.*, **38(80)**, 431–450 (Russian).

Birman, M. S. 1962. Perturbations of the continuous spectrum of a singular elliptic operator by varying the boundary and the boundary conditions. *Vestnik Leningrad. Univ.* **17**, 22–55. English translation in *Spectral theory of differential operators*, Amer. Math. Soc. Transl. Ser. 2, **225**. Amer. Math. Soc. Providence, R.I. 2008, 19–53.

Birman, M. S., and Solomyak, M. Z. 1977. Asymptotic behavior of the spectrum of pseudodifferential operators with anisotropically homogeneous symbols. *Vestnik Leningrad. Univ.*, **13**, 13–21. English translation in *Vestn. Leningr. Univ. Math.*, **10** (1982), 237–247.

Birman, M. S., and Solomyak, M. Z. 1980. Asymptotics of the spectrum of variational problems on solutions of elliptic equations in unbounded domains. *Funkts. Analiz Prilozhen.*, **14**, 27–35. English translation in *Funct. Anal. Appl.*, **14** (1981), 267–274.

Boutet de Monvel, L. 1971. Boundary problems for pseudodifferential operators, *Acta Math.*, **126**, 11–51.

Brown B. M., Grubb G., and Wood, I. G. 2009. *M*-functions for closed extensions of adjoint pairs of operators with applications to elliptic boundary problems. *Math. Nachr.* **282**, 314–347.

Brown, B. M., Marletta M., Naboko S., and Wood, I. G. 2008. Boundary triplets and M-functions for non-selfadjoint operators, with applications to elliptic PDEs and block operator matrices. *J. Lond. Math. Soc.*, **77**, 700–718.

Brüning, J. 1974. Zur Abschätzung der Spektralfunktion elliptischer Operatoren. *Math. Z.*, **137**, 75–85.

Brüning, J., Geyler V., and Pankrashkin, K. 2008. Spectra of self-adjoint extensions and applications to solvable Schrödinger operators. *Rev. Math. Phys.*, **20**, 1–70.

Calderón, A. P., and Zygmund, A. 1957. Singular integral operators and differential equations. *Amer. J. Math.*, **79**, 901–921.

Courant, R., and Hilbert, D. 1953. *Methods of Mathematical Physics, Vol. I.* Interscience Publishers, Inc., New York, N.Y.

Courant, R., and Hilbert, D. 1962. *Methods of Mathematical Physics, Vol. II: Partial Differential Equations* (by R. Courant). Interscience Publishers (a division of John Wiley & Sons), New York-London.

Derkach, V., Hassi, S., Malamud, M. M., and de Snoo, H.S.V. 2006. Boundary relations and their Weyl families. *Trans. Amer. Math. Soc.*, **358**, 5351–5400.

Derkach, V. A., and Malamud, M. M. 1991. Generalized resolvents and the boundary value problems for Hermitian operators with gaps. *J. Funct. Anal.*, **95**, 1–95.

Friedrichs, K. 1934. Spektraltheorie halbbeschränkter Operatoren und Anwendung auf die Spektralzerlegung von Differentialoperatoren. *Math. Ann.*, **109**, 465–487.

Gesztesy, F., and Mitrea, M. 2008. Generalized Robin boundary conditions, Robin-to-Dirichlet maps, and Krein-type resolvent formulas for Schrödinger operators on bounded Lipschitz domains. *Perspectives in Partial Differential Equations, Harmonic Analysis and Applications: A Volume in Honor of Vladimir G. Maz'ya's 70th Birthday,* Proceedings of Symposia in Pure Mathematics (eds. Mitrea, D., and Mitrea, M.) **79**, Amer. Math. Soc., Providence, R.I., 105–173.

Gesztesy, F., and Mitrea, M. 2011. A description of all selfadjoint extensions of the Laplacian and Kreĭn-type resolvent formulas in nonsmooth domains. *J. Analyse Math.*, **113**, 53–172.

Gorbachuk, M. L., and Mikhailets, V. A. 1976. Semibounded selfadjoint extensions of symmetric operators. *Dokl. Akad. Nauk SSSR*, **226**, 765-767. English translation in *Soviet Math. Doklady*, **17**, 185–187.

Gorbachuk, V. I., and Gorbachuk, M. L. 1991. *Boundary value problems for operator differential equations.* Kluwer, Dordrecht.

Grubb, G. 1968. A characterization of the non-local boundary value problems associated with an elliptic operator. *Ann. Scuola Norm. Sup. Pisa*, **22**, 425–513.

Grubb, G. 1970. Les problèmes aux limites généraux d'un opérateur elliptique, provenant de la théorie variationnelle. *Bull. Sc. Math.*, **94**, 113–157.

Grubb, G. 1971. On coerciveness and semiboundedness of general boundary value problems. *Israel J. Math.*, **10**, 32–95.

Grubb, G. 1973. Weakly semibounded boundary problems and sesquilinear forms. *Ann. Inst. Fourier Grenoble*, **23**, 145–194.

Grubb, G. 1974. Properties of normal boundary problems for elliptic even-order systems. *Ann. Scuola Norm. Sup. Pisa (4)*, **1**, 1–61.

Grubb, G. 1983. Spectral asymptotics for the "soft" selfadjoint extension of a symmetric elliptic differential operator. *J. Operator Theory*, **10**, 9–20.

Grubb, G. 1984a. Singular Green operators and their spectral asymptotics. *Duke Math. J.*, **51**, 477–528.

Grubb, G. 1984b. Remarks on trace extensions for exterior boundary problems. *Comm. Partial Diff. Equ.*, **9**, 231–270.

Grubb, G. 1996. *Functional Calculus of Pseudodifferential Boundary Problems*, Progress in Math. vol. 65, Second Edition. Birkhäuser, Boston.

Grubb, G. 2008. Krein resolvent formulas for elliptic boundary problems in nonsmooth domains. *Rend. Sem. Mat. Univ. Pol. Torino*, **66**, 13–39.

Grubb, G. 2009. *Distributions and operators.* Graduate Texts in Mathematics **252**. Springer, New York 2009.

Grubb, G. 2011a. Perturbation of essential spectra of exterior elliptic problems. *Applicable Analysis*, **90**, 103–123.

Grubb, G. 2011b. Spectral asymptotics for Robin problems with a discontinuous coefficient. *J. Spectral Theory*, **1**, 155–177.

Grubb, G. 2011c. The mixed boundary value problem, Krein resolvent formulas and spectral asymptotic estimates. *J. Math. Anal. Appl.*, **382**, 339–263.

Grubb, G. 2012. Krein-like extensions and the lower boundedness problem for elliptic operators on exterior domains. *J. Differential Equations*, **252**, 852–885.

Hörmander, L. 1963. *Linear Partial Differential Operators,* Grundlehren Math. Wiss. vol. 116. Springer Verlag, Berlin.

Hörmander, L. 1965. Pseudo-differential operators. *Comm. Pure Appl. Math.*, **18**, 501–517.

Hörmander, L. 1968. The spectral function of an elliptic operator. *Acta Math.*, **121**, 193–218.

Hörmander, L. 1971. Fourier integral operators I. *Acta Math.*, **127**, 79–183.

Hormander, L. 1985. *The Analysis of Linear Partial Differential Operators III, Pseudo-differential Operators,* Grundlehren Math. Wiss. vol. 274. Springer Verlag, Berlin.

Ivrii, V. Ja. 1982. Accurate spectral asymptotics for elliptic operators that act in vedtor bundles. *Functional Analysis Prilozhen*, **16**, 30–38, English translation in *Functional Analysis Appl.*, **16**, (1983), 101–108.

Ivrii, V. 1984. *Precise spectral asymptotics for elliptic operators acting in fiberings over manifolds with boundary.* Lecture Notes in Mathematics, 1100. Springer-Verlag, Berlin.

Ivrii, V. 1991. *Microlocal Analysis and Precise Spectral Asymptotics.* Springer-Verlag, Berlin.

Kohn, J. J., and Nirenberg, L. 1965. An algebra of pseudo-differential operators. *Comm. Pure Appl. Math.*, **18**, 269–305.

Kočubeĭ, A. N. 1975. Extensions of symmetric operators and symmetric binary relations. *Math. Notes* (1), **17**, 25–28.

Kopachevskiĭ, N.D., and Kreĭn, S.G. 2004. Abstract Green formula for a triple of Hilbert spaces, abstract boundary value and spectral problems. *Ukr. Math. Bull.*, **1**, 77–105.

Kreĭn, M. G. 1947. The theory of self-adjoint extensions of semi-bounded Hermitian transformations and its applications. I. *Mat. Sb.*, **20**, 431–495 (Russian).

Ladyzhenskaya, O. 1985. *The Boundary Value Problems of Mathematical Physics.* Springer-Verlag, New York.

Lions, J.-L., and Magenes, E. 1968. E. *Problèmes aux limites non homogènes et applications, 1.* Éditions Dunod, Paris.

Lopatinskiĭ, Ya. B. 1953. On a method of reducing boundary problems for a system of differential equations of elliptic type to regular integral equations. *Ukrain. Mat. Zb.*, **5**, 123–151.

Lyantze, V. E., and Storozh, O. G. 1983. *Methods of the Theory of Unbounded Operators.* Naukova Dumka, Kiev (Russian).

Malamud, M. M. 2010. Spectral theory of elliptic operators in exterior domains. *Russian J. Math. Phys.*, **17**, 96–125.

Malamud, M. M., and Mogilevskii, V. I. 2002. Kreĭn type formula for canonical resolvents of dual pairs of linear relations. *Methods Funct. Anal. Topology*, (4) **8**, 72–100.

Mihlin, S. G. 1948. Singular integral equations. *Uspehi Matem. Nauk (N.S.)*, **3**, 29–112. English translation in Amer. Math. Soc. Translations **24** (1950), 116 pp.

von Neumann, J. 1929. Allgemeine Eigenwerttheorie Hermitescher Funktionaloperatoren. *Math. Ann.*, **102**, 49–131.

Nirenberg, L. 1955. Remarks on strongly elliptic partial differential equations. *Comm. Pure Appl. Math.*, **8**, 649–675.

Posilicano, A., and Raimondi, L. 2009. Krein's resolvent formula for self-adjoint extensions of symmetric second-order elliptic differential operators. *J. Phys. A* **42** 015204, 11 pp.

Rempel, S. and Schulze, B.-W. 1982 *Index Theory of Elliptic Boundary Problems.* Akademie-Verlag, Berlin.

Ryzhov, V. 2007. A general boundary value problem and its Weyl function. *Opuscula Math.*, **27**, 305–331.

Safarov, Yu., and Vassiliev, D. 1997. *The Asymptotic Distribution of Eigenvalues of Partial Differential Operators.* Translated from the Russian manuscript by the authors. Translations of Mathematical Monographs, **155**. American Mathematical Society, Providence, R.I.

Schwartz, L. 1950-51. *Théorie des distributions I–II.* Hermann, Paris.

Seeley, R. T. 1965. Refinement of the functional calculus of Calderon and Zygmund. *Proc. Koninkl. Nederl. Akad. Wetensch.*, **68**, 521–531.

Seeley, R. T. 1978. A sharp asymptotic remainder estimate for the eigenvalues of the Laplacian in a domain of R^3. *Adv. in Math.*, **29**, 244–269.

Shapiro, Z. Ya. 1953. On general boundary value problems of elliptic type. *Isz. Akad. Nauk, Math. Ser.*, **17**, 539–562.

Sobolev, S. L. 1950. *Some applications of functional analysis in mathematical physics.* Izdat. Leningrad. Gos. Univ., Leningrad. English translation by Browder, F. E. Translations of Mathematical Monographs, **7**, American Mathematical Society, Providence, R.I. 1963.

Treves, F. 1980. *Introduction to Pseudodifferential and Fourier Integral Operators, 1-2.* Plenum Press, New York.

Vainerman, L. I. 1980. On extensions of closed operators in Hilbert space. *Math. Notes*, **28**, 871–875.

Vishik, M. I. 1952. On general boundary value problems for elliptic differential operators. *Trudy Mosc. Mat. Obsv.*, **1**, 187–246. English translation in *Amer. Math. Soc. Transl. (2)* , **24**, (1963), 107–172.

Weyl, H. 1912. Das asymptotische Verteilungsgesetz der Eigenwerte linearer partieller Differentialgleichungen (mit einer Anwendung auf die Theorie der Hohlraumstrahlung). *Math. Ann.*, **71**, 441–479.

9

Dirac structures and boundary relations

Seppo Hassi, Arjan van der Schaft, Henk de Snoo and Hans Zwart

Abstract It is shown how Dirac structures, boundary control and conservative state/signal system nodes, and unitary boundary relations are connected via proper transforms between the underlying Kreĭn spaces that are involved in all of these notions.

9.1 Introduction

The present chapterer explores the connections between the concepts of Dirac structures, boundary control systems, and boundary relations. The notion of a Dirac structure has its origin in the Hamiltonian approach to systems theory. It is used to formalize the concept of interconnection of port-Hamiltonian systems, see [Dalsmo and van der Schaft, 1999; van der Schaft, 2000; van der Schaft and Maschke, 2002]. For the notion of boundary control system, see [Curtain and Zwart, 1995; Staffans, 2005; Malinen and Staffans, 2007].

The notions of a boundary relation and a corresponding Weyl family were introduced in [Derkach et al., 2006], as generalizations of the notions of boundary triplets introduced by A.N. Kochubei and V.M. Bruck (see [Gorbachuk and Gorbachuk, 1991]), and Weyl functions introduced by [Derkach and Malamud, 1991, 1995].

Dirac structures, boundary control systems, and boundary relations are related to unitary relations in Kreĭn spaces [Derkach et al., 2006, 2009; Kurula at al., 2006, 2010]. It is the purpose of the present chapter to explain the basic connections between these different notions in simple terms, using suitable transforms between the underlying Kreĭn spaces.

9.2 Linear relations in Hilbert and Kreĭn spaces

Let $\mathfrak{H}$ and $\mathfrak{K}$ be Hilbert spaces; the notation for the inner products will be suppressed where ever possible. The algebra of all bounded linear operators from $\mathfrak{H}$ to $\mathfrak{K}$ is denoted by $\boldsymbol{B}(\mathfrak{H}, \mathfrak{K})$; in case $\mathfrak{K} = \mathfrak{H}$ the notation $\boldsymbol{B}(\mathfrak{H})$ will be used.

Let $j \in \boldsymbol{B}(\mathfrak{H})$ be selfadjoint and boundedly invertible. The operator j provides the Hilbert space $\mathfrak{H}$ with a Kreĭn space structure by means of the inner product $[\cdot, -] = (j\,\cdot, -)$; the notation $(\mathfrak{H}, j)$ is used to indicate the Kreĭn space. Very often the operator j is taken to be a *signature* operator j: $j = j^* = j^{-1}$, in which case j is called a *fundamental symmetry.* See [Azizov and Iokhvidov, 1989; Bognar, 1974] for further details. A subspace $\mathfrak{X}$ of a Kreĭn space $(\mathfrak{H}, [\cdot, -])$ is said to be

(i) nonpositive if $[h, h] \leq 0$, $h \in \mathfrak{X}$;
(ii) nonnegative if $[h, h] \geq 0$, $h \in \mathfrak{X}$;
(iii) neutral if $[h, h] = 0$, $h \in \mathfrak{X}$, or $\mathfrak{X} \subset \mathfrak{X}^{[\perp]}$;
(iv) hyper-maximal neutral if $\mathfrak{X} = \mathfrak{X}^{[\perp]}$,

where $[\perp]$ refers to the Kreĭn space inner product.

A *linear relation* T from a linear space $\mathfrak{H}$ to a linear space $\mathfrak{K}$ is defined as a subspace of the product space $\mathfrak{H} \times \mathfrak{K}$, i.e., $T \subset \mathfrak{H} \times \mathfrak{K}$. The linear space of all linear relations from $\mathfrak{H}$ to $\mathfrak{K}$ (with component-wise sum) is denoted by $\boldsymbol{L}(\mathfrak{H}, \mathfrak{K})$; and by $\boldsymbol{L}(\mathfrak{H})$ when $\mathfrak{K} = \mathfrak{H}$. When $\mathfrak{H}$ and $\mathfrak{K}$ are Banach spaces, then the relation $T \in \boldsymbol{L}(\mathfrak{H}, \mathfrak{K})$ is said to be *closed* if T is a closed subspace of the Banach space $\mathfrak{H} \times \mathfrak{K}$. For linear relations and their decompositions, see [Hassi et al., 2009].

Linear relations in Hilbert spaces Let $\mathfrak{H}$ and $\mathcal{H}$ be Hilbert spaces and let T be a linear relation between $\mathfrak{H}$ and $\mathcal{H}$. The adjoint T^* of T is defined by

$$T^* = \{\, \{f, f'\} \in \mathcal{H} \times \mathfrak{H} : \, (f', h) = (f, h'), \ \{h, h'\} \in T \,\},$$

which is a closed linear relation from $\mathcal{H}$ to $\mathfrak{H}$. A relation T in a Hilbert space $\mathfrak{H}$ is said to be

(i) dissipative if $\operatorname{Im}(f', f) \geq 0$;
(ii) accumulative if $\operatorname{Im}(f', f) \leq 0$;
(iii) symmetric if $\operatorname{Im}(f', f) = 0$, or $T \subset T^*$;
(iv) selfadjoint if $T = T^*$.

A selfadjoint relation is automatically the orthogonal sum of (the graph of) a selfadjoint operator and a pure multivalued part, being selfadjoint in its own subspace. A relation T in a Hilbert space $\mathfrak{H}$ is said to be

(i) accretive if $\mathrm{Re}\,(f', f) \geq 0$;
(ii) anti-accretive if $\mathrm{Re}\,(f', f) \leq 0$;
(iii) skew symmetric if $\mathrm{Re}\,(f', f) = 0$, or $T \subset -T^*$;
(iv) skew selfadjoint if $T = -T^*$.

A skew-selfadjoint relation is automatically the orthogonal sum of (the graph of) a skew-selfadjoint operator and a pure multivalued part.

Kreĭn space interpretations For a Hilbert space $\mathfrak{H}$ the product space $\mathfrak{H} \times \mathfrak{H}$ is turned into a Kreĭn space denoted by $(\mathfrak{H}^2, J_{\mathfrak{H}})$, by means of the fundamental symmetry

$$J_{\mathfrak{H}} := \begin{pmatrix} 0 & -\mathrm{i}I_{\mathfrak{H}} \\ \mathrm{i}I_{\mathfrak{H}} & 0 \end{pmatrix}. \tag{9.1}$$

This leads to the identity

$$(J_{\mathfrak{H}}\{f, f'\}, \{g, g'\}) = -\mathrm{i}[(f', g) - (f, g')], \quad \{f, f'\}, \{g, g'\} \in \mathfrak{H}^2,$$

and in particular

$$(J_{\mathfrak{H}}\{f, f'\}, \{f, f'\}) = 2\,\mathrm{Im}\,(f', f), \quad \{f, f'\} \in \mathfrak{H}^2. \tag{9.2}$$

A linear relation T in the Hilbert space $\mathfrak{H}$ can be seen as a subspace of the Kreĭn space $(\mathfrak{H}^2, J_{\mathfrak{H}})$. It follows from (9.2) that T is selfadjoint, symmetric, dissipative, accumulative in $\mathfrak{H}$ if and only if T (as a subspace) is hyper-maximal neutral, neutral, nonnegative, nonpositive in $(\mathfrak{H}^2, J_{\mathfrak{H}})$, respectively.

Likewise the product $\mathfrak{H} \times \mathfrak{H}$ is turned into a Kreĭn space denoted by $(\mathfrak{H}^2, J_{\mathfrak{H}})$, by means of the fundamental symmetry

$$\iota_{\mathfrak{H}} := \begin{pmatrix} 0 & I_{\mathfrak{H}} \\ I_{\mathfrak{H}} & 0 \end{pmatrix}. \tag{9.3}$$

Now this leads to the identity

$$(\iota_{\mathfrak{H}}\{f, f'\}, \{g, g'\}) = (f', g) + (f, g'), \quad \{f, f'\}, \{g, g'\} \in \mathfrak{H}^2,$$

and in particular

$$(\iota_{\mathfrak{H}}\{f, f'\}, \{f, f'\}) = 2\,\mathrm{Re}\,(f', f), \quad \{f, f'\} \in \mathfrak{H}^2. \tag{9.4}$$

It follows from (9.4) that T is skew selfadjoint, skew symmetric, accretive, anti-accumulative in $\mathfrak{H}$ if and only if T (as a subspace) is hypermaximal neutral, neutral, nonnegative, nonpositive in $(\mathfrak{H}^2, J_{\mathfrak{H}})$, respectively.

For applications it is sometimes useful to consider product spaces of the form $\mathfrak{H}_1 \times \mathfrak{H}_2$ instead of $\mathfrak{H}^2$. Let $\mathfrak{H}_1$ and $\mathfrak{H}_2$ be Hilbert spaces with the same dimension and let $\mathfrak{u}$ be a unitary operator from $\mathfrak{H}_2$ onto $\mathfrak{H}_1$. Then the block operator defined on the space $\mathfrak{H}_1 \times \mathfrak{H}_2$ via

$$J_{\mathfrak{u}} = \begin{pmatrix} 0 & \mathfrak{u} \\ \mathfrak{u}^* & 0 \end{pmatrix} : \begin{matrix} \mathfrak{H}_1 \\ \times \\ \mathfrak{H}_2 \end{matrix} \to \begin{matrix} \mathfrak{H}_1 \\ \times \\ \mathfrak{H}_2 \end{matrix} \tag{9.5}$$

satisfies $J_{\mathfrak{u}}^2 = J_{\mathfrak{u}} = J_{\mathfrak{u}}^*$, i.e., it is a fundamental symmetry. Hence $J_{\mathfrak{u}}$ generates a Kreĭn space structure on $\mathfrak{H}_1 \times \mathfrak{H}_2$ with the indefinite inner product

$$[h, k]_{\mathfrak{H}_1 \times \mathfrak{H}_2} = (J_{\mathfrak{u}} h, k)_{\mathfrak{H}_1 \times \mathfrak{H}_2}, \quad h, k \in \mathfrak{H}_1 \times \mathfrak{H}_2.$$

Linear relations in Kreĭn spaces Let the Hilbert spaces $\mathfrak{H}$ and $\mathcal{H}$ be provided with the signature operators $j_{\mathfrak{H}}$ and $j_{\mathcal{H}}$, respectively, and let $(\mathfrak{H}, j_{\mathfrak{H}})$ and $(\mathcal{H}, j_{\mathcal{H}})$ be the corresponding Kreĭn spaces with fundamental symmetries $j_{\mathfrak{H}}$ and $j_{\mathcal{H}}$. Then the adjoint $T^{[*]}$ of a linear relation T from the Kreĭn space $(\mathfrak{H}, j_{\mathfrak{H}})$ to the Kreĭn space $(\mathcal{H}, j_{\mathcal{H}})$ is given by

$$T^{[*]} = \{ \{j_{\mathcal{H}} f, j_{\mathfrak{H}} f'\} : \{f, f'\} \in T \},$$

which is equal to $j_{\mathfrak{H}} T^* j_{\mathcal{H}}$. Let T be a linear relation from $(\mathfrak{H}, j_{\mathfrak{H}})$ to $(\mathcal{H}, j_{\mathcal{H}})$. Then T is said to be

(i) *contractive* if $[f', f'] \leq [f, f]$, $\{f, f'\} \in T$;
(ii) *expansive* if $[f', f'] \geq [f, f]$, $\{f, f'\} \in T$;
(iii) *isometric* if $[f', f'] = [f, f]$, $\{f, f'\} \in T$, or $T^{-1} \subset T^{[*]}$;
(iv) *unitary* if $T^{-1} = T^{[*]}$.

A *standard unitary operator* is an isometric mapping from $(\mathfrak{H}, j_{\mathfrak{H}})$ onto $(\mathcal{H}, j_{\mathcal{H}})$, which is defined everywhere; cf. [Azizov and Iokhvidov, 1989; Derkach et al., 2006]. Note that a unitary relation T in a Hilbert space $\mathfrak{H}$ is automatically (the graph of) a unitary operator in $\boldsymbol{B}(\mathfrak{H})$. In contrast to the Hilbert space situation, unitary operators and relations may be nondensely defined, unbounded, and multivalued operators; cf. [Arens, 1961; Shmul'yan, 1976]. The following facts on unitary relations are taken from Derkach et al. [2006]. Let U be a unitary relation from the Kreĭn space $(\mathfrak{H}, j_{\mathfrak{H}})$ to the Kreĭn space $(\mathcal{H}, j_{\mathcal{H}})$. Then

$$\ker U = (\operatorname{dom} U)^{[\perp]}, \quad \operatorname{mul} U = (\operatorname{ran} U)^{[\perp]}$$

and, moreover, $\operatorname{dom} U$ is closed if and only if $\operatorname{ran} U$ is closed. Observe that, if $\mathcal{L} \subset \mathfrak{H}$ and $\mathcal{M} \subset \mathcal{H}$, then

$$\ker U \subset \mathcal{L} \subset \overline{\operatorname{dom}} U \quad \Leftrightarrow \quad \ker U \subset \mathcal{L}^{[\perp]} \subset \overline{\operatorname{dom}} U,$$

$$\operatorname{mul} U \subset \mathcal{M} \subset \overline{\operatorname{ran}} U \quad \Leftrightarrow \quad \operatorname{mul} U \subset \mathcal{M}^{[\perp]} \subset \overline{\operatorname{ran}} U.$$

In Kreĭn spaces those unitary relations whose domain, and hence equivalently, whose range is closed map orthogonal complements of subspaces in a similar way as unitary operators in Hilbert spaces.

Lemma 9.1 *Let U be a unitary relation from the Kreĭn space $(\mathfrak{H}, j_{\mathfrak{H}})$ to the Kreĭn space $(\mathcal{H}, j_{\mathcal{H}})$ and assume that* $\operatorname{dom} U$ *is closed or, equivalently, that* $\operatorname{ran} U$ *is closed. If $\mathcal{L} \subset \mathfrak{H}$ is a subset such that* $\ker U \subset \mathcal{L} \subset \operatorname{dom} U$, *then*

$$U(\mathcal{L}^{[\perp]}) = U(\mathcal{L})^{[\perp]}.$$

Proof Denote $\mathcal{M} = U(\mathcal{L})$. Then $\operatorname{mul} U \subset \mathcal{M} \subset \operatorname{ran} U$, since $\operatorname{ran} U$ is closed. Moreover, by the isometry of U one has $U(\mathcal{L}^{[\perp]}) \subset \mathcal{M}^{[\perp]}$. Since U is unitary also U^{-1} is isometric and, therefore, $U^{-1}(\mathcal{M}^{[\perp]}) \subset \mathcal{L}^{[\perp]}$. This implies that

$$U(\mathcal{L}^{[\perp]}) \supset U(U^{-1}(\mathcal{M}^{[\perp]})) = \mathcal{M}^{[\perp]} + \operatorname{mul} U = \mathcal{M}^{[\perp]},$$

which combined with $U(\mathcal{L}^{[\perp]}) \subset \mathcal{M}^{[\perp]}$ proves the assertion. □

When this result is specialized to unitary operators U with closed domain from the Kreĭn space $(\mathfrak{H}^2, J_{\mathfrak{H}})$ to the the Kreĭn space $(\mathfrak{H}^2, J_{\mathcal{H}})$ with $J_{\mathfrak{H}}$ and $J_{\mathcal{H}}$ of the form (9.1), the results in Lemma 9.1 applied to a linear relation T from $\mathfrak{H}$ to $\mathcal{H}$ takes the form

$$U(T^*) = U(T)^*, \quad \ker U \subset T \subset \operatorname{dom} U \, (= \overline{\operatorname{dom}} U).$$

9.3 Linear relations in product spaces

Let $\mathfrak{H}$ and $\mathcal{H}$ be linear spaces. There are different ways to form product spaces involving $\mathfrak{H}$ and $\mathcal{H}$. One way is to form the product of the product spaces $\mathfrak{H}^2 = \mathfrak{H} \times \mathfrak{H}$ and $\mathcal{H}^2 = \mathcal{H} \times \mathcal{H}$, denoted by $\mathfrak{H}^2 \times \mathcal{H}^2$. Another way is to define the product $\mathfrak{H} \times \mathcal{H}$ and then to form the product of this space with itself, denoted by $(\mathfrak{H}\times\mathcal{H})^2$. These different ways will be now explored.

Mappings from $\boldsymbol{L}(\mathfrak{H}^2,\mathcal{H}^2)$ to $\boldsymbol{L}(\mathfrak{H}^2,\mathcal{H}^2)$ Define the linear mapping $\mathcal{I}$ from $\mathfrak{H}^2\times\mathcal{H}^2$ to $\mathfrak{H}^2\times\mathcal{H}^2$ by

$$\mathcal{I}:\left\{\begin{pmatrix} f\\ f'\end{pmatrix},\begin{pmatrix} h\\ h'\end{pmatrix}\right\}\mapsto\left\{\begin{pmatrix} f\\ -\mathrm{i}f'\end{pmatrix},\begin{pmatrix} h\\ -\mathrm{i}h'\end{pmatrix}\right\},\quad f,f'\in\mathfrak{H},\ h,h'\in\mathcal{H}.$$

Clearly this defines a unitary operator between these Hilbert spaces. Observe that $\mathcal{I}$ may be considered as a mapping from $\boldsymbol{L}(\mathfrak{H}^2,\mathcal{H}^2)$ to $\boldsymbol{L}(\mathfrak{H}^2,\mathcal{H}^2)$.

Let $\Gamma\in\boldsymbol{L}(\mathfrak{H}^2,\mathcal{H}^2)$, so that Γ is a subset of $\mathfrak{H}^2\times\mathcal{H}^2$; a typical element in Γ has the form

$$\left\{\begin{pmatrix} f\\ f'\end{pmatrix},\begin{pmatrix} h\\ h'\end{pmatrix}\right\}\in\Gamma,\quad f,f'\in\mathfrak{H},\quad h,h'\in\mathcal{H}. \tag{9.6}$$

The relation $\mathcal{I}(\Gamma)$ is then defined by

$$\mathcal{I}(\Gamma)=\left\{\left\{\begin{pmatrix} f\\ -\mathrm{i}f'\end{pmatrix},\begin{pmatrix} h\\ -\mathrm{i}h'\end{pmatrix}\right\}:\left\{\begin{pmatrix} f\\ f'\end{pmatrix},\begin{pmatrix} h\\ h'\end{pmatrix}\right\}\in\Gamma\right\},$$

and it belongs to $\boldsymbol{L}(\mathfrak{H}^2,\mathcal{H}^2)$. Clearly, as a unitary operator $\mathcal{I}$ preserves linearity and closedness of subspaces in $\boldsymbol{L}(\mathfrak{H}^2,\mathcal{H}^2)$. In addition, it is straightforward to check that

$$\mathcal{I}(\Gamma)^{-1}=\mathcal{I}(\Gamma^{-1}),\quad \mathcal{I}(\Gamma)^*=\mathcal{I}(\Gamma^*), \tag{9.7}$$

for any linear relation $\Gamma\in\boldsymbol{L}(\mathfrak{H}^2,\mathcal{H}^2)$.

Now provide the product spaces $\mathfrak{H}^2$ and $\mathcal{H}^2$ with the fundamental symmetries as in (9.1) and (9.3). Then the transform $\mathcal{I}$ preserves certain properties relative to the fundamental symmetries.

Lemma 9.2 *Consider $\mathcal{I}$ as a mapping between $\boldsymbol{L}((\mathfrak{H}^2,J_{\mathfrak{H}}),(\mathcal{H}^2,J_{\mathcal{H}}))$ and $\boldsymbol{L}((\mathfrak{H}^2,\iota_{\mathfrak{H}}),(\mathcal{H}^2,\iota_{\mathcal{H}}))$ and let Γ be in $\boldsymbol{L}((\mathfrak{H}^2,J_{\mathfrak{H}}),(\mathcal{H}^2,J_{\mathcal{H}}))$. Then*

$$\mathcal{I}(J_{\mathfrak{H}}\Gamma^*J_{\mathcal{H}})=\iota_{\mathfrak{H}}\mathcal{I}(\Gamma)^*\iota_{\mathcal{H}}. \tag{9.8}$$

Moreover, the following identity

$$\begin{aligned}\left[\begin{pmatrix} h\\ -\mathrm{i}h'\end{pmatrix},\begin{pmatrix} h\\ -\mathrm{i}h'\end{pmatrix}\right]_{\iota_{\mathcal{H}}}&-\left[\begin{pmatrix} f\\ -\mathrm{i}f'\end{pmatrix},\begin{pmatrix} f\\ -\mathrm{i}f'\end{pmatrix}\right]_{\iota_{\mathfrak{H}}}\\ &=\left[\begin{pmatrix} h\\ h'\end{pmatrix},\begin{pmatrix} h\\ h'\end{pmatrix}\right]_{J_{\mathcal{H}}}-\left[\begin{pmatrix} f\\ f'\end{pmatrix},\begin{pmatrix} f\\ f'\end{pmatrix}\right]_{J_{\mathfrak{H}}}\end{aligned} \tag{9.9}$$

holds for all elements of the form (9.6).

Proof Let $\Gamma \in \boldsymbol{L}((\mathfrak{H}^2, J_{\mathfrak{H}}), (\mathcal{H}^2, J_{\mathcal{H}}))$. Then the product $J_{\mathfrak{H}}\Gamma^* J_{\mathcal{H}}$ is given, by definition, by

$$\begin{aligned} J_{\mathfrak{H}}\Gamma^* J_{\mathcal{H}} &= \left\{ \left\{ J_{\mathcal{H}} \begin{pmatrix} \varphi \\ \varphi' \end{pmatrix}, J_{\mathfrak{H}} \begin{pmatrix} \psi \\ \psi' \end{pmatrix} \right\} : \left\{ \begin{pmatrix} \varphi \\ \varphi' \end{pmatrix}, \begin{pmatrix} \psi \\ \psi' \end{pmatrix} \right\} \in \Gamma^* \right\} \\ &= \left\{ \left\{ \begin{pmatrix} -\mathrm{i}\varphi' \\ \mathrm{i}\varphi \end{pmatrix}, \begin{pmatrix} -\mathrm{i}\psi' \\ \mathrm{i}\psi \end{pmatrix} \right\} : \left\{ \begin{pmatrix} \varphi \\ \varphi' \end{pmatrix}, \begin{pmatrix} \psi \\ \psi' \end{pmatrix} \right\} \in \Gamma^* \right\}. \end{aligned} \tag{9.10}$$

Likewise, the product $\iota_{\mathfrak{H}}\mathcal{I}(\Gamma)^*\iota_{\mathcal{H}}$ is given by

$$\begin{aligned} \iota_{\mathfrak{H}}\mathcal{I}(\Gamma)^*\iota_{\mathcal{H}} &= \left\{ \left\{ \iota_{\mathcal{H}} \begin{pmatrix} \varphi \\ \varphi' \end{pmatrix}, \iota_{\mathfrak{H}} \begin{pmatrix} \psi \\ \psi' \end{pmatrix} \right\} : \left\{ \begin{pmatrix} \varphi \\ \varphi' \end{pmatrix}, \begin{pmatrix} \psi \\ \psi' \end{pmatrix} \right\} \in \mathcal{I}(\Gamma)^* \right\} \\ &= \left\{ \left\{ \begin{pmatrix} \varphi' \\ \varphi \end{pmatrix}, \begin{pmatrix} \psi' \\ \psi \end{pmatrix} \right\} : \left\{ \begin{pmatrix} \varphi \\ \mathrm{i}\varphi' \end{pmatrix}, \begin{pmatrix} \psi \\ \mathrm{i}\psi' \end{pmatrix} \right\} \in \Gamma^* \right\} \\ &= \left\{ \left\{ \begin{pmatrix} -\mathrm{i}\varphi' \\ \varphi \end{pmatrix}, \begin{pmatrix} -\mathrm{i}\psi' \\ \psi \end{pmatrix} \right\} : \left\{ \begin{pmatrix} \varphi \\ \varphi' \end{pmatrix}, \begin{pmatrix} \psi \\ \psi' \end{pmatrix} \right\} \in \Gamma^* \right\}, \end{aligned}$$

where in the second equality the second result in (9.7) has been used. The last equality together with (9.10) leads to (9.8). Finally, the identities

$$\left[\begin{pmatrix} f \\ f' \end{pmatrix}, \begin{pmatrix} f \\ f' \end{pmatrix} \right]_{J_{\mathfrak{H}}} = \left[\begin{pmatrix} f \\ \mathrm{i}f' \end{pmatrix}, \begin{pmatrix} f \\ -\mathrm{i}f' \end{pmatrix} \right]_{\iota_{\mathfrak{H}}},$$

and

$$\left[\begin{pmatrix} h \\ h' \end{pmatrix}, \begin{pmatrix} h \\ h' \end{pmatrix} \right]_{J_{\mathcal{H}}} = \left[\begin{pmatrix} h \\ -\mathrm{i}h' \end{pmatrix}, \begin{pmatrix} h \\ -\mathrm{i}h' \end{pmatrix} \right]_{\iota_{\mathcal{H}}},$$

lead to the equality (9.9). □

Lemma 9.3 *The transform $\mathcal{I}$ gives a one-to-one correspondence between the unitary, isometric, contractive, and expansive relations in* $\boldsymbol{L}((\mathfrak{H}^2, J_{\mathfrak{H}}), (\mathcal{H}^2, J_{\mathcal{H}}))$ *and the unitary, isometric, contractive, and expansive relations in* $\boldsymbol{L}((\mathfrak{H}^2, \iota_{\mathfrak{H}}), (\mathcal{H}^2, \iota_{\mathcal{H}}))$.

Proof Let Γ be a unitary relation in $\boldsymbol{L}((\mathfrak{H}^2, J_{\mathfrak{H}}), (\mathcal{H}^2, J_{\mathcal{H}}))$, so that $\Gamma^{-1} = J_{\mathfrak{H}}\Gamma^* J_{\mathcal{H}}$. Due to the first identity in (9.7) and the identity in (9.8), it follows that

$$\mathcal{I}(\Gamma)^{-1} = \mathcal{I}(\Gamma^{-1}) = \mathcal{I}(J_{\mathfrak{H}}\Gamma^* J_{\mathcal{H}}) = \iota_{\mathfrak{H}}\mathcal{I}(\Gamma)^*\iota_{\mathcal{H}},$$

so that $\mathcal{I}(\Gamma)$ is unitary in $\boldsymbol{L}((\mathfrak{H}^2, \iota_{\mathfrak{H}}), (\mathcal{H}^2, \iota_{\mathcal{H}}))$. The remaining assertions follow from (9.9). □

Mappings between $\boldsymbol{L}(\mathfrak{H}^2, \mathcal{H}^2)$ and $\boldsymbol{L}(\mathfrak{H}\times\mathcal{H}, \mathfrak{H}\times\mathcal{H})$ The transform $\mathcal{J}$ from $\mathfrak{H}^2 \times \mathcal{H}^2$ to $(\mathfrak{H}\oplus\mathcal{H})^2$ defined by

$$\mathcal{J} : \left\{ \begin{pmatrix} f \\ f' \end{pmatrix}, \begin{pmatrix} h \\ h' \end{pmatrix} \right\} \mapsto \left\{ \begin{pmatrix} f \\ h \end{pmatrix}, \begin{pmatrix} f' \\ -h' \end{pmatrix} \right\}, \quad f, f' \in \mathfrak{H}, \; h, h' \in \mathcal{H},$$

was introduced in [Derkach et al., 2006]. Consider the transform $\mathcal{J}$ as a mapping between $\boldsymbol{L}(\mathfrak{H}^2, \mathcal{H}^2)$ and $\boldsymbol{L}(\mathfrak{H}\times\mathcal{H}, \mathfrak{H}\times\mathcal{H})$: for $\Gamma \in \boldsymbol{L}(\mathfrak{H}^2, \mathcal{H}^2)$ the relation $\mathcal{J}(\Gamma)$, defined by

$$\mathcal{J}(\Gamma) = \left\{ \left\{ \begin{pmatrix} f \\ h \end{pmatrix}, \begin{pmatrix} f' \\ -h' \end{pmatrix} \right\} : \left\{ \begin{pmatrix} f \\ f' \end{pmatrix}, \begin{pmatrix} h \\ h' \end{pmatrix} \right\} \in \Gamma \right\},$$

belongs to $\boldsymbol{L}(\mathfrak{H}\times\mathcal{H}, \mathfrak{H}\times\mathcal{H})$. This mapping establishes a one-to-one correspondence between the (closed) linear relations $\boldsymbol{L}(\mathfrak{H}^2, \mathcal{H}^2)$ (i.e., (closed) subspaces of $\mathfrak{H}^2\times\mathcal{H}^2$) and the (closed) linear relations in $\boldsymbol{L}(\mathfrak{H}\times\mathcal{H}, \mathfrak{H}\times\mathcal{H})$ (i.e., (closed) subspaces of $(\mathfrak{H}\times\mathcal{H})^2$). It is straightforward to observe that

$$\left\{ \begin{pmatrix} \alpha \\ \beta \end{pmatrix}, \begin{pmatrix} \gamma \\ \delta \end{pmatrix} \right\} \in (\mathcal{J}\Gamma)^* \text{ if and only if } \left\{ \begin{pmatrix} \delta \\ \beta \end{pmatrix}, \begin{pmatrix} -\gamma \\ \alpha \end{pmatrix} \right\} \in \Gamma^*. \tag{9.11}$$

The transform $\mathcal{J}$ preserves certain properties when the spaces $\mathfrak{H}^2$ and $\mathcal{H}^2$ are provided with fundamental symmetries as in (9.1) and (9.3). For the following result, see [Derkach et al., 2006].

Lemma 9.4 *Assume that* $\Gamma \in \boldsymbol{L}((\mathfrak{H}^2, J_{\mathfrak{H}}), (\mathcal{H}^2, J_{\mathcal{H}}))$. *Then*

$$\mathcal{J}(J_{\mathfrak{H}}\Gamma^* J_{\mathcal{H}})^{-1} = (\mathcal{J}\Gamma)^*.$$

Moreover, the following identity

$$\left[\begin{pmatrix} f \\ f' \end{pmatrix}, \begin{pmatrix} f \\ f' \end{pmatrix} \right]_{J_{\mathfrak{H}}} - \left[\begin{pmatrix} h \\ h' \end{pmatrix}, \begin{pmatrix} h \\ h' \end{pmatrix} \right]_{J_{\mathcal{H}}} = 2\,\mathrm{Im} \left(\begin{pmatrix} f' \\ -h' \end{pmatrix}, \begin{pmatrix} f \\ h \end{pmatrix} \right) \tag{9.12}$$

holds for all elements of the form (9.6).

Proof Use the identity in (9.10):

$$J_{\mathfrak{H}}\Gamma^* J_{\mathcal{H}} = \left\{ \left\{ \begin{pmatrix} \varphi' \\ -\varphi \end{pmatrix}, \begin{pmatrix} \psi' \\ -\psi \end{pmatrix} \right\} : \left\{ \begin{pmatrix} \varphi \\ \varphi' \end{pmatrix}, \begin{pmatrix} \psi \\ \psi' \end{pmatrix} \right\} \in \Gamma^* \right\}.$$

Hence

$$(J_{\mathfrak{H}}\Gamma^* J_{\mathcal{H}})^{-1} = \left\{ \left\{ \begin{pmatrix} \psi' \\ -\psi \end{pmatrix}, \begin{pmatrix} \varphi' \\ -\varphi \end{pmatrix} \right\} : \left\{ \begin{pmatrix} \varphi \\ \varphi' \end{pmatrix}, \begin{pmatrix} \psi \\ \psi' \end{pmatrix} \right\} \in \Gamma^* \right\},$$

and therefore

$$\mathcal{J}(J_{\mathfrak{H}}\Gamma^* J_{\mathcal{H}})^{-1} = \left\{ \left\{ \begin{pmatrix} \psi' \\ \varphi' \end{pmatrix}, \begin{pmatrix} -\psi \\ \varphi \end{pmatrix} \right\} : \left\{ \begin{pmatrix} \varphi \\ \varphi' \end{pmatrix}, \begin{pmatrix} \psi \\ \psi' \end{pmatrix} \right\} \in \Gamma^* \right\}.$$

It follows from (9.11) that the righthand side is equal to $(\mathcal{J}\Gamma)^*$. The equation (9.12) follows by writing out the inner products. □

Lemma 9.5 *Assume that* $\Gamma \in \boldsymbol{L}((\mathfrak{H}^2, \iota_{\mathfrak{H}}), (\mathcal{H}^2, \iota_{\mathcal{H}}))$. *Then*

$$\mathcal{J}(\iota_{\mathfrak{H}}\Gamma^*\iota_{\mathcal{H}})^{-1} = -(\mathcal{J}\Gamma)^*.$$

Moreover, the following identity

$$\left[\begin{pmatrix} f \\ f' \end{pmatrix}, \begin{pmatrix} f \\ f' \end{pmatrix}\right]_{\iota_{\mathfrak{H}}} - \left[\begin{pmatrix} h \\ h' \end{pmatrix}, \begin{pmatrix} h \\ h' \end{pmatrix}\right]_{\iota_{\mathcal{H}}} = 2\,\mathrm{Re}\left(\begin{pmatrix} f' \\ -h' \end{pmatrix}, \begin{pmatrix} f \\ h \end{pmatrix}\right) \tag{9.13}$$

holds for all elements of the form (9.6).

Proof It follows from the identity

$$\iota_{\mathfrak{H}}\Gamma^*\iota_{\mathcal{H}} = \left\{\left\{\begin{pmatrix} \varphi' \\ \varphi \end{pmatrix}, \begin{pmatrix} \psi' \\ \psi \end{pmatrix}\right\} : \left\{\begin{pmatrix} \varphi \\ \varphi' \end{pmatrix}, \begin{pmatrix} \psi \\ \psi' \end{pmatrix}\right\} \in \Gamma^*\right\},$$

that

$$(\iota_{\mathfrak{H}}\Gamma^*\iota_{\mathcal{H}})^{-1} = \left\{\left\{\begin{pmatrix} \psi' \\ \psi \end{pmatrix}, \begin{pmatrix} \varphi' \\ \varphi \end{pmatrix}\right\} : \left\{\begin{pmatrix} \varphi \\ \varphi' \end{pmatrix}, \begin{pmatrix} \psi \\ \psi' \end{pmatrix}\right\} \in \Gamma^*\right\}.$$

Therefore

$$\mathcal{J}(\iota_{\mathfrak{H}}\Gamma^*\iota_{\mathcal{H}})^{-1} = \left\{\left\{\begin{pmatrix} \psi' \\ \varphi' \end{pmatrix}, \begin{pmatrix} \psi \\ -\varphi \end{pmatrix}\right\} : \left\{\begin{pmatrix} \varphi \\ \varphi' \end{pmatrix}, \begin{pmatrix} \psi \\ \psi' \end{pmatrix}\right\} \in \Gamma^*\right\},$$

and the righthand side is equal to $-(\mathcal{J}\Gamma)^*$, since by (9.11) one has

$$\left\{\begin{pmatrix} \varphi \\ \varphi' \end{pmatrix}, \begin{pmatrix} \psi \\ \psi' \end{pmatrix}\right\} \subset \Gamma^* \quad \text{if and only if} \quad \left\{\begin{pmatrix} \psi' \\ \varphi' \end{pmatrix}, \begin{pmatrix} -\psi \\ \varphi \end{pmatrix}\right\} \in (\mathcal{J}\Gamma)^*.$$

The equation (9.13) follows by writing out the inner products. □

Lemma 9.6 *Let* $\mathcal{J}$ *be the mapping between the spaces* $\boldsymbol{L}(\mathfrak{H}^2, \mathcal{H}^2)$ *and* $\boldsymbol{L}(\mathfrak{H} \times \mathcal{H}, \mathfrak{H} \times \mathcal{H})$. *Then*

(i) $\mathcal{J}$ *gives a one-to-one correspondence between the unitary, isometric, contractive, and expansive relations from* $(\mathfrak{H}^2, J_{\mathfrak{H}})$ *to* $(\mathcal{H}^2, J_{\mathcal{H}})$ *and the selfadjoint, symmetric, dissipative, and accumulative relations in* $\mathfrak{H} \times \mathcal{H}$;

(ii) $\mathcal{J}$ *gives a one-to-one correspondence between the unitary, isometric, contractive, and expansive relations from* $(\mathfrak{H}^2, \iota_{\mathfrak{H}})$ *to* $(\mathcal{H}^2, \iota_{\mathcal{H}})$ *and the skew-selfadjoint, skew-symmetric, accretive, and anti-accretive relations in* $\mathfrak{H} \times \mathcal{H}$.

Proof The results follow from Lemma 9.4 and Lemma 9.5. □

Based on Lemmas 9.3 and 9.6 the following result is clear.

Proposition 9.7 *The following diagram*

$$\begin{array}{ccc} \boldsymbol{L}((\mathfrak{H}^2, J_{\mathfrak{H}}), (\mathcal{H}^2, J_{\mathcal{H}})) & \stackrel{\mathcal{J}}{\rightarrow} & \boldsymbol{L}(\mathfrak{H} \times \mathcal{H}, \mathfrak{H} \times \mathcal{H})) \\ \mathcal{I} \downarrow & & \downarrow -\mathrm{i} \\ \boldsymbol{L}((\mathfrak{H}^2, \iota_{\mathfrak{H}}), (\mathcal{H}^2, \iota_{\mathcal{H}})) & \underset{\mathcal{J}}{\rightarrow} & \boldsymbol{L}(\mathfrak{H} \times \mathcal{H}, \mathfrak{H} \times \mathcal{H})) \end{array}$$

is commutative: $\mathcal{J} \circ \mathcal{I} = (-\mathrm{i})\mathcal{J}$.

The commutativity of the above diagram expresses the fact that for boundary relations one uses the usual form of Green's identity whereas for Dirac structures one uses a modified form of Green's identity. The interpretation of this diagram is that, roughly speaking, Dirac structures and boundary relations are the same objects, but formulated in different languages; this is made more clear in the next section.

9.4 The connections between various structures

In this section some basic connections of general boundary triplets and relations to Dirac and Tellegen structures as well as to boundary control and state/signal systems are given.

Dirac and Tellegen structures For Dirac structures and, more generally, Tellegen structures the following terminology is often used; cf. [Dalsmo and van der Schaft, 1999; van der Schaft, 2000; van der Schaft and Maschke, 2002]. The *space of efforts* is a Hilbert space denoted here by $\mathfrak{E}$ and the *space of flows* another Hilbert space denoted here by $\mathfrak{F}$. In what follows it is assumed that these space are complex Hilbert spaces with the same dimension and that $\mathfrak{u}$ is a unitary operator from $\mathfrak{E}$ onto $\mathfrak{F}$. As in (9.5) the block operator $J_{\mathfrak{u}}$ defined on the product Hilbert space $\mathfrak{F} \times \mathfrak{E}$ via

$$J_{\mathfrak{u}} = \begin{pmatrix} 0 & \mathfrak{u} \\ \mathfrak{u}^* & 0 \end{pmatrix} : \begin{matrix} \mathfrak{F} \\ \times \\ \mathfrak{E} \end{matrix} \rightarrow \begin{matrix} \mathfrak{F} \\ \times \\ \mathfrak{E} \end{matrix} \tag{9.14}$$

generates a Kreĭn space structure on $\mathfrak{F} \times \mathfrak{E}$ with the indefinite inner product $[h, k]_{\mathfrak{F} \times \mathfrak{E}} = (J_{\mathfrak{u}} h, k)_{\mathfrak{F} \times \mathfrak{E}}$, $h, k \in \mathfrak{F} \times \mathfrak{E}$. The Kreĭn space

$$\mathcal{B} := (\mathfrak{F} \times \mathfrak{E}, J_{\mathfrak{u}})$$

is often called a *bond space*. This terminology and the following definition is taken from [Kurula at al., 2010, Definition 2.1].

Definition 9.8 A linear subspace $\mathcal{D}$ of the bond space $\mathcal{B}$ is called a Tellegen structure on $\mathcal{B}$ if $\mathcal{D} \subset \mathcal{D}^{[\perp]}$ and $\mathcal{D}$ is called a Dirac structure on $\mathcal{B}$ if $\mathcal{D} = \mathcal{D}^{[\perp]}$.

To connect Dirac and Tellegen structures to (operator) colligations appearing in system theory assume that $\mathfrak{H}$, $\mathcal{H}_0$, and $\mathcal{H}_1$ are Hilbert spaces such that $\dim \mathcal{H}_0 = \dim \mathcal{H}_1$ and let

$$\mathfrak{w} : \mathcal{H}_0 \to \mathcal{H}_1$$

be a unitary operator. Now consider the product Hilbert space

$$\mathfrak{F} \times \mathfrak{E}, \quad \mathfrak{F} = \mathfrak{H} \oplus \mathcal{H}_1, \quad \mathfrak{E} = \mathfrak{H} \oplus \mathcal{H}_0.$$

Clearly, the operator

$$\mathfrak{u} = I_{\mathfrak{H}} \oplus -\mathfrak{w} : \mathfrak{E} \to \mathfrak{F} \tag{9.15}$$

is unitary in the Hilbert space sense and thus the corresponding bond space with a so-called *power inner product* is now obtained by putting

$$\mathcal{B} := (\mathfrak{F} \times \mathfrak{E}, J_{\mathfrak{u}}), \tag{9.16}$$

where $J_{\mathfrak{u}}$ is the fundamental symmetry as in (9.14) generated by the unitary operator $\mathfrak{u}$ in (9.15). After permutation of the elements

$$\{f, h_1\} \times \{f', h_0\} \in \mathfrak{F} \times \mathfrak{E}$$

as follows

$$\left\{ \begin{pmatrix} f \\ f' \end{pmatrix}, \begin{pmatrix} h_0 \\ h_1 \end{pmatrix} \right\}, \quad f, f' \in \mathfrak{H}, \quad h_0 \in \mathcal{H}_0, \, h_1 \in \mathcal{H}_1, \tag{9.17}$$

one can identify the Kreĭn space $\mathcal{B}$ with the Kreĭn space

$$\mathfrak{H}_{\mathfrak{E}\mathfrak{F}} = (\mathfrak{H} \times \mathfrak{H}) \times (\mathcal{H}_0 \times \mathcal{H}_1), \quad J_{\mathfrak{F}\mathfrak{E}} = \iota_{\mathfrak{H}} \oplus -J_{\mathfrak{w}^*}. \tag{9.18}$$

In view of (9.17) one can interpret the subspaces $\mathcal{D}$ of $(\mathfrak{H}_{\mathfrak{E}\mathfrak{F}}, J_{\mathfrak{F}\mathfrak{E}})$, hence also of the bond space $\mathcal{B} \sim (\mathfrak{H}_{\mathfrak{E}\mathfrak{F}}, J_{\mathfrak{F}\mathfrak{E}})$ in (9.16), as multi-valued linear mappings form the Kreĭn space $(\mathfrak{H}^2, \iota_{\mathfrak{H}})$ to the Kreĭn space $(\mathcal{H}_0 \times \mathcal{H}_1, J_{\mathfrak{w}^*})$. In particular, Dirac and Tellegen structures on $\mathcal{B}$ can now be considered as multi-valued operators from $(\mathfrak{H}^2, \iota_{\mathfrak{H}})$ to $(\mathcal{H}_0 \times \mathcal{H}_1, J_{\mathfrak{w}^*})$.

Proposition 9.9 *The mapping*

$$W : \left\{ \begin{pmatrix} f \\ f' \end{pmatrix}, \begin{pmatrix} h \\ h' \end{pmatrix} \right\} \to \left\{ \begin{pmatrix} f \\ -\mathrm{i}f' \end{pmatrix}, \begin{pmatrix} h \\ -\mathrm{i}\mathfrak{w}h' \end{pmatrix} \right\}, \tag{9.19}$$

is a standard unitary operator from the Kreĭn space $(\mathfrak{H}^2 \oplus \mathcal{H}_0^2, J_{\mathfrak{H}} \oplus -J_{\mathcal{H}_0})$ *onto the Kreĭn space* $(\mathfrak{H}_{\mathfrak{E}\mathfrak{F}}, J_{\mathfrak{F}\mathfrak{E}})$.

The linear relation $\Gamma : (\mathfrak{H}^2, J_{\mathfrak{H}}) \to (\mathcal{H}_0^2, J_{\mathcal{H}_0})$ *is a unitary (isometric, contractive, expansive) relation with* $\ker \Gamma = S$ *and* $\operatorname{dom} \Gamma = T$ *if and only if* $\mathcal{D} = W(\Gamma) : (\mathfrak{H}^2, \iota_{\mathfrak{H}}) \to (\mathcal{H}_0 \times \mathcal{H}_1, J_{\mathfrak{w}^*})$ *is a unitary (isometric, contractive, expansive) relation with* $\ker \mathcal{D} = -\mathrm{i}S$ *and* $\operatorname{dom} \mathcal{D} = -\mathrm{i}T$.

Proof The isometry of the mapping W in (9.19) is straightforward to check by writing down the equality between the inner products; cf. (9.1), (9.18). Moreover, W is clearly everywhere defined and surjective, hence it is a standard unitary operator.

The remaining assertions concerning correspondence between various linear relations are now immediate. □

If $\mathcal{H}_0 = \mathcal{H}_1$ and $\mathfrak{u} = I$, then W coincides with $\mathcal{I}$. Now one can extend Proposition 9.7 by replacing $\mathcal{I}$ with W and $-\mathrm{i}$ by $-\mathrm{i}(I_{\mathcal{H}_0} \oplus \mathfrak{w})$ in the diagram given therein, and then its commutativity reads as $\mathcal{J} \circ \mathcal{I} = -\mathrm{i}(I_{\mathcal{H}_0} \oplus \mathfrak{w}) \circ \mathcal{J}$.

Proposition 9.9 shows, in particular, that Dirac structures correspond to unitary boundary relations, and that Tellegen structures correspond to isometric boundary relations; see Chapter 7.

In the special case that $\Gamma = \{\Gamma_0, \Gamma_1\}$ is an ordinary boundary triplet for a densely defined symmetric operator $S = \ker \Gamma$ the statement in Proposition 9.9 was proved in [Kurula at al., 2010, Theorem 4.8].

Remark Since W in Proposition 9.9 is a standard unitary operator between Kreĭn spaces, it preserves all basic geometric and algebraic properties of the subspaces Γ and $W(\Gamma)$. Hence, for instance, the results proved in [Derkach et al., 2009, Sections 4, 5] for certain transformed unitary and isometric boundary triplets $\Gamma = (\Gamma_0, \Gamma_1)$ (associated with some intermediate extensions) immediately carry over to similar transformation results for Dirac and Tellegen structures $\mathcal{D} = W(\Gamma) = (\Gamma_0, -\mathrm{i}\mathfrak{w}\Gamma_1)$, and conversely; see for instance [Kurula at al., 2010, Sections 4, 5].

Boundary control and conservative state/signal system nodes
To make a connection with boundary control systems assume that $\Gamma = \{\mathcal{H}_0, \Gamma_0, \Gamma_1\}$ in Proposition 9.9 is a boundary triplet for a densely defined

symmetric operator S in $\mathfrak{H}$. Then the mappings Γ_0, Γ_1 can be interpreted to be mappings from $\operatorname{dom} S^*$ onto $\mathcal{H}_0 \times \mathcal{H}_0$, as in the original definition of boundary triplet; see [Gorbachuk and Gorbachuk, 1991]. Then by Proposition 9.9 $W(\Gamma)$ is a Dirac structure that, after rearranging the elements (cf. (9.17)), coincides with a so-called conservative boundary control system; for further details, see [Malinen and Staffans, 2007] and Theorem 4.8. If $\Gamma = \{\mathcal{H}_0, \Gamma\}$ in Proposition 9.9 is a unitary boundary relation for the adjoint of a symmetric relation in $\mathfrak{H}$, then $W(\Gamma)$ is a Dirac structure that (after rearranging the elements) with two additional conditions on Γ, which guarantee a proper dynamics for an associated state/signal system, coincides with a so-called impedance-conservative i/s/o system node $W(\Gamma)$; see Theorem 5.35 and Propositions 5.36, 5.38.

9.5 Weyl families and transfer functions

Weyl families Each linear relation from $\mathfrak{H}^2$ to $\mathcal{H}^2$ generates a family of relations in $\mathcal{H}$. This family is the abstract analog of the Titchmarsh-Weyl m function from the theory of singular Sturm-Liouville operators; see [Derkach et al., 2006].

Definition 9.10 Let $\mathfrak{H}$ and $\mathcal{H}$ be Hilbert spaces and let Γ be a relation in the space $\boldsymbol{L}(\mathfrak{H}^2, \mathcal{H}^2)$. The *Weyl family* $M(\lambda)$ corresponding to the relation Γ is defined by

$$M(\lambda) := \left\{ \widehat{h} \in \mathcal{H}^2 : \{\widehat{f}_\lambda, \widehat{h}\} \in \Gamma \text{ for some } \widehat{f}_\lambda = \{f, \lambda f\} \in \mathfrak{H}^2 \right\}, \quad \lambda \in \mathbb{C}.$$

Thus, for each $\lambda \in \mathbb{C}$, $M(\lambda)$ belongs to $\boldsymbol{L}(\mathcal{H})$. $M(\lambda)$ is called a Weyl function when it is operator-valued.

Lemma 9.11 *Let $\mathfrak{H}$ and $\mathcal{H}$ be Hilbert spaces and let Γ be a relation in $\boldsymbol{L}(\mathfrak{H}^2, \mathcal{H}^2)$. Let $M(\lambda)$ be the Weyl family corresponding to Γ and let $N(\lambda)$ be the Weyl family corresponding to $\mathcal{I}(\Gamma)$. Then*

$$N(\lambda) = -\,\mathrm{i}M(\mathrm{i}\lambda), \quad \lambda \in \mathbb{C}.$$

Proof The definition of $M(\lambda)$ can be rewritten as

$$M(\lambda) = \left\{ \begin{pmatrix} h \\ h' \end{pmatrix} : \left\{ \begin{pmatrix} f \\ \lambda f \end{pmatrix}, \begin{pmatrix} h \\ h' \end{pmatrix} \right\} \in \Gamma \right\}.$$

Observe that

$$\left\{ \begin{pmatrix} f \\ \lambda f \end{pmatrix}, \begin{pmatrix} h \\ h' \end{pmatrix} \right\} \in \Gamma \quad \text{if and only if} \quad \left\{ \begin{pmatrix} f \\ -\,\mathrm{i}\lambda f \end{pmatrix}, \begin{pmatrix} h \\ -\,\mathrm{i}h' \end{pmatrix} \right\} \in \mathcal{I}(\Gamma).$$

Hence it follows that

$$-\,\mathrm{i}M(\lambda) = \left\{ \begin{pmatrix} h \\ -\,\mathrm{i}h' \end{pmatrix} : \left\{ \begin{pmatrix} f \\ -\,\mathrm{i}\lambda f \end{pmatrix}, \begin{pmatrix} h \\ -\,\mathrm{i}h' \end{pmatrix} \right\} \in \mathcal{I}(\Gamma) \right\}.$$

Clearly, the righthand side is equal to $N(\mathrm{i}\lambda)$. □

Definition 9.12 A family of relations $M(\lambda)$, $\lambda \in \mathbb{C}_+ \cup \mathbb{C}_-$, in a Hilbert space $\mathcal{H}$ is said to be a *Nevanlinna family on* $\mathbb{C}_+ \cup \mathbb{C}_-$ if

(i) $\operatorname{Im} M(\lambda) \geq 0$, $\lambda \in \mathbb{C}_+$;
(ii) $M(\lambda)^* = M(\bar{\lambda})$;
(iii) for each $\nu \in \mathbb{C}_+(\mathbb{C}_-)$ the operator-valued function $(M(\lambda)+\nu)^{-1} \in \boldsymbol{B}(\mathcal{H})$ is holomorphic for $\lambda \in \mathbb{C}_+(\mathbb{C}_-)$.

Definition 9.13 A family of relations $M(\lambda)$, $\lambda \in \mathbb{C}_r \cup \mathbb{C}_l$, in a Hilbert space $\mathcal{H}$ is said to be a *Carathéodory family on* $\mathbb{C}_r$ if

(i) $\operatorname{Re} M(\lambda) \geq 0$, $\lambda \in \mathbb{C}_r$;
(ii) $M(\lambda)^* = M(-\bar{\lambda})$;
(iii) for each $\nu \in \mathbb{C}_r(\mathbb{C}_l)$ the operator-valued function $(M(\lambda)+\nu)^{-1} \in \boldsymbol{B}(\mathcal{H})$ is holomorphic for $\lambda \in \mathbb{C}_r(\mathbb{C}_l)$.

Clearly, $N(\lambda)$ is a Carathéodory family if and only if in

$$N(\lambda) = -\,\mathrm{i}M(\mathrm{i}\lambda), \quad \lambda \in \mathbb{C},$$

the family $M(\lambda)$ is a Nevanlinna family.

Lemma 9.14 *Let $\mathfrak{H}$ and $\mathcal{H}$ be Hilbert spaces and let Γ be a relation in the space $\boldsymbol{L}(\mathfrak{H}^2, \mathcal{H}^2)$. Then:*

(i) *if $\Gamma \in \boldsymbol{L}((\mathfrak{H}^2, J_{\mathfrak{H}}), (\mathcal{H}^2, J_{\mathcal{H}}))$ is unitary, then the Weyl family $M(\lambda)$ corresponding to Γ is a Nevanlinna family;*
(ii) *if $\Gamma \in \boldsymbol{L}((\mathfrak{H}^2, \iota_{\mathfrak{H}}), (\mathcal{H}^2, \iota_{\mathcal{H}}))$ is unitary, then the Weyl family $M(\lambda)$ corresponding to Γ is a Carathéodory family.*

For proofs, see [Derkach et al., 2006]. Under suitable minimality conditions each Nevanlinna family or Carathéodory family is a Weyl family of a unitary boundary relation or of a Dirac structure.

Let $\mathfrak{H}$ be a Hilbert space over $\mathbb{C}$ and define for $\mu, \rho \in \mathbb{C}$ the *Cayley transforms* C_μ and D_ρ as mappings from $\mathfrak{H} \times \mathfrak{H}$ to $\mathfrak{H} \times \mathfrak{H}$ by:

$$C_\mu\{f, f'\} = \{f' - \mu f, f' - \bar{\mu} f\}, \quad \{f, f'\} \in \mathfrak{H} \times \mathfrak{H}, \tag{9.20}$$

$$D_\rho\{f, f'\} = \{f' - \rho f, f' + \bar{\rho} f\}, \quad \{f, f'\} \in \mathfrak{H} \times \mathfrak{H}. \tag{9.21}$$

Then it is clear that

$$\|f' - \bar{\mu}f\|^2 - \|f' - \mu f\|^2 = 4(\mathrm{Im}\,\mu)\,\mathrm{Im}\,(f', f), \quad \{f, f'\} \in \mathfrak{H} \times \mathfrak{H}, \tag{9.22}$$

and that

$$\|f' + \bar{\rho}f\|^2 - \|f' - \rho f\|^2 = 4(\mathrm{Re}\,\rho)\,\mathrm{Re}\,(f', f), \quad \{f, f'\} \in \mathfrak{H} \times \mathfrak{H}. \tag{9.23}$$

The mappings C_μ and D_ρ in (9.20) and (9.21) belong to $\boldsymbol{B}(\mathfrak{H} \times \mathfrak{H})$. They are versions of the transformers introduced in [Shmuljan, 1980]. The identities (9.22) and (9.23) give rise to isometries by normalizing the inner product.

Lemma 9.15 *For $\mu \in \mathbb{C} \setminus \mathbb{R}$ the Cayley transform C_μ gives a one-to-one correspondence between the selfadjoint and symmetric relations, and the unitary and isometric operators. For $\mu \in \mathbb{C}_+(\mathbb{C}_-)$ the Cayley transform C_μ gives a one-to-one correspondence between the dissipative (accumulative) relations, and the contractive operators.*

Lemma 9.16 *For $\rho \in \mathbb{C} \setminus i\mathbb{R}$ the Cayley transform D_ρ gives a one-to-one correspondence between the skew-selfadjoint and skew symmetric relations, and the unitary and isometric operators. For $\rho \in \mathbb{C}_r(\mathbb{C}_l)$ the Cayley transform D_ρ gives a one-to-one correspondence between the accretive (anti-accretive) relations, and the contractive operators.*

Under the Cayley transform C_μ the selfadjoint (in the Hilbert space sense, see Lemma 9.6) image $\mathcal{J}(\Gamma)$ of a unitary boundary relation Γ (unitary in the Kreĭn space sense) is transformed into a unitary colligation (unitary in the Hilbert space sense), see [Behrndt et al., 2009, Section 5]. Hence, this composition of the transforms behaves in a similar way as the Potapov-Ginzburg transform, see [Azizov and Iokhvidov, 1989, Chapter 5: §1] for further details. The *transfer function* of the unitary colligation is a Schur function $\Theta(z)$. The connection with the Weyl (Nevanlinna) family of the boundary relation Γ is given by

$$\Theta(z) = I - (\mu - \bar{\mu})(M(\lambda) + \mu)^{-1}, \quad \lambda \in \mathbb{C}_+,$$

where $\mu \in \mathbb{C}_+$ and $z(\lambda) = (\lambda - \mu)/(\lambda - \bar{\mu})$; cf. [Behrndt et al., 2009]. Similarly, Dirac systems have corresponding scattering functions which can be seen as transforms of the Weyl (Carathéodory) family. For isometric boundary relations and Tellegen structures, there are corresponding minimality conditions (relative to a half-plane).

References

Arens, R. 1961. Operational calculus of linear relations. *Pacific J. Math.*, **11**, 9–23.

Azizov, T. Ya., and Iokhvidov, I.S. 1989. *Linear operators in spaces with indefinite metric*, John Wiley and Sons, New York.

Behrndt, J., Hassi, S., and de Snoo, H.S.V. 2009. Boundary relations, unitary colligations, and functional models. *Complex Analysis Operator Theory*, **3**, 57–98.

Bognar, J. 1974. *Indefinite inner product spaces*, Ergebnisse der Mathematik und ihrer Grenzgebiete, **78**, Springer-Verlag, New York-Heidelberg.

Curtain, R.F., and Zwart, H.J. 1995. *An introduction to infinite-dimensional linear systems theory*, Springer-Verlag.

Dalsmo, M., and van der Schaft, A.J. 1999. On representations and integrability of mathematical structures in energy-conserving physical systems. *SIAM J. Control Optimization*, **37**, 54–91.

Derkach, V.A., Hassi, S., Malamud, M.M., and de Snoo, H.S.V. 2006. Boundary relations and Weyl families. *Trans. Amer. Math. Soc.*, **358**, 5351–5400.

Derkach, V.A., Hassi, S., Malamud, M.M., and de Snoo, H.S.V. 2009. Boundary relations and generalized resolvents of symmetric operators. *Russ. J. Math. Phys.*, **16**, 17–60.

Derkach, V.A., and Malamud, M.M. 1991. Generalized resolvents and the boundary value problems for hermitian operators with gaps. *J. Funct. Anal.*, **95**, 1–95.

Derkach, V.A., and Malamud, M.M. 1995. The extension theory of hermitian operators and the moment problem. *J. Math. Sciences*, **73**, 141–242.

Gorbachuk, V.I., and Gorbachuk, M. L. 1991. *Boundary value problems for operator differential equations*, Mathematics and its Applications (Soviet Series), **48**, Kluwer Academic Publishers, Dordrecht.

Hassi, S., de Snoo, H.S.V., and Szafraniec, F.H. 2009. Componentwise and canonical decompositions of linear relations. *Dissertationes Mathematicae*, **465** (59 pages).

Kurula, M., van der Schaft, A.J., and Zwart, H.J. 2006. Composition of infinite-dimensional linear Dirac-type structures. *17th International Symposium on Mathematical Theory of Networks and Systems*, 24–28 July 2006, Kyoto, Japan.

Kurula, M., van der Schaft, A.J., Zwart, H.J., and Behrndt, J. 2010. Dirac structures and their composition on Hilbert Spaces. *J. Math. Anal. Appl.*, **372**, 402–422.

Malinen, J., and Staffans, O.J. 2007. Impedance passive and conservative boundary control systems, *Complex Analysis Operator Theory*, **1**, 279–300.

van der Schaft, A.J. 2000. *L_2-gain and passitivity techniques in nonlinear control*, 2nd revised and enlarged edition, Springer Communications and Control Engineering series, **218**, Springer-Verlag.

van der Schaft, A.J., and Maschke, B. 2002. Hamiltonian formulation of distibuted-parameter systems with boundary energy flow. *J. Geometry Physics*, **42**, 166–194.

Shmul'jan, Yu. L. 1976. Theory of linear relations, and spaces with indefinite metric. *Funkcional. Anal. i Priložen*, **10**, 67–72 (Russian).

Shmuljan, Yu.L. 1980. Transformers of linear relations in *J*-spaces. *Funkts. Analiz i Prilogh.*, **14**, 39–44 (Russian). English translation in *Functional Anal. Appl.*, **14**, (1980), 110–113.

Staffans, O.J. 2005. *Well-posed linear systems*, Cambridge University Press.

10

Naĭmark dilations and Naĭmark extensions in favour of moment problems

Franciszek Hugon Szafraniec [a]

Abstract Making use of Naĭmark extensions of a symmetric operator arising from an indeterminate Hamburger moment sequence we manufacture a machinery for providing representing measures with the following properties

1° the support of each of them is in arithmetic progression;
2° the supports of all the measures together partition $\mathbb{R}$;
3° none of them is N-extremal;
4° all of them are of infinite order

All this is based on and, in fact, is a kind of guide to [Cichoń, Stochel and Szafraniec, 2010].

Moment problems and their close relatives, orthogonal polynomials were pretty often treated by the same means, mostly continuous fractions. Then, starting from [Stone, 1932, Chapter X] operator theory was used to handle the problem, look at [Landau, 1980] and [Fuglede, 1983, especially p. 51] for further references as well as for a fairly decent introduction to the subject, in a contemporary language.

Sustaining this idea in what refers to the single real variable case, sooner or later one has to deal with a symmetric operator which, as a matter of course, has deficiency indices $(0,0)$ or $(1,1)$. The latter is of interest here as it corresponds to an indeterminate moment problem and the routine procedure is to pass to extensions in the same Hilbert space as they always have to exist. The other way, which seems to be much less exploited with some exceptions, like [Kreĭn and Krasnoselskiĭ, 1947; Gil de Lamadrid, 1971; Langer, 1976; Simon, 1998] and references therein, is to go for extensions beyond the space. Those we propose to call the Naĭmark extensions regardless of the way they are constructed. The Naĭmark extensions which appear in this chapter come in the most

[a] Research supported by the MNiSzW grant NN201 1546438

natural way as differential operators in L^2-spaces over a finite interval. From this point of view our paper may be seen as a kind of continuation of [Simon, 1998] in those respects which take into account the Naĭmark extensions. As a consequence we get the properties 1° up to 4° (cf. Abstract) avoiding rude calculation which happens from time to time when using orthogonal polynomial techniques.

Referring to 4° let us remember that the so-called N-extremal measures have the property that the polynomials are dense in their L^2 spaces; this can be rephrased by saying that they are always of order 0 [1].

The example (or rather a class of examples) we are going to build up here has a representing measure equivalent to the Lebesgue measure on $\mathbb{R}$ as well as an uncountable family of representing measures with discrete closed supports. The Lebesgue-equivalent measure [2] is produced with the help of the Fourier transform, whereas the discrete measures are obtained with the aid of Naĭmark extensions of a symmetric first order differential operator.

In fact, we are able to construct a large class of Hamburger moment sequences depending on, what is called here, a test function. Among the parameters describing the representing measures of such a sequence the most significant role is played by the interval over which the corresponding extension spaces are built; in each case the underlying interval has to contain the support of the test function. If the interval is not minimal with respect to the aforesaid support, then the property 4° (and consequently 3°) holds, cf. Theorem 10.22. Otherwise the issue becomes much more subtle: under some circumstances the same conclusion can be derived (cf. Theorems 10.22 and 10.21) while in general it is left as an open problem.

A comprehensive historical attribution of classical results on moment problems may be found in [Shohat and Tamarkin, 1943; Akhiezer, 1965] and also in Stone [1932] which seems to be underestimated presumably due to its intense operator theory approach.

10.1 Dilations and extensions

In this section we collect some facts which are scattered in the literature, sometimes written in a slighly old-fashioned way. For some it would be

[1] Note that 1° always implies 3°.

[2] Its density, when ranging over the Paley-Wiener space, may be viewed as a way of identifying the whole class of examples we are going to present here.

nothing but an occasion to refresh their memory, for others a moment of getting new thoughts.

Some ideology This part is rather scratchy, its only purpose is to evoke some forgotten or overlooked associations rather than provide precise statements; they will follow. Consider two Hilbert spaces $\mathcal{H}$ and $\mathcal{K}$ with $\mathcal{H} \subset \mathcal{K}$ isometrically. Suppose $\boldsymbol{A}$ and $\boldsymbol{B}$ are families of linear densely defined operators in $\mathcal{H}$ and $\mathcal{K}$ resp. Moreover, suppose both $\boldsymbol{A}$ and $\boldsymbol{B}$ are composed of pairwise commuting operators which is meant pointwise where ever executable. The two notions can be defined along with: for any finite collection $A_1, \ldots, A_n$ in $\boldsymbol{A}$ there correspond $B_1, \ldots, B_n$ in $\boldsymbol{B}$ such that

$$A_1 \cdots A_n \subset \begin{cases} PB_1 \cdots B_n & \text{DILATION} \\ \text{or} & \\ B_1 \cdots B_n & \text{EXTENSION} \end{cases} \tag{10.1}$$

where P stands for the orthogonal projection of $\mathcal{K}$ onto $\mathcal{H}$.. An operator $B \in \boldsymbol{B}$ which is a dilation or an extension of $A \in \boldsymbol{A}$ has a matrix decomposition with respect to $\mathcal{K} = \mathcal{H} \bigoplus \mathcal{H}^{\perp}$ respectively

$$\begin{pmatrix} A & * \\ \& & \# \end{pmatrix} \text{ DILATION,} \qquad \begin{pmatrix} A & * \\ 0 & \# \end{pmatrix} \text{ EXTENSION.} \tag{10.2}$$

It is clear that extensions are dilations; it is also clear for extensions singletons in (10.1) are in a sense enough to be considered. The recipe (10.1) can be decoded as follows

$$\mathcal{D}(A_1 \cdots A_n) \subset \mathcal{D}(B_1 \cdots B_n) \text{ and for } f \in \mathcal{D}(A_1 \cdots A_n),\ g \in \mathcal{H}$$
$$\langle A_1 \cdots A_n f, g \rangle = \langle B_1 \cdots B_n f, g \rangle \text{ DILATION}$$
$$\text{or}$$
$$A_1 \cdots A_n f = B_1 \cdots B_n f \text{ EXTENSION}$$

Thus dilations can be thought of as weak extensions.

The main reason for considering dilations and extensions is to deduct some properties of operators in the ground space, say $\mathcal{H}$, from those acting in a superspace, like $\mathcal{K}$ here. Therefore, the problem consists in

given $\boldsymbol{A}$, determine a class $\boldsymbol{B}$ with properties well-suited to those of the class $\boldsymbol{A}$ we are particulary interested in.

A simple though instructive example of such deduction is the case when operators inherit closedness or closability from that of their extensions.

Assorted particular cases

operator	dilation	extension	designation
single contraction	single unitary		Julia Halmos
powers of contractions	powers of unitaries		Sz.–Nagy
semispectral measure	spectral measure		Naĭmark
symmetric		selfadjoint	von Neumann Naĭmark
subnormal		normal	Halmos Stochel-Szafraniec
			Naĭmark GNS Stinespring Sz.-Nagy ⋮ *others* ⋮
PD functions	$*$–representations		Szafraniec

For a fairly general treatment look at [Sz.-Nagy, 1955; Mlak, 1978; Szafraniec, 1993], for synthesis, updates and further generalizations see [Szafraniec, 2010].

Representations (10.2) show that dilatability is $*$-invariant while extendibility is not unless the operators in $\boldsymbol{A}$ are symmetric. They also suggest the rough proof of the following statement.

Fact 10.1 *B is an extension of A if and only if B is a dilation of A and B^*B is a dilation of A^*A and, equivalently, if and only if $B^{*j}B^i$ is a dilation of $A^{*j}A^i$ for $i, j = 0, 1, \ldots$.*

This if made precise can be proved in much the way as in [Sz.-Nagy, 1955, p. 20] for bounded operators.

Fact 10.1 makes it clear why the difference between dilations and extensions in the case of symmetric operators becomes invisible.

In general we do not exclude the case $\mathcal{H} = \mathcal{K}$. Then the notion of dilation becomes irrelevant though that of extension is flourishing. On the other hand, for the extension questions everywhere defined operators do not bring in any novelty; the unbounded operators are the only players left – as also this volume shows very rewarding players.

However, there is one important and beautiful instance when dilations and extensions (or, that is to say: bounded and unbounded) go together – that is Naĭmark's.

For the preamble If $\mathcal{H}$ is a Hilbert space $\boldsymbol{B}(\mathcal{H})$ stands for the C^*-algebra of all bounded operators on $\mathcal{H}$. Referring to unbounded operators we want to consider them always densely defined and closable. We keep up another sort of notation; domain, range and kernel of an operator A are denoted by $\mathcal{D}(A)$, $\mathcal{R}(A)$ and $\mathcal{N}(A)$ resp. $\mathcal{D}^\infty(A) \stackrel{\text{def}}{=} \bigcap_{n=0}^\infty \mathcal{D}(A^n)$.

If $\mathcal{E}$ is a subspace of $\mathcal{D}(A)$ as usual $A|_\mathcal{E}$ stands for the *restriction* of A, i.e. $\mathcal{D}(A|_\mathcal{E}) = \mathcal{E}$ and $A|_\mathcal{E} f = Af$ for $f \in \mathcal{E}$; in this case the restriction has to be densely defined. If a subspace $\mathcal{E}$ of $\mathcal{D}(A)$ is *invariant* for A, i.e. $A(\mathcal{E}) \subset \mathcal{E}$ then $A|_\mathcal{E}$ is understood as an operator in $\overline{\mathcal{E}}$. The latter includes the following case. If $\mathcal{L}$ is a closed subspace of $\mathcal{H}$ such that $A(\mathcal{D}(A) \cap \mathcal{L}) \subset \mathcal{L}$ (we again say that $\mathcal{L}$ is *invariant* for A), the operator $A{\restriction}_\mathcal{L} \stackrel{\text{def}}{=} A|_{\mathcal{D}(A)\cap\mathcal{L}}$ considered in $\mathcal{L}$ is called the *restriction* of A to $\mathcal{L}$ (in case $\mathcal{E} \subset \mathcal{D}(A)$ and $\mathcal{E}$ is closed, both these versions of invariance coincide and $A|_\mathcal{E} = A{\restriction}_\mathcal{E}$). The space $\mathcal{L}$ is invariant for A if and only if $PAP = AP$, where P is the orthogonal projection of $\mathcal{H}$ onto $\mathcal{L}$. We say that $\mathcal{L}$ is a *reducing* subspace for A (or $\mathcal{L}$ *reduces* A) if $P(\mathcal{D}(A)) \subset \mathcal{D}(A)$ and both $\mathcal{L}$ and $\mathcal{L}^\perp$ are invariant for A; then we call $A{\restriction}_\mathcal{L}$ the *part* of A in $\mathcal{L}$. The space $\mathcal{L}$ reduces A if and only if $PA \subset AP$, cf. [Stone, 1932, p. 150]. The space $\mathcal{D}^\infty(A) \stackrel{\text{def}}{=} \bigcap_{n=0}^\infty \mathcal{D}(A^n)$ is always invariant for A.

In principle we do not suppose the operators to be closed, unless stated explicitly. This is related to the fact that closed operators with invariant domain, and this happens here, are very likely to be bounded, certainly when an operator is symmetric, cf. [Ôta, 1984; Okazaki, 1986]. Under these circumstances another notion takes care of an operator: we say $\mathcal{D} \subset \mathcal{D}(A)$ is a *core* of A if $\overline{(A|_\mathcal{D})} = \bar{A}$.

Typographically A stands for symmetric, B for selfadjoint operators

Naĭmark's dilations Let $(T, \mathfrak{M})$ be a measure space and the mapping $F\colon \mathfrak{M} \to \boldsymbol{B}(\mathcal{H})$ be such that $\langle E(\cdot)f, f\rangle$ is a positive (=nonnegative) measure on T for every $f \in \mathcal{H}$ and $E(T) = I_\mathcal{H}$; call E a *semispec-*

tral measure[3] in $\mathcal{H}$. Applying the polarization formula to $\langle E(\cdot)f, f\rangle$ one gets complex measures $\langle E(\cdot)f, g\rangle$, $f, g \in \mathcal{H}$. If $\mathfrak{M} = \mathfrak{B}(T)$, a collection of Borel subsets of a topological space T, we additionally require the measure $\langle E(\cdot)f, f\rangle$ to be regular. A semispectral measure E is said to be *spectral* if the operators $E(\sigma)$ are idempotents for $\sigma \in \mathfrak{M}$, therefore orthogonal projections; this is equivalent to $E(\sigma \cap \tau) = E(\sigma)E(\tau)$, $\sigma, \tau \in \mathfrak{M}$, see [Mlak, 1978].

Theorem 10.2 (Naĭmark) *Let F be a semispectral measure in $\mathcal{H}$. Then there is a Hilbert space $\mathcal{K}$ containing $\mathcal{H}$ isometrically and a spectral measure in $\mathcal{K}$, the dilation of F, i.e. such that*[4] $F(\sigma) = PE(\sigma)|_{\mathcal{H}}$, $\sigma \in \mathfrak{M}$, *satisfying the minimality condition*

$$\mathcal{K} = \operatorname{clolin}\{E(\sigma)f\colon \sigma \in \mathfrak{M},\ f \in \mathcal{H}\}. \tag{10.3}$$

Moreover, for $(E_i, \mathcal{K}_i)$, $i = 1, 2$, minimal dilations of F, i.e. satisfying (10.3), there is a unitary operator U between $\mathcal{K}_1$ and $\mathcal{K}_2$ such that[5] $U|_{\mathcal{H}} = I_{\mathcal{H}}$ *and* $UE_1(\sigma) = E_2(\sigma)U$, $\sigma \in \mathfrak{M}$.

In fact in [Naĭmark, 1943] this theorem is proved for F finitely additive, in [Mlak, 1978, Theorem 4, p.30] a complete proof based on [Sz.-Nagy, 1955, Théorème principal] is provided. Notice that dropping minimality the only conclusion one can get from Sz.–Nagy theory is the dilating operators $E(\sigma)$ are orthogonal projections; to have $E(\cdot)$ a measure requires minimality.

Naĭmark's extensions The very classical result of von Neumann completely characterizes symmetric operators with equal deficiency indices[6], as those having selfadjoint extensions in the same Hilbert space; call the latter *von Neumann extensions*. Naĭmark's extensions go much further if one allows selfadjoint extensions to act in a larger space. He classifies these external extensions into two types depending on whether or not[7]

$$A = B \cap (\mathcal{H} \times \mathcal{H}) \subsetneqq B \text{ (type II)}. \tag{10.4}$$

Notice that closedness of A is necessary for (10.4) to hold.

[3] Another name tagged to it pretty often is a positive operator valued measure, in short POV-measure

[4] Whenever the positioning $\mathcal{H} \subset \mathcal{K}$ appears P stands for the orthogonal projection onto $\mathcal{H}$.

[5] Which means they are *$\mathcal{H}$–unitarily equivalent.*

[6] Some authors, like [Weidmann, 1980, p. 230], call them defect indices.

[7] We often identify an operator with its graph without any warning.

The basic result of [Naĭmark, 1940] in this matter says the following.

Theorem 10.3 *Let A be a symmetric operator in $\mathcal{H}$. Then*

(α) *there exists a Hilbert space $\mathcal{K}$ with $\mathcal{H} \subsetneqq \mathcal{H}$ isometrically and a selfadjoint operator B in $\mathcal{K}$ such that $A \subset B$;*

(β) *if A closed then B can always be chosen to satisfy* (10.4).

The main step in the proof of (α) is to consider in $\mathcal{K} \overset{\text{def}}{=} \mathcal{H} \bigoplus \mathcal{H}$ the operator $\tilde{B} \overset{\text{def}}{=} A \oplus -A$ and to show it has equal deficiency indices; for an accessible proof look at [Akhiezer and Glazman, 1981, vol. II, Theorem 1, p. 127]. For details of the proof that a selfadjoint extension can always be chosen to be of type II, i.e. to satisfy (10.4), cf. [Stochel and Szafraniec, 1991].

Notice that

$$\mathcal{D}(A) \subset \mathcal{D}(B) \cap \mathcal{H} \subset P\mathcal{D}(B) \subset \mathcal{D}(A^*) \tag{10.5}$$

with equality instead of the second inclusion when B is von Neumann.

Call those appearing in (α) of Theorem 10.3 *Naĭmark extensions*. A symmetric operator besides having Naĭmark extensions (which always exist) may have also von Neumann ones, the interplay between these two is what we are going to consider in more detail in this paper.

Example 10.4 It is well known that the operator

$$\mathcal{D}(A) \overset{\text{def}}{=} \{f \in \mathcal{L}^2[0,+\infty) \colon f \text{ absolutely continuous in } [0,+\infty),$$
$$f' \in \mathcal{L}^2[0,+\infty) \text{ and } f(0)=0\},$$
$$Af \overset{\text{def}}{=} -\mathrm{i}f', \quad f \in \mathcal{D}(A)$$

is maximal symmetric, cf. [Akhiezer and Glazman, 1981, vol. I, p. 111] with deficiency indices $(0,1)$. The main step of the proof of Theorem 10.3 suggests, see [Akhiezer and Glazman, 1981, vol. II, p. 138], to consider the operator $\tilde{B} \overset{\text{def}}{=} -A \oplus A \subset B$ in

$$\mathcal{K} \overset{\text{def}}{=} \mathcal{L}^2[0,+\infty) \bigoplus \mathcal{L}^2[0,+\infty) = \mathcal{L}^2(-\infty,0] \bigoplus \mathcal{L}^2[0,+\infty)$$
$$= \mathcal{L}^2(-\infty,0) \bigoplus \mathcal{L}^2[0,+\infty) = \mathcal{L}^2(\mathbb{R}).$$

It is not too difficult to show that

$$\mathcal{D}(\tilde{B}) \overset{\text{def}}{=} \{f \in \mathcal{L}^2(\mathbb{R}) \colon f \text{ absolutely continuous in } \mathbb{R},$$
$$f' \in \mathcal{L}^2(\mathbb{R}) \text{ and } f(0)=0\},$$
$$\tilde{B}f \overset{\text{def}}{=} -\mathrm{i}f', \quad f \in \mathcal{D}(\tilde{B})$$

Then B defined by

$$\mathcal{D}(B) \stackrel{\text{def}}{=} \{f \in \mathcal{L}^2(\mathbb{R})\colon f \text{ absolutely continuous in } \mathbb{R}, f' \in \mathcal{L}^2(\mathbb{R}),$$
$$Bf \stackrel{\text{def}}{=} -\mathrm{i}f', \quad f \in \mathcal{D}(B)$$

is a selfadjoint extension of A. Moreover $\tilde{B}$ has deficiency indices $(1,1)$ [8] and, referring to (10.4) and (10.5), we have

$$\mathcal{D}(A) = \mathcal{D}(\tilde{B}) \cap \mathcal{H} \subsetneqq \mathcal{D}(B) \cap \mathcal{H} = P\mathcal{D}(B) = \mathcal{D}(A^*).$$

Around the spectral theorem Instead of stating the theorem formally we rather would like to describe it in a more detailed and slightly more cautious way. Given a selfadjoint operator B in $\mathcal{H}$, there are several ways of proving existence of a unique spectral measure E in $\mathcal{H}$ such that

$$\langle Bf, g\rangle = \int_{\mathbb{R}} t \, \langle E(\mathrm{d}t)f, g\rangle, \quad f \in \mathcal{D}(B), \, g \in \mathcal{H} \tag{10.6}$$

and the domain $\mathcal{D}(B)$ is equal to

$$\mathcal{D}_E \stackrel{\text{def}}{=} \{f \in \mathcal{H}\colon \int_{\mathbb{R}} t^2 \, \langle E(\mathrm{d}t)f, f\rangle < +\infty\}.$$

For the spectrum of B we have $\mathrm{sp}(B) = \operatorname{supp} E$ and $E(\sigma)B \subset BE(\sigma)$ for every $\sigma \in \mathfrak{B}(\mathbb{R})$. Moreover, if the set $\mathcal{T}$ is total in $\mathcal{H}$, because E is a regular measure, we get

$$\mathcal{B}_E \stackrel{\text{def}}{=} \operatorname{span}\{E(\sigma)f\colon f \in \mathcal{T}, \, \sigma \in \mathfrak{B}(\mathbb{R}) \text{ compact}\} \text{ is a core of } B;$$

it is evidently invariant.

The converse goes as follows: given a spectral measure E such that the linear space $\mathcal{D}_E$ is dense in $\mathcal{H}$, the operator B defined by (10.6) with $\mathcal{D}(B) = \mathcal{D}_E$ is selfadjoint.

Remark 10.5 Notice that with $f \in \mathcal{D}^\infty(B)$ the formula (10.6) holds for each B^n with the integrand t^n; this is due to multiplicativity of spectral integrals.

One of the unnoticed features of the theory, which does not carry over

[8] Because of this the symmetric operator $\tilde{B}$ is a proper target for constructions of von Neumann and even Naĭmark extensions. However going down, to the initial space $\mathcal{H}$ all this ends up with the unique semispectral measure representing A as it is maximal symmetric, cf. [Akhiezer and Glazman, 1981, vol. II, Theorem 2, p. 135].

to the multivariate case, is that the restriction of a selfadjoint operator to its invariant subspace remains symmetric; this procedure is a kind of converse to extension. Therefore the question of how much of the spectral representation symmetric operators inherit from that of their selfadjoint extensions is of immense importance.

Combining Theorem 10.2, Theorem 10.3 and the spectral theorem for selfadjoint operators we get an extension free characterization of symmetric operators by means of integral representation.

Corollary 10.6 *An operator A in $\mathcal{H}$ is symmetric if and only if there is a semispectral measure F in $\mathcal{H}$ such that*

$$\langle A^m f, A^n g\rangle = \int_{\mathbb{R}} t^{m+n}\langle F(\mathrm{d}t)f, g\rangle, \quad m, n = 0, 1,\ f, g \in \mathcal{D}(A). \tag{10.7}$$

Moreover,

$$\mathcal{D}(A) \subset \{f \in \mathcal{H} \colon \int_{\mathbb{R}} t^2 \langle F(\mathrm{d}t)f, f\rangle < +\infty\}$$

with equality when the minimal dilation of F stems from a selfadjoint extension satisfying (10.4) or when A is maximal symmetric.

The right hand side of (10.7) defines a symmetric operator A_F acting within $\mathcal{H}$, which is an extension of A, by

$$\mathcal{D}(A_F) \overset{\mathrm{def}}{=} \{f \in \mathcal{H} \colon \int_{\mathbb{R}} t^2\langle F(\mathrm{d}t)f, f\rangle < +\infty\},$$
$$\langle A_F f, g\rangle \overset{\mathrm{def}}{=} \int_{\mathbb{R}} t\langle F(\mathrm{d}t)f, g\rangle,\ f \in \mathcal{D}(A_F),\ g \in \mathcal{H}.$$

If A is not essentially selfadjoint the set of measures F satisfying (10.7) is rather sizable.

Minimality Call a selfadjoint extension B of A *minimal* if the only closed subspace of $\mathcal{K}$ including $\mathcal{H}$ and reducing B is $\mathcal{K}$ itself[9].

Proposition 10.7 *Let A be symmetric in $\mathcal{H}$ and let B be is a selfadjoint extension in $\mathcal{K}$ with E its spectral measure. The closed subspace*

$$\mathcal{K}_E \overset{\mathrm{def}}{=} \mathrm{clolin}\{E(\sigma)f \colon \sigma \in \mathfrak{M},\ f \in \mathcal{H}\}$$

of $\mathcal{K}$ reduces B and $B{\restriction}_{\mathcal{K}_E}$ is a minimal selfadjoint extension of A.

[9] In [Stochel and Szafraniec, 1989] such an extension is called minimal of spectral type just to distinguish this kind of minimality from other kinds, not present here.

Proof Make use of $E(\sigma)B \subset BE(\sigma)$. □

As an immediate consequence of the above proposition we get that minimal selfadjoint extensions always exist; the restrictions $B\restriction_{\mathcal{H}_E}$ are such. This observation can serve as an alternative definition of minimality: B is minimal if $\mathcal{K} = \mathcal{K}_E$. Notice that von Neumann extensions are always minimal due to the definition.

An important fact concerning this notion is that

$$\operatorname{sp}(B) \subset \operatorname{sp}(A) \tag{10.8}$$

as long as B is a minimal selfadjoint extension of A, look at [Stochel and Szafraniec, 1989, Theorem 1] for this as well as for other spectral relations. If $(B_i, \mathcal{K}_i)$, $i = 1, 2$, are minimal extensions of A there is a unitary operator U between $\mathcal{K}_1$ and $\mathcal{K}_2$ such that $U|_{\mathcal{H}} = I_{\mathcal{H}}$ and $UB_1 = B_2U$, we say B_1 and B_2 are *$\mathcal{H}$–unitarily equivalent.* The spectral inclusion (10.8) suggests at once how to find minimal normal extensions which are not $\mathcal{H}$-unitarily equivalent, see [Stochel and Szafraniec, 1989, Example 1]. In the sequel we develop this idea extensively along the lines of [Cichoń, Stochel and Szafraniec, 2010].

Cyclicity Notice that for selfadjoint operators $\mathcal{B}_E \subset \mathcal{D}^\infty(A)$ while for some symmetric operators the situation may be even worse, for instance $\mathcal{D}(A^2) = \{0\}$, cf. [Naĭmark, 1940a]. However, sometimes it happens that A has an invariant core. The most comfortable situation is when this core is generated by a single vector, that is when A is *cyclic*[10], which means there is a vector $e \in \mathcal{D}^\infty(A)$ (called a *cyclic vector* of A) such that $\mathcal{D}_e \overset{\text{def}}{=} \{p(A)e \colon p \in \mathbb{C}[X]\}$ is a core of A. Thus a cyclic operator has an invariant subspace $\mathcal{D}_e$ which is its core.

The following simple result shows the strength of cyclicity; it is somehow in the flavour of [Gorbachuk and Gorbachuk, 1997, Theorem 7.1, p. 91].

Proposition 10.8 *A cyclic symmetric operator must necessarily have equal deficiency indices or, equivalently, von Neumann extensions.*

Proof Indeed, due to

$$\|\sum\nolimits_n p_n A^n e\| = \|(\sum\nolimits_n p_n A^n)^* e\| = \|\sum\nolimits_n \bar{p}_n A^n e\|,$$

the operator $V\colon \sum_n p_n A^n e \to \sum_n \bar{p}_n A^n e$ is a well-defined conjugation.

[10] There is another notion: $*$–cyclicity which is designed for formally normal (hence normal) operators. For symmetric operators they coincide.

Because $\mathcal{D}_e$ is a core of A the transparent inclusion $V\mathcal{D}_e \subset \mathcal{D}_e$ extends to $V\mathcal{D}(A) \subset \mathcal{D}(A)$. Therefore A is V–real and, consequently, it has a selfadjoint extension within $\mathcal{H}$, cf. [Weidmann, 1980, p. 235]. □

Proposition 10.9 *Suppose A is a symmetric operator such that*

$$\mathcal{D}_e \text{ is dense in } \mathcal{H} \tag{10.9}$$

with some $e \in \mathcal{D}^\infty(A)$ and B is a minimal selfadjoint extension of A. If $P\mathcal{D}(B) \subset \mathcal{D}(B)$ then B is a von Neumann extension of A. Consequently, if B is a Naĭmark extension then[11] $P\mathcal{D}(B) \not\subset \mathcal{D}(B)$.

Proof Because B is an extension of A, $\mathcal{D}_e = \{p(A)e\colon p \in \mathbb{C}[X]\} = \{p(B)e\colon p \in \mathbb{C}[X]\}$, the closed subspaces $\overline{\mathcal{D}_e}$ and $\overline{\mathcal{D}_e}^\perp$ of $\mathcal{K}$ are invariant for B. If $P\mathcal{D}(B) \subset \mathcal{D}(B)$ the closure $\overline{\mathcal{D}_e}$ reduces B. Because $\mathcal{H} = \overline{\mathcal{D}_e}$, minimality of B enforces $\mathcal{H} = \mathcal{K}$. □

Let μ be a positive measure on $\mathbb{R}$ with finite moments[12] and such that

$$\text{the polynomials } \mathbb{C}[X] \text{ are dense in } \mathcal{L}^2(\mu). \tag{10.10}$$

Define the operator M of *multiplication* by the independent variable in $\mathcal{L}^2(\mu)$ as follows

$$\mathcal{D}(M) \overset{\text{def}}{=} \{f \in \mathcal{L}^2(\mu)\colon g_f \in \mathcal{L}^2(\mu)\},\ Mf \overset{\text{def}}{=} g_f,\ g_f(t) \overset{\text{def}}{=} tf(t)\ \mu\text{-a.e.}$$

Because $\mathbb{C}[X] \subset M$ the operator A is densely defined. It is an easy task to show that A is symmetric and $\mathcal{D}(A) = \mathcal{D}(A^*)$, therefore A is selfadjoint. Notice (10.10) is too little for $\mathbb{C}[X]$ to be a core of M, cf. Spectral Theorem 10.12.

Remark 10.10 Let A be a symmetric operator such that (10.9) holds with some $e \in \mathcal{D}^\infty(A)$[13]. If B is any selfadjoint extension of A with the spectral measure E then

$$\|p(A)e\|^2 = \|p(B)e\|^2 = \int_{\mathbb{R}} |p(t)|^2 \langle E(\mathrm{d}t)e, e\rangle = \int_{\mathbb{R}} |p(t)|^2 \mu(\mathrm{d}t)$$
$$\mu \overset{\text{def}}{=} \langle E(\cdot)e, e\rangle. \tag{10.11}$$

Thus the mapping

$$U\colon \mathcal{D}_e \ni p(A)e \to p \in \mathcal{L}^2(\mu) \tag{10.12}$$

[11] Cf. (10.5).
[12] That is $\int_{\mathbb{R}} t^{2n} \mu(\mathrm{d}t) < +\infty$, $n = 0, 1, \ldots$
[13] This is still less than to say $\mathcal{D}_e$ is a core of A, cyclicity may not be present.

is an isometry. In any case, with M as above

$$UAp(A)e = MUp(A)e.$$

If A is cyclic, this implies $U\overline{A} \subset MU$.

Proposition 10.11 *Suppose A is a symmetric operator satisfying* (10.9) *for some $e \in \mathcal{D}^\infty(A)$. Then B is a von Neumann extension of A if and only if* (10.10) *holds with μ defined by* (10.11). *If this happens U defined by* (10.12) *must necessarily be unitary.*

Proof Suppose B is a von Neumann extension. With E being a spectral measure of B and μ as (10.11), for $p, q \in \mathbb{C}[X]$ and $\rho, \sigma \in \mathfrak{B}(\mathbb{R})$ multiplicative properties of the spectral integral give us

$$\langle p(B)E(\rho)e, q(B)E(\sigma)e\rangle = \int_{\mathbb{R}} \chi_\rho p \overline{\chi_\sigma q}\, \mathrm{d}\mu. \tag{10.13}$$

Putting $\sigma = \rho = \mathbb{R}$, due to (10.13), the mapping $p(B)E(\sigma)e \to p\chi_\sigma$ coincides on $\mathcal{D}_e$ with U given by (10.12). This makes U onto and, consequently, the polynomials are dense in $\mathcal{L}^2(\mu)$.

Conversely, if (10.10) holds U is a unitary mapping between $\mathcal{H}$ and $\mathcal{L}^2(\mu)$. The inverse U^{-1} sends p to $p(B)e$ with the latter necessarily in $\mathcal{H}$. Thus B is a von Neumann extension. □

Proposition 10.11 is a version of [Buchwalter and Cassier, 1984, Proposition 1.4]. It can be restated roughly that von Neumann extensions are those for which the conclusion of the well-known result of [Riesz, 1923] holds, *ergo* they are precisely N-extremal (actually this serves in [Buchwalter and Cassier, 1984] for a definition of N-extremal measures).

'Diagonal' version of the spectral theorem This is an enhancement of what is known for bounded operators; if the current version were somewhere in the literature, it might be hidden deeply.

Theorem 10.12 (Spectral Theorem) *A selfadjoint operator B satisfying* (10.9) *is unitary equivalent to the operator M of multiplication by the independent variable in $\mathcal{L}^2(\mu)$ with μ defined as in* (10.11). *The unitary mapping U is that given by* (10.12).

If B is cyclic, then $UB \subset MU$ and the polynomials $\mathbb{C}[X]$ are dense in $\mathcal{L}^2((1+X^2)\mu)$.

Proof of the extras Because B is its von Neumann extension most of the theorem stems from Proposition 10.11.

Denseness of the polynomials in $\mathcal{L}^2((1+X^2)\mu)$ comes out from the graph topology interpretation of $\mathbb{C}[X]$ being a core of B. □

The last conclusion of Theorem 10.12 warns that not every measure with finite moments makes its multiplication operator cyclic, *vide* N-extremal measures (cf. Remark 10.10 as well).

If A is a cyclic symmetric operator then (10.7) holds for all $m, n = 0, 1, \ldots$ with $f, g \in \mathcal{D}_e$. In fact we have much more, with $\mu = \langle F(\cdot)e, e\rangle = \langle E(\cdot)e, e\rangle$ for any spectral measure E dilating F,

$$\langle A^m e, A^n e\rangle = \int_{\mathbb{R}} t^{m+n}\mu(\mathrm{d}t), \quad m, n = 0, 1, \ldots$$

If this happens we say μ is a *representing measure* of A.

For a cyclic symmetric operator A in $\mathcal{H}$ with the cyclic vector e let B be its selfadjoint extension (regardless the Hilbert space it acts in) with its spectral measure E. Consider the positive measure

$$\mu_B \stackrel{\text{def}}{=} \langle E(\cdot)e, e\rangle. \tag{10.14}$$

Denote by $\mathfrak{S}$ the collection of all measures μ_B defined by (10.14) when E ranges over the spectral measures of all selfadjoint extensions B of A.

Corollary 10.13 *Suppose A is a cyclic symmetric operator with a cyclic vector e. Then*

- (a) *for each $B \in \mathfrak{S}$, μ_B is a representing measure of A;*
- (b) *if $B_1, B_2 \in \mathfrak{S}$, then $\mu_{B_1} = \mu_{B_2}$ if and only if B_1 and B_2 are $\mathcal{H}$-unitarily equivalent;*
- (c) *for every representing measure μ of A there exists $B \in \mathfrak{S}$ such that $\mu = \mu_B$;*
- (d) *for every $B \in \mathfrak{S}$, the spectrum of B is equal to* $\operatorname{supp} \mu_B$.

10.2 The example

Fulfilling our promise we give in this section merely a sketch of [Cichoń, Stochel and Szafraniec, 2010] where further details like proofs, references, comments and, above of all, subtle considerations concerning the Fourier transform are included.

The moment problem . . . Recall that $(a_n)_{n=0}^{\infty}$ is said to be a *Hamburger moment* sequence if there exists a positive Borel measure μ on the real line $\mathbb{R}$ such that

$$a_n = \int_{\mathbb{R}} x^n \mathrm{d}\mu(x), \quad n = 0, 1, 2, \ldots$$

The measure μ is called a *representing* measure of $(a_n)_{n=0}^{\infty}$. A Hamburger moment sequence is said to be *indeterminate* if it has more than one representing measure; otherwise we call it *determinate*. In particular (cf. [Fuglede, 1983] for instance), every Hamburger moment sequence which has a compactly supported representing measure is determinate.

Our central point is the moment sequence $(a_n)_{n=0}^{\infty}$ defined by means of a fixed nonzero *test function* ω on $\mathbb{R}$, i.e. a complex function of class $\mathcal{C}^{\infty}$ whose closed support supp ω is compact, by

$$a_n \overset{\mathrm{def}}{=} (-\mathrm{i})^n \int_{\mathbb{R}} \frac{\mathrm{d}^n \omega}{\mathrm{d}x^n}(x)\, \overline{\omega(x)}\, \mathrm{d}x, \quad n = 0, 1, 2, \ldots \tag{10.15}$$

Applying the Plancherel theorem (cf. [Rudin, 1973, Theorem 7.9]) to (10.15), we see that

$$a_n = \int_{\mathbb{R}} x^n\, |\widehat{\omega}(x)|^2 \mathrm{d}x, \quad n = 0, 1, 2, \ldots, \tag{10.16}$$

where $\widehat{\omega}$ stands for the Fourier transform of ω, i.e.

$$\widehat{\omega}(y) = \frac{1}{\sqrt{2\pi}} \int_{\mathbb{R}} \omega(x) \mathrm{e}^{-\mathrm{i}\, xy} \mathrm{d}x, \quad y \in \mathbb{R}.$$

We get from (10.16) that all the numbers a_n are real. Moreover,

$$a_{2n} > 0, \quad n = 0, 1, 2, \ldots \tag{10.17}$$

Indeed, otherwise $a_{2n} = 0$ for some integer $n \geqslant 0$, which implies that $\widehat{\omega}(y) = 0$ for almost every y with respect to the Lebesgue measure on $\mathbb{R}$. Then, by the Plancherel theorem and continuity of ω, $\omega(x) = 0$ for every $x \in \mathbb{R}$, which contradicts our assumption.

It follows from (10.16) that $\{a_n\}_{n=0}^{\infty}$ is a Hamburger moment sequence and that $|\widehat{\omega}(x)|^2 \mathrm{d}x$ is its representing measure. This measure turns out to be equivalent to the Lebesgue measure on the real line $\mathbb{R}$. Indeed, by the Paley-Wiener theorem (cf. [Rudin, 1973, Theorem 7.22]), $\widehat{\omega}$ extends to an entire function on $\mathbb{C}$, which by the Plancherel theorem is nonzero (as $\omega \neq 0$). Hence, the set $\{x \in \mathbb{R} \colon \widehat{\omega}(x) = 0\}$ has no accumulation points and consequently is at most countable. This gives the desired equivalence.

... and its operators A sequence is a Hamburger moment one if and only if it is positive definite; this is the classical, very old fact. A typical way, at least when operator theory gets involved, is to attach to a positive definite sequence $(a_n)_{n=0}^{\infty}$ a Hilbert space $\mathcal{H}$, a cyclic symmetric operator A in $\mathcal{H}$ and a vector $e \in \mathcal{D}^{\infty}(A)$ such that

$$a_n = \langle A^n e, e\rangle, \quad n = 0, 1, \dots \tag{10.18}$$

The *moment triplet* $(\mathcal{H}, A, e)$ is uniquely determined up to unitary equivalence; the operator A itself is sometimes called the *moment operator*[14]. Recall, two triplets $(\mathcal{H}_i, A_i, e_i)$, $i = 1, 2$ with A_i's cyclic are isomorphic with a unique unitary $U\colon \mathcal{H}_1 \to \mathcal{H}_2$ such that $Ue_1 = e_2$ and $UA_1 = A_2U$. If a moment operator has deficiency indices $(0, 0)$ the Spectral Theorem 10.12 takes care of it in full; otherwise one may to refer to Proposition 10.11 and Corollary 10.13. In any case they are a convenient way of transferring moment problems to operator theory.

Regarding our example we have just got in a natural way two such triplets, due to the Plancherel Theorem being unitary equivalent via the Fourier transform.

① Formula (10.16), which is in fact already an integral representation of the sequence $(a_n)_{n=0}^{\infty}$, determines the operator A_1 as multiplication by the independent variable in $\mathcal{L}^2(|\hat{\omega}(t)|^2\mathrm{d}t)$ with $\mathcal{D}(A_1)$ equal to the restriction of $\mathbb{C}[X]$ to $\mathcal{L}^2(|\hat{\omega}(t)|^2\mathrm{d}t)$, the Hilbert space $\mathcal{H}_1$ equal to the closure of $\mathbb{C}[X]$ in $\mathcal{L}^2(|\hat{\omega}(t)|^2\mathrm{d}t)$ and $e_1 = 1$. Apparently (10.16) corresponds to (10.18).

② On the other hand, formula (10.15) gives rise to the operator $A_2 \stackrel{\text{def}}{=} \frac{\mathrm{d}}{\mathrm{d}x}$, with $\mathcal{D}(A_2) \stackrel{\text{def}}{=} \operatorname{span}\left(\frac{\mathrm{d}^n\omega}{\mathrm{d}x^n}\right)_{n=0}^{\infty}$, the Hilbert space $\mathcal{H}_2 \stackrel{\text{def}}{=} \operatorname{clolin}\left(\frac{\mathrm{d}^n\omega}{\mathrm{d}x^n}\right)_{n=0}^{\infty}$, the closure being in $\mathcal{L}^2(\mathrm{d}x)$, and $e_2 \stackrel{\text{def}}{=} \omega$. Now (10.18) becomes (10.15).

The case ② embodies the working model which helps us to promote the use of Naĭmark extensions in describing solutions of an indeterminate moment problem. Therefore referring to it we drop the subscript 2 for all its ingredients.

So as to make the initial space tight put for an arbitrary but fixed nonzero test function ω on $\mathbb{R}$

$$\alpha \stackrel{\text{def}}{=} \inf \operatorname{supp} \omega \quad \text{and} \quad \beta \stackrel{\text{def}}{=} \sup \operatorname{supp} \omega. \tag{10.19}$$

14 In fact moment triplets (or moment operators when the other members are clear) are in one-to-one correspondence with moment sequences, so they are replaceable.

Define the operator $\tilde{A}$ in $\mathcal{L}^2[\alpha,\beta]$ by

$$\mathcal{D}(\tilde{A}) = \{f \in \mathcal{L}^2[\alpha,\beta]\colon f \text{ absolutely continuous in } [\alpha,\beta],\ f' \in \mathcal{L}^2[\alpha,\beta] \\ \text{and } f(\alpha) = f(\beta) = 0\},$$

$$\tilde{A}f = -\mathrm{i}\, f', \quad f \in \mathcal{D}(\tilde{A}).$$

Then $\tilde{A}$ is a closed symmetric operator in $\mathcal{L}^2[\alpha,\beta]$ with the deficiency indices $(1,1)$ (cf. [Stone, 1932, Theorem 10.7]). Let $\mathcal{H}$ be the closure in $L^2[\alpha,\beta]$ of span $\{\tilde{A}^n\omega\}_{n=0}^{\infty}$ and let A be the operator in $\mathcal{H}$ defined on $\mathcal{D}(A) \overset{\text{def}}{=} \text{span}\,\{\tilde{A}^n\omega\}_{n=0}^{\infty}$ by $Af = \tilde{A}f$ for $f \in \mathcal{D}(A)$. Clearly, A is a symmetric cyclic operator in $\mathcal{H}$ (with a cyclic vector ω) and the operator $\tilde{A}$ is its extension that acts in the Hilbert space $L^2[\alpha,\beta]$, which, due to [Cichoń, Stochel and Szafraniec, 2010, Corollary 2.11], is always a proper superspace of $\mathcal{H}$. The formula (10.15) can now be written as (10.18) which is a good starting point for the extension theory to be performed.

von Neuman extensions of the moment operator and N-extremal solutions An analytical description of all selfadjoint extesions of moment operators (=moment problems) is an old affair, cf. [Stone, 1932; Akhiezer, 1965]; it is given by means of Nevanlinna matrices of entire functions. More precisely, there is a bijective correspondence between moment operators and quadruples (A, B, C, D) of suitable entire functions, called the Nevannlina matrices, such that the Stieltjes transform (or the so-called generalized resolvent) of any representing measure, say μ is of the form

$$\int_{\mathbb{R}} \frac{\mu(\mathrm{d}t)}{z-t} = \frac{A(z)\varphi(z) - C(z)}{B(z)\varphi(z) - D(z)}, \tag{10.20}$$

where φ ranging over the Nevanlinna class $\cup\,\{\infty\}$ is a parameter determining μ.

Dealing with indeterminacy one pretty often exposes N-extremality of representing measures. *N–extremal* measures (cf. Proposition 10.11 and the paragraph following its proof) correspond to the constant $\varphi \in \mathbb{R} \cup \{\infty\}$ in (10.20), see [Buchwalter and Cassier, 1984] for a contemporary, well-balanced approach as well as [Simon, 1998]. Therefore our preparatory work profits in having the one-parameter family of selfadjoint extensions B_s, $s \in \mathbb{R} \cup \{\infty\}$, of the moment operator as well as the one-parameter family of measures μ_s coming from (10.14) via the spectral measures E_s of the operators B_s. By [Simon, 1998, Theorem 4.11], $\{\mu_s \overset{\text{def}}{=} \mu_{B_s} \colon s \in \mathbb{R} \cup \{\infty\}\}$ is a family of discrete representing measures of $\{a_n^\omega\}_{n=0}^\infty$ whose supports form a partition of $\mathbb{R}$. These supports are described as the zero sets of appropriate entire functions coming from the Nevanlinna parametrization (cf. [Simon, 1998, Theorem 4.10 and the proof of Theorem 4.11]), far from being explicit. Lastly, $\{\mu_s : s \in \mathbb{R} \cup \{\infty\}\}$ is the set of all *N-extremal* representing measures of $(a_n)_{n=0}^\infty$.

N-extremal measures must necessarily satisfy

$$\sum_{\lambda \in \operatorname{supp} \mu \setminus \{0\}} |\lambda|^{-p} < +\infty \quad \text{for all real } p > 1, \tag{10.21}$$

see [Simon, 1998, Theorem 4.11]. Condition (10.21) is a convenient criterion which helps to exclude straight away N-extremality of some other measures. In particular, if the support of a representing measure has an accumulation (cluster) point, then the measure can never be N-extremal as (10.21) fails to hold.

Remark 10.14 The support of an N-extremal measure of an indeterminate Hamburger moment sequence is never in arithmetic progression. In particular 1° always implies 3°

Naĭmark extensions of the moment operator and their representing measures In the sequel $\mathtt{d}$ is a shorthand notation for the $\mathtt{data}$ [15] $([a,b],t)$ such that $[\alpha,\beta] \subset [a,b]$ and $t \in [0,2\pi)$; each time when $[a,b]$ or t appears one has to admit they are hidden in the closest $\mathtt{d}$ around and as such are represented implicitly by $\mathtt{d}$; be careful and suspicious, satisfaction guaranteed.

Let $B_{\mathtt{d}}$ be a selfadjoint operator defined in the complex Hilbert space $\mathcal{L}^2[a,b]$ by

$$\mathcal{D}(B_{\mathtt{d}}) \stackrel{\text{def}}{=} \{f \in \mathcal{L}^2[a,b] : f \text{ absolutely continuous in } [a,b], \\ f' \in \mathcal{L}^2[a,b] \text{ and } f(a) = \mathrm{e}^{\mathrm{i}\,t} f(b)\},$$

$$B_{\mathtt{d}} f \stackrel{\text{def}}{=} -\mathrm{i}\, f', \quad f \in \mathcal{D}(B_{\mathtt{d}}),$$

where f' is the derivative of f (cf. [Stone, 1932, Theorem 10.7] or

15 There are two kinds of parameters for extended objects and their follow-ups: $s \in \mathbb{R} \cup \{\infty\}$ refers to von Neumann's and $\mathtt{d} = ([a,b],t)$ does to Naĭmark 's; no confusion possible.

[Akhiezer and Glazman, 1981, vol. I, Section 55]). We set [16]

$$\lambda_{\mathsf{d},k} \overset{\text{def}}{=} \frac{2k\pi - t}{b-a}, \quad f_{\mathsf{d},k}(x) \overset{\text{def}}{=} \frac{1}{\sqrt{b-a}} \exp\big(\mathrm{i}\, x \lambda_{\mathsf{d},k}\big), \ k \in \mathbb{Z},\ x \in [a,b]. \tag{10.22}$$

The sequence $(f_{\mathsf{d},k})_{k\in\mathbb{Z}}$ is an orthonormal basis of $\mathcal{L}^2[a,b]$ which consists of eigenvectors of B_{d}, cf. [Stone, 1932, Theorem 10.7]. The spectral measure E_{d} of B_{d} can be described as follows

$$E_{\mathsf{d}}(\sigma) f = \sum_{k\in\mathbb{Z}} \langle f, f_{\mathsf{d},k}\rangle\, \chi_\sigma\big(\lambda_{\mathsf{d},k}\big)\, f_{\mathsf{d},k}, \quad f \in \mathcal{L}^2[a,b],\ \sigma \in \mathfrak{B}(\mathbb{R}),$$

where $\langle \cdot, \text{-}\rangle$ is the inner product of $\mathcal{L}^2[a,b]$ and χ_σ is the characteristic function of σ. Therefore for every Borel subset σ of $\mathbb{R}$,

$$\mu_{\mathsf{d}}(\sigma) \overset{\text{def}}{=} \langle E_{\mathsf{d}}(\sigma)\omega, \omega\rangle = \frac{2\pi}{b-a} \sum_{k\in\mathbb{Z}} |\widehat{\omega}(\lambda_{\mathsf{d},k})|^2\, \chi_\sigma(\lambda_{\mathsf{d},k}). \tag{10.23}$$

Since $\omega \in \bigcap_{n=1}^\infty \mathcal{D}((B_{\mathsf{d}})^n)$, we obtain

$$\begin{aligned} a_n = \langle (B_{\mathsf{d}})^n \omega, \omega\rangle &= \Big\langle \int_{\mathbb{R}} x^n E_{\mathsf{d}}(\mathrm{d}x)\omega, \omega \Big\rangle \\ &= \int_{\mathbb{R}} x^n \mu_{\mathsf{d}}(\mathrm{d}x), \ n = 0, 1, \ldots, \end{aligned} \tag{10.24}$$

which means that μ_{d} is a representing measure of the Hamburger moment sequence $(a_n)_{n=0}^\infty$. This fact and (10.23) imply that [17]

$$a_n = \frac{2\pi}{b-a} \sum_{k\in\mathbb{Z}} (\lambda_{\mathsf{d},k})^n\, |\widehat{\omega}(\lambda_{\mathsf{d},k})|^2, \quad n = 0, 1, 2, \ldots,$$

where the series is absolutely convergent. Put $Y_{\mathsf{d}} \overset{\text{def}}{=} \{\lambda_{\mathsf{d},k} \colon k \in \mathbb{Z}\}$. It is clear that [18]

$$\text{the family } \{Y_{\mathsf{d}}\}_{t\in[0,2\pi)} \text{ is a partition of } \mathbb{R}, \tag{10.25}$$

$$\operatorname{supp} \mu_{\mathsf{d}} \overset{(10.23)}{=} \{\lambda_{\mathsf{d},k} \colon k \in \mathbb{Z},\ \widehat{\omega}(\lambda_{\mathsf{d},k}) \neq 0\} \subset Y_{\mathsf{d}}. \quad t \in [0, 2\pi), \tag{10.26}$$

What is more, by (10.17) and (10.24), $\operatorname{supp} \mu_{\mathsf{d}} \neq \varnothing$ for every $t \in [0, 2\pi)$. Hence, combining (10.24), (10.25) and (10.26), we conclude that $(a_n)_{n=0}^\infty$

[16] $\mathbb{Z} \overset{\text{def}}{=} \{\ldots, -2, -1, 0, 1, 2, \ldots\}$.

[17] Use the Lebesgue monotone convergence theorem for measures and series.

[18] Notice that, by (10.22) and (10.23), Y_{d} and μ_{d} depend in fact on $b-a$ with the other parameters being fixed; moreover, any real number greater than or equal to the diameter of $\operatorname{supp}\omega$ can be written as $b-a$ with $\operatorname{supp}\omega \subset [a,b]$.

is an indeterminate Hamburger moment sequence which has the continuum of representing measures μ_{d}, $t \in [0, 2\pi)$, with pairwise disjoint supports. It is clear that

$$\text{each set supp } \mu_{\mathsf{d}} \text{ is infinite,} \tag{10.27}$$

otherwise the moment sequence would be determinate. Taking into account all possible values of the difference $b - a$ (cf. footnote [18]) we get more, namely that there are uncountably many families of representing measures, each of which contains measures with pairwise disjoint supports. This follows from (10.27) and the ensuing observation:

$$\begin{array}{l}\text{the set supp } \mu_{\mathsf{d}} \cap \text{supp } \mu_{\mathsf{d}'} \text{ has at most one point when-}\\ \text{ever the number } (b-a)(b'-a')^{-1} \text{ is irrational.}\end{array} \tag{10.28}$$

In turn, (10.28) can be deduced from the fact that the distance between two different points of supp μ_{d} is an integer multiple of $2\pi(b-a)^{-1}$.

Minimality of B_{d} Our next goal is to describe the part of A_{d} which is a minimal selfadjoint extension of A. This allows us to get rid of 'ghost' eigenspaces.

Proposition 10.15 [Cichoń, Stochel and Szafraniec, 2010, Proposition 2.2] *Let $(f_{\mathsf{d},k})_{k\in\mathbb{Z}}$, given by (10.22), be the orthonormal basis of $\mathcal{L}^2[a,b]$. Then the space*

$$\text{clolin}\,(f_{\mathsf{d},k})_{k\in\mathbb{Z},\,\widehat{\omega}(\lambda_{\mathsf{d},k})\neq 0} \tag{10.29}$$

reduces the operator B_{d} to a minimal selfadjoint extension of A. The operator B_{d} is itself a minimal selfadjoint extension of A if and only if $\widehat{\omega}(\lambda_{\mathsf{d},k}) \neq 0$ for all $k \in \mathbb{Z}$. This happens in particular if the function $\widehat{\omega}$ has no real zeros.

From now on keep the same notation B_{d} for the restriction of the operator B_{d} to its reducing subspace (10.29). In view of Proposition 10.15, this definition is correct and B_{d} is <u>now</u> a minimal selfadjoint extension of A.

Corollary 10.16 *For all real t except at most a countable number, the operator B_{d} is a minimal selfadjoint extension of A. For such t $\mathcal{K}_{\mathsf{d}} = L^2[a,b]$.*

If the selfadjoint extension B_{d} of A is minimal, then, in view of Proposition 10.15, we have

$$\mathcal{L}^2(\sigma) \subset \mathcal{L}^2[a,b] = \mathcal{K}_{\mathsf{d}} \quad \text{for any Borel subset } \sigma \text{ of } [a,b].$$

The case when B_{d} is not minimal is in the following.

Proposition 10.17 *If the selfadjoint extension B_{d} of S_ω is not minimal, then $L^2(\sigma) \not\subset \mathcal{K}_{\mathrm{d}}$ for each Borel subset σ of $[a,b]$ with nonempty interior.*

It follows from (10.23) that μ_{d} is a measure given by (10.14) with $B = B_{\mathrm{d}}$ and $e = \omega$. A direct consequence of Corllary 10.13 and Proposition 10.15 is that the closed support of the representing measure μ_{d} of $(a_n)_{n=0}^\infty$ coincides with the spectrum of B_{d}. Below we give a condition characterizing N-extremality of measures μ_{d} by means of the spaces $\mathcal{K}_{\mathrm{d}}$.

On the other hand, Corollary 10.16 is somehow aligned with the following (cf. Theorem 10.22).

Lemma 10.18 *$\mathcal{H} = \mathcal{K}_{\mathrm{d}}$ if and only if μ_{d} is N-extremal.*

Order of representing measures Our next goal is to show that, in general, the representing measure μ_{d} of $(a_n)_{n=0}^\infty$ is of infinite order. Given a positive Borel measure ρ on $\mathbb{R}$ with all moments finite, we define the *order* $\operatorname{ord}(\rho)$ of ρ as

$$\operatorname{ord}(\rho) = \dim \mathcal{L}^2(\rho) \ominus \mathcal{P}^2(\rho),$$

where $\mathcal{P}^2(\rho)$ is the closure in $\mathcal{L}^2(\rho)$ of $\mathbb{C}[X]$ and "dim" stands for the Hilbert space dimension. According to [Riesz, 1923] a representing measure of an indeterminate Hamburger moment sequence is N-extremal if and only if its order is equal to 0.

As shown below, the spaces appearing in the definition of $\operatorname{ord}(\mu_{\mathrm{d}})$ can be unitarily identified with the spaces $\mathcal{K}_{\mathrm{d}}$ and $\mathcal{H}$ respectively. Extra care is required when constructing a unitary isomorphism of $\mathcal{K}_{\mathrm{d}}$ onto $\mathcal{L}^2(\mu_{\mathrm{d}})$ which at the same time identifies $\mathcal{H}$ with $\mathcal{P}^2(\mu_{\mathrm{d}})$.

Proposition 10.19 *There exists a unitary isomorphism U between $\mathcal{K}_{\mathrm{d}}$ and $\mathcal{L}^2(\mu_{\mathrm{d}})$ such that $U(\omega)1 = 1$, $U(\mathcal{H}) = \mathcal{P}^2(\mu_{\mathrm{d}})$ and $UB_{\mathrm{d}} = MU$, where M is the operator of multiplication by the independent variable X in $\mathcal{L}^2(\mu_{\mathrm{d}})$. Moreover, $\operatorname{ord}(\mu_{\mathrm{d}}) = \dim \mathcal{K}_{\mathrm{d}} \ominus \mathcal{H}$.*

[Cichoń, Stochel and Szafraniec, 2010, Proposition 3.2, Corollary 3.3 and Theorem 3.4] is a source of further information.

Proposition 10.20 *The following equality holds:*

$$\dim \mathcal{L}^2[\alpha,\beta] \ominus \mathcal{H} = \#\{k \in \mathbb{Z}\colon \widehat{\omega}(\lambda_{\mathrm{d},k}) = 0\} + \operatorname{ord}(\mu_{\mathrm{d}}), \quad t \in \mathbb{R}.$$

For every $t \in \mathbb{R}$, either $\widehat{\omega}(\lambda_{\mathsf{d},k}) = 0$ for some $k \in \mathbb{Z}$ or the representing measure μ_{d} is not N-extremal. Moreover, the set of all $t \in \mathbb{R}$ for which the representing measure μ_{d} is N-extremal is at most countable.

Theorem 10.21 *Assume that there exists a nonempty open interval $\varDelta \subset [\alpha, \beta]$ and a nonzero polynomial $p \in \mathbb{C}[X]$ such that $\big[p(\frac{\mathrm{d}}{\mathrm{d}x})\omega\big](x) = 0$ for all $x \in \varDelta$. Then $\operatorname{ord}(\mu_{\mathsf{d}}) = \aleph_0$ for all $t \in \mathbb{R}$. In particular, $\dim \mathcal{L}^2[\alpha, \beta] \ominus \mathcal{H} = \aleph_0$.*

Concerning the assumptions imposed upon ω in Theorem 10.21, it is possible to indicate a class of test functions which obey the condition $\big[p(\frac{\mathrm{d}}{\mathrm{d}x})\omega\big](x) = 0$ for x ranging over an open nonempty interval $\varDelta \subset [\alpha, \beta]$. It is well known that the restriction of such ω to the interval $\varDelta$ must be of the form

$$\omega(x) = \sum_{j=1}^{m} p_j(x)\mathrm{e}^{a_j x}, \quad x \in \varDelta, \tag{10.30}$$

where $p_j \in \mathbb{C}[X]$ and $a_j \in \mathbb{C}$. Hence, if ω_0 is any test function taking value 1 in the interval $\varDelta$, then by multiplying it by the right-hand side of (10.30) we obtain a test function ω satisfying the assumptions of Theorem 10.21.

More on $\operatorname{supp}\omega$ **versus** $[a, b]$ Appeal first to [Cichoń, Stochel and Szafraniec, 2010, Theorem 2.7].

Theorem 10.22 *If $\operatorname{supp}\omega \subsetneq [a, b]$, then $\dim \mathcal{K}_{\mathsf{d}} \ominus \mathcal{H} = \aleph_0$ for all $t \in \mathbb{R}$.*

It may be interesting to see how $[\alpha, \beta]$ situated against the parameter intervals $[a, b]$ makes the difference concerning the order. Consider the following localizations (recall (10.19))

(i) $[\alpha, \beta] \neq [a, b]$,
(ii) $[\alpha, \beta] = [a, b]$ and $\operatorname{supp}\omega \neq [\alpha, \beta]$,
(iii) $[\alpha, \beta] = [a, b]$ and $\operatorname{supp}\omega = [\alpha, \beta]$.

Owing to Proposition 10.19 and Theorem 10.22, the order $\operatorname{ord}(\mu_{\mathsf{d}})$ for every $t \in \mathbb{R}$ is infinite if either (i) or (ii) holds. Theorem 10.21 goes even further and says that $\operatorname{ord}(\mu_{\mathsf{d}}^{min})$ happens to be infinite for every $t \in \mathbb{R}$ in the case of (iii); apparently the class of such ω's is pretty large. Under these circumstances, none of the measures μ_{d} is N-extremal. On the other hand, Corollary 10.20 implies that the representing measures

μ_{d} may be N-extremal at most for a countable number of $t \in \mathbb{R}$; this may happen exclusively when (iii) is satisfied. The above discussion makes it legitimate to pose the following open questions for $\mathrm{d} = ([\alpha, \beta], t)$.

Question 1 Is the order of μ_{d} infinite for all $t \in \mathbb{R}$ and for all nonzero test functions ω?

Question 2 Do there exist a nonzero test function ω and $t \in \mathbb{R}$ such that the measure μ_{d} is N-extremal?

Question 1 is related to the property 4° (see Abstract) whereas Question 2 is connected with the property 3°. We may modify any nonzero test function without affecting its support so that its Fourier transform has no real zeros (however this argument is not constructive). If the Fourier transform $\widehat{\omega}$ of ω has no real zeros, then the indeterminate Hamburger moment sequence $(a_n)_{n=0}^{\infty}$ has properties 1°, 2° and 3° (cf. Remark 10.14).

[Derkach, Hassi, Malamud, and de Snoo, 2009, Proposition 7.5] upholds Question 1 in the case of Naĭmark extensions satisfying (10.4); on this occasion we would like to call attention to Theorem 10.22. However, it has to be mentioned that our notion of the order of a measure is more refined than that of finite dimensionality of extensions [Derkach, Hassi, Malamud, and de Snoo, 2009, p. 54] though the latter has been contemplated under more general circumstances; even the present excerpts from [Cichoń, Stochel and Szafraniec, 2010] provide arguments for the case.

Another thing, which is the point in [Derkach, Hassi, Malamud, and de Snoo, 2009] and left apart in [Cichoń, Stochel and Szafraniec, 2010], is whether the extensions in question are of type II or III. This can be assessed by looking at the boundary conditions determining the domain of A and its Naĭmark extensions B_{d}'s.

References

Akhiezer, N. I. 1965. *The Classical Moment Problem.* Hafner, New York.

Akhiezer, N. I. and Glazman, I. M. 1981. *Theory of linear operators in Hilbert space* vol. I and II. Pitman, Boston-London-Melbourne.

Buchwalter, H. and Cassier, G. 1984. La paramétrisation de Nevanlinna dans le problème des moments de Hamburger. *Expo. Math.* **2**, 155-178.

Cichoń, D., Stochel, J. and Szafraniec F. H. 2010. Naimark extensions for indeterminacy in the moment problem. An example. *Indiana Univ. Math. J.*, **59**, 1947-1970.

Fuglede, B., 1983. The multidimensional moment problem. *Expo. Math.*. **1**, 47-65.

Gil de Lamadrid, J. 1971. Determinacy theory for the Livšic moments problem. *J. Math. Anal. Appl.* **34**, 429-444.

Gorbachuk, M. L. and Gorbachuk, V. I. 1997. *M. G. Kreĭn's lectures on entire operators*. Operator Theory: Advances and Applications, **97**. Birkhäuser Verlag, Basel.

Derkach, V.A., Hassi, S., Malamud, M.M. and de Snoo, H.S.V. 2009. Boundary relations and generalized resolvents of symmetric operators. *Russ. J. Math. Phys.*, **16**, 17–60.

Kreĭn, M. G. and Krasnoselskiĭ, M. A. 1947. Fundamental theorems on the extension of Hermitian operators and certain of their applications to the theory of orthogonal polynomials and the problem of moments. (*Russian*) *Uspehi Matem. Nauk* (*N. S.*) **2**, no. 3(19), 60-106.

Landau, H. J. 1980. The classical moment problem: Hilbertian proofs. *J. Funct. An.*, **38**, 255-272.

Langer, R. W. 1976. More determinacy theory for the Livšic moments problem. *J. Math. Anal. Appl.* **56**, 586-616.

Mlak, W. 1978. Dilations of Hilbert space operators (general theory). *Dissertationes Math.* (*Rozprawy Mat.*) **153**, 61 pp.

Naĭmark, M. A. 1940. Self-adjoint extensions of the second kind of a symmetric operator (in Russian), *Bull. Acad. Sci. URSS. Ser. Math.* [*Izvestii Akad. Nauk SSSR*] **4**, 53-104.

Naĭmark, M. A. 1940. On the square of a closed symmetric operator. *C. R.* (*Doklady*) *Acad. Sci. URSS* (*N.S.*) **26**, 866-870.

Naĭmark, M. A. 1943. On a representation of additive operator set functions. *C. R.* (*Doklady*) *Acad. Sci. URSS* (*N.S.*) **41**, 359-361.

Okazaki, Y. 1986. Boundedness of closed linear operator T satisfying $\mathcal{R}(T) \subset \mathcal{D}(T)$, *Proc. Japan Acad.*, **62**, 294-296.

Ôta, S. 1984. Closed linear operators with domain containing their range, *Proc. Edinburgh Math. Soc.*, **27**, 229–233.

Riesz, M. 1923. Sur le problème des moments. Troisième Note, *Ark. för mat., astr. och fys.* **17**, (16).

Rudin, W. 1973. *Functional analysis*. McGraw-Hill Series in Higher Mathematics, McGraw-Hill Book Co., New York-Düsseldorf-Johannesburg.

Shohat, J. A. and Tamarkin, J. D. 1943. *The Problem of Moments*, American Mathematical Society Mathematical Surveys, vol. II, American Mathematical Society, New York.

Simon, B. 1998. The classical moment problem as a self-adjoint finite difference operator. *Advances Math.* **137**, 82-203.

Stochel, J. and Szafraniec, F. H. 1989. On normal extensions of unbounded operators. III. Spectral properties. *Publ. RIMS, Kyoto Univ.* **25**, 105-139.

Stochel, J. and Szafraniec F. H. 1991. A few assorted questions about unbounded subnormal operators. *Univ. Iagel. Acta Math.*, **28**, 163-170; available at `http://www2.im.uj.edu.pl/actamath/issues.php`.

Stone, M. H. 1932. *Linear transformations in Hilbert space and their applications to analysis.* Amer. Math. Soc. Colloq. Publ. 15, Amer. Math. Soc., Providence, R.I.

Szafraniec, F. H. 1993. The Sz.-Nagy "théorème principal" extended. Application to subnormality. *Acta Sci. Math.* (*Szeged*), **57**, 249-262.

Szafraniec, F. H. 2010. Murphy's *Positive definite kernels and Hilbert C*-modules* reorganized. Pages 275-295 of *Noncommutative harmonic analysis with applications to probability II*, Banach Center Publ., vol. 89, Polish Acad. Sci. Inst. Math., Warsaw.

Sz.-Nagy, B. 1955. Prolongements des transformations de l'espace de Hilbert qui sortent de cet espace. Appendice au livre "Leçons d'analyse fonctionnelle" par F. Riesz et B. Sz.-Nagy. (French) Akadémiai Kiadó, Budapest, 36 pp.

Weidmann, J. 1980. *Linear operators in Hilbert spaces.* Springer–Verlag, Berlin, Heidelberg, New York.